CURRENT TOPICS IN BIOTECHNOLOGY

CURRENT TOPICS IN BIOTECHNOLOGY

Edited By
C P MALIK
BHAVNEET KAUR
CHITRA WADHWANI

MD Publications Pvt Ltd
New Delhi

Published by
MD Publications Pvt Ltd
"MD House" 11, Darya Ganj
New Delhi-110002
Tel. : +91-11-41563325, 41562846
E-mail : contact@mdppl.com
Website : www.mdppl.com

ISBN : 978-81-7533-102-0

Price : Rs. 1495/-

Published and Printed by **MD Publications Pvt Ltd** at Times Press, 910, Jatwara Street, Darya Ganj, New Delhi.

PREFACE

The book entitled Current Topics in Biotechnology is a humble dedication to an incredible person, Mr Sandeep Bakshi, on the happy occasion of his birthday and for academic service to the world. Apart from making meritorious contributions in modern facets of education, Mr Bakshi has rendered Yeoman's service to the subject of Life Sciences especially Biotechnology in several ways. A gifted administrator and an emblem of self confidence and maturity, Mr Bakshi is unique for his clarity, eloquence and understanding.

Biotechnology has woven the elements of novel ideas and innovations into the fabric of Life Sciences. The subject undoubtedly has triggered intellectual minds into a more productive mode, thereby, making the IMPOSSIBLE – – – now POSSIBLE.

The field of biotechnology has revolutionized the concepts in agriculture, food, industrial feed stocks and health care in the past three decades. It has furnished new tools and techniques to enhance the agricultural productivity, raise value-added products and health care systems and has ensured better environments. Over the years it has exploited biological systems at all levels of complexity or organisations for improving the life of common man. Scientific breakthroughs in gene expression, protein engineering, and drug designing, animal health care, industrial chemistry, preservation of biodiversity and cell fusion are being increasingly translated into new products and services. Rapid advances in different areas of biotechnology have ushered tremendous changes in agriculture, medicine and cell biology.

Truly speaking biotechnology has affected all spheres of human endeavour; genetic engineering of plants and animals has created fears amongst sections of common man and is currently hotly debated.

The present book *Current Topics in Biotechnology* incorporates review articles on varied aspects of biotechnology. The volume contains 10 chapters written by experts in the disciplines who have reviewed recent developments in selected areas.

Dr Edison T. Liu, Executive Director, Genome Institute of Singapore has briefly discussed, What is next in Biotechnology?

Drs Weichang Yu and James A. Birchler, Division of Biological Sciences, University of Missouri-Columbia USA have brought out a succinct discussion on Minichromosomes: The next generation technology for plant genetic engineering.

Dr VS Hanur, Scientist (Senior Scale), Division of Biotechnology, Indian Institute of Horticultural Research, Hesaraghatta, Bangalore, has reviewed the pertinent information on Plant stem cells: Molecular biology and Biotechnology.

Networks of sophisticated, delicately but stringently balanced and hierarchically regulated genes largely govern the body plan architecture throughout the life cycle of angiosperms. Pluripotent stem cells are the eternal source of all the organs below and above the ground. At every stage of specific organ formation, genes and cell pools finely interact to determine morphological events systematically. Genes are regulated quantitatively in the stem cells and their products. Powerful molecular genetic, biochemical and bioinformatics tools have unravelled the magnificent genetic programs underlying stem cells, meristems, organs and organogenesis at different levels. Biotechnological applications of the rich information can lead to the rational engineering of the field of plant stem cell research and agricultural biotechnology through a systems biology approach.

Sonali Chadha discusses Cloning vectors: the vehicles of genetic engineering. Besides describing general features, vectors based on Ti and Ri plasmids of *Agrobacterium*, bacteriophage, virus vectors,' phagemids, YAC, transformation techniques using *Agrobacterium*, prerequisites for production of transgenic plants are also described. The chapter is profusely illustrated to aid understanding by a graduate student.

In India, vigorous efforts have been initiated for development of GM crops, in view of their potential and importance. Cotton is an important cash crop in India and plays a significant role in the national economy, contributing about Rs. 360 billion (US $8 billion) towards export income and 4% of GDP. It is estimated to support about 60 million people, including farmers who cultivate the crop and those involved in the cotton industry for processing and trading. All the four species of cultivated cotton, *Gossypium herbaceum*, *G. arboreum* (both called 'Desi,' meaning local, or Asian cotton), *G. hirsutum* (American cotton) and *G. barbadense* (Egyptian cotton) are grown in India. The cotton area has fluctuated between 8 and 9 million hectares in the last decade of which, in recent years, about 70% constituted hybrid cotton. Among hybrids, *G. hirsutum* represents about 90%. Maharashtra, Andhra Pradesh, Gujarat, Madhya Pradesh, Punjab, Haryana, Rajasthan, Karnataka and Tamil Nadu are the important cotton growing states in our country.The information on various aspects of *Bt*-cotton has been unfolded in the form of answers to several questions, divided into several sections such as bollworms, *Bacillus thuringiensis* (*Bt*), development of *Bt*-cotton, efficacy, safety, insect resistance management, field performance and adoption, costs and benefits, opposition to *Bt*-cotton, legal and illegal seeds, regulation etc. As of now, *Bt* cotton is the only transgenic crop approved for commercial cultivation in the country. Since its introduction in 2002, there has been a phenomenal increase in the area under *Bt* cotton which presently stands at 3.5 million hectare. Till now, 111 hybrids and 4 events viz. Bollgard 1 (BG-1), Bollgard 2 (BG-2), Event 1 and GFM event of *Bt* cotton are approved for commercial cultivation in the country. The first GM food crop of India i.e. *Bt* brinjal is in pipeline for introduction in the market. In fact, with the release of *Bt* brinjal, India is likely to take global lead in introduction of first GM eggplant. Further, various other GM crops including rice, potato, tomato, mustard, groundnut and several other food crops are under different stages of development and trials in the country. The genetic modification is primarily targeted at introduction of crop protection traits which include: resistance to pests, disease & herbicides; abiotic stresses such as drought, heat, cold or salinity and quality traits for enhanced nutrition, shelf life, improved taste, colour, fragrance & increased productivity. The results of socio-economic impact assessment studies on *Bt* cotton are encouraging in terms of economic and environmental benefits. Drs Dapinder Bakshi and JK Arora (Add Director, Biotechnology, PSCST, Chandigarh) have made excellent attempt to present a succinct review.

Considering that cotton fiber biology is important for describing, defining and refining fundamental questions in the fields of cell and developmental biology, Prof. C P Malik, Dr Bhavneet Kaur, Chitra Wadhwani and Aman Verma have discussed The Mechanism of cotton fiber development. In order to understand the regulation of the development and functioning of cotton fibers, and ultimately perhaps to manipulate fiber properties, we need more fundamental information on the developmental and environmental signals and mechanisms associated with signal transduction, leading to changes in fiber gene expression. Characterization of the cotton fiber transcriptome has immediate applications in agricultural biotechnology and molecular breeding programs geared toward the genetic improvement of yield and fiber quality, The wealth of genomic tools and resources now available to the cotton community are paving the way to manipulating the fiber transcriptome using molecular approaches to develop enhanced germplasm. Genetic control of fiber size and shape, and hence agronomic properties, is governed by the rate and duration of fiber elongation, which in turn, positively correlates to fiber length. Breeding for long staple genotypes has been successful, resulting in lines that possess improved yield and fiber quality but without unduly delaying crop maturity Improved fiber yield and maturity would result from the ability to control secondary wall cellulose deposition quantitatively, including making the process less sensitive to environmental stress. Both genetic engineering and marker-assisted breeding are possible avenues for the control of fiber morphogenesis once key regulatory genes and gene products have been identified. In addition, future cotton fiber biotechnology will require strong fiber-specific promoters that drive targeted gene expression in order to avoid pleiotropic effects on plant development.

Prof. C.P. Malik, Aman Verma, Chitra Wadhwani and Bhavneet Kaur (Dr) have made in depth survey and reviewed the pertinent literature on Bt cotton and its economic impact. India, the second largest producer of fruits and vegetables in the world, aims to double its production to 300 million tones by 2012. They contribute nearly 28 per cent of GDP and 54 per cent of export share in Agriculture from the cultivated area share of 8.5% only. This sector could significantly benefit from India's new infrastructural technology support like cold storage chains, refrigerated transport facilities and centralized food processing facilities. In fact Post-harvest operations are assuming greater importance due to higher yields and increased cropping intensity indicating that there is tremendous scope of increasing production and exports in Horticulture. Fruits and vegetables earn 20-30 times more foreign exchange per unit area than cereals due to higher yields and higher price available in the international market. But India loses about 35 - 45% of the harvested fruits and vegetables during handling, storage, transportation, etc., a loss estimated at Rs 40,000 crores per year! In India, fruits and vegetables wastage every year is equivalent to the annual consumption of the United Kingdom. Post Harvest management of fresh produce using emerging techniques and technology will be one of the biggest business opportunities in India. Prof. C P Malik and his colleagues have compiled recent literature on the techniques especially molecular biology of fruit ripening. The chapter on Prions by Chitra Wadhwani and C P Malik is aimed at looking at the Prion diseases from a molecular point of view, taking into account the genetic as well as the transmissible and sporadic aspect of the diseases. It attempts to first fully identify these novel diseases to give an exact idea of what the research is about and why it is taking place. The molecular side starts by looking at the molecular and genetic evidence from the beginning of the discovery of the 'prion' hypothesis. Then, by following a fascinating path of logical deductions and discovery with exciting consequences that were revolutionary for molecular biology and resulted in uncertainties and fierce scientific debates, the properties of the prion diseases are investigated. The chapter critically analyzes the 'prion' hypothesis, and compares it to the viral and bacterial hypotheses as well, ending in identifying the progress that needs to be made in the future, and above all, the major uncertainty of variant CJD.

Drs. S.P. Vyas and Amit K. Goyal have done extremely fascinating job to review the available literature on the recent aspects of edible vaccines. Evidence has accumulated in support of plant-derived antigens as effective, safe and inexpensive means of acquiring or delivering a vaccine. The success of human trials with edible transgenic plant vaccines and successful challenge in animal studies has demonstrated the potential of plant-derived vaccines to expand the armament of preventive medicine. Vaccines produced using plants as bioreactors have been successful when intranasal administered. However, oral immunization has only been successful when the vaccine was delivered in plant. This suggests that additional protection afforded by plant tissues permits successful oral immunization.

We are thankful to all the learned contributors for their unstinted cooperation in compiling the useful information on different facets of biotechnology. Our heartfelt thanks go to them for presenting up-to-date information. Some tasks are too great; necessitate dependence on the available discussions and we acknowledge them fully. We would like to thank all those whose original investigations are the basis of this volume. We are indebted to Dr. Edison T. Liu and Drs. James A. Birchler and Weichang Yu for their excellent contributions. We apologise unreservedly for any mistakes, omissions or failure to acknowledge fully. Finally we would like to thank our families (CPM his spouse, Surakhsha; BK and CW their parents) for their continued support and encouragement throughout the completion of this book.

Thanks are due to Mrs Saxena MD Publishers, New Delhi, who supported us fully for completing this project.

Jaipur
1.10.07

CP Malik
Bhavneet Kaur
Chitra Wadhwani

We (BK & CW) as beginners, feel privileged to be associated with the living legend of Life Sciences, Prof. C.P. Malik and wish to go miles with this icon of knowledge. We owe an unrepayable debt of gratitude for his encouragement and unquestioning support over time.

BK & CW

CONTENTS

Page

List of Authors (xiii)

1. **WHAT IS NEXT IN BIOTECHNOLOGY?** 1
Edison T Liu

2. **MINICHROMOSOMES: THE NEXT GENERATION TECHNOLOGY FOR PLANT GENETIC ENGINEERING** 7
Weichang Yu and James A. Birchler

3. **PLANT STEM CELLS: MOLECULAR BIOLOGY AND BIOTECHNOLOGY** 13
Vageeshbabu S Hanur

4. **CLONING VECTORS: THE VEHICLES OF GENETIC ENGINEERING** 41
Sonali Chadha

5. **GENETICALLY MODIFIED CROPS IN INDIAN PERSPECTIVE** 87
Dapinder K Bakshi and JK Arora

6. **THE MECHANISMS OF COTTON FIBER DEVELOPMENT** 101
C P Malik, Bhavneet Kaur, Chitra Wadhwani and Aman Verma

7. **TRANSGENIC *BT* COTTON AND ITS COMMERCIAL USE** 147
C P Malik, Aman Verma, Chitra Wadhwani and Bhavneet Kaur

8. **PRIONS: THE LETHAL PROTEINS** 227
Chitra Wadhwani and CP Malik

9. **PLANT DERIVED EDIBLE VACCINES** 311
SP Vyas and Amit K Goyal

10. **THE ROLE OF BIOTECHNOLOGY IN PROLONGING THE SHELF LIFE OF FRUITS** 353
C P Malik, Aman Verma, Bhavneet Kaur and Chitra Wadhwani

LIST OF AUTHORS

ARORA JK
PUNJAB STATE COUNCIL OF SCIENCE AND TECHNOLOGY, CHANDIGARH

BAKSHI DAPINDER
PUNJAB STATE COUNCIL OF SCIENCE AND TECHNOLOGY, CHANDIGARH

BIRCHLER JAMES A
DIVISION OF BIOLOGICAL SCIENCES, UNIVERSITY OF MISSOURI, COLUMBIA, USA

CHADHA SONALI
SEEDLING INSTITUTE FOR INTEGRATED LEARNING AND ADVANCED STUDIES, JAGATPURA, JAIPUR

GOYAL AMIT K
DRUG DELIVERY RESEARCH LABORATORY, DEPARTMENT OF PHARMACEUTICAL SCIENCES, DR. HARISINGH GOUR VISHWAVIDYALAYA, SAGAR [MP]

HANUR VAGEESHBABU S
SCIENTIST (SENIOR SCALE), DIVISION OF BIOTECHNOLOGY, INDIAN INSTITUTE OF HORTICULTURAL RESEARCH, HESARAGHATTA LAKE POST, BANGALORE

KAUR BHAVNEET
INSTITUTE OF BIOTECHNOLOGY AND ALLIED SCIENCES, SEEDLING ACADEMY OF DESIGN, TECHNOLOGY AND MANAGEMENT, JAGATPURA, JAIPUR

LIU EDISON T
THE EXECUTIVE DIRECTOR OF THE GENOME INSTITUTE OF SINGAPORE, SINGAPORE

MALIK C P
SEEDLING ACADEMY OF DESIGN, TECHNOLOGY AND MANAGEMENT, JAGATPURA, JAIPUR

VERMA AMAN

SEEDLING INSTITUTE FOR INTEGRATED LEARNING AND ADVANCED STUDIES, JAGATPURA, JAIPUR

VYAS SP

DRUG DELIVERY RESEARCH LABORATORY, DEPARTMENT OF PHARMACEUTICAL SCIENCES, DR. HARISINGH GOUR VISHWAVIDYALAYA, SAGAR [MP]

WADHWANI CHITRA

INSTITUTE OF BIOTECHNOLOGY AND ALLIED SCIENCES, SEEDLING ACADEMY OF DESIGN, TECHNOLOGY AND MANAGEMENT, JAGATPURA, JAIPUR

YU WEICHANG

DIVISION OF BIOLOGICAL SCIENCES, UNIVERSITY OF MISSOURI, COLUMBIA, USA

SANDEEP BAKSHI

Man of
Vision, Values and Vitality

Traversing an incredible journey ....

FROM

SEEDLING PUBLIC SCHOOL

TO

JAIPUR NATIONAL UNIVERSITY

1

WHAT IS NEXT IN BIOTECHNOLOGY?

EDISON T LIU

Throughout the world, biotechnology is synonymous with the likes of Amgen, Vertex Pharmaceuticals, and Genentech, companies centered on the development of therapeutics; and companies such as Caliper Life Sciences, Affymetrix, Celera (the original company) and Illumina who provide technology platforms for research and pharmaceutical development. In Asia, as governments and investors push to develop local and regional biotechnology industries, the focus has been on establishing entities based on these existing models. On the whole, the majority of Asian biotechnology companies emerging in the first wave of investment from Taiwan to New Delhi have followed this format. This is all well and good, and both money and societal benefit can still be extracted from this model, but a new business opportunity is arising that is being overlooked by investors. It is one that will inevitably rise in clinical importance and in terms of profit margins. This emerging sector will be in medical informatics, a discipline defined as information technology dedicated to the collection, storage, indexing, processing, retrieval, and analysis of medical information. The prediction is that this underserved sector will see an expansive growth in the next two decades as the need and demand for efficiency, portability, and safety in medical care rises in both the developed and emerging economies.

What is the origin of this need?

The cost of medical care has become a significant burden to most countries both advanced and developing. This trend of increased expenditures in health care is seen in every developed nation with the US showing the most dramatic increases. In the US, the national health spending as a percentage of the GDP exceeds 15 per cent and is projected to rise to 18 per cent by 2012. These levels are unsustainable and are already affecting the competitiveness of US industries such as the automotive industry. Though the pharmaceutical industry has been able to thrive relying on the US market, there are political signs that these progressive increases will not be tolerated in the future. Therefore, in the beginning of the 21st century, there will be much pressure to limit the rise in health care spending from all governments. This of course will have ramifications to the biotechnology and pharmaceutical industries which rely on government buyers and pricing policies.

Earlier, the main driver for medical information technology was for billing, but now, medical informatics is needed for directly medical reasons. A series of reports from the US Institute of Medicine described the dire state of affairs: 44,000-98,000 Americans die because of preventable medical errors, US one-third of the $1.6 trillion health care spending in 2003 went to inappropriate or duplicate procedures, and 190,000 hospitalizations per year could have been prevented by better information technologies (Kohn *et al.*, 1999; Leape & Berwick, 2005). Across North America and Europe, hospitals are progressively implementing barcode-tracked drug dispensing systems which significantly reduces system errors in dispensing drugs. Any of us who has worked in a hospital system knows the frustration of not being able to obtain a medical chart, X-ray, or lab data of a patient in the emergency room or in the clinic simply because of lost medical records or X-ray files. Whereas health economists have pondered the costs of medical innovation, these reports raise the possibility that major improvement in health care, and major cost savings have been achieved simply by better management of medical information.

Today, in an era of electronic medical records and digitalized radiographs, the only rational medium for medical information is digital and electronic. Yet, this domain has not been exploited by the IT sector as a key business opportunity. It may be because of the earlier disappointments with the cost and inefficiencies of the first medical IT systems, and the "boutique" nature of some of the systems. Moreover, the first systems were geared solely for components within a hospital and not integrated as a health delivery system. The key challenges now are in data integration and interoperability, and in indexing and retrieval.

What kind of innovation are these?

Instead of the medical records locked in hospital vaults, the individual patient will carry his or her complete medical records potentially in a portable memory device, or more likely, in virtual electronic medical record warehouses that will be under the complete control of the patient who can give access to doctors anywhere in the world. This will change the focus of control of medical records to the patient rather than the hospital. Online services will emerge that will allow physicians and patients to intelligently search their medical records for the appropriate information - just like current web search engines do. The digital radiographs/X-rays can be rendered searchable and automated alert systems will be developed that will notify physician and patient of an unusual change in patterns. Online medical "concierge" services will be developed whereby a patient can access medical information about a disorder, or a medical product pertinent to his/her medical concerns. When linked to whole genome sequencing, the individual will have the option of archiving his or her genetic code in a very private vault that can be automatically interrogated as new information arises. The client will be given feedback as to the ramifications of these findings.

I can imagine that medical educational systems, annotated medical literature, and textbook references will be linked to this medical information access platform. In this manner, organized medical educational content can be employed by healthcare workers to build a culture of continuous information expansion as much as continuous quality improvement (CQI) has become an important business process in the manufacturing industry. Business opportunities would be open to those servicing the infrastructure for these medical informatics systems. Clinically relevant information needs to be accessed with the widest bandwidth possible, but also with stringent requirements for data security and confidentiality. The need for medical information to simultaneously have open access to a distributed community but to be highly secure is a difficult problem to resolve. However, solutions are available and drawn from banking and defense IT experience. Those who resolve these security and confidentiality problems will probably sweep the field in as much as IT data security systems are exceedingly profitable today.

By my estimate, medical informatics will be the next "big thing" in biomedicine. I predict that like what happened in the computer industry, where the search engines like Google have achieved market capitalization that exceeded that of the hardware makers, and that rivals that of the software giants, the same will happen in medical IT as compared to medical information hardware.

How big is this market?

It is very premature to estimate, but it could be in the billions of dollars once mature. There are already signs that the big companies are not only thinking about the issues but acting on them. IBM recently launched a system for their US employees whereby all of their medical information can be digitalized and made accessible to the employee wherever he or she is transferred to within the international company (www.informationweek.com). In addition to employee private medical information, it is envisaged that the service will later evolve so that these health records will be co-populated by other data sources, such as detailed information about their health plans, and catalogues from drug mail-order companies providing prescription and pharmacy services to their employees – thus, metamorphosing to a medical "concierge" system.

A most intriguing development has been the establishment of Dossia (www.economist.com; Lopes, 2006) by Omnimedix Institute, a non-profit organization based in Portland, Oregon (USA) (www.omnimedix.org) at the behest of a consortium of five major companies: Applied Materials Inc., BP America Inc., Intel, Pitney Bowes Inc. and Wal-Mart Stores. Each are spending "seven figure" amounts to fund this web-based framework for 2.5 million workers, dependents and retirees to keep their standardized, portable, and interoperable personal health record (including X-ray data) in a secure database that is available only to those health care providers approved by the employee. The role of Omnimedix is not only as a service provider, but also as a "trusted third party" that has official firewalls to ensure employee confidentiality especially from his/her employer. Thus, the parent companies cannot access personalized information that can be linked to individual identifiers. In this manner, aggregate information about employee health can be provided to the company for financial and strategic planning without jeopardizing patient identities. Already IT spin-off products and projects have emerged from the Dossia effort. Angeldocs is a web-based application that allows health care centers, social services workers, and emergency room staff match uninsured patients with physicians willing to free services. The advantage to the medical provider is a "paper" trail of activity that can be used for tax purposes, and, in the future, a system to validate the truly needy patient.

Clearly the need is there and the human and technical capabilities are available to answer this need. If so, then why has there not been more activity? In fact, much of the activities in medical informatics in the past have been either premature, misplaced or hampered by confidentiality issues and regulatory uncertainties. Earlier efforts in artificial intelligence to mimic the medical expert did not address any core needs from the medical community. Medical IT during the dot com boom was simply too expensive and cumbersome focusing only on medical billing. Moreover, most systems were neither flexible nor scalable. A number of medical information service providers failed because they were too small or too focused on one narrow area of medical informatics such as patient referrals, or on archiving only the printed medical records. This fragmentation of the medical information provide only local departmental solutions (i.e. procurement or billing), but did not address the need for integrated and often heterogeneous information. Finally, security of comprehensive systems dealing with private medical information is a very sensitive issue and trust by the subscriber is paramount. In dealing with confidential health information, most for-profit business models have not developed adequate strategies to gain and then ensure that trust. This is why Omnimedix may be a very important first effort in the US: it is structured as a not-for-profit organization which is more likely to be structured for public-good. However, for the first time, the need is recognized by organizations with the resources to implement change, and by governments desperate to control medical care costs. So there are reasons to be optimistic.

Why should Asia be interested in this projected medical informatics boom?

The first should be the obvious follow through from the IT miracle model of India and Greater China. The human resources, technical and business savvy is resident in this region. The second, however, is less well-known and is based on the significant economic expansion of Asia. With the rise of the middle class in all of Asia, the demand for quality medical services both for local consumption and for medical tourism is very high. New hospitals and medical systems are being built at an accelerated pace in China, India, and Southeast Asia. Each of these systems will strive to have the most advanced information systems to enhance efficiency and safety. The business opportunities not only lie in the provision of services, but also in developing novel IT systems tailored to specific needs and conditions. Therefore, I predict that much of the innovation in medical informatics will come from Asia.

Biotechnology as we knew it in the past must and will change to remain relevant to the medical market place. Here, we explored one new form of biotechnology and that is in informatics. In future essays, we will explore other trends, some obvious, and others more subtle but equally important. In all cases, what you will get will be an honest opinion coming from an Asian source, with a global perspective. Stay tuned.

REFERENCES

Kohn LT Corrigan JM Donaldson MS (1999) Eds Committee on Quality of Health Care in America. To Err Is Human: Building a Safer Health System Institute of Medicine

Leape LL Berwick DM (2005) "Five Years After To Err Is Human: What Have We Learned?" *Journal of the American Medical Association* **293**: 2384–2390

"IBM Provides Its US Workers With Digital Medical Records" http://www.informationweek.com/showArticle.jhtml?articleID=172302570

"Health care in America: Bit by bit, Wal-Mart and other big firms are pushing for electronic medical records". The Economist print edition. December 7, 2006. http://www.economist.com/business/displaystory.cfm?story_id=8381465

Gregory Lopes "5 firms fund network to fight medical costs". The Washington Times. December 7, 2006. http://washingtontimes.com/business/20061206-100002-6877r.htm

http://www.omnimedix.org/dossia.html

The author is the Executive Director of the Genome Institute of Singapore, Dr Edison T Liu, is a physician-scientist who trained with Nobel Laureate, J Michael Bishop. Dr Liu has authored over 220 papers, reviews and book chapters. He sits on a number of scientific boards for biotechnology and pharmaceutical entities.

2

MINICHROMOSOMES: THE NEXT GENERATION TECHNOLOGY FOR PLANT GENETIC ENGINEERING

WEICHANG YU AND JAMES A. BIRCHLER

Genetic transformation occurs frequently in nature in prokaryotes. The transfer of genes from one organism to another is termed horizontal gene transfer(Richardson & Palmer 2007). For example, bacteria can acquire virulencefactors, as well as antibiotic resistance genes, which may lead to the breakdown of the efficacy of antibiotics. Horizontal gene transfers are rare in higher eukaryotes, but years ago scientists found that a pathogenic bacterium, Agrobacterium, could transfer genes from its genome to its plant hosts, where expression of the transferred genes caused crown gall disease (Chilton *et al.* 1977).

The development of Agrobacterium-mediated genetic transformation, and direct transformation bybiolistics, i.e., the high velocity delivery of DNA attached to metal particles, led to the first generation of transgenic plants and the rapid application of this technology to crop improvement.

Genetic engineering as a driving force for modern agriculture

Genetic engineering is a powerful tool for improving crop quality and productivity, and reducing information systems for biotechnologyPlant ReseaRch labor and resource utilization of farming. For example, farmers saved up to an estimated 60% of costs for pest control by growing Bt (insect resistance) cotton in certain regions of the US in 1997, according to a USDA report (Price *et al.* 2003). The reduction of pesticide spray has other safety benefits for both the environment and humans. Because of these and other benefits, the adoption of three primary GE crops (corn, cotton and soybean) for two traits, HT (herbicide tolerance) and Bt, has dramatically increased since the first commercial introduction in 1996. According to recent USDA survey data tp://www.ers.usda.gov/Data/biotechcrops/), the acreage of GE soybeans reached 89% in the US in 2006, followed by 83% for cotton, and 61% for corn. Although most of the rrent GE crops are either herbicide tolerant or insect resistant, there is a trend toward a small but steadily increasing percentage of GE crops with stacked genes (containing both HT plus Bt traits). For example, 39% of GE cotton planted last year were arieties with stacked genes, compared with only 20% in 2000; 15% of GE corn had stacked genes, compared with 1% in 2000.

The next generation – engineered minichromosomes

Despite many successes, genetic engineering does have limitations. For example, the technology used to create stacked genevarieties is not efficient. When stacking genes in inbred lines, the two genes located on different chromosomes will segregate at a ratio of 1:2:1 following first a genetic cross, and then self-pollinate to make each homozygous. This means that only one-sixteenth of the progeny will maintain the two homozygous stacked transgenes in the next generation. Gene stacking or pyramiding is also widely used in conventional crop breeding programs to combine multiple characteristic traits.

Another difficulty is linkage drag, which refers to the reduction in fitness of a cultivar due to deleterious genes introduced along with the beneficial gene during backcrossing. Linkage drag may occur because genes closely linked to the desired genes or transgenes are usually difficult to recombine during the transfer from one line to another. Many generations of backcrosses are usually required to introduce a good character gene to an elite cultivar without linkage drag. This problem has created a bottleneck that limits the advancement of plant genetic engineering technology from the first generation of GE crops, characterized as 'input traits', to the next generation, characterized as 'output traits' (Halpin, 2005).

In a recent paper, Yu *et al.* (Yu W *et al.* 2007) describe an engineered minichromosome system that can be used as a genetic engineering platform for the expression of an almost unlimited number of genes that can be inherited together as an independent unit. This engineered minichromosome technology should remove the bottleneck described above, and may lead to the second generation of genetic engineered crops.

Chromosome engineering and minichromosomes

Minichromosomes have been produced in yeast and mammalian cells through either de novo construction using the minimum constituent parts of chromosomes or telomere-mediated chromosomal truncation of existing Chromosome (Farr, 1991; Harrington *et al.*, 1997).

Arrows denote minichromosomes. B chromosome centromeres are labeled green with a B specific repeat probe. Transgenes carrying the truncating sequences and site specific recombination cassettes, as detected with the transgene probe, are red. Chromosomes are stained blue with DAPI. A) A minichromosome without the transgene (a copy of a stably integrated transgene is present on an A chromosome). B) Two minichromosomes with the truncating terminal transgenes (red) occurring in the same cell are shown, as well as two normal B chromosomes

However, these technologies for minichromosome construction were not successful in plants until recently when telomere-mediated chromosomal truncation in maize was reported8, which causes chromosome fracture at the telomere sequence insertion site. An array consisting of 2.6 kb telomere repeats efficiently produced chromosomal truncations in maize. However, due to deficiencies in truncated chromosomes, recovery of chromosomes with large truncations is difficult to attain through the genetic transformation process. Moreover, genetic transmission of truncated chromosomes is unlikely; they are non-viable in the haploid gametophyte generation because they lack essential genes. However, the lethality of truncated chromosomes can be overcome either by using tetraploid materials in which the gametes will be diploid and the deficiency of truncated chromosomes would be compensated by additional copies of the homologs, or by using a non-essential chromosome, such as the maize B chromosome5.

Yu *et al.* (2007) recovered a minichromosome formed by the truncation of chromosome 7 in a tetraploid plant. This minichromosome was transferred to a diploid background by repeated backcrossing and as stably maintained. By using the same set of constructs, they targeted the maize B chromosome with biolistic-mediated gene transformation. Truncated B chromosomes were recovered with much greater efficiency. The sizes of the mini B chromosomes ranged from very small to the size of the large B chromosome. These chromosomes from A and B chromosomes of maize were studied for their utility as utility as artificial chromosome for future genetic engineering in plants.

Attaching genes to minichromosomes

Minichromosomes generated by telomere truncation can have different configurations: those with transgenes and those without. This difference depends on the orientation of construct integration relative to the centromere of the recipient chromosome, which is subsequently broken at the site of telomere integration. The transgene can segregate with either fragment of the broken chromosome. However, if the transgene is retained on the chromosomal fragment without a functional centromere, which is essential for normal chromosome segregation during cell division, it will be lost after cell division. In this circumstance, minichromosomes lacking a transgene will be formed and could be selected if there are other transgenes in the same cell able to produce resistance to the selection agent in the transformation process (Fig. 1A). If the transgene is present on the fragment containing a centromere, it is then present on a minichromosome (Fig. 1B). In fact, both types of minichromosomes were recovered. Minichromosomes with attached transgenes were characterized in detail for mitotic and meiotic behaviors, transmission frequencies, gene expression, and the ability to use site-specific recombination.

All minichromosomes in the experiment were transmissible from generation to generation. They never paired with other chromosomes during meiosis, which minimized the chance of recombination with other normal chromosomes in the genome; thus all transgenes on the minichromosome can transmit as a unit. In addition, the research demonstrated that foreign genes could be expressed from maize B chromosomes, which otherwise are inert chromosomes with no essential genes.

Minichromosomes from maize B chromosomes have other properties that make them exceptionally useful as artificial chromosome platforms. For example, mini-B chromosomes can accumulate to high copy numbers in the presence of normal B chromosomes, but remain stable in the absence of normal B chromosomes. This property might allow one to manage the copy number of the chromosome in order to increase gene expression from transgenes on the chromosome under appropriate circumstances.

Manipulation of genes on minichromosomes

Yu *et al.* also demonstrated that additional transgenes could be targeted to the minichromosome by site-specific recombination, which allowed the addition of subsequent genes. Yu *et al.*(2007) crossed a minichromosome containing a promoterless lox-DsRed gene, which encodes a red fluorescent protein, with another transgenic plant that expresses Cre, the recombinase for Cre/lox recombination, in a 35S promoter-lox-Cre cassette. Recombination occurred between the two chromosomes at the lox site, thus activating DsRed gene expression, and at the same time placing all genetic elements distal to the 35Slox- Cre gene onto the minichromosome5. The demonstration of the Cre/lox recombination system in the minichromosome facilitates further manipulation of foreign genes onto minichromosomes. By using other site-specific recombination systems (Yu W *et al.* 2006; Ow *et al.*, 2007; Wright DA *et al.* 2005). Yu *et al.* predict they will be able to add, delete, or replace genes on minichromosomes in order to redesign existing engineered chromosome platforms. This new technology has numerous advantages over the current genetic engineering technology.

As noted above, an immediate application of minichromosome technology is to enable stacking of genes involved with herbicide tolerance and pest resistances. Future developments could facilitate any application that requires the inheritance of multiple foreign genes as a unit. One can now contemplate adding entire biochemical pathways to plants to confer new properties or to synthesize novel metabolites in mass quantities. With mini-B chromosomes, the quantity can be increased to maximize.

PLATE 1

1

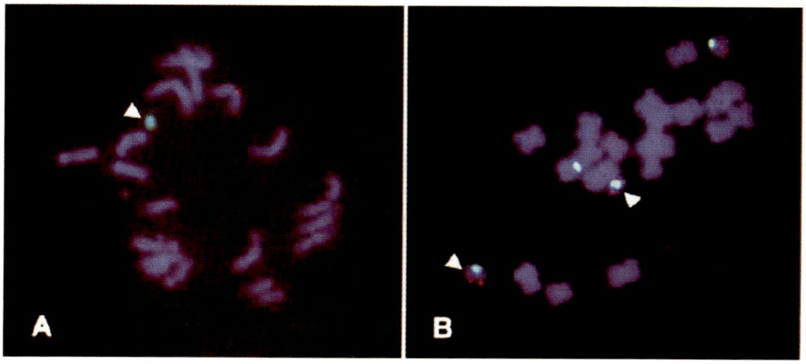

2

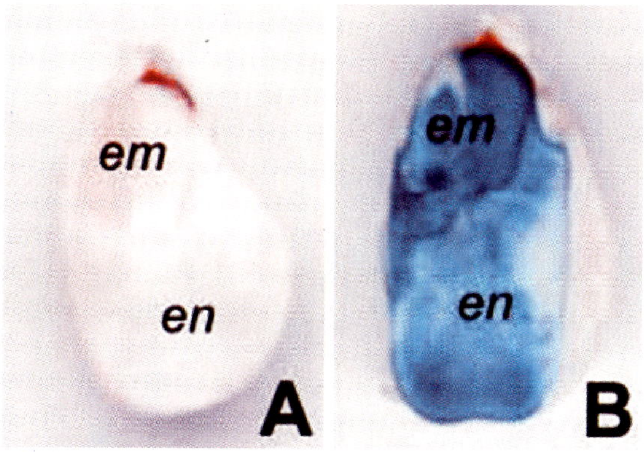

Fig.1. Minichromosomes produced from maize B chromosome truncation. Arrows denote minichromosomes. B chromosome centromeres are labeled green with a B specific repeat probe. Transgenes carrying the truncating sequences and site specific recombination cassettes, as detected with the transgene probe, are red. Chromosomes are stained blue with DAPI. **A)** A minichromosome without the transgene (a copy of a stably integrated transgene is present on an A chromosome). **B)** Two minichromosomes with the truncating terminal transgenes (red) occurring in the same cell are shown, as well as two normal B chromosomes.

Fig. 2. GUS gene expression from a targeted gene on a miniB chromosome. **A)** shows no GUS expression in the absence of the transgene. **B)** shows GUS gene expression in both embryo (em) and endosperm (en) of a mature kernel with a copy of the transgene on a miniB chromosome.

REFERENCES

Richardson AO & Palmer JD (2007) Horizontal gene transfer in plants *J Exp Bot* **58**: 1-9

Chilton MD *et al* (1977) Stable incorporation of plasmid DNA into higher plant cells: the molecular basis of crown gall tumorigenesis *Cell* **11** 263-71

Price GK *et al* (2003) The size and distribution of marker benefits from adopting agricultural biotechnology *U S Department of Agriculture Economic Research Service Technical Bulletin* No 1906

Halpin C (2005) Gene stacking in transgenic plants - the challenge for 21st century plant biotechnology *Plant Biotechnol J* **3** 141-55

Yu W *et al* (2007) Construction and behavior of engineered minichromosomes in maize *Proc Natl Acad Sci U S A* **104** 8924-9

Farr C *et al* (1991) Functional reintroduction of human telomeres into mammalian cells *Proc Natl Acad Sci USA* **88** 7006-10

Harrington JJ *et al* (1997) Formation of de novo centromeres and construction of first-generation human artificial microchromosomes *Nat Genet* **15** 345-55

Yu W *et al* (2006) Telomere-mediated chromosomal truncation in maize *Proc Natl Acad Sci USA* **103** 17331-6

Ow DW (2007) GM maize from site-specific recombination technology what next? *Curr Opin Biotechnol* **18** 115-20

Wright DA *et al* (2005) High-frequency homologous recombination in plants mediated by Zinc-Finger Nucleases *Plant J* **44** 693-705

3

PLANT STEM CELLS: MOLECULAR BIOLOGY AND BIOTECHNOLOGY

VAGEESHBABU S. HANUR

INTRODUCTION

Continuous production of new organs and postembryonic elaboration of architecture throughout the life cycle extending often even decades is a unique ability of the higher plants. This requires the steady availability and maintenance of a reservoir of undifferentiated stem cells at the apical meristems. Along the axis, the upper shoot apical meristem (SAM) produces all the aerial organs like stem, leaves and flowers whereas the root apical meristem below the ground level produces the primary and lateral root systems, thus playing a central role in body plan and organ development. The SAM, which acts as a self-renewing source of pluripotent stem cell populations, becomes a source of initiation of new organs. It is amazing to understand that all the parts of a plant, from the dwarfest plants to the giant trees, are continuously developed systematically from a very few cells, the stem cells, residing in the meristematic region. Rate of (asymmetric) stem cell division is balanced with the rate of loss of stem cells due to consumption of cells towards differentiation and this balance results in the maintenance of critical mass of stem cell pool. In plants, fine mechanisms exist to establish the stem cell pool immediately during embryogenesis as well as to maintain this pool throughout the life cycle. Termination of the stem cell maintenance occurs with the formation of inner whorls of the flower particularly the carpels (gynoecium) in the inflorescence meristems. How this critical mass of stem cells is delicately maintained throughout the life cycle of a plant in SAM has been the subject of intense research in the field of plant biology (Clark, 2001, Baurle and Laux, 2003). Availability of mutants coupled with genetic, molecular and biochemical investigations has opened up this extremely interesting field of plant stem cells with equally stimulating questions and challenges (Hanur, 2007).

Origin of stem cells

Angiosperms typically have a highly evolved layered structure of SAM. The establishment of stem cells and their maintenance in the SAM involves the coordinate orchestration of several genes connected by interlinked signal transduction pathways in different cellular zones as studied mainly in *Arabidopsis* (Irish and Sussex, 1992; Lyndon, 1998, Vernoux and Benfey, 2005). The stem cells are maintained mainly in the central zone of the meristem dome in three layers in *tunica corpus*. The outer layer, the tunica, consists of one or two layers (L1 and L2 layers) of cells. Cells in these layers divide anticlinically (new cell walls form perpendicular to the plane of surface) so that progeny cells will remain in their layer of origin and establish a clone. The cells within one clone can have diverse fates. Corpus is enclosed by tunica in L3 layer. In corpus, cell divisions take place in various directions. Apart from the tunica corpus structure, the SAM can also be histologically subdivided into three distinct zones: peripheral, central and rib zones. The central zone is situated at the summit and provides undifferentiated cells to the surrounding peripheral zone and the rib zone. The peripheral and rib zones in which cells divide more rapidly generate cells that are subsequently used in lateral organs and stem tissues. The central zone therefore serves as a type of stem cell source, replenishing both peripheral and rib zones while continuing its role as the source of stem cells.

KNOX (KNOTTED-like HOMEOBOX) over expression experiments have shown that these genes interact with the cytokinin, auxin and gibberellin pathways. The L1 layer has been shown to be necessary for the maintenance of indeterminacy in the underlying meristem layers. A maize gene, *EXTRA CELL LAYERS1* (*XCL1*), when mutated leads to the production of multiple epidermal layers by overproduction of a normal gene product (Kessler *et al.*, 2006). Meristem size is reduced in mutant plants and more cells are incorporated into the incipient leaf primordium. *Xcl1* may provide a link between L1 division patterns, hormonal pathways and meristem maintenance. *Xcl1* suppresses the *KN1* phenotype but has a synergistic interaction with *GNARLEY1* and *ROUGH SHEATH1 (RS1)*, possibly correlated with changes in gibberellin and auxin signaling. Double mutants with *xcl1* and *crinkly4* results in defective SAM maintenance. Thus proper L1 development is essential for meristem function and *XCL1* may act to coordinate hormonal effects with KNOX gene function at the shoot apex. The maize *KNOTTED1* (*KN1*) gene acts non-cell autonomously in the maize leaf and *KN1* was the first plant protein shown to traffic cell-to-cell, presumably through plasmodesmata. *GFP-KN1* fusion protein was able to traffic from the inner layers of the leaf to the epidermis, but not in the opposite direction, from epidermis to mesophyll and also able to traffic out of the L1 epidermal layer in the SAM indicating that *KN1* movement out of the L1 is developmentally regulated (Kim *et al.*, 2003). *GFP-KNAT1/BREVIPEDICELLUS* and *GFP-STM* fusions could also traffic from the L1 to the L2/L3 layers of the meristem. The activity of *KN1* and related homeobox proteins is maintained following intercellular trafficking, and that trafficking may be required for their normal developmental function. Microsurgical and laser ablation analysis of interactions between the zones and layers of the tomato SAM has shown that ablation of the central zone leads to ectopic expression of the WUS gene at the periphery, followed by the establishment of a new meristem center but organ formation continues without a lag suggesting that the central zone does not participate in organogenesis except as the ultimate source of founder cells (Reinhardt *et al.*, 2003). Microsurgical removal of the external L1 layer induces periclinal cell divisions and terminal differentiation in the subtending layers. No organs initiate in areas devoid of L1 demonstrating an important role of the L1 in organogenesis. Regeneration of L1 cells never occurs after ablation showing that while zones of the meristem show a remarkable capacity to regenerate after interference, elimination of the L1 layer is irreparable and causes terminal differentiation. Meanwhile, light is known to inhibit expression of *DE-ETIOLATED1* (*DET1*) gene, which in turn negatively regulates KNOX genes.

Plant stem cells are therefore not specific for certain cell lineages, but they serve as progenitors of the individual cell layers. Despite this early separation into distinct layers, cells in the meristem act in a coordinated manner to produce the plant body. Both leaves and flowers originate on the flanks of the meristem from the peripheral zone that surrounds the central zone, where the stem cells reside. Cell loss in the peripheral zone due to organ formation is compensated by divisions of stem cells in the central zone and cell displacement to the periphery.

Various morphological events occur in sequential and systematic manner from embryogenesis along the apical-basal axis. During the initiation of SAM, the embryo initially establishes the axis with clear-cut polarity followed by differentiation of the embryonic organs and meristems at appropriate positions and times along the axis. Higher plant embryogenesis can be viewed as the initial differentiation of the shoot and root fates and their maintenance.

Molecular genetics of stem cells

Formation and maintenance of polarity along the embryo axis and embryo proper

The SAM development can be simplistically classified in different steps: formation and maintenance of the polarity along the apical-basal axis, fate specification along the axis (containing embryo proper and the suspensor) and elaboration of the apical region of the embryo. Although little is known about how the initial polarity is generated, recent molecular genetic and physiological analyses suggest that the fixation of the pre-existing axis involves directional vesicle transport and polar auxin transport. The direction of the apical-basal axis of the egg and zygote is invariably along the chalaza-micropyle axis of the ovule, largely arising from a patterning process during female embryogenesis. Even though plant cells can establish this polarity without maternal tissues as demonstrated by somatic embryogenesis, maternal information does provide a positional cue that promotes a preferential polarity of the egg in angiosperms. One maternal effect mutant, *short integument1* (*sin1*) of *Arabidopsis*, originally reported to affect ovule formation and flowering time, also affects embryo development in mutant ovules (Ray *et al.*, 1996). The phenotype of the *sin1* embryos is dependent on the genotype of the maternal tissue, but not on the zygotic tissue, showing a maternal effect of the *sin1* mutations, although function of the *SIN1* gene is not properly known.

The *GNOM/EMBRYO DEFECTIVE30* (*GN/EMB30*) gene of *Arabidopsis* is required for the fixation of the apical-basal polarity with the earliest phenotype of the *gn/emb30* mutant detected at the first division of the zygote (Mayer *et al.*, 1993). This altered polarity is revealed by the variable expression of an apical-specific marker gene, *Arabidopsis thaliana Lipid Transfer Protein1* (*AtLTP1*) (Vroemen *et al.*, 1996). Cellular organization in *gn/emb30* embryos can be examined using a protein marker for cell polarity, PIN-FORMED1 (PIN1) (Steinmann *et al.*, 1999) indicating that the direction of auxin transport is along the apical-basal axis promoting the same polarity. *GN/EMB30* and *PIN1* can generate a positive feedback loop in which *GN/EMB30* promotes coordinated polar localization of *PIN1* whereas coordinated *PIN1* localization strengthens the polarity by promoting polar auxin transport along the stochastic axis.

The *GN/EMB30* gene encodes a protein homologous to yeast Gea1p and Gea2p, which are large guanine nucleotide exchange factors (GEFs) for G proteins of the ADP-ribosylation factor (ARF) family required for vesicle transport. Gea1p and Gea2p can be specifically inhibited by the fungal metabolite brefeldin A (BFA). BFA treatment can completely block polar localization of PIN1 and results in its even distribution along the entire cell surface suggesting that BFA-sensitive ARF GEF is indeed involved in polar localization of PIN1. Mutations in the *PIN1* gene cause variable alterations in the number or position of cotyledons as well as fusion of cotyledons. Also, *pin1* embryos show variable expression patterns of two genes, *CUC1* and *CUC2*, which are discussed below.

As embryogenesis proceeds, specification of embryonic organs along the apical basal axis occurs sequentially viz; separation of the early proembryo into two different fates, the embryo proper and the suspensor and subdivision of the embryo proper into the SAM, cotyledons, hypocotyl, root and the root apical meristem. In *Arabidopsis*, the *twin* mutants (*twn1* and *twn2*) frequently develop embryos from the basal cell derivatives suggesting that the basal cell retains the capacity to change its fate to the embryo proper (Vernon and Meinke, 1994). In *twn2*, the primary defect occurs in the apical cells, which stops development immediately after the first division of the zygote, and a housekeeping gene, *Valyl-tRNA synthetase*, apparently causes this developmental arrest. Similar situation can be observed in the *suspensor* (*sus*) and *raspberry* mutants, in which the embryo proper arrests at certain stages of development and embryo-like structures develop from the basal lineage after the arrest (Zhang and Somerville, 1997). In *twn1* mutant, twinning can occur without developmental arrest of the embryo proper. *Arabidopsis LEAFY COTYLEDON1* (*LEC1*) and *FUSCA3* (*FUS3*), which regulate many aspects of embryogenesis such as cotyledon identity, desiccation tolerance and repression of precocious germination, are also involved in preventing suspensor cells from adopting an embryo fate (Lotan *et al.*, 1998; Parcy *et al.*, 1997). *LEC1* and *FUS3* have overlapping functions that limit the basal cell fate. *LEC1* encodes a protein homologous to the CCAAT box binding transcription factor, HAP3.

Fate specification within the embryo proper involves a number of interacting genes. In *fass* (*fs*) of *Arabidopsis*, the regularity of cell division pattern is completely disrupted leading to the formation of extremely compressed embryos (Torres-Ruiz and Jurgens, 1994). The *topless* (*tpl*) mutant shows defects in formation of the SAM and cotyledons, which are mainly derived from the apical region (Long *et al.*, 1999). The phenotype of the *tpl* mutation is variable, and embryos with the most severly-affected mutants develop a root instead of cotyledons and the SAM. Embryos of *tpl* fail to express the SAM-specific genes *SHOOT MERISTEMLESS* (*STM*) and *UNUSUAL FLORAL ORGANS* (*UFO*). *TPL* is required both to specify the fate of the apical region of the embryo and to suppress the root fate. Mutations in the *MONOPTEROS* (*mp*) gene cause deletion of the hypocotyl and root that are derived from the central and basal regions of the axis (Berleth and Jurgens, 1993). Occasionally, *mp* mutants produce adventitious roots from seedlings or tissue cultures, suggesting that the *MP* gene is not required for the formation of the root *per se*. *MP* gene is required to promote vascular strand continuity both in embryonic and postembryonic development. The *MP* gene encodes AUXIN RESPONSE FACTOR5 (ARF5), a possible transcription factor that can-bind to short conserved auxin response elements, which are found in a certain class of auxin inducible genes and mediate the auxin signaling required for cell axialization. AUXIN RESPONSE FACTOR (ARF)-mediated signaling conveys positional information during embryonic and postembryonic organogenesis and mutations in *MONOPTEROS* (*MP/ARF5*) result in severe patterning defects during embryonic and postembryonic development. MP patterning activity is largely dispensable when the presumptive carboxypeptidase ALTERED MERISTEM PROGRAM 1 (AMP1) is not functional, indicating that *MP* is primarily necessary to counteract *AMP1* activity (Vidaurre *et al.*, 2007). In the absence of *MP* activity, cells in apical meristems and along the paths of procambium formation acquire differentiated identities and this is largely dependent on differentiation-promoting *AMP1* activity. Positions of antagonistic interaction between MP and AMP1 coincide with MP expression domains within the larger AMP1 expression domain. These observations suggest a model in which auxin-derived positional information through *MP* carves out meristematic niches by locally overcoming a general differentiation-promoting activity involving *AMP1*. Other auxin response mutants, *bodenlos* (*bdl*) (Hamann *et al.*, 1999) and *auxin resistant6* (*axr6*) (Hobbie *et al.*, 2000) also show defects both in the cell axialization process and formation of the embryo axis. The *hobbit* (*hbt*) mutant in *Arabidopsis* fails to develop a functional root apical meristem and shows irregular cell division patterns in the basal region from the quadrant stage onwards giving rise to seedlings that lack a quiescent center and columella root cap. *hbt* mutants fail to establish an active root apical meristem and also fail to differentiate lateral root cap cells derived from the central region (Willemsen *et al.*, 1998).

Initiation and maintenance of stem cells in the SAM

Genetic and molecular analyses of shoot meristem development in maize, *Arabidopsis* and other plants have led to the identification of genes critical to the development of various zonations. Maize *KNOTTED1* (*KN1*) was the first gene to be identified as specifically expressed in the SAM during embryo development (Smith *et al.*, 1992) and in the growing tip, but not in the leaf primordia (Jackson *et al.*, 1994). The *KN1* gene encodes a homeodomain protein. Ectopic expression of maize *KN1* in tobacco or its homologs in *Arabidopsis* cause induction of adventitious shoot formation from the transgenic leaf tissues. Loss-of-function mutations of *KN1* in maize are defective in shoot meristem maintenance. The homeobox gene family plays a crucial role during the development of multicellular organisms. The *KNOTTED*-like genes from *Arabidopsis*, especially *KNAT6* and *KNAT2* are close relatives of *STM* and another gene *BREVIPEDICELLUS*. *KNAT6* contributes redundantly with *STM* to the maintenance of the SAM and organ separation (Belles-Boix *et al.*, 2006). The expression domain of *KNAT6* in the SAM marks the boundaries between the SAM and cotyledons. The lack of meristematic activity in the *knat6 stm-2* double mutant and the fusion of cotyledons are linked to the modulation of *CUC* activity. During embryogenesis, *KNAT6* is expressed later than *STM* and *CUC* and therefore *CUC1* and *CUC2* are redundantly required for *KNAT6* expression. Therefore, *KNAT6* contributes to SAM maintenance and boundary establishment in the embryo via the *STM/CUC* pathway.

In *Arabidopsis*, isolation of *DORNROSCHEN* (*drn-D*) mutants using transposon activation tagging that prematurely arrest shoot meristem activity with the formation of radicalized lateral organs show that the mutated gene (*DRN/ESR1*) which encodes an *AP2/ERF* protein, is expressed in a subdomain of meristem stem cells, in lateral organ and transiently in the distal domain of organ primordial (Kirch *et al.*, 2003). During the development of *drn-D* mutants, expression of *STM* is downregulated and later reactivated in an altered domain and expression of *CLV3* and *WUS* is increased. It suggests that *DRN/ESR1* is involved in the regulation of gene expression patterns in meristems and specific misexpression of *DRN* in meristem stem cells affects organ polarity and outgrowth in the meristem periphery and that *DRN/ESR1* itself, or a process regulated by *DRN/ESR1*, can act non-cell-autonomously.

In plants, KNOX proteins have been known to interact specifically with the *BEL1*-like (*BELL*) class of homeodomain proteins through a domain conserved between plants and animals. Mutation in a *BELL* homeobox gene, *VAAMANA* (*VAN*) causes a dwarf phenotype and *van* inflorsescence have clusters of cauline leaves; typically three are produced at each node (Bhatt *et al.*, 2004). *VAN* specifically interacts with *STM*, *BREVIPEDICELLUS* (*BP*) and *KNAT6* and nuclear localization of a *VAN-GFP* fusion depends on co-expression of *STM* or *BP* in tobacco leaves, suggesting that localization of VAN, like that of the animal PBC homeodomain protein is also regulated by interaction with a partner homeodomain protein. While *STM* and *BP* are needed for meristem function but excluded from leaves, members of the *HD-ZipIII* (class III homeodomain leucine zipper) protein family function to promote both meristem activity and adaxial leaf fate. Another gene, *ARGONAUTE1* (*AGO1*) is essential for biogenesis of miRNAs. An *ago1* mutation has been shown to perturb the adaxial-abaxial symmetry of lateral organs and toinduce ectopic expression of *HD-ZipIII* in the leaves (Kidner and Martienssen, 2004, Juarez *et al.*, 2004). The zinc finger protein *SERRATE* acts in a microRNA gene-silencing pathway to regulate expression of the *HD-ZipIII* gene *PHABULOSA* while also limiting the competence of shoot tissue to respond to *KNOX* expression (Grigg *et al.*, 2005). *SERRATE* acts to coordinately regulate meristem activity and leaf axial patterning.

Two paralogous *BELL* homeobox genes, *PENNYWISE* (*PNY*) and *POUND-FOOLISH* (*PNF*), expressed in SAM, encode DNA-binding proteins that are essential for specifying floral primordial and establishing early internode patterning events during inflorescence development (Kanrar *et al.*, 2006). PNY associates with STM and BP. PNY-BP heterodimers are essential for establishing early internode patterning events while PNY-STM heterodimers are critical for SAM function. PNF interacts with STM and BP indicating that PNY and PNF are redundant functioning proteins. Inflorescence development but not vegetative development is sensitive to the dosage levels of PNY, PNF and STM. *STM* is also found to be involved in floral specification and internode development.

ClassI *KNOTTED1-like homeobox* (*KNOX*) transcription factors promote meristem function partly through repression of biosynthesis of the growth regulator gibberellin (GA). *KNOX* function is also mediated by cytokinin and this function is sufficient to rapidly activate both cytokinin biosynthetic gene expression and a SAM-localized cytokinin-response regulator (Jasinski *et al.*, 2005). Cytokinin signaling is necessary for SAM function in a weak hypomorphic allele of the *KNOX* gene *STM*. A combination of constitutive GA signaling and reduced cytokinin levels is detrimental to SAM function. Cytokinin activity is both necessary and sufficient for stimulating GA catabolic gene expression, thus reinforcing the low-GA regime established by KNOX proteins in the SAM. KNOX proteins may act as general orchestrators of growth regulator homeostasis at the shoot apex of *Arabidopsis* by simultaneously activating cytokinin and repressing GA biosynthesis thus promoting meristem activity. In *Arabidopsis*, the *REGULATOR OF AXILLARY MERISTEMS1* (*RAX1*) locus, encoding MYB-like transcription factor *MYB37* and homologous to the tomato *BLIND* gene, is transiently expressed in a small central domain within the boundary zone (Keller *et al.*, 2006). *RAX1* genetically interacts with *CUC* genes and is required for the expression of *CUC2* in the *RAX1* expression domain. *RAX1* also affects the timing of developmental phase transitions by negatively regulating gibberellic acid levels in the shoot apex. *RAX1* thus defines a novel activity that links the specification of SAM formation with the modulation of the rate of progression through developmental phases.

PASTICCINO (*PAS*) genes are required for coordinated cell division and differentiation during plant development. In loss-of-function *pas* mutants, plant aerial tissues show ectopic cell division that is specifically enhanced by cytokinins, leading to disorganized tumour-like tissue. Ectopic and enlarged expression of *CYCLIN DEPENDENT KINASE A* and *CYCLIN B1* results in *pas* mutants (Harrar *et al.*, 2003). The expression of *KNAT2*, *KNAT6* and *STM* is also increased in *pas* mutants. The loss of meristem function caused by *stm* mutation can be suppressed by *pas2* mutation. Cytokinin induction of the primary response markers, *ARABIDOPSIS RESPONSE REGULATOR5* (*ARR5*) and *ARR6* is enhanced and lasts longer in pas mutants suggesting that *PAS* genes repress cytokinin responses and modify auxin response indicating that *PAS* genes are involved in the hormonal control of cell division and differentiation.

In *Arabidopsis*, the *STM* gene encodes a KN1-type homeodomain protein. During the elaboration of the apical region of the globular embryo, formation of the SAM and two cotyledonary primordia takes place and *CUP-SHAPED COTYLEDON1* (*CUC1*), *CUC2* and *STM* genes interact in this process (Aida *et al.*, 1997, 1999). *CUC1* and *CUC2* seem to be redundant genes. *cuc1cuc2* double mutants fail to develop the SAM. Boundaries between cotyledons, where growth is normally suppressed, grow ectopically and lead to the formation of a fused cup-shaped cotyledon. *CUC1* and *CUC2* are required for the SAM formation at the center of the apical region and also for the suppression of growth at the cotyledon boundaries to separate cotyledons. Similarly, mutations in the *STM* gene cause loss of a SAM and slight fusion of cotyledons at the base. *STM* is absolutely required for the SAM formation but only partially required for cotyledon separation. Even though the expression of all the three genes begins at the globular stage, they soon overlap in the stripe region between cotyledonary primordial by the early heart stage of the embryo. This stripe largely corresponds to the presumptive SAM and the boundaries between cotyledons, where the three genes resolve to express complementarily. *STM* is down regulated at the peripheries and becomes restricted to the central SAM, whereas *CUC1* and *CUC2* are down regulated at the center and become restricted to the region surrounding the center. *STM* is required for the restriction of *CUC1* and *CUC2* expression. Activation of *STM* expression is a critical step in SAM formation during embryogenesis and *CUC1* and *CUC2* genes play an important role in this activation. In *Arabidopsis*, a mutation called *early extra petals1* (*eep1*) has identified a role for microRNA *miR164c* in regulating petal number (Baker *et al.*, 2005).

Transgenic plants over-expressing *CUC1* (*35S::CUC1*) had two basal lobes in the cotyledons, small and round epidermal cells between sinuses and adventitious SAMs on the adaxial surface suggesting that *CUC1* promotes adventitious SAM formation by maintaining epidermal cells in an undifferentiated state (Hibara *et al.*, 2003). In these plants, *STM* and *BP* genes were ectopically expressed before adventitious SAM formation. In *stm* mutants, ectopic expression of *CUC1* could not induce adventitious SAMs, whereas they continued to be observed in *bp* mutants suggesting that *STM* but not *BP* is necessary for the formation of adventitious SAMs in these transgenic lines. *CUC1* positively regulates SAM formation not only through *STM* but also through an *STM*-independent pathway that is negatively regulated by *AS1* and *AS2*.

The *Arabidopsis FERTILIZATION-INDEPENDENT ENDOSPERM* (*FIE*) polycomb group (*PcG*) protein, a *WD40* homologue of *Drosophila EXTRA SEX COMB* (*ESC*), regulates endosperm and embryo development and represses flowering during embryo and seedling development. In *FIE* co-suppressed plants morphological aberrations are found to be similar to those exhibited by plants over-expressing *AGAMOUS* or *CURLY LEAF* (*clf*) mutants and this aberrant leaf morphology of *FIE* silenced and *clf* mutants correlates with derepression of class I *KNOX* genes including *KNAT2* and *STM* whereas *BP* is upregulated in *FIE*-silenced plants but not in the *clf* mutant (Katz *et al.*, 2004). This indicates that *FIE* is essential for the control of shoot and leaf development and it interacts with *CLF*. It is suggested that in plants, as in mammals and insects, *PcG* proteins control expression of homeobox genes especially the *FIE* which is derived from association with different *SET (SU (VAR) 3-9, E(Z), Trithorax*) domain *PcG* proteins, resulting in differential regulation of gene expression throughout the plant life cycle.

MicroRNAs (miRNAs) are small 20-25 nucleotide nonprotein-coding RNAs that negatively regulate expression of genes in many organisms ranging from plants to humans. The *MIR164* family of miRNAs in *Arabidopsis* consists of three members that share sequence complementarity to transcripts of NAC (*NAM/ATAF1/CUC2*) family transcription factors, including *CUC1* and *CUC2*. The phenotype of *eep1* is the formation of extra petals in early-arising flowers. Double mutants of *miR164c* and *eep1 mir164b* reveal that *miR164c* controls petal number in a nonredundant manner by regulating the transcript accumulation of *CUC1* and *CUC2*. The microRNA *miR164* guides the cleavage of endogenous and transgenic *NAC1* mRNA, producing 3′ fragments and this cleavage is blocked by *NAC1* mutations that disrupt base pairing with *miR164* (Guo-HuiShan *et al.*, 2005). Mutant plants of *mir164a* and *mir164b* expressed less *miR164* and more *NAC1* mRNA and produced more lateral roots while inducible expression of *miR164* in wild type plants led to decreased *NAC1* mRNA levels and reduced lateral root emergence. Auxin induction of *miR164* was mirrored by an increase in the *NAC1* mRNA 3′ fragment, which was not observed in the auxin-insensitive mutants *auxin resistant1* (*axr1-12*), *axr2-1* and *transport inhibitor response1* suggesting that auxin induction of *miR164* provides a homeostatic mechanism to clear *NAC1* mRNA to down regulate auxin signals.

Expression of a *miR164*-resistant version of *CUC1* mRNA from the *CUC1* promoter causes alterations in *Arabidopsis* embryonic, vegetative and floral development, including cotyledon orientation defects, reduction of rosette leaf petioles, dramatically misshapen rosette leaves, one to four extra petals and one to two missing sepals (Mallory *et al.*, 2004). Reciprocally, constitutive overexpression of *miR164* recapitulates *cuc1 cuc2* double mutant phenotypes, including cotyledon and floral organ fusions. This demonstrates that *miR164*-directed regulation of *CUC1* is necessary for normal embryonic, vegetative and floral organs thus implicating *miR164* as a common regulatory component of the molecular circuitry that controls the separation of different developing organs and thereby exposes a posttranscriptional layer of NAC-domain gene regulation during plant development.

The *jabba-1D* (*jba-1D*) mutant of *Arabidopsis* displays multiple enlarged shoot meristems, radicalized leaves, reduced gynoecia and vascular defects in *Arabidopsis* (Williams *et al.*, 2005). The *jba1-D* meristem phenotypes are caused by over expression of *miR166g* microRNA and require the activity of RNAase III helicase *DCL1*. Over expression of *miR166g* in *jba-1D* plants affects the transcripts of several class III homeodomain leucine zipper (*HD-ZIP*) family target genes. The expression of *PHABULOSA*, *PHAVOLUTA* and *CORONA* is significantly reduced in a *jba-1D* background while *REVOLUTA* expression is elevated and *ATHB8* is unchanged. In wild type and *jba1-D* embryos, *miR166* has a dynamic expression, all revealing a compex regulation of the class III *HD-ZIP* family and *WUS* by miRNAs.

In rice, *OsNAC1* and *OsNAC2* are classified in the *NAM* subfamily, which includes *NAM* and *CUC2*, while *OsNAC5* and *OsNAC6* fall into the *ATAF* subfamily (Kikuchi *et al.*, 2000). Besides the rice genome contains *NAC* genes *NAC* genes *OsNAC3*, *OsNAC4* (both in the *OsNAC3* subfamily), *OsNAC7* and *OsNAC8*, indicating that *OsNAC* genes constitute a large gene family in the rice genome. Each *OsNAC* gene is expressed in a specific pattern in different organs suggesting that this family has diverse and important roles in rice development.

In *Antirrhinum, cupuliformis* mutants are defective in SAM formation but *cup* plants overcome this early barrier to development to reach maturity (Weir *et al.*, 2004). *CUP* encodes a *NAC*-domain transcription factor homologous to Petunia *NAM* and *Arabidopsis CUC* proteins. The mutant exhibits dramatic organ fusion throughout development, severe lateral organ fusion in leaves and inflorescences and highly fasciated apical meristem. *CUP* directly interacts with a *TCP*-domain transcription factor, members of which are known to regulate organ growth.

Numerous highly conserved *RING-H2* domains are found in *Arabidopsis.* During the characterization of potential *RING-H2* protein interactions, Greve *et al.* (2003) used the small *RING-H2* protein RHA2a as bait in a yeast two-hybrid screen. *RHA2a* interacted with one of the plant-specific NAC transcription factors namely *ANAC* (abscisic acid-responsive NAC). The core *RING-H2* was sufficient for the multispecific interaction. A short divergent region in *RING-H2* domains is able to modulate interaction specificity. *ANAC* has a degenerate bipartite nuclear localization signal (NLS), while *RHG1a,* also identified as an ANAC interaction partner, contains a basic NLS.

Several genes are expressed in cotyledon primordial with varying expression domains such as throughout the primordial or subregions like adaxial and abaxial surfaces. Most genes that express in cotyledonary primordial also express in leaf primordial suggesting that cotyledons and leaves are homologous organs. The *Arabidopsis AINTEGUMENTA* (*ANT*) gene, which encodes a transcription factor that belongs to the *AP2/EREBP* family, is expressed in primordial of all shoot lateral organs including cotyledons. Loss or gain of *ANT* function results in the decrease or increase in size of lateral organs, respectively, suggesting that *ANT* is required for growth of the shoot lateral organs by promoting cell division. The *ant* mutation can cause complete loss of most floral organs when combined with *apetala2* mutation. Similarly, *ASYMMETRIC LEAVES1* (*AS1*) is another regulator of shoot lateral organ formation. It encodes a MYB transcription factor and is closely related to *PHANTASTICA* (*PHAN*) of *Antirrhinum* and *ROUGH SHEATH2* (*RS2*) of maize. In a groundbreaking work, the *PHAN* gene was shown to regulate organ polarity in snpadragon. *PHAN* orthologues were later isolated from many species including maize and *Arabidopsis* and shown to retain conserved developmental functions (Waites *et al.*, 1998). All these genes are negative regulators of knotted-like homeobox (*knox*) genes. In *as1* mutant cotyledons, *KNAT1,* a class I *knox* gene, is ectopically expressed implying that *AS1* negatively regulates *KNAT1* but the expression of *STM* is unaffected. Another gene, *PINOID* (*PID*), encoding a serine-threonine kinase, is also expressed in the presumptive cotyledons at the globular stage. The *pid* mutant develops a pin-formed stem. Over expression of *PID* results in altered sensitivity to auxin, suggesting that *PID* mediates auxin signaling and promotes formation of primordia in response to auxin. In the cells of the abaxial side of cotyledon primordia, another gene, *FILAMENTOUS*

FLOWER (*FIL*) is expressed. *FIL* encodes a putative transcription factor with a C2C2 zinc finger-like domain and a helix-loop-helix domain. Along with another redundant gene, *YABBY3, FIL* specifies abaxial cell fate in cotyledons and other lateral shoot organs. The *fil* mutant does not show an obvious phenotype related to adaxial-abaxial polarity unless it is combined with *yabby3* mutation. Also, *fil yabby3* double mutants exhibit ectopic meristem formation on the adaxial surfaces of cotyledons and leaf blades (Kumaran *et al.*, 2002). In addition to abaxial specification, lateral organ development requires *YABBY* function to downregulate *KNOX* genes so that meristem initiation and growth are restricted to the apex. Molecular and genetic analysis of the *rotunda2* (*ron2*) mutants in *Arabidopsis*, isolated based on their wide and serrated vegetative leaf lamina, indicate that *RON2* acts at later stages of leaf development by restricting cell expansion during leaf growth (Cnops *et al.*, 2004). *RON2* is identical to *LEUNIG* (*LUG*), which encodes a transcriptional co-repressor that affects floral development. Quantification of expression of *KNOX, WUS, YABBY3, LEAFY, ASYMMETRIC LEAVES* and *GIBBERELLIN OXIDASE* genes by real-time reverse-transcription PCR in expanding and fully expanded rosette leaf laminas of the wild type and *ron2* and lug mutants show that *RON2* has a function in the later stages of leaf development.

The *WUSCHEL* (*WUS*, "bushy" or "tousled-looking" in German) gene is responsible for the continuous production of stem cells and is regarded as the master regulator. *WUS* encodes a novel subtype of homeodomain-containing, nuclear transcription factor that belongs to a different class from the KNOX family. Expression of *WUS* starts as early as the dermatogen stage of embryogenesis. In the active SAM, *WUS* is expressed not in stem cells *per se* but in a small group of cells underneath the presumed stem cell population in the OC, affecting the stem cell fate in a non-cell-autonomous fashion. The *CLAVATA* (*CLV*, "club-like" in Latin) genes, *CLV1, CLV2* and *CLV3* expressed as a receptor protein complex, negatively interact with *WUS* gene in the stem cells thereby controlling the effective stem cell population. CLV3 positively regulates *CLV1* (coupled with CLV2). (*WUS* is epistatic over *CLV1* and *CLV3*, which in turn are epistatic over *CLV2*). WUS positively regulates *CLV3* but CLV3 represses *WUS*. This positive-negative feedback loop mechanism is an extremely delicately maintained equilibrium where the most important regulation on stem cells is exercised (Schoof *et al.*, 2000).

CLV signaling restricts the size of the organizing center by repressing *WUS* transcription in the neighboring cells whereas *WUS* induces the expression of *CLV3*, thereby dynamically adjusting the size of the stem cell population in real time. Down regulation of *WUS* leads to down regulation of *CLV3*; this in turn leads to up regulation of *WUS*. Up regulation of *WUS* leads to up regulation of *CLV3*, which in turn down regulation of *WUS*. This system eventually reaches an equilibrium point at which the expression of *WUS* and *CLV3* would be stable. Imbalances in the expression of either of the two tend to return to equilibrium. CLV3 expression is spatially separated from WUS expressing cells in the OC and therefore regulates *WUS* expression by CLAVATA complex. CLAVATA3 protein functions as a mobile intercellular signal in the SAM that spreads laterally from the stem cells and acts both on their lateral neighbours and on the stem cells themselves to repress WUS transcription (Lenhard *et al.*, 2003). The spread and range of action of CLAVATA3 can be limited by binding to its receptor CLAVATA1, which offers an explanation for how CLAVATA3 is prevented from entering the organizing centre and repressing *WUS* transcription. This regulated spread of a secreted signaling molecule enables the SAM to permit the onset of cell differentiation in the periphery, but at the same time to maintain a stable niche for its stem cells in the centre. Characterization of CLV1 receptor protein complexes in wild type and mutant *Arabidopsis* plants reveals the nature of inactive and active CLV1 complexes, the association of a putative Rho/Rac-GTPase and the requirement for CLV3 in receptor activation (Trotochaud *et al.*, 1999). Increasing the CLV3 signal can very rapidly repress *WUS* expression during development, which in turn causes a fast reduction of *CLV3* expression (Muller *et al.*, 2006). Increased *CLV3* signaling restricts meristem growth and promotes allocation of peripheral meristem cells into organ primordia. Besides, meristem homeostasis tolerates variation in CLV3 levels over a 10-fold range and that high level CLV3 signaling can be partially compensated with time, indicating that the level of CLV3 expression communicates only limited information on stem cell number to the underlying OC cells.

STIMPY (*STIP*; also called *WOX9*), a homeobox gene required for the growth of the vegetative SAM and in several aerial organs and root (Wu *et al.*, 2005). What sets *STIP* apart from *STM* and *WUS* is that *stip* mutants can be fully rescued by stimulating the entry into the cell cycle with sucrose. Therefore, *STIP* is likely to act in all these tissues by maintaining cell division and preventing premature differentiation and this gene identifies a new genetic pathway integrating developmental signals with cell cycle control.

The *CLV3/ESR* (*CLE*) motif, together with the adjacent C-terminal sequence, is sufficient to execute *CLV3* function when fused behind an N-terminal sequence of another gene, *ERECTA*. Most of the sequences flanking the *CLE* motif of *CLV3* can be deleted without affecting *CLV3* function (Fiers *et al.*, 2006). A mechanistic link between *CLV/WUS* network and hormonal control shows that *WUS*, a positive regulator of stem cells, directly represses the transcription of several two-component *ARABIDOPSIS RESPONSE REGULATOR* genes (*ARR5*, *ARR6*, *ARR7* and *ARR15*), which act in the negative feedback loop of cytokinin signaling (Leibfried *et al.*, 2005). *ARR* genes might negatively influence meristem size and that their repression by *WUS* might be necessary for proper meristem function. Supportably, a mutant *ARR7* allele which mimics the active, phosphorylated form, causes the formation of aberrant shoot apical meristems and conversely, a loss-of-function mutation in a maize ARR homologue causes enlarged meristems. Another important finding is that *WUS* induces shoot stem cell activity and developmental plasticity in the root meristem (Gallois *et al.*, 2004). *WUS* alone can specify a shoot stem cell niche when it is expressed in the root induces the expression of shoot stem cell markers and the formation of shoot lateral organs.

The *Hg-SYV46* parasitism gene is expressed exclusively in the dorsal oesophageal gland cell of parasitic stages of the soybean cyst nematode, *Heterodera glycines,* and it encodes a secretory protein that contains a C-terminal motif of the *CLAVATA3/ESR*-related (*CLE*) protein family. Constitutive misexpression of *Hg-SYV46* in *Arabidopsis* results in a *wus*-like phenotype including premature termination of the SAM and the development of flowers lacking the central gynoecium and the *wus*-like phenotype being similar to that of over expression of *CLV3* and *CLV40* (Wang *et al.*, 2005). In the transgenic plants a short root phenotype similar to over expression of *CLV3, CLV40* and *CLE19* in roots suggesting that a functional similarity of *Hg-SYV46* to plant secreted *CLE* ligands that may play a role in the differentiation or division of feeding cells induced in plant roots by parasitic nematodes.

The cellular organization is disrupted in post-embryonic meristems both in the root and in the shoot and their meristematic activity is reduced or become abnormal in the *halted root* (*hlr*) mutant of *Arabidopsis* (Ueda *et al.*, 2004). In the post-embryonic SAM, the expression pattern of *WUS* is disturbed in the mutant indicating that the *HLR* gene is essential to maintain the cellular organization and normal nature of the root and shoot apical meristems. The *HLR* gene encodes RPT2a, which is a subunit of the 26 proteasome that degrades key proteins in diverse cellular processes. The activity of proteasomes is reduced in the mutant. Proteasome-dependent programmed proteolysis is required to maintain the meristem integrity both in the shoot and in the root.

Meristem plasticity requires the *ZWILLE* (*ZLL*) gene. In *zll* mutant embryos, the apical cells are misspecified causing a variability of the meristems size and function (Moussian *et al.*, 2003). ZLL is solely localized in vascular tissue. In *zll* loss-of-function background, *wus* has a negative effect whereas *clv* mutations have a positive effect on meristem size. *ZLL* therefore buffers meristem stability non-cell-autonomously by ensuring the critical number of apical cells required for proper meristem function.

In a screen for mutations modifying *clv1* mutants, Green *et al.*, (2005) identified a novel regulator of meristem development, *CORONA* (*CNA*). Whereas *cna* single mutant plants exhibit subtle defects in meristem development, *clv cna* double mutants develop massively enlarged apices that display early loss of organogenesis, mis-expression of

WUS and *CLV3*, and eventual differentiation of the entire apex. *CNA* encodes a class III homeodomain leucine zipper protein. *CNA* is expressed in developing vascular tissue, diffusely through shoot and flower meristems and within developing stamens and carpels and functions in parallel to the *CLV* loci to promote organ formation. Contrary to current models, results of Green *et al.*, revealed that *WUS* is neither necessary nor sufficient for stem cell specification and that neither *WUS* nor *CLV3* is a marker for stem cell identity. *CNA* has been proposed to function in parallel to the *CLV* loci to promote organ formation. Transcriptome analysis of cytokinin, benzylaminopurine (BAP) treated *Arabidopsis* populations indicate that expression of *CLV1* is significantly reduced by the exogenous cytokinin (Lindsay *et al.*, 2006). Time course analysis by RT-PCR of transcript levels in BAP-treated plants during flower development show a decline and subsequent recovery of *CLV1* and a concurrent increase in *WUS*. The temporal coincidence of aberrant floral phenotype with changes in transcript levels of *CLV1* and *WUS* suggests that cytokinins regulate flower development through genes controlling SAM activity.

In tandem, the premature differentiation of the stem cells is suppressed by the gene *SHOOTMERISTEMLESS* (*STM*). *STM* encodes a homeodomain KNOX transcription factor. This gene is expressed throughout the meristematic region except where organ initiation takes place and it promotes meristematic cell identity. The *STM* gene acts in parallel with *WUS* to prevent the stem cells from being differentiated and consumed by lateral organs in the SAM and therefore maintains the stem cell pool and provides meristem identity. Transcriptional control of *WUS* gene shows that the *WUS* promoter contains distinct regulatory regions that control tissue specificity and levels of transcription in a combinatorial manner. A minimal 57 basepair regulatory region is the only required region to control the boundaries of *WUS* transcription in the shoot meristem stem cell niche and this activity can be further assigned to two adjacent short sequence motifs within this region (Baurle and Laux, 2005) and also indicating that the diverse regulatory pathways that control the stem cells in the shoot meristem converge at these two short sequence elements of the *WUS* promoter thereby suggesting that the integration of regulatory signals takes place at the level of a central transactivating complex.

Characterization of a meristem maintenance mutant in *Antirrhinum majus* shows that it arises from a defect in the *WUS* orthologue *ROSULATA* (*ROA*) (Kieffer *et al.*, 2006). The semidominant *roa* allele revealed an essential role for the conserved C-terminal domain. Expression of either *ROA* or *WUS* lacking this domain causes a failure of meristem maintenance suggesting that the conserved domain mediates an interaction between *WUS* and two members of a small family of corepressor-like proteins in *Arabidopsis*. *WUS* is demonstrated to function by recruiting transcriptional corepressors to repress target genes that promote differentiation, thereby ensuring stem cell maintenance.

Table 1. Some important genes that regulate SAM

Gene	*Mutant Phenotype*	*Protein*	*Function*
KNOTTED1 (KN1)	Defective SAM	Homeodomain transcription factor	Maintenance of SAM
WUSCHEL (WUS)	Reduced SAM and floral organ number	Homeodomain transcription factor	Generation of stein cell daughter cell population
CLAVATA 1, 2 (CLV1, 2)	Excess SAM cells and enlarged meristems	LRR receptor like proteins	Component of receptor kinase signaling complex
CLAVATA 3 (CL V3)	Excess SAM cells and enlarged meristems	Extracellular secretary protein of CLE/ESR1 family	Negatively regulates WUS as a component of CLV1 CLV2 complex
SHOOT MERISTEMLESS (S TM)	Lack SAM and stem cells	Homeodomain transcription factor	Prevents meristem differentiation and organ initiation at SAM
FASCIA TA (FAS)	Enlarged and fasciated SAM	Component of chromatin assembly factor-1	Maintain WUS expression domain and stable WUS transcription
A GA MO US (A G)	Defective stamens and carpels	MADS-box transcription factor	Promotes floral meristem identity
ULTRAPETALA (UL T)	Shoot and floral meristem enlargement	Small family of transcriptional regulator with B box-like motif and SAND domain	Negatively regulates size of WUS domain
SPLAYED (SYD)	Altered SAM	SNF2 chromatin-remodeling ATPase	Regulation of stem cell pool maintenance by transcriptionally controlling WUS
POLTERGEIST and POL-LIKE1 (POL, PLL1)	Suppressed clv mutant phenotypes	Protein phosphatase	Stem cell specification and suppression of stem cell differentiation
HAIRY MERISTEM (HA M)	meristem develops as shoot axis-like tissue	GRAS family transcription factor	Maintenance of SAM
APETALA3 (AP3)	Affected SAM and flower development	MADS-box transcription factor	Specification of pistils and stamens; regulation of WUS and CL V3
EXTRA CELL LA YERS I (ECL 1)	Reduced meristem size		Proper LI development
BREVIPEDICELL us (BP)	Defective pedicel development, cell	KNA T1 homeodomain	Defines growth and differentiation of

	differentiation and radial stylar growth	protein	inflorescence stem, pedicel and style
SHORT INTEGUMENT] (SINI)	Improper ovule formation, flowering time and embryo development		Formation of embryo proper
GNOM/EMBRYO DEFECTIVE30 (GN/EMB30)	Altered zygote polarity	Guanine nucleotide exchange factor	Fixation of apical-basal polarity
PIN-FORMED] (PIN])	Altered number or position of cotyledons and fusion of cotyledons		Polar auxin transport along embryo axis
LEAFY COTYLEDON] (LEC1) FUSCA3	Suspensor adopts embryo fate	CCAAT box binding transcription factor	Cotyledon identity, desiccation tolerance and repression of precocious germination of embryo
MONOPTEROS (MP)	Deleted hypocotyls, adventitious roots from tissue cultures	Auxin response factor 5	Promotes vascular strand continuity in embryonic and postembryonic development
DRN/ESRI	Premature arrest of SAM activity	AP21ERF protein	Regulation of gene expression in meristems
VAMANA (VAN)	Dwarf phenotype and cauline leaves in inflorescence	BELL-homeobox gene	Promotes meristem activity
SERRA TE	Altered meristem and leaf formation	Zinc finger protein	Regulates meristem activity and leaf axial patterning
REG ULATOR OF AXILLARY MERISTEMS1 (RAXI)	Altered timing of meristem development	MYB transcription factor	Links specification of SAM formation with modulation of rate of progression through developmental phases
PASTICCINO (PAS)	Disorganized tumour-like tissue development of aerial parts		Hormonal control of cell division and differentiation
FERTILIZATION-INDEPENDENT ENDOSPERM (FIE)	Altered aerial development	WD homologue of *Drosophila* polycomb protein	Control of shoot and leaf development

CUPULLFORMIS (CUP)	Defective SAM formation	NAC-domain transcription factor	Regulation of organ growth
AINTEGUMENTA (ANT)	Altered size of lateral organs	AP21EREBP transcription factor	Required for growth of shoot lateral organs
ASYMMETRIC LEA VES1 (AS1)	Affected shoot lateral organ formation	MY13 transcription factor	Negative regulation of KNOX genes
PHANTASTICA (PHAN)	-do-		-do
ROUGH SHEATH2 (RS2)	-do-		-do
PINOID (PID)	Pin formed stem	Serine threonine protein kinase	Mediates auxin signaling
FILAMENTOUS FLOWER (FIL)	Abaxial-adaxial polarity is affected	Putative transcription factor with a C2C2 zinc finger domain and a helix-loop-helix domain	Abaxial specification
ROTUNDA2 (RON2)	Altered leaf lamina phenotype	Transcriptional corepressor	Regulates cell expansion during leaf growth.
S TIMP Y (s TP)	Abnormal SAM development	Homeobox gene	Required for growth of vegetative SAM and aerial organs
HALTED ROOT (HLR)	Abnormal or reduced meristematic activity	RPT2a subunit of proteasome	Maintains cellular organization of root and shoot apical meristems
ZWILLE (ZLL)	Variable meristem size and function		Buffers meristem stability
CORONA (CNA)	Altered apical meristem size	Class III homeodomain leucine zipper protein	Promotes organ formation
BRAHMA (BRM)	Dwarfism, altered vegetative and reproductive development	SNF2 homologue of SWI/SNF complex	Regulates chromatin remodeling during. transcription
PRETTY FEW SEEDS2 (PFS2)	Failed embryonic development and reduced fecundity	Homeodomain protein	Regulates ovule patterning or differentiation
TOPLESS (TPL)	Defective SAM	Transcriptional corepressor	Mediates the repressive effects of AUX/IAA proteins

There are other genes that profoundly interact with *WUS*, *CLV* and *STM* with additional levels of regulation (Table 1). For example, expression of *STM* requires the activity of two genes, *CUP-SHAPED COTYLEDON1* (*CUC1*) and *CUC2*, which are positive regulators of *STM* in SAM formation. *STM* suppresses organ initiation by inhibiting the expression of *ASYMMETRIC LEAVES1* (*AS1*) and *AS2* genes that promote organ formation. *AS1* and *AS2*, in turn, repress the expression of *KNOX* genes. Therefore, indirectly, *STM* prevents meristem differentiation in the meristem by allowing the expression of *KNOX* genes. *STM* and *WUS* therefore encode two major regulators of meristem formation and maintenance in *Arabidopsis* and perform independent and complementary roles. Over expression of *AS2* results in the repression of the *KNOX* genes, *BP*, *KNAT2* and *KNAT6* but not *STM* and this over expression also leads to a perturbation of normal adaxial-abaxial asymmetry in lateral organs, resulting in the replacement of abaxial cell types with adaxial cell types, suggesting that *AS2* is sufficient to induce adaxial cell fate and repress *KNOX* gene expression (Lin *et al.*, 2003). *AS2* is a member of plant-specific *LATERAL ORGAN BOUNDARIES* (*LOB*)-domain gene family. In a recent study, Song *et al* (2006) demonstrated that *POLTERGEIST* (*POL*) and *POLTERGEIST-LIKE1* (*PLL1*) are integral components of the *CLV1* signaling pathway; essential for stem cell specification and suppression of stem cell differentiation and are central players in the stem cells by regulating WUS. POL encodes a protein phosphatase 2C (PP2C) with a predicted nuclear localization sequence indicating that it has a role in signal transduction downstream of the CLV1 receptor (Yu *et al.*, 2003). The POL protein represents a unique subclass of plant PP2Cs. POL is ubiquitously expressed through the plant. Apart from POL and PLL1, WUS is an important target of the SNF2-class ATPases SPLAYED (SYD) and BRAHMA (BRM) in the *Arabidopsis* SAM. SNF2 chromatin remodeling ATPases play a vital role in ensuring proper organ development in higher eukaryotes by modulating the accessibility of cis-regulatory DNA regions to transcription factors and to the transcriptional machinery. However, little is known about the biological targets of these regulatory ATPases. Similarly, chromatin assembly factor CAF-1 facilitates the formation of nucleosomes on newly replicated DNA in vitro. Two genes, FASCIATA-1 (FAS-1) and FASCIATA-2 (FAS-2), which regulate genome replication and plant development, are required for the formation of CAF-1 and are implicated in maintaining WUS expression domain by promoting stable WUS transcription by facilitating the appropriate chromatin conformation.

BRM is a SNF2 family protein homologue of BRAHMA, the ATPase of the *Drosophila* SWI/SNF complex involved in chromatin remodeling during transcription. Homozygous BRM loss of function mutants are viable but exhibit numerous defects including dwarfism, altered leaf and root development and several reproduction defects including both male and female gametophyte transmission (Hurtado *et al.*, 2006). The brm mutants present reduced levels of APETALA2 (AP2), APETALA3 (AP3), PISTILLATA and NAC-LIKE activated by AP3/PI. BRM strongly interacts with *Arabidopsis* SWI3C gene but interacts weakly with SWI3B but not with SWI3A or SWI3D genes. A dominant negative allele of the AP2 gene, I28, as suggested by allele competition experiments, isolated from *Arabidopsis*, affects WUS and CLV3 expression (Wurschum *et al.*, 2006). Unlike in floral patterning, AP2 mRNA is active in the center of the SAM and acts via a mechanism independent of AG, which is a repressor of WUS and stem cell maintenance in the floral meristem. Genetic analysis shows that termination of the primary shoot meristem in I28 mutants requires an active CLV signaling pathway, indicating that AP2 functions in stem cell maintenance by modifying the WUS-CLV3 feedback loop.

In contrast to the indeterminate shoot meristem, the floral meristem terminates at the end of flower development, specifically with the formation of gynoecium where stem cell differentiation takes place. This is a result of the overcoming of the self-regulatory WUS-CLV3 feedback loop as discussed earlier. Floral regulatory genes like AGAMOUS (AG), LEAFY (LFY), APETALA1 (AP1) and ULTRAPETALA (ULT) have now been shown to regulate WUS expression. AG ensures floral meristem termination by repressing WUS transcription while WUS activates AG transcription in the centre of the floral meristem. LFY is another key floral meristem gene whose regulation is critical to the control of flower development. AG is directly activated by the transcription factor LFY. LFY and WUS act in a well-coordinated manner in regulating AG expression; while LFY provides flower specificity to AG, WUS provides the regional specificity for AG induction in the central region of floral meristems. Another dimension of temporal specificity for AG induction is provided by ULT. ULT interestingly has a DNA-binding motif previously reported only in animal transcription factors. ULT is a key negative regulator of cell accumulation and size of WUS-expressing OC in *Arabidopsis*. Transgenic petunia in which the E-type and D-type MADS-box genes FLORAL BINDING PROTEIN2 (FBP2) and FBP11, respectively, are simultaneously overexpressed show an early arrest in development at the cotyledon stage and reveal that the possible combined action of FBP1 and FBP11 repress the petunia WUS homolog, TERMINATOR (Ferrario *et al.*, 2006). Ectopic upregulation of the C-type and D-type homeotic genes FBP6 and FBP7 show that they participate in a complex which causes the determinacy supporting a model that a transcription factor complex consisting of C-, D- and E-type MADS-box proteins controls the stem cell population in the floral meristem. In the Petunia mutant hairy meristem (ham), shoot meristems differentiate post-embryonically as continuations of the subtending stem (Stuurman *et al.*, 2002). HAM encodes a putative transcription factor of the GRAS family, which acts non-cell-autonomously from L3-derived tissue of lateral organ primordia and stem provasculature. HAM acts in parallel with TERMINATOR (PhWUS) and is required for continued cellular response to TERMINATOR and PhSTM, thus revealing a novel mechanism by which signals from differentiating tissues extrinsically control stem cell fate in the shoot apex.

Histone acetyltransferases, which are able to acetylate histone and non-histone proteins, play important roles in gene regulation. Many histone acetyltransferases are related to yeast Gcn5, a component of two transcription regulatory complexes SAGA and ADA. Characterization of a mutation in the *Arabidopsis* GCN5 gene showed that in addition to pleiotropic effects on plant development, this mutation also leads to the production of terminal flowers (Bertrand *et al.*, 2003). The flowers show homeotic transformations of petals into stamens and sepals into filamentous structures and produce ectopic carpels. The phenotypes correlate to an expansion of the expression domains within floral meristems of the key regulatory genes WUS and AG, suggesting that the *Arabidopsis* GCN5 is required to regulate floral meristem activity through the WUS/AG pathway.

The PRETTY FEW SEEDS2 (PFS2) gene in *Arabidopsis* encodes a homeodomain protein that regulates ovule development and shares high homology with the PRESSED FLOWER and WUS proteins. In the pfs2-1 allele, the integuments display morphological abnormalities and most embryo sacs fail to develop properly resulting in reduced fecundity (Park *et al.*, 2005). Ectopic PFS2 expression interferes with differentiation of meristem primordia resulting in fascinated stems, altered phyllotaxy, a cessation of primordia differentiation or a combination of these. PFS2 activity also alters AG expression, which accounts for some of the gain- and loss-of-function phenotypes, suggesting that PFS2 affects either ovule patterning or differentiation.

Similarly, along the proximal-distal axis of the outgrowing radially symmetrical ovule primordium, three elements, funiculus, chalaza and nucellus can be distinguished. The asymmetric initiation of the outer integument marks the switch to adaxial-abaxial development, which leads to the formation of a bilaterally symmetrical ovule. The putative transcriptional regulator, NOZZLE (NZZ) plays a role in mediating this transition by controlling the timing of expression of the putative transcriptional regulator INNER NO OUTER (INO) in an abaxial domain of the chalaza, from where the outer integument initiates. Integument formation depends on WUS, which is expressed in the nucellus and is sufficient to induce integuments non-cell autonomously from a region adjacent to its expression domain (Sieber *et al.*, 2004). NZZ is required to restrict the expression pattern of the leucine zipper homeobox gene PHABULOSA (PHB) expression to the distal chalaza from where the inner integument initiates. PHB expression is not established in the distal chalaza of two mutants, aintegumenta (ant) and wus which fail to form integuments. WUS may control integument formation by establishing the chalaza and that outer and inner integument identity determination depends on additional region-specific factors. NZZ is also essential for the normal nucellar expression pattern of WUS suggesting that both WUS and PHB affect processes downstream of NZZ action during the transition from proximal-distal to adaxial-abaxial ovule development.

Biotechnological applications of stem cells

Molecular genetic analysis of SAM regulation has now highlighted the complex circuitry of a plethora of genes (Sablowski, 2007). Several genes like are involved in the different levels of SAM maintenance, differentiation and organ initiation (Fig. 1). A number of other equally important genes which profoundly regulate the SAM maintenance and development have not been discussed in this review,. What about equally fascinating area of root apical meristems with all the interesting genes like SCARECROW, SHORTROOT, PLETHORA and RETINOBLASTOMA etc? (Wildwater *et al.*, 2005, Vernoux and Benfey, 2005). What other genes and microRNAs (Zhou *et al.*, 2007) that directly or indirectly regulate the master regulators, WUS, CLV and STM etc? How exactly WUS and other multiple signals the overlying cells to establish stem cell fate and the circuitry shaping the shoot (Vandenbussche and van der Straeten, 2004)? Which genes are required within stem cells per se to maintain their identity? What down stream signal transduction pathways are involved in SAM formation? What is the exact nature of the "stemness" signature of the stem cells (Mikkers and Frisen, 2005)? How can we select interesting QTLs and mutations through the use of sensitive genetic screens, genome-wide transcriptional profiling and in vivo biochemical experiments along with techniques like laser microdissection? What are the roles of hormones, especially auxins, cytokinins and gibberellic acid in the SAM development both in planta and in vitro? In the present exciting research in the field of animal stem cell research, how stem cells *per se* can be studied in plants and animals on a common theme (Benfey, 1999).

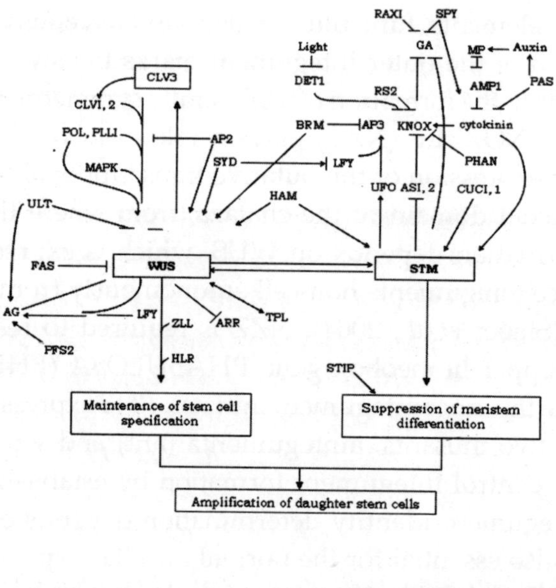

Fig. 1 : Simplified interaction map of genes of SAM Development

Obviously, micropropagation is one of the first major areas benefited by advances in stem cell research. Application of knowledge of stem cell biology can help in the isolation of critical mass of stem cell explant and meristem for virus-free clones and multiply faster and without introducing the unwarranted problems of somaclonal variations. Germplasm conservation of important species, endangered species and commercially exploitable species can be visualized by this knowledge. Identification and utilization of vigorous planting material in elite cultivars may be aided by the knowledge of stem cells. More importantly, rational development of transgenics for engineering yield, marketable quality and specific plant phenotypes and ideotypes using targeted genes in the nebulous metabolic network of stem cell interactions can lead to fruitful application of stem cell research through a systems biology approach. Plant stem cell research may throw, unexpectedly albeit, windfall solutions to many of the presently nagging problems in cultivar development and mprovement. These are some of the questions, vistas and paradigms that emerge in the studies on plant stem cell research particularly in the induction and maintenance of stem cells and organs. A "picturesque denouement" of the plant stem cell research is on the horizon based on the deluge of information that is being generated in this field. It also simply translates into the fact that the science of plant stem cells is as exciting or more than the animal counterpart with equally interesting expectations in store!

REFERENCES

Aida M Ishida T Fukaki H Fujisawa H Tasaka M (1997) Genes involved in organ separation in *Arabidopsis*: an analysis of the *cup-shaped cotyledon* mutant *Plant Cell* **9**: 84-857

Aida M Ishida T Tasaka M (1999) Shoot apical meristem and cotyledon formation during *Arabidopsis* embryogenesis: interaction among the *CUP-SHAPED COTYLEDON* and *SHOOT MERISTEMLESS* genes *Development* **126**: 1563-1570

Baker CC Sieber P Wellmer F Meyerowitz EM (2005) The *early extra petals1* mutant uncovers a role for microRNA miR164c in regulating petal number in *Arabidopsis Current Biol* **15**: 303-315

Baurle I Laux T (2005) Regulation of *WUSCHEL* transcription in the stem cell niche of the *Arabidopsis* shoot meristem *Plant Cell* **17**: 2271-2280

Baurle I Laux T (2003) Apical meristems: the plant's fountain of youth *BioEssays* **25**: 961-970

Belles-Boix E Hamant O Witiak SM Morin H Traas J Pautot V (2006) *KNAT6*: an *Arabidopsis* homeobox gene involved in meristem activity and organ formation *Plant Cell* **18**: 1900-1907

Benfey PN (1999) Stem cells: A tale of two kingdoms *Curr Biol* **9**: 171-172

Berleth T Jurgens G (1993) The role of the *monopteros* gene in organizing the basal body region of the *Arabidopsis* embryo *Development* **118**: 575-587

Bertrand C Bergounioux C Domenichini S Delarue M Zhou-DaoXiu (2003) *Arabidopsis* histone acetyltransferase *AtGCN5* regulates the floral meristem activity through the *WUSCHEL/AGAMOUS* pathway *J Biol Chem* **278**: 28246-28251

Bezhani S Winter C Hershman S Wagner JD Kennedy JF Kwon CS Pfluger J Su Y Wagner D (2007) Unique shared and redundant roles for the *Arabidopsis SWI/SNF* chromatin remodeling ATPases *BRAHMA* and *SPLAYED Plant Cell* **19**: 403-416

Bhatt AM Etchells JP Canales C Lagodienko A Dickinson H (2004) *VAAMANA*-a *BEL1*-like homeodomain protein interacts with KNOX proteins *BP* and *STM* and regulates inflorescence stem growth in *Arabidopsis Gene* **328**: 103-111

Carles CC Choffnes-Inada D Reville K Lertpiriyapong K Fletcher JC (2005) *ULTRAPETALA1* encodes a SAND domain putative transcriptional regulator that controls shoot and floral meristem activity in *Arabidopsis Development* **132**: 897-911

Clark SE (2001) Cell signalling at the shoot meristem *Nature Rev* **2**: 276-284

Cnops G Jover-Gil S Peters JL Neyt P Block S de Robles P Ponce MR Gerats T Micol JL Lijsebettens M van (2004) The *rotunda2* mutants identify a role for the *LEUNIG* gene in vegetative leaf morphogenesis *J Exp Bot* **55**: 1529-1539

DeCook R Lall S Nettleton D Howell SH (2006) Genetic regulation of gene expression during shoot development in *Arabidopsis Genetics* **172**: 1155-1164

Exner V Taranto P Schonrock N Gruissem W Hennig L (2006) Chromatin assembly factor *CAF-1* is required for cellular differentiation during plant development *Development* **133**: 4163-4172

Ferrario S Shchennikova AV Franken J Immink RGH Angenent GC (2006) Control of floral meristem determinacy in petunia by MADS-box transcription factors *Plant Physiol* **140**: 890-898

Fiers M Golemiec E Schors R van der Geest L van der Li KW Stiekema WJ Liu CM (2006) The *CLAVATA3/ESR* motif of *CLAVATA3* is functionally independent from the nonconserved flanking sequences *Plant Physiol* **141**: 1284-1292

Gallois JL Nora FR Mizukami Y Sablowski R (2004) *WUSCHEL* induces shoot stem cell activity and developmental plasticity in the root meristem *Genes Dev* **18**: 375-380

Green KA Prigge MJ Katzman RB Clark SE (2005) *CORONA* a member of the class III homeodomain leucine zipper gene family in *Arabidopsis* regulates stem cell specification and organogenesis *Plant Cell* **17**:691-704

Greve K Cour T la Jensen MK Poulsen FM Skriver K (2003) Interactions between plant RING-H2 and plant-specific NAC (NAM/ATAF1/CUC2) proteins: RING-H2 molecular specificity and cellular localization *Biochem J* **371**: 97-108

Grigg SP Canales C Hay A Tsiantis M (2005) *SERRATE* coordinates shoot meristem functionand leaf axial patterning in *Arabidopsis Nature* **437**: 1022-1026

Hamann T Mayer U Jurgens G (1999) The auxin-insensitive *bodenlos* mutation affects primary root formation and apical-basal patterning in the *Arabidopsis* embryo *Development* **126**: 1387-1395

Hanur VS (2007) Exciting developments in plant stem cell research *Curr Sci* **92**: 1668-1670

Harrar Y Bellec Y Bellini C Faure JD (2003) Hormonal control of cell proliferation requires *PASTICCINO* genes *Plant Physiol* **132**: 1217-1227

Hibara K Takada S Tasaka M (2003) *CUC1* gene activates the expression of SAM-related genes to induce adventitious shoot formation *Plant J* **36**: 687-696

Hobbie L McGovern M Hurwitz LR Pierro A Liu NY Bandyopadhyay A Estelie M (2000) the *axr6* mutants of *Arabidopsis thaliana* define a gene involved in auxin response and early development *Development* **127**: 23-32

Hurtado L Farrona S Reyes JC (2006) The putative *SWI/SNF* complex subunit *BRAHMA* activates flower homeotic genes in *Arabidopsis thaliana Plant Mol Biol* **62**: 291-304

Irish VF Sussex IM (1992) A fate map of the *Arabidopsis* embryonic shoot apical meristem *Development* **115**: 745-753

Jackson D Veit B Hake S (1994) Expression of maize *KNOTTED1* related homeobox genes in the shoot apical meristem predicts patterns of morphogenesis in the vegetative shoot *Development* **120**: 405-413

Jasinski S Piazza P Craft J Hay A Woolley L Rieu I Phillips A Hedden P Tsiantis M (2005) KNOX action in *Arabidopsis* is mediated by coordinate regulation of cytokinin and gibberellin activities *Curr Biol* **15**: 1560-1565

Juarez MT Kul JS Thomas J Heller BA Timmermans MC (2004) microRNA-mediated repression of *rolled leaf* specifies maize leaf polarity *Nature* **428**: 84-88

Kanrar S Onguka O Smith HMS (2006) *Arabidopsis* inflorescence architecture requires the activities of KNOX-BELL homeodomain heterodimers *Planta* **224**: 1163-1173

Katz A Oliva M Mosquna A Hakim O Ohad N (2004) *FIE* and *CURLY LEAF* polycomb proteins interact in the regulation of homeobox gene expression during sporophyte development *Plant J* **37**: 707-719

Keller T Abbott J Moritz T Doerner P (2006) *Arabidopsis* REGULATOR OF AXILLARY MERISTEMS controls a leaf axil stem cell niche and modulates vegetative development *Plant Cell* **18**: 598-611

Kessler S Townsley B Sinha N (2006) L1 division and differentiation patterns influence shoot apical meristem maintenance *Plant Physiol* **141**: 1349-1362

Kidner CA Martienssen RA (2004) Spatially restricted microRNA directs leaf polarity through *ARGONAUTE1 Nature* **428**: 81-84

Kieffer M Stern Y Cook H Clerici E Maulbetsch C Laux T Davies B (2006) Analysis of the transcription factor WUSCHEL and its functional homologue in Antirrhinum reveals a potential mechanism for their roles in meristem maintenance *Plant Cell* **18**: 560-573

Kikuchi K Ueguchi-Tanaka M Yoshida KT Nagato Y Matsusoka M Hirano HY (2000) Molecular analysis of the NAC gene family in rice *Mol Gen Genet* **262**: 1047-1051

Kim JY Yuan Z Jackson D (2003) Developmental regulation and significance of KNOX protein trafficking in *Arabidopsis Development* **130**: 4351-4362

Kirch T Simon R Grunewald M Werr W (2003) The *DORNROSCHEN/ENHANCER OF SHOOT REGENERATION1* gene of *Arabidopsis* acts in the control of meristem cell fate and lateral organ development *Plant Cell* **15**: 694- 705

Kumaran MK Bowman JL Sundaresan V (2002) *YABBY* polarity genes mediate the repression of KNOX homeobox genes in *Arabidopsis Plant Cell* **14**: 2761-2770

Leibfried A To JPC Busch W Stehling S Kehle A Demar M Kieber JJ Lohmann JU (2005) *WUSCHEL* controls meristem function by direct regulation of cytokinin-inducible response regulators *Nature* **438**: 1172-1175

Lenhard M Jurgens G Laux T (2002) The *WUSCHEL* and *SHOOT MERISTEMLESS* genes fulfil complementary roles in *Arabidopsis* shoot meristem regulation *Development* **129**: 3195-3206

Lin WC Shuai B Springer PS (2003) The *Arabidopsis LATERAL ORGAN BOUNDARIES*-domain gene *ASYMMETRIC LEAVES2* functions in the repression of KNOX gene expression and in adaxial-abaxial patterning *Plant Cell* **15**: 2241-2252

Lindsay DL Sawhney VK Bonham-Smith PC (2006) Cytokinin-induced changes in *CLAVATA1* and *WUSCHEL* expression temporally coincide with altered floral development in *Arabidopsis. Plant Sci* **170**: 1111-1117

Lohmann JU Hong RL Hobe M Busch MA Parcy F Simon R Weigel D (2001) A molecular link between stem cell regulation and floral patterning in *Arabidopsis. Cell* **105**: 793-803

Long JA Woody S Barton MK (1999) A mutation at the *TOPLESS* locus causes shoot to root transformation during embryogenesis In 10th International Conference on *Arabidopsis Research* (Melbourne)

Lotan T Ohto M Matsudaira K West MAL Lo R Kwong RK Yamagishi K Fischer RL Goldberg RB Harada JJ (1998) *Arabidopsis* LEAFY COTYLEDON1 is sufficient to induce embryo development in vegetative cells *Cell* **93**: 1195-1205

Lyndon RF (1998) The shoot apical meristem: Its growth and development Cambridge University Press

Mallory AC Dugas DV Bartel DP Bartel B (2004) MicroRNA regulation of NAC-domain targets is required for proper formation and separation of adjacent embryonic vegetative and floral organs *Curr Biol* **14**: 1035-1046

Mayer U Buttner G Jurgens G (1993) Apical-basal pattern formation in the *Arabidopsis* embryo: Studies on the role of the *gnom* gene *Development* **117**: 149-162

Mikkers P Frisen J (2005) Deconstructing stemness *EMBO J* **24**: 2715-2719

Moussian B Haecker A Laux T (2003) *ZWILLE* buffers meristem stability in *Arabidopsis thaliana Development Genes Evol* **213**: 534-540

Muller R Borghi L Kwiatkowska D Laufs P Simon R (2006) Dynamic and compensatory responses of *Arabidopsis* shoot and floral meristems to CLV3 signaling *Plant Cell* **18**: 1188-1198

Parcy F Valon C Kohara A Misera S Giraudat J (1997) The *ABSCISIC ACID-INSENSITIVE3 FUSCA3* and *LEAFY COTYLEDON1* loci act in concert to control multiple aspects of *Arabidopsis* seed development *Plant Cell* **9**: 1265-1277

Park SO Zheng Z Oppenheimer DG Hauser BA (2005) The *PRETTY FEW SEEDS2* gene encodes an *Arabidopsis* homeodomain protein that regulates ovule development *Development* **132**: 841-849

Ray S Golden T Ray A (1996) Maternal effects of the *short integument* mutation on embryo development in *Arabidopsis Dev Biol* **180**: 365-369

Reinhardt D Frenz M Mandel T Kuhlemeier C (2003) Microsurgical and laser ablation analysis of interactions between the zones and layers of the tomato shoot apical meristem *Development* **130**: 4073-4083

Sablowski R (2007) Flowering and determinacy in *Arabidopsis*. *J Exp Bot* **58:** 899-907

Schoof H Lenhard M Haecker A Mayer KFX Jurgens G Laux T (2000) The stem cell population of *Arabidopsis* shoot meristems is maintained by a regulatory loop between the *CLAVATA* and *WUSCHEL* genes *Cell* **100**: 635-644

Sieber P Gheyselinck J Gross-Hardt R Laux T Grossniklaus U Schneitz K (2004) Pattern formation during early ovule development in *Arabidopsis thaliana Dev Biol* **273**: 321-334

Smith L Greene GB Veit B Hake S (1992) A dominant mutation in the maize homeobox gene Knotted-1 causes its ectopic expression in leaf cells with altered fates *Development* **116**: 21-30

Song SK Lee MM Clark SE (2006) *POL* and *PLL1* are *CLAVATA1* signaling intermediates required for *Arabidopsis* shoot and floral stem cells *Development* **133**: 4691-4698

Steinmann T Geldner N Grebe M Mangold S Jackson CL Paris S Galweiler L Palme K Jurgens G (1999) Coordinated polar localization of auxin efflux carrier *PIN1* by *GNOM ARF GEF Science* **286**: 316-318

Stuurman J Jaggi F Kuhlemeier C (2002) Shoot meristem maintenance is controlled by a GRAS-gene mediated signal from differentiating cells *Genes Dev* **16**: 2213-2218

Torres-Ruiz RA Jurgens G (1994) Mutations in the *FASS* gene uncouple pattern formation and morphogenesis in *Arabidopsis* development *Development* **120**: 2967-2978

Trotochaud AE Hao T Guang W Yang Z Clark SE (1999) The *CLAVATA1* receptor-like kinase requires *CLAVATA3* for its assembly into a signaling complex that includes KAPP and a Rho-related protein *Plant Cell* **11**: 393-405

Ueda M Matsui K Ishiguro S Sano R Wada T Paponov I Palme K Okada K (2004) The *HALTED ROOT* gene encoding the 26S proteasome subunit *RPT2a* is essential for the maintenance of *Arabidopsis* meristems *Development* **131**: 2101-2111

Vandenbussche F van der Straeten D (2004) Shaping the shoot: a circuitry that integrates multiple signals *Trends Plant Sci* **9**: 499-506

Vernon DM Meinke DW (1994) Embryonic transformation of the suspensor in *twin* a polyembrynic mutant of *Arabidopsis Dev Biol* **165**: 566-573

Vernoux T Benfey PN (2005) Signals that regulate stem cell activity during plant development *Curr Opin Genet Dev* **15**: 388-394

Vidaurre DP Ploense S Krogan NT Berleth T (2007) *AMP1* and *MP* antagonistically regulate embryo and meristem development in *Arabidopsis Development* **134**: 2561-2567

Vroeman CW Mordhorst AP Albrecht C Kwaaitaal MACJ Vries SC de (2003) The *CUP-SHAPED COTYLEDON3* gene is required for boundary and shoot meristem formation in *Arabidopsis. Plant Cell* **15**: 1563-1577

Waites R Selvadurai HR Oliver IR Hudson A (1998) The *PHANTASTICA* gene encodes a MYB transcription factor involved in growth and dorsoventrality of lateral organs in *Antirrhinum Cell* **93**: 779-789

Wang XH Mitchum MG Gao BL Li CY Diab H Baum TJ Hussey RS Davis EL (2005) A parasitism gene from a plant-parasitic nematode with function similar to *CLAVATA3/ESR* (*CLE*) of *Arabidopsis thaliana Mol Plant Pathol* **6**: 187-191

Weir I Lu JP Cook H Causier B Schwarz-Sommer Z Davies B (2004) *CUPULIFORMIS* establishes lateral organ boundaries in *Antirrhinum Development* **131**: 915-922

Wildwater M Campilho A Perez-Perez JM Heidstra R Billou I Korthout H Chatterjee J Mariconti L Gruissem W Scheres B (2005) The *RETINOBLASTOMA-RELATED* gene regulates stem cell maintenance in *Arabidopsis* roots *Cell* **123**: 1337-1349

Willemsen V Wolkenfelt H de Vrieze G Weisbeek P Scheres B (1998) The *HOBBIT* gene is required for formation of the root meristem in the *Arabidopsis* embryo *Development* **125**: 521-531

Williams L Grigg SP Xie MT Christensen S Fletcher JC (2005) Regulation of *Arabidopsis* shoot apical meristem and lateral organ formation by microRNA miR166g and its AtHD-ZIP target genes *Development* **132**: 3657-3668

Wu XL Dabi T Weigel D (2005) Requirement of homeobox gene *STIMPY/WOX9* for *Arabidopsis* meristem growth and maintenance *Curr Biol* **15**: 436-440

Wurschum T Gross-Hardt R Laux T (2006) *APETALA2* regulates the stem cell niche in the *Arabidopsis* shoot meristem *Plant Cell* **18**: 295-307

Yu LP Miller AK Clark SE (2003) *POLTERGEIST* encodes a protein phosphatase 2C that regulates *CLAVATA* pathways controlling stem cell identity at *Arabidopsis* shoot and flower meristems *Curr Biol* **13**: 179-188

Zhang JZ Somerville CR (1997) Suspensor-derived polyembryony caused by altered expression of valyl-tRNA synthetase in the *twn2* mutant of *Arabidopsis Proc Natl Acad Sci USA* **94**: 7349-7355

Zhou GK Kubo M Zhong R Demura T Ye ZH (2007) Overexpression of miR165 affects apical meristem formation organ polarity establishment and vascular development in *Arabidopsis. Plant Cell Physiol* **48**: 391-404

4

CLONING VECTORS: THE VEHICLES OF GENETIC ENGINEERING

SONALI CHADHA

INTRODUCTION

The technique of genetic engineering became possible because vectors like plasmid and phages reproduce in a host (e.g. *E.coli*) in their usual manner even after insertion of foreign DNA, so that the inserted DNA will also replicate faithfully with the parent DNA. This technique is known as Gene Cloning and the vectors used for this purpose are known as cloning vectors. A vector can thus be defined as a DNA molecule capable of replication in a host organism into which a gene is inserted to construct a recombinant DNA molecule. With this technique, using a variety of vectors, genes can be isolated, cloned and characterized so that the technique has lead to significant progress in all areas of molecular biology.

This chapter will deal with the various vectors for cloning in bacteria and subsequently at the vectors needed for cloning in eukaryotes.

Properties of a Good Vector

A good vector must have the following properties:

1. It should be able to replicate autonomously. When the objective of cloning is to obtain a large number of colonies of the DNA insert, the vector replicon must be under relaxed control so that it can generate multiple copies of itself in a single host cell.
2. It should be easy to isolate and purify.
3. It should have a low molecular weight. Low molecular weight of the plasmid can be beneficial firstly because it becomes much easier to handle i.e. becomes more resistant to damage and is readily isolated from the host cells. Secondly, low molecular weight plasmids occur as multicopies which not only facilitate

their isolation but also lead to gene dosage effects. Finally, with a low molecular weight there is less chance that the vector will have multiple substrate sites for any restriction endonuclease.

4. It should be easily introduced into the host cells, i.e., transformation of the host with the vector should be easy.
5. The vector should have suitable marker genes that allow easy detection and/or selection of the transformed host cells.
6. It should have the ability to integrate itself or the DNA insert it carries into the genome of the host cell.
7. The cells transformed with the vector molecules containing the DNA insert should be identifiable or selectable from the host molecules only.
8. A vector should contain unique restriction sites into which the DNA insert can be integrated without disrupting an essential function.
9. When expression of the DNA insert is desired, the vector should contain at least suitable control elements, e.g. promoter, operator and ribosome binding sites.

PLASMID VECTORS

Plasmids are the most widely used versatile and easily manipulated vectors which occur widely in nature and are found in most bacterial species. They are naturally occurring extra chromosomal DNA molecules. (Plate I-A) Most plasmids exist as double stranded circular DNA molecules. If both strands of DNA are intact circles, the molecule are described as covalently closed circular DNA or CCC DNA. If only one strand is intact, then the molecules are described as open circles or OC DNA.

Not all the plasmids exist as circular molecules. Linear plasmids have been found in a variety of bacteria; e.g. Streptomyces sps. and *Borrelia burdoferi*. To prevent the action of nuclease enzyme on plasmids, two general mechanisms have evolved.

- Either there are repeated sequences ending in a terminal DNA hairpin loop (*Borrelia*)
- The ends protected by covalent attachment of a protein (Streptomyces)

General properties of plasmids

a) They often carry one or more genes and often these genes are responsible for a useful characteristic displayed by the host bacterium eg. plasmid often carries antibiotic resistance genes such as for chloramphenicol, ampicillin, kanamycin, tetracycline etc. Due to the presence of these antibiotic resistance genes bacterium can survive in toxic concentrations of antibiotics. Antibiotic resistance is generally used in the laboratory as selectable marker to ensure that the bacteria contain a particular plasmid.

b) Ability to replicate in the host bacterium is the most fundamental property of the plasmid. Most, or all, of the enzymes and other products needed for this replication are already present in the host cell. The amount of information that the plasmid has to supply may be only a few hundred base pairs. This region of the plasmid that is necessary for replication is referred to as the 'ori' region of replication.

c) Another interesting feature that makes a plasmid an efficient cloning vector is a cloning site. It is a unique restriction site, so that the enzyme concerned will cut the plasmid only once. If a circular molecule is broken at one position, it is converted into a linear molecule and it is relatively simple to join the ends together to reform an intact circle. If the enzyme cuts more than once, the plasmid will be cut in two or more pieces, and joining them up again to make an intact plasmid becomes a much more complicated process. Thus an intact plasmid must contain two or more such unique restriction sites located at a position not essential for replication and any other functions. (Fig. 1)

d) Plasmid Copy Number: Each plasmid is maintained in their bacterial cell at a characteristic copy number mainly due to its replication control system. Thus we can categorize plasmids into two types:

- **Single copy plasmids**: The replication control of single copy plasmids is the same as that of their bacterial host cells so that they replicate and segregate with the bacterial chromosome; this is called *Stringent replication* and the plasmids carrying this kind of replication are known as stringent plasmids.
- **Multi-copy plasmids**: The replication of multi copy plasmids is different from that of their bacterial host genome so that they undergo more than one replication for each replication of their host genome; this is referred to as *relaxed replication* and the plasmids carrying this kind of replication is know as relaxed plasmid.

e) Size of plasmids: The size of the plasmid varies from less than 1 kb to more than 200 kb. Smaller the size of the plasmid more they become easy for gene cloning experiments. Larger plasmids are less in number whereas smaller ones are more in number.

f) Conjugation and Compatibility: Plasmids can be classified into one of the two major types depending upon whether or not they carry a set of transfer genes called the *tra* genes which promote bacterial conjugation. Conjugative plasmids contain these *tra* genes but these genes are absent from the non conjugative plasmids. However, a non conjugative plasmid may under some circumstances, be cotransferred along with a conjugative plasmid when both are present in the same plasmid.

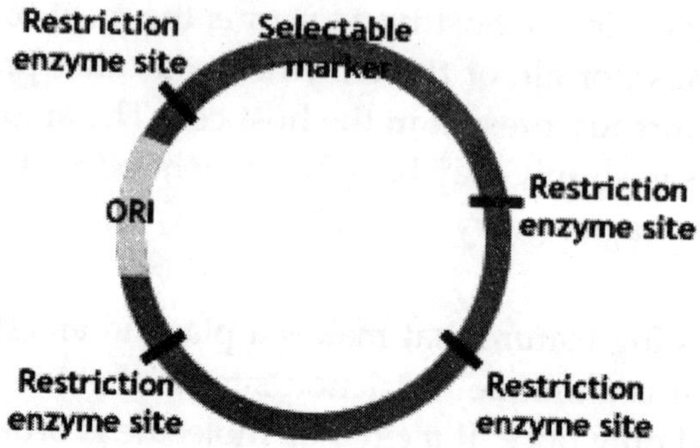

Fig. 1: Plasmid showing the main features as a vector.

Several different kinds of plasmids may be found in a single cell, including more than one different conjugative plasmid at any one time. Cells of *E. coli* have been known to contain upto seven different plasmids at once. In order to be able to exist in the same cell, different plasmids must be compatible. If two plasmids are incompatible then one or the other will be quite rapidly lost from the cell. Different types of plasmids can therefore be assigned to different incompatibility groups on the basis of whether or not they coexist together. Plasmids from a single incompatibility group are often related to each other in various ways.

Development of Plasmid Cloning Vectors

Development of plasmid cloning vectors proceeded in the following three phases:

In the I phase, vectors that were made available during 1973- 1976 included pSC101, ColE1 and pCR1. All of these had the following drawbacks:

a) They showed poor replication (Additional reading*)

b) There was an absence of suitable selectable markers

c) They carried not more than two restriction sites for cloning.

These drawbacks were taken care of in pBR313, but this plasmid was too large, half of its DNA being nonessential. This was reduced in size giving rise to pBR322. (4.362 Kb) both pBR313 and pBR322 were produced in 1977 and remained popular for many years.

In the II phase, size of plasmid vectors was greatly reduced, because the efficiency of transformation is inversely related to size of vector. Also larger plasmids have lower copy number. A number of derivatives of pBR322 were obtained like pAT153, pXf3, pBR327 etc. they made use of antibiotic resistance for selecting recombinant plasmids.

The III phase of development of plasmid vectors involved:

a) Incorporation of Hind III fragment of β galactosidase gene to prevent selection through blue/white colonies due to a- complementation e.g. pUC8 vectors.

b) Incorporation of DNA sequences from single stranded M13 phage for the generation of ss DNA template for sequencing e.g. phagemid vectors

c) Incorporation of promoter DNA sequence for transcription *in vitro* e.g. pGEM vectors

ADDITIONAL READING

Replication of ColE1 based plasmids

ColE1 is a small circular plasmid which codes for a 57 KDal protein toxin (colicin E1) which can kill other ***E. coli*** cells by depolarizing the bacterial membrane.

It carries an origin of replication (oriV) and a region (bom) at which it can be mobilized for transfer to other bacteria. It also contains some genes mentioned in Table 1.

The ColE1 plasmid carries the following genes:

Table 1 : Table showing the various genes present in Col E1

gene	function
cea	the colicin toxin
imm	immunity protein
kil	lysis protein
inc	RNA I - incompatibility determinant
RNA II	primer for replication
rop	protein that regulates priming and copy number
mob	proteins for mobilization during conjugation
cer	maintains plasmid as monomers
exc	exclusion protein

Replication of ColE1 proceeds unidirectionally from oriV and depends on RNA II which is transcribed from a promoter 555 bp away from oriV. When this transcript reaches oriV, one of two possible fates is possible:

- The RNA-DNA hybrid is cleaved by RNaseH to generate a free 3'-OH end which serves as a primer for DNA polymerase I to initiate replication. DNA polymerase III takes over after a short distance.
- The RNA-DNA hybrid is not cleaved by RNase H and transcription continues nonproductively.

The choice between these fates depends on the formation of a specific secondary structure in the nascent RNA II transcript. If the structure forms correctly then RNaseH processing occurs correctly; if not, it does not. Clearly, anything that interferes with the formation of the correct secondary structure will interfere with plasmid replication.

There is a specific RNA whose function is to do just that! RNA I is a 108 nt molecule which is transcribed — in the opposite direction to RNA II — from a promoter located between the RNA II promoter and the origin of replication:

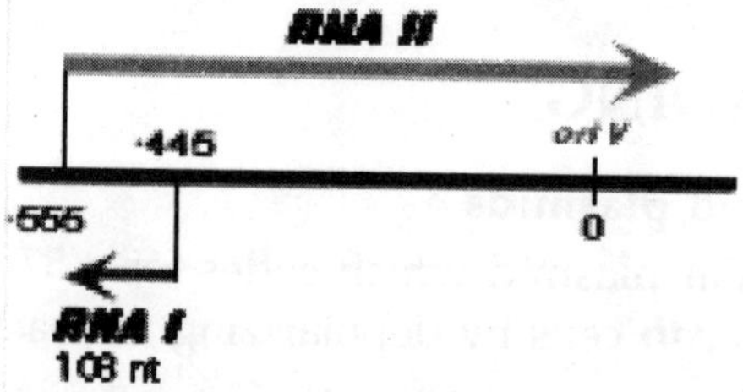

Fig 2: Showing Antisense RNA regulation with RNA I and RNA II transcribing in the opposite direction.

Since RNA I is perfectly complementary to RNA II, it is able to form a stable RNA-RNA hybrid with it. If it does, then RNA II will not be able to function as a primer. This is an example of **antisense RNA** regulation (Fig. 2).

Replication of ColE1 also depends on the Rop protein.

Rop is a small protein (63 aas) which increases the rate of binding of RNA I to RNA II. If Rop is absent, plasmid replication will be more frequent.

The relative copy numbers of pBR322 and pUC18 illustrate the role of Rop perfectly. There are approximately 15 copies of pBR322 per cell; however, there are 50-100 copies of pUC18, which is derived from pBR322. The relevant difference between the two plasmids is the fact that pBR322 contains the rop gene but pUC18 does not.

Turnover of the RNAI molecule also plays a role in the control of replication of these plasmids. RNAI is a rare example of a prokaryotic RNA that is polyadenylated. It contains a short, 25-50 nt, long polyA tail. This tail acts as a starting site for exonucleolytic degradation of RNAI by polynucleotide phosphorylase, a degradation that is facilitated if RNA I has been cleaved by RNAse E.

pBR322

Nomenclature of plasmids:

In the name pBR322 :

- 'p' stands for plasmid
- BR stands for the laboratory in which the vector was originally constructed (BR stands for Bolivar and Rodriguez, the two researchers who developed pBR322)
- 322 distinguish this plasmid from others developed in the same laboratory (there are also plasmids called pBR325, pBR327, pBR328, etc.)

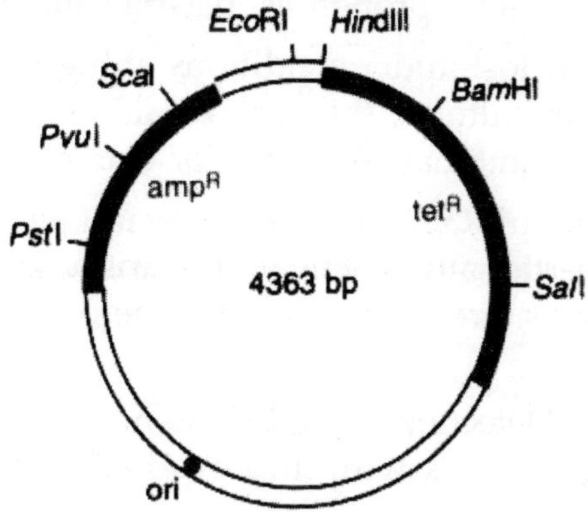

Fig 3: Diagram showing features of Plasmid pBR322

The plasmid pBR322, an ideal cloning vector, is the most common plasmid used in gene cloning experiments. It has the following advantages:

1. The plasmid is much smaller than the natural plasmid (4363 bp). (Fig. 3). Low size avoids the problems encountered during handling and purification of plasmid. Thus the recombinant molecule can be constructed with ease.
2. Since this plasmid is small, the length of foreign DNA that can be incorporated increases. Foreign DNA of upto 6 kb length can be inserted into the plasmid.
3. Another feature of plasmid pBR322 is that it carries two sets of antibiotic resistance genes namely amp^r and tet^r with unique restriction sites. Insertion of a new gene at a site in any one of these marker genes will inactivate the marker genes. This helps in the selection of *E. coli* cells harboring recombinant plasmids.
4. pBR322 has a relaxed origin of replication i.e. its replication is not tightly linked to cell division. The usual copy number of pBR322 is 15 molecules per *E. coli* cell. But this number can be increased upto 1000-3000, by plasmid amplification in the presence of a protein synthesis inhibitor such as chloramphenicol.

   Some plasmids have the ability to replicate in the absence of protein synthesis. In the presence of protein synthesis inhibitor the main bacterial chromosome cannot replicate. During this time the plasmid can replicate. This results in high copy number.
5. This plasmid is used as a raw material to construct several other useful plasmids or other cloning vectors.
6. Single restriction sites for various enzymes are present on the plasmid. This can be used to open the plasmid at a specific point in order top insert the foreign DNA.

Besides having so many useful properties, plasmid pBR322 has few disadvantages.

1. *Instability*: pBR322 is lost under conditions of high pressure (e.g. in continuous culture and in batch culture.) this may create a problem in the large scale fermentation of the recombinant bacteria for commercial purposes.
2. *Cloning sites*: Though pBR322 has twenty one unique restriction sites, only eleven can be used to insertionally inactivate the antibiotic resistance genes. Another three restriction sites, are present within the origin of replication region and thus cannot be used.
3. *Detection technique*: Detection of pBR322 is not convenient and reliable. Thus a number of pBR322 vectors with improved selectable markers have been designed.

Pedigree of Plasmid pBR322 : Plasmid pBR322 was designed in order to posses the following properties.

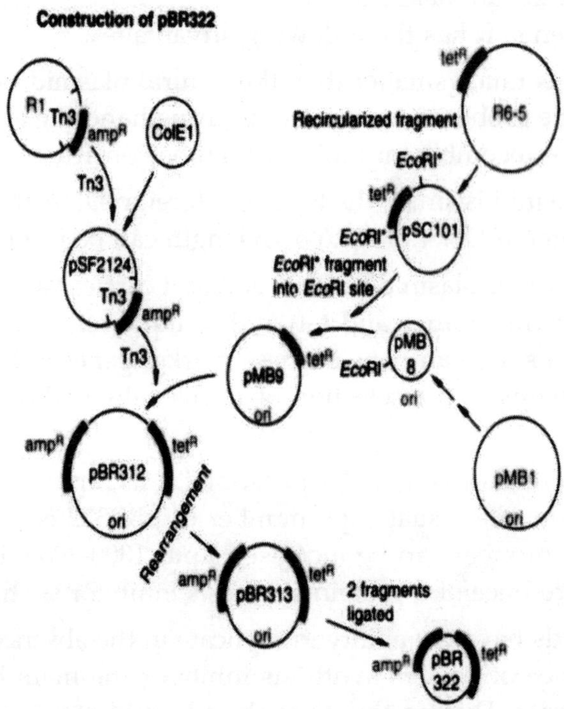

Fig. 4 : Pedigree of pBR322 (After Brown, 2001)

pBR322 comprises DNA derived from three different naturally occurring plasmids. The amp^r gene originally resided on the plasmid R1. R1 is an antibiotic resistance plasmid which occurs in the natural population of *E.coli*. The tet^r gene is derived from R6-5 which again is an antibiotic resistant plasmid.

The amp^r transposon, Tn3 from the plasmid R1was transposed onto pMB1 to form pMB3. This plasmid was reduced in size by EcoRI* rearrangement to form a tiny plasmid, pMB8, which carries only colicin immunity. EcoRI* fragments from pSC101 were combined with pMB8, opened at its unique EcoRI site and the resulting the resulting chimeric molecule rearranged by EcoRI* activity to generate pMB9. In a separate event, Tn3 of R1 was transposed to ColE1 to form pSF2124. The Tn3 element was then transposed to pMB9 to form pBR312. EcoRI* rearrangement of pBR312 lead to the formation of pBR313 from which two different fragments were isolated and ligated together to form pBR322. (Fig. 4)

pBR327: This plasmid was constructed by removing a 1089 bp segment from pBR322 (Fig. 5). This deletion left the amp^r and tet^r genes intact, but changed the replicative and conjugative abilities of the resulting plasmid. This made pBR327 to differ from pBR322 in two ways:

a) pBR327 has a higher copy number than pBR322 i.e. 30–35 molecules are present per *E.coli*. This number can be amplified to 100 molecules per cell. Such high numbers makes this vector more suitable to study the function of cloned gene. If more copies of cloned gene are present in the host cell, then it becomes more likely to detect the effect of cloned gene.

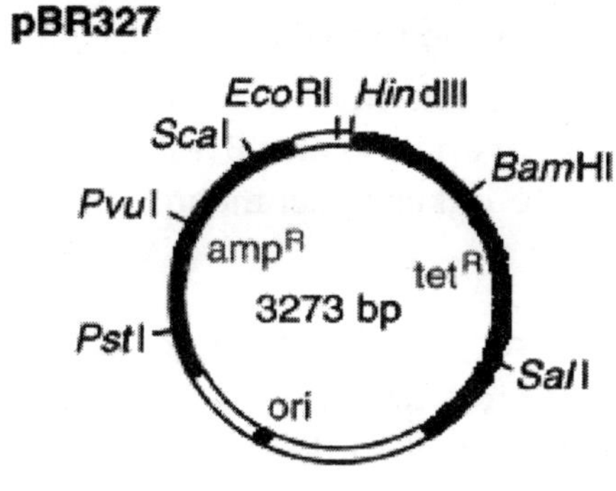

Fig. 5: Diagram showing features of pBR327

b) The deletion leads to destruction of the conjugative ability of the plasmid thereby making it non conjugative. A non conjugative plasmid thus cannot mediate its own transfer to other *E. coli*. This becomes important biological containment, preventing the possibility of recombinant pBR327 molecule from escaping from the test tube and colonizing bacteria in the gut of the worker. This becomes favorable when the cloned gene is harmful.

pUC8: The pUC series of plasmids were all constructed by **Jeffrey Vieira** and **Joachim Messing** and all are derived from pBR322. It is a 2750 bp plasmid derived from pBR322 although only the ori region and amp^r gene have remained. The nucleotide sequence of the amp^r has been changed so that it no longer contains the unique restriction sites. In fact all the cloning sites are clustered into a short segment of the lacZ′ gene carried by the pUC8.

Advantages of pUC8 over pBR322

a) The manipulations in the pBR322 lead to a mutation in the ori region. This resulted in the plasmid of very high copy number of about 500-700 even before amplification.

b) This was significant in the yield of cloned DNA.

c) Identification of the recombinant cells can achieved in a single step i.e. plating the cells onto the agar medium containing ampicillin and X- gal. The selection time for pUC8 is just half that for pBR322 and pBR327. The lacZ′ gene codes for the enzyme β-galactosidase. Cloning with pUC8 involves insertional inactivation of the lacZ′ gene, with recombinants identified because of their inability to synthesize β-galactosidase. This enzyme is involved in the breakdown of lactose to glucose and galactose. X- gal (5-bromo 4 βchloro 3 β indolyl- β-D galactopyranoside) is broken down by β -galactosidase to a product that is colored deep blue. If X- gal (plus an inducer of enzyme such as IPTG) is added to agar, along with ampicillin, then nonrecombinant colonies, the cells of which synthesize β-galatosidase will be colored blue, whereas recombinants with a disrupted lacZ′ gene unable to make β–galactosidase will appear white.

d) The clustering of restriction sites, allows DNA fragment with two different sticky ends, eg. EcoR1 at one end and BamH1 at the other, to be cloned without any manipulations like linker attachment.

Shuttle Vectors

Plasmids with narrow host ranges will only replicate in certain bacteria. In particular, most of the modern cloning vectors are derived from the Col E1 plasmid which has a narrow host range. Thus if you want to do molecular genetics in strains outside this range, you have to either (a) construct a new set of vectors suitable for general cloning in your chosen strain, or (b) modify one of the common vectors so that it will also replicate in your chosen organism.

The second option is the easier option since all it requires is the addition of an origin of replication which will function in your chosen organism. Such vectors, with two different origins of replication suitable for two different bacteria, are called shuttle vectors.

BACTERIOPHAGE:

Bacteriophages, phages, bacterial viruses or virions are small particles of protein coated nucleic acid molecules. The genetic material may be linear or circular, double or single stranded RNA or DNA. As the name implies, phages infect suitable bacterial hosts. Coliphages are particularly interesting as they are *E. coli* specific, which is scientists' favorite experimental microorganism.

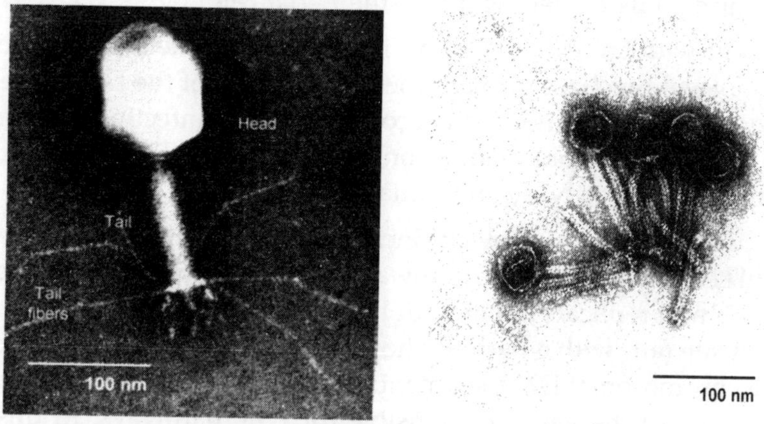

Fig. 6: Electron Microscopic images of
(a) T4 phage (b) Lambdoid phage

A number of such viruses of different genetic material have been reported eg. φ × 174, M13, R209 and phage λ. Unlike plasmid vectors, the phage vectors are required for cloning of large DNA fragments and, therefore, the gene bank or the genomic library can be constructed. (Figs. 6 and 7)

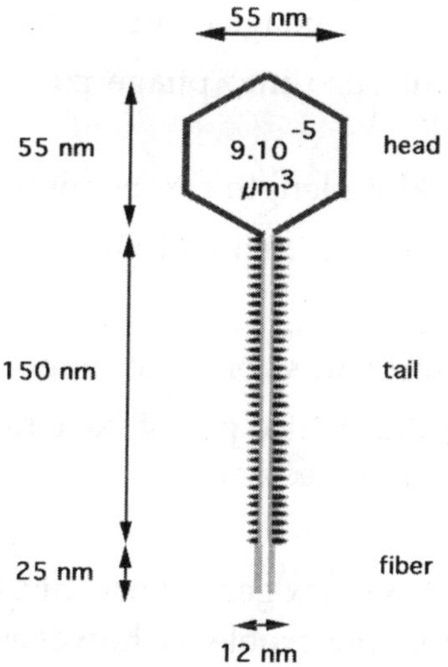

Fig. 7: Bacteriophage lambda with head and tail structure

Bacteriophage Lambda (λ) is a tadpole shaped coliphage consisting of an isometric head and a non-contractile tail. It is a typical head and tail phage. The DNA is contained in the polyhedral head structure and the tail serves to attach the phage to the bacterial surface and to inject the DNA into the cell. In the head it possesses 50 genes in its 49kb genome. The genome has 12 nucleotide single stranded complementary 5′-ends (cosL and cosR). These sticky ends are joined together by host ligase forming a circular molecule, once the virion injected its nucleic acid into the *E. coli* cell.

On attachment with tail to cell wall of *E. coli* it injects linear DNA into the cell. The linear ds DNA molecule cyclizes through the single strand of the DNA (See Plate Fig. B). As a temperate virus, it can undergo a productive lytic infection or form a dormant lysogen, hence producing turbid plaques on an *E. coli* lawn. The decision, whether the infection becomes virulent or lysogenic is made very early in λ's lifecycle.

In the lytic pathway phage λ replicate independently and phage particles are released into the bacterial cell. In the lysogenic pathway phage λ stably integrate into the bacterial host chromosomes and replicate along with host chromosome. Phage genome integrates by an attachment site (att) with a partially homologous site of *E. coli* chromosome, where it replicates as a chromosomal DNA segment. In this case a protein is produced by the cI genes which repress all the genes responsible for lytic pathways. In this pathway no phage structural proteins are synthesized. The interaction of two proteins, the cI genes expressed proteins (by phage genome) and 'cro' genes expressed protein (by *E. coli* chromosomes) decides the events of the lytic and lysogenic cycle. Phage λ has about 50 genes where only about 50% of the genes are necessary for growth and phage formation. Non- essential DNA sequence, therefore is replaced by donor DNA and the inserted DNA propagated as phage λ recombinant with no detrimental to its replication or packaging.

Following are the advantages of phage cloning system over the plasmids:

- DNA can be packed in vitro into phage particles and transcribed into *E.coli* with high efficiency,
- Foreign DNA upto 25kb in length can be inserted into phage vector, and
- Screening and storage of recombinant DNA is easier.
- Stable propagation
- Multiple copies of insert and strong promoters

Although there are many different types of bacteriophage, only phage λ and M13 are found to be potential as cloning vectors.

Phage λ as a vector:

The large genome size and complex genetic organization of lambda had posed initial problems with its use as a vector. The problems, however, were surmounted through the sustained efforts of researchers, and lambda has been developed into an efficient vector.

The broad objectives in constructing various phage vectors are:

(i) The presence of cloning sites only in the dispensable fragments,

(ii) The capacity to accommodate foreign DNA fragments of various sizes,

(iii) The presence of multiple cloning sites,

(iv) An indication of incorporation of DNA fragments by a change in the plaque type,

(v) The ability to control transcription of a cloned fragment from promoters on the vector,

(vi) The possibility of growing vectors and clones in high yield in addition to easy and ready recovery of cloned DNA, and

(vii) Introduction of features contributing to better biological containment.

There are several difficulties in the use of lambda as a vector. Some of them are:

1. Plaques are produced not colonies.
2. Useful for cloning DNA but no use for gene expression, eg. production of gene products in biotechnology.
3. 20 kb is still not large enough for many eukaryotic genes. For larger inserts cosmids or eukaryotic vectors can be used.

Manipulation of Restriction Sites

The major obstacle to the use of phage lambda as a cloning vector was essentially the presence of multiple recognition sites for a number of restriction enzymes in its genome. Initially, all attempts were directed towards minimizing the number of EcoRI sites. Murray and Murray (1974) were able to construct derivatives of lambda with only one or two EcoRI sites. Similarly, Rambach and Toillais constructed lambda derivatives with EcoRI sites only in the nonessential region of the genome by repeated transfer on restrictive and nonrestrictive hosts. After several cycles of digestion, packaging, and growth, phage derivatives with desirable restriction sites and full retention of infectivity were obtained.

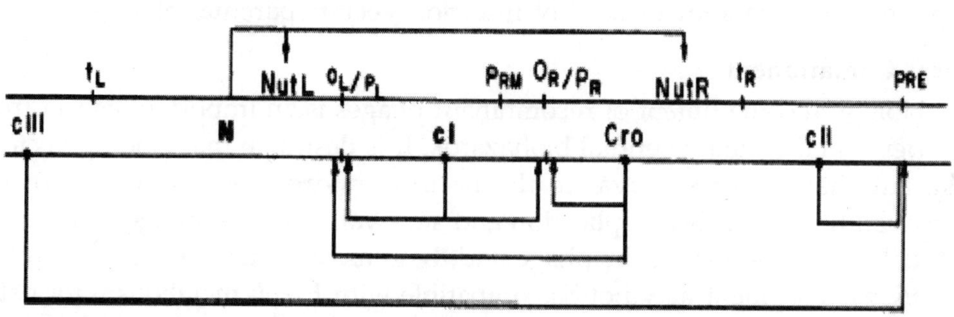

Fig. 8: Regulatory region of lambda genome. Binding sites of specific gene products are indicated by the arrowheads.

Size Limitation for Packaging

The second problem was the requirement of a minimum and maximum genome length (38 and 53 kbp, respectively) for the efficient packaging and for the production of viable phage particles. The viability of the bacteriophage decreases when its genome length is greater than 105% or less than 78% of that of wild-type lambda. Genetic studies of specialized transducing bacteriophages showed, however, that the central one-third of the genome is not essential for lytic growth. (Fig. 8)

However, too much DNA cannot be deleted because there is a minimum 38-kbp requirement essential for efficient packaging.

Transfection of Recombinant Molecules

The problem of transfection of recombinant molecules constructed in vitro was overcome by the successful in vitro assembly of viable and infectious phage particles Two types of in vitro packaging systems have been developed so far, i.e., two-strain packaging and single-strain packaging.

Two-strain packaging: The basis of the two-strain in vitro packaging system is the complementation of two amber mutations. Two lambda lysogens, each carrying a single amber mutation in a distinctly different gene, are induced and grown separately so that they can synthesize the necessary proteins. Neither of the lysogens alone is capable of packaging the phage DNA and producing infectious phage particles. However, when the extracts are mixed, mature phage particles are produced by complementation.

Single-strain packaging: There were successful development of a single-strain packaging system by a deletion in the cos region of prophage, rendering the prophage.

DNA unpackagable because cos is the packaging origin. Induction of the lysogen results in the intracellular accumulation of all protein components needed for packaging. However, packaging of phage DNA is prevented by the lack of cos sites on the prophage DNA. On the other hand, exogenous DNA with cos sites is packaged efficiently to produce an infectious bacteriophage particle. The single-strain system is superior to two-strain system in having a lower need of Insertion Vectors parental phages.

Biological Containment

The biological containment of recombinant phages is an important aspect from the point of view of ethics and eventual biohazards. It is desirable that cloning vectors and recombinants have poor survival in the natural environment and require special laboratory conditions for their replication and survival. The lytic phages offer a natural advantage in this respect since the phage and the sensitive bacteria coexist only briefly. A newly inserted segment may not be compatible with *E. coli* metabolism for extended periods.

Many phage vectors have been constructed in the recent past, each with its own special features. However, there is no universal lambda vector which can fulfill all the desirable objectives of the cloning experiments. The choice of a vector depends mainly on (a) the size of the DNA fragm ent to be inserted, (b) the restriction enzymes to be used, (c) the necessity of the expression of the cloned fragment, and (d) the method of screening to be used to select the desired clones.

Bacteriophage lambda vectors can be broadly classified into two types:

(i) Insertion vectors and (ii) Replacement vectors.

Insertion vectors

Since the maximum packageable size of the lambda genome is 52 kb, small DNA fragments can be inserted without the removal of the non essential (stuffer) fragment.

These vectors are therefore called insertion vectors. Cloning of foreign DNA in these vectors exploits the insertional inactivation of the biological function which differentiates recombinants from non-recombinants. When cloning into an insertion vector the phage DNA is cleaved with restriction enzyme that cuts it only once and the target DNA is inserted into this site. (Fig. 9)

Insertion vectors are particularly useful in cloning small DNA fragments such as cDNA, λgt10, λgt11 and λ ZAPII are examples of this type of vector.

λgt10: It can carry upto 8kb of new DNA inserted into a unique EcoRI restriction site located in the cI gene.

λ ZAPII: in this , 10 kb of DNA can be inserted in any of the 6 restriction sites within a polylinker. This inactivates the lacZ′ gene and thus the non-recombinants can be distinguished as clear plaques in X-gal Agar.

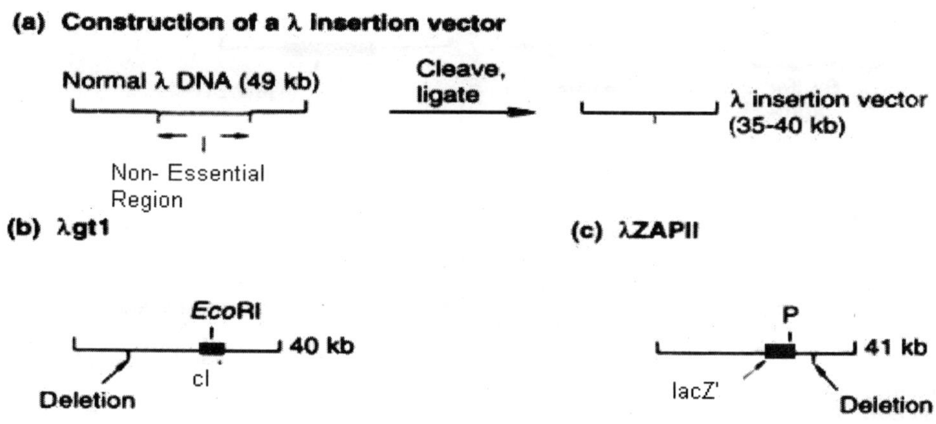

Fig.9: (a) Construction of λ insertion vector. (b, c). P = polylinker, in the lacZ′ gene of λ ZAPII, containing unique restriction sites EcoRI, NotI, XbaI, XhoI, ScaI, SpeI. (After Brown, 2001)

Replacement Vectors: Taking advantage of the maximum and minimum genome size essential for efficient packaging and the presence of the nonessential central fragment, it is possible to remove the stuffer fragment and replace it with a foreign DNA in the desired size range. This forms the basis of lambda derived replacement vectors. Cloning of a foreign DNA in these vectors involves :-

(i) Preparation of left and right arms by physical elimination of the nonessential region,

(ii) Ligation of the foreign DNA fragment between the arms, and

(iii) In vitro packaging and infection.

The replacement vectors contain a pair of restriction sites to excise the central stuffer fragments, which can be replaced by a desired DNA sequence with compatible ends. The presence of identical sites within the stuffer fragment but not in the arms facilitates the separation of the arms and the stuffer on density gradient centrifugation. In many vectors, sets of such sites are provided on attached polylinkers so that an insert can be easily excised.(Fig. 10) Two purified arms cannot be packaged despite their being ligated to each other, because they fall short of the minimum length required for packaging. This provides positive selection of recombinants. The replacement vectors are convenient for cloning of large (in some cases up to 24 kbp) DNA fragments and are useful in the construction of genomic libraries of higher eukaryotes.

Two popular replacement vectors are:

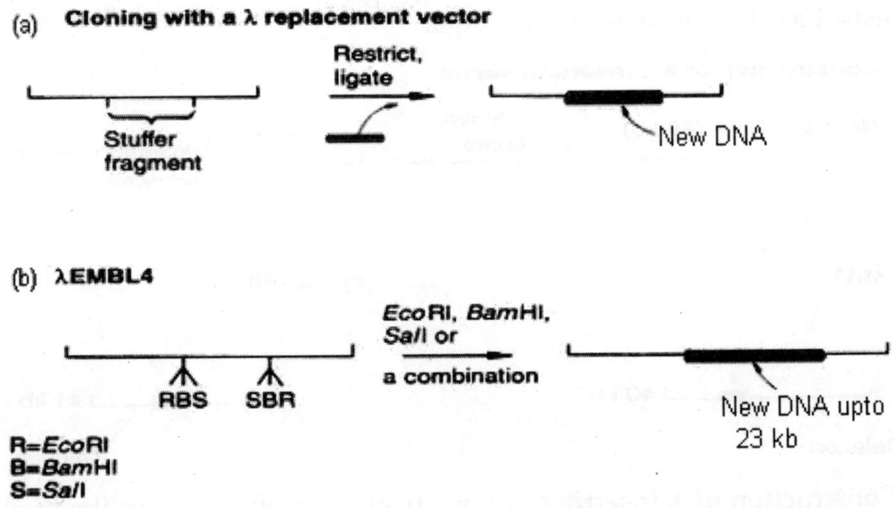

Fig.10: λ replacement vectors. (a) Showing cloning with a λ replacement vector (b) Construction of λEMBL4 (After Brown, 2001)

- ***λEMBL4*** : The λEMBL series of phages is derived from λ1059 by replacing the region of the ColEl plasmid by a fragment carrying the trpE gene of *E. coli*. The removal of the ColEl plasmid region facilitates screening of recombinants by using DNA probes cloned in ColEl-derived plasmid vectors. This vector is named λEMBL1. EcoRI sites are removed by repeated cycling on restricting and non modifying hosts to derive λEMBL2. A new polylinker is introduced into λEMBL2 to construct λEMBL3, which has SalI, BamHI, and EcoRI sites. λEMBL4 has the same sites as λEMBL3 in the reverse orientation. λ EMBL4 can carry upto 20kb of inserted DNA.

- ***λGEM11 and λGEM12:*** Each of these has the capacity of 23kb with the stuffer fragment flanked by polylinkers that contain seven different restriction sites. The polylinkers are slightly different in the two vectors, but in most cases the outermost sites are for the restriction enzyme SfiI, which recognizes the sequence GGCCNNNNNGGCC. This sequence is very rare and unlikely to be present in the DNA fragment that has been cloned. Restriction of the recombinant vector with SfiI can therefore be used to cut out the cloned fragment, with a high likelihood that the fragment will be recovered intact. As with λEMBL4, size or the Spi phenotype is used to select the recombinants.

Filamentous Phages as cloning vectors: Due to their unique biology, M13 and related single-stranded phages (f1, fd) are useful cloning vectors. These f virions are long and thin and contain a closed loop of single stranded DNA. Because the phages readily accept inserts of foreign DNA and they supply one strand of that DNA in an easily isolated form, vectors based on Ff phages have become standard choice.

M13 Bacteriophage:

M13 is a filamentous, male-specific coliphage composed of a single-stranded circular DNA genome and a protein coat made up of three kinds of capsomeres. (Fig. 11) It is 870nm long and 6nm wide. Because of its filamentous structure additional DNA can inserted into its genome and packaged into an extended.

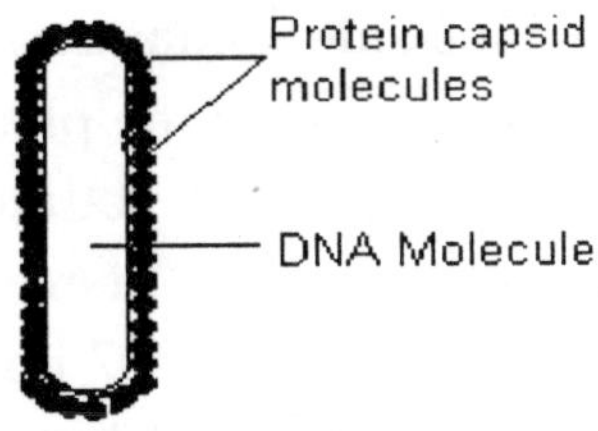

Fig. 11: Filamentous structure of M13

During infection, the single-stranded DNA is converted to the double-stranded replicative form (RF) from which the mature single-stranded form is synthesized. M13 does not lyse its host, but instead replicates until over 100 copies are present in the cell. When the bacterium divides, each daughter cell receives copies of the phage genome, which continues to replicate, thereby maintaining its overall numbers per cell. The phage particles containing single-stranded DNA are continuously secreted into the media by the cells. On agar plates, turbid plaques form because infected cells grow 2-3 times more slowly than uninfected cells. In liquid culture, as many as 5 X 10^{12} phage/ml can be obtained. Since the virus particles are very small relative to the size of the host bacteria the host cells can be easily removed by low-speed centrifugation.

Phage M13 is widely used in nucleotide sequencing and site-directed mutagenesis since its genome can exist either in a single-stranded form inside a phage coat or as a double stranded replicative form within the infected cell. During replication, only the plus strand of the replicative form is selectively packaged by the phage proteins. The replicative form is a covalently closed circular molecule and hence can be used as a plasmid vector and transformed into the host by the usual transformation procedures. The vectors derived from M13, e.g., M13mp18 and M13mpl9, have the same polylinker as that of pUC18 and pUC19, respectively. The DNA fragments having non complementary ends can be directionally cloned in this pair of vectors, and the two strands of DNA can be sequenced independently.

M13 as Vectors

There are several features of M13 which make it an attractive cloning vector. Firstly, the small size of M13 i.e. 10kb makes it easy to handle. Secondly, the double stranded replicative form (RF) of the M13 genome behaves very much like a plasmid and can be treated like a one for cloning experiments. M13 can thus be easily prepared from a culture of infected *E. coli* cells and can be reintroduced by transfection. Thirdly, the genes cloned with an M13 based vector can be easily obtained on the form of single- stranded DNA which is useful for the techniques of in vitro mutagenesis, DNA sequencing.

As the wild type M13 DNA molecule contains several genes, for its replication, genes coding for its coat protein and genes for the phage specific replicative enzymes, any modification in the portion of phage DNA coding for the above will impair the replicative ability of the resulting molecule. Thus there is very little scope for alterations in the phage genome. In case of M13 genome (6407 bp) only 507 nucleotide intergenic sequences is free for insertion of any foreign DNA but this region contains the origin of replication which must remain intact. Thus, there is very less space for the incorporation of foreign DNA.

Inspite of the above mentioned hassles, M13 acts an excellent cloning vehicle as it offers single stranded versions of cloned DNA.

M13mp1 and M13mp2

M13mp1 is constructed by incorporation of lacZ'gene into the intergenic sequence of the wild type M13 phage. This resulted into blue plaques on X-gal agar plates. The lacZ' gene does not have any unique restriction site, however it has a hexa nucleotide sequence, GGATTC near the start. If the second G residue was placed by a residue then this becomes the site for EcoRI. (GAATTC). This conversion was carried out by in vitro mutagenesis thereby resulting in M13mp2. M13mp2 has a slightly different lacZ' gene base sequence. Due to this, the enzyme β- galactosidase encoded by the lacZ' gene has aspargine instead of aspartic acid in the fifth position. The change does not affect the activity of β - galactosidase enzyme (See Plate Fig. E). This can be tested on X-gal Agar plates where non recombinants give blue plaques where as recombinants form clear plaques.

M13 mp7

Messing, Crea and Seeburg modified M13mp2 by introducing BamHI, Hind III, Sal I PstI sites into the b-galactosidase gene. This phage is termed M13mp7. In simpler words a polylinker was inserted into the EcoRI site of the lacZ' gene. (See Plate Fig. F) Thus M13mp7 is a more complex vector with four possible insertional sites. Foreign DNA with the corresponding sites can be inserted into these sites. The recombinant M13mp7 will form clear plaques whereas the non recombinants will form blue plaques on X-gal agar plates.

Insertion of the DNA fragment into the polylinker will inactivate the lacZ' gene which in turn prevents the production of β galactosidase. This is shown by the formation of clear plaques on X-gal agar by the recombinant phage DNA.

ADDITIONAL READING

High level expression of recombinant proteins in bacteria is a common process in Biotechnology. The production of human insulin in bacteria is one such example. Another example is the production of plant calreticulin protein in bacteria as a means to generate anti-calreticulin antibodies in rabbits.

Probably the most widely used bacterial protein expression is the **pET vector system** which was developed in 1984 by Bill Studier and colleagues at Brookhaven National Labs. The vector system is now sold commercially by **Novagen.** One of the most innovative pET vectors is **pETBlue** which utilizes a lacZ gene (alpha region) on the opposite coding strand such that blue-white screening can be done but the recombinant protein does not contain any lacZ fusion sequences.

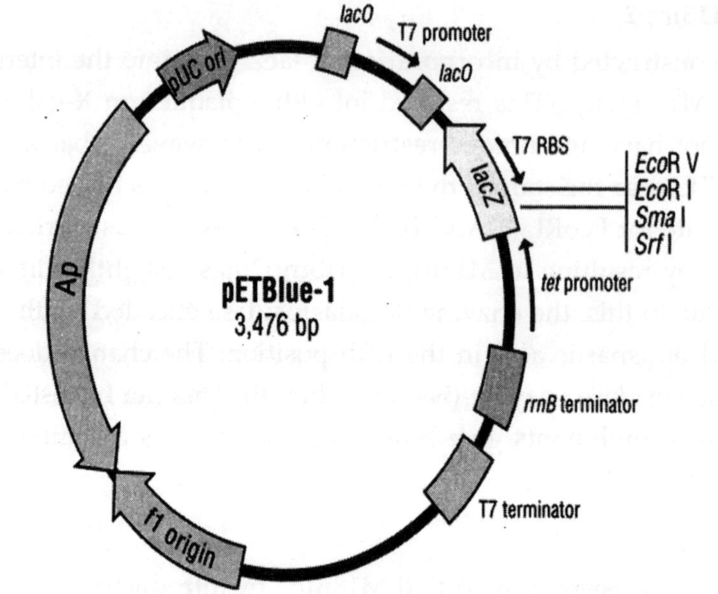

Fig. 12: Showing the pETBlue pasmid.

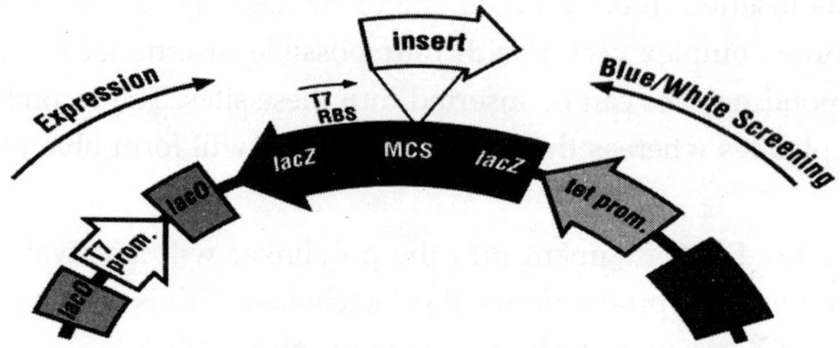

Fig. 13: Showing the pETBlue pasmid which utilizes a lacZ gene (alpha region) on the opposite coding strand but the recombinant protein does not contain any lacZ fusion sequences.

The Cloning Strategy

The two essential components of this T7 bacteriophage RNA polymerase-based expression system are the pET plasmid containing the coding sequence of a DNA insert downstream of the T7 promoter (Tpr), and an *E. coli* strain containing a genomic copy of the T7 RNA polymerase (T7 pol) gene under the control of the lac UV5promoter.

Also shown in this example is a T7 lysozyme encoding plasmid containing the chloramphenicol-resistance gene (cat), which provides a means to inhibit the small amount of T7 RNA polymerase that is expressed in the absence of IPTG. High level expression of the histidine-tagged calreticulin protein is obtained by blocking lacI repressor function with the addition of IPTG to the media. (See Plate Fig. D)

This IPTG-inducible expression system is based on the finding that bacteriophage T7 RNA polymerase is highly specific, and efficient, in transcribing genes containing a T7 promoter. By transforming a recombinant calreticulin pET vector into an *E. coli* strain containing the T7 RNA polymerase gene under lac control, it will be possible to induce calreticulin expression by treating the cells with IPTG. Since low level expression of the T7 RNA polymerase gene occurs in this system in the absence of IPTG, and appreciable levels of the plant calreticulin protein may inhibit bacterial growth, a compatible plasmid containing the T7 lysozyme gene will also be introduced into the host cells. T7 lysozyme is a bifunctional protein that not only cleaves peptidoglycan linkages in the bacterial cell wall but also stoichiometrically inhibits the function of T7 RNA polymerase. Therefore, low level expression of T7 lysozyme is able to inhibit the activity of T7 RNA polymerase molecules present in cells even before IPTG treatment.

Regulated expression of genes in *E. coli* using the pET vector system:

The two essential components of this T7 bacteriophage RNA polymerase-based expression system are the pET plasmid containing the coding sequence of a DNA insert downstream of the T7 promoter (Tpr), and an *E. coli* strain containing a genomic copy of the T7 RNA polymerase (T7 pol) gene under the control of the lac UV5 promoter. Also shown in this example is a T7 lysozyme encoding plasmid containing the chloramphenicol-resistance gene (cat), which provides a means to inhibit the small amount of T7 RNA polymerase that is expressed in the absence of IPTG. High level expression of the histidine-tagged calreticulin protein is obtained by blocking lacI repressor function with the addition of IPTG to the media.

PHAGEMIDS :

Phagemids (e.g., pUC118 and pUC119 and pBS vectors combine desirable properties of both plasmids and filamentous phages. They carry (i) the ColEl origin of replication, (ii) a selectable marker such as antibiotic resistance, and (iii) the major intergenic region of a filamentous phage. The segments of foreign DNA cloned in these vectors can be propagated as plasmids. When cells harboring these plasmids are infected with a suitable helper bacteriophage, these plasmids are infected with a suitable helper bacteriophage, the mode of replication of the plasmid changes under the influence of the gene II product of the incoming virus. Interaction of the intergenic region of the plasmid with the gene II protein initiates the rolling-circle replication to generate copies of one strand of the plasmid DNA, which are packaged into progeny bacteriophage particles. The single-stranded DNA purified from these particles is used as a template to determine the nucleotide sequence of one strand of the foreign DNA segment, for site-directed mutagenesis or as a strand-specific probe. Phagemids provide high yields of double-stranded DNA and render the time-consuming process of subcloning DNA fragments from plasmids to filamentous bacteriophages unnecessary.

pUC118 and pUC119

These phagemids (plasmid- phage hybrid) were named so for plasmid, University of California. The numbers 118 and 119 were assigned in order to differentiate them from other vectors constructed from the same series.

The phagemid vector pUC118 and pUC119 (Fig. 14) is a pair of vectors that are essentially identical except that the regions into which foreign DNAs are inserted are present in opposite orientations i.e turned end to end, relative to the rest of the genes of the vector. Thus if a foreign DNA is inserted into a specific restriction sites in both vectors, one vector will package one strand of the gene and the other will package the complementary strand of the gene. Therefore both the strands can be isolated and sequenced.

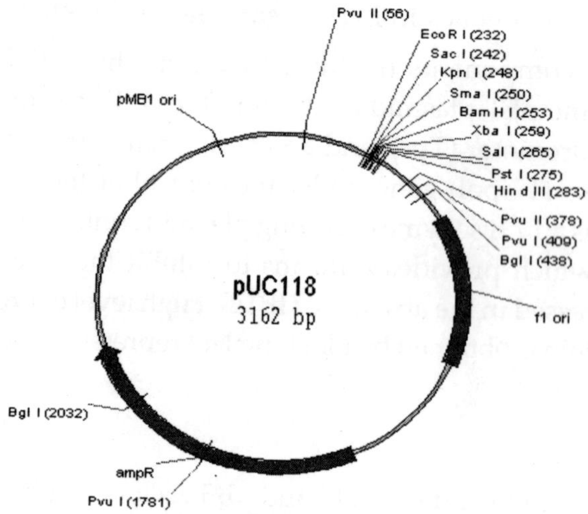

Fig. 14: Showing the genome of pUC118

The vector pUC118 and pUC119 differ from other vectors of the same series in having the origin of replication of M13 phage. As a result of this pUC118 and pUC119 can replicate either as a double stranded plasmid when the helper phage in not present on they can replicate as single stranded DNA that is packaged in M13 phage coats and extruded from the cell in the presence of "Helper phage". In the former case, replication is controlled by the plasmid origin of replication whereas in the latter case it is directed by the M13 phage origin of replication that has been added to these plasmids.

The following properties of pUC118 and pUC119 make them superior to the other vectors:

1. **Easy isolation:** Since these vectors are small, supercoiled and covalently closed circular DNA, thus they can be easily isolated and manipulate further in vitro.

2. **Easy selection:** These vectors contain amp^r gene as selectable marker which makes their selection easier.
3. **High copy number:** These phagemids have a copy number of 500 copies per bacterium which is more than that of other vectors. This helps in obtaining large yields of DNA from small cell cultures.
4. **Presence of polyclonal region:** The presence of a polyclonal region containing a variety of restriction enzyme cleavage sites helps in the insertion of many different types of restriction fragments without any modification. The polyclonal region also contains the lacZ′ gene. Thus the recombinants can be easily isolated by a simple color test, as the insertion of any foreign DNA will inactivate the lacZ′ gene.
5. **Easy sequencing of inserted DNA:** The polyclonal regions are present in the opposite orientations. Thus, if pUC118 packages one strand of the cloned gene, pUC119 will package the opposite strand.

pEMBL8 : This phagemid was made by transferring into pUC8 a 1300 bp fragment of the M13 genome. This fragment of M13 DNA contains the signal sequence recognized by the enzymes that convert the normal double stranded M13 molecule into single stranded DNA before secretion of new phage particles. Though detached from the rest of the M13 genome, this fragment is still functional.

Thus pEMBL8 molecules are also converted into single stranded DNA and secreted as defective phage particles.

When pEMBL8 vector is used, the *E. coli* cells should also be infected with an intact M13 particle which acts as a helper phage providing the essential enzymes and the phage coat proteins. Since pEMBL8 is derived from pUC8 it has the polylinker at the necessary lacZ′ gene.

pEMBL8 has thus got the following advantages over other plasmid and M13 vectors:

1. Foreign DNA fragments with two different sticky ends can be inserted at the multiple cloning sites on the lacZ′ gene.
2. Recombinant pEMBL8 genome can be packaged into the capsid of M13 phage.
3. The phage particles containing the recombinant pEMBL8 genome are as efficient as the wild type M13 phages in their ability to infect *E .coli* cells and then in the expression of the foreign gene.
4. Recombinant pEMBL8 genome can be easily screened using the medium containing X-gal.

COSMIDS:

Cosmids are plasmids that contain a minimum of 250 bp of lambda DNA which helps in whole genome in circularization and ligation. The term cosmid is a combination of two words: 'cos' (cos site of lambda phage also known as cohesive ends) and 'mid' from the word plasmids. This is because the cosmids are made by joining a plasmid and cos site of lambda phage to give an efficient hybrid cloning vector. They consist of:

1. The cos site
2. Sequences needed fro binding of and cleavage by *terminase* so that under appropriate conditions they are packaged in vitro into empty – phage particles.

A typical cosmid consists of (a) replication origin (b) a suitable selectable marker and (c) unique restriction sites

Cosmid vectors are constructed using recombinant DNA techniques by Collins and Hohn in 1978.

Cosmid lack genes encoding viral proteins; therefore neither viral particle is formed within the host cell nor does cell lysis occur. They have a length of about 50 – 60 kb. The upper size limit of the foreign DNA that can be inserted in cosmids and packaged into phage particles is therefore, approximately 45 kb, much larger than it would be possible to clone in phage.

According to the size of cos sites and upper limit in the head of phage, the recombinant DNA molecule can be packed into bacteriophage particles in *in vitro* packaging system consisting of packaging enzymes, head and tail protein.

There are four essential features of cosmids that make it a remarkable cloning vehicle:

a) The linear cosmid molecule upon injecting a bacterial cell, becomes circular through cohesive cos sites and then replicates just as a plasmid.
b) Cosmids have selective markers (such as those of antibiotic resistance or the ability to utilize certain sugars) which facilitate easy selection of cosmids carrying foreign DNA.
c) The cos site of lambda phage DNA is present. ("cos" sites are required for packaging DNA into lambda virions)
d) Cosmids have unique restriction sites making easy the insertion of foreign DNA.

Procedure of DNA Cloning by using Cosmids as vectors:

After constructing a cosmid, foreign DNA to be cloned is spliced into this unique cloning vehicle. For this, cosmid DNA is first linearized- straightened out by cutting it with an appropriate restriction endonuclease. It is further treated with another enzyme, namely the "calf intestinal phosphatase." This enzyme removes the 5′ group from ends and thus prevents recircularization of cosmid DNA. Meanwhile, the foreign DNA, usually a high molecular weight DNA is also treated with restriction enzyme. Now, the enzyme treated cosmid and foreign DNA are mixed together in the presence of T4 DNA ligase.

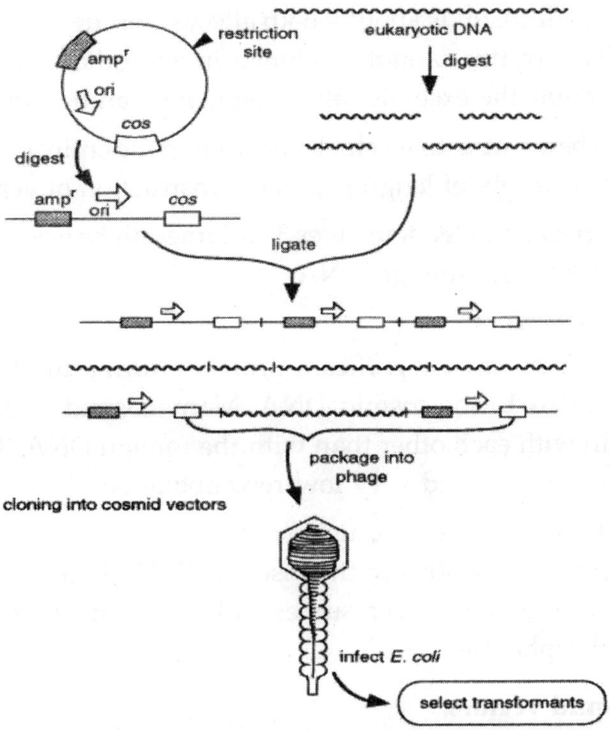

Fig. 15: Showing a cosmid and the way it is used to clone large fragments of DNA.

During this reaction, several combinations of the cloning vehicle and foreign DNA are generated. Among the products, it is possible by chance that in some cases, foreign DNA gets bound to the cos site of cosmid which are with the same orientation. When such molecules are incubated in an *in vitro* packaging reaction mixture (containing adequate amount of phage head and tail protein), the cos sites are cleaved and intervening DNA is hence forth packaged into bacteriophage particles. When susceptible bacteria infected with such by such phage carrying the foreign DNA, this hybrid DNA gets injected into the cell upon entry. The DNA circularize due to the presence of 'cos' site. Further, it replicates as a plasmid and express the antibiotic resistance gene (Fig. 15).

The infected bacteria are then plated onto selective medium for isolation of those host colonies, which harbor cloning vehicles as they survive antibiotic present in the medium. These bacterial colonies are then picked up and the plasmid DNA is isolated from them. The purified plasmid DNA is then analyzed for the presence of foreign DNA.

Advantages over other vectors:

a) The packaging process imposes a direct selection for cosmids carrying the foreign DNA as those without foreign DNA are not packaged due to their small size.

b) The use of small cosmids about 4-6 kb allows cloning of DNA fragment as long as 45 kb. DNA of this cannot be cloned in any other existing cloning vehicle. Thus, in cosmids the exceptionally long genes can be cloned.

Cosmids seem to be an attractive vehicle for cloning the entire genome of an organism cut by enzyme's into a variety of lengths i.e. the construction of genome libraries.

c) The in vitro packaging system gives 3- 10 times higher yield of cosmids carrying the insertion of these foreign DNA.

Drawbacks of cosmid vectors:

- Genetic engineers face a significant problem during the reaction of combining the foreign DNA to the cosmid DNA. Many cosmid vector molecules pose a threat to join with each other than with the foreign DNA. This results in a high parental background and very low recombinants.
- Recombination can occur among two or more cosmid molecules. This results in rearrangement or deletions in the inserted DNA. In a significant development the former drawback has been successfully overcome by using the enzyme calf intestinal phosphatase.

Applications of cosmid vectors:

Cosmid vectors are used in the construction of new breeds of plants.

They are used in the replacement of defective genes in humans suffering from genetic diseases.

YEAST VECTORS :

Saccharomyces cerevisiae, commonly known as yeast, is a eukaryote with 34 chromosomes. Its haploid DNA content is only three times that of *E.coli*. Till now only a single yeast plasmid of 2 μm has been discovered which has been used to construct some useful vectors. The copy number of the yeast plasmid is between 70–200. The plasmid contains an origin of replication, two genes REP1 and REP2 which are involved in the replication of the plasmid. Besides this another gene FLP codes for a protein that can convert the A form of plasmid to the B form, in which the gene order has been rearranged by the intermolecular recombination (Fig. 16).

The above mentioned genes are not sufficient for the yeast plasmid to act as a cloning vector. A selectable marker is needed. Generally a normal yeast gene which codes for an enzyme involved in amino acid synthesis is is taken as a selectable marker.

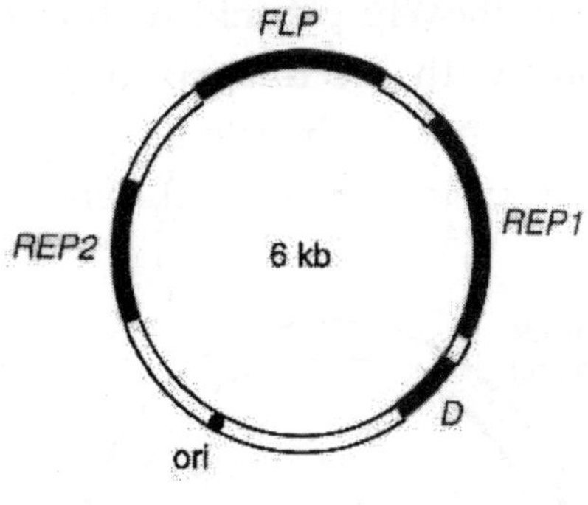

Fig. 16. Showing Yeast Plasmid

Yeast Vectors

A wide range of vectors are available to meet various requirements for insertion, deletion alteration and expression of genes in yeast. Most plasmids used for yeast studies are shuttle vectors, which contain sequences permitting them to be selected and propagated in *E. coli,* thus allowing for convenient amplification and subsequent alteration in vitro. The most common yeast vectors originated from pBR322 and contain an origin of replication (ori), promoting high copy-number maintenance in *E. coli,* and the selectable antibiotic markers, the β-lactamase gene, bla (or AmpR), and sometime to tetracycline-resistance gene, tet or (TetR), conferring resistance to, respectively, ampicillin and tetracycline.

In addition, all yeast vectors contain markers that allow selection of transformants containing the desired plasmid. The most commonly used yeast markers include URA3, HIS3, LEU2, TRP1 and LYS2, which complement specific auxotrophic mutations in yeast, such as ura3-52, and lys2-201. These complementable yeast mutations have been chosen because of their low-reversion rate. Also, the URA3, HIS3, LEU2 and TRP1 yeast markers can complement specific *E. coli* auxotrophic mutations.

The URA3 and LYS2 yeast genes have an additional advantage because both positive and negative selections are possible, as discussed below. Although there are numerous kinds of yeast shuttle vectors, those used currently can be broadly classified in either of following three types: integrative vectors, YIp; autonomously replicating high copy-number vectors, YEp; or autonomously replicating low copy-number vectors, YCp.

YIp Vectors

The YIp integrative vectors do not replicate autonomously, but integrate into the genome at low frequencies by homologous recombination. Integration of circular plasmid DNA by homologous recombination leads to a copy of the vector sequence flanked by two direct copies of the yeast sequence (Fig. 17). The site of integration can be targeted by cutting the yeast segment in the YIp plasmid with a restriction endonuclease and transforming the yeast strain with the linearized plasmid. The linear ends are recombinogenic and direct integration to the site in the genome that is homologous to these ends. In addition, linearization increases the efficiency of integrative transformation from 10- to 50-fold.

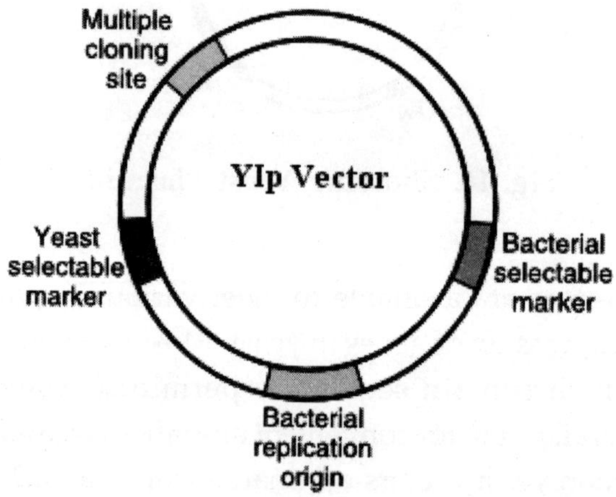

Fig. 17: Yeast Integrative plasmid Cloning Vector

The YIp vectors typically integrate as a single copy. However multiple integration do occur at low frequencies, a property that can be used to construct stable strains overexpressing specific genes. YIp plasmids with two yeast segments, such as YFG1 and URA3 marker, have the potential to integrate at either of the genomic loci, whereas vectors containing repetitive DNA sequences, such as Ty elements or rDNA, can integrate at any of the multiple sites within genome. Strains constructed with YIp plasmids should be examined by PCR analysis, or other methods, to confirm the site of integration.

Strains transformed with YIp plasmids are extremely stable, even in the absence of selective pressure. However, plasmid loss can occur at approximately 10-3 to 10-4 frequencies by homologous recombination between tandemly repeated DNA, leading to looping out of the vector sequence and one copy of the duplicated sequence.

YEp Vectors

The YEp yeast episomal plasmid vectors replicate autonomously because of the presence of a segment of the yeast 2 µm plasmid that serves as an origin of replication (2 µm ori). The 2 µm ori is responsible for the high copy-number and high frequency of transformation of YEp vectors.

YEp vectors contain either a full copy of the 2 µm plasmid, or, as with most of these kinds of vectors, a region which encompasses the ori and the REP3 gene. The REP3 gene is required in cis to the ori for mediating the action of the trans-acting REP1 and REP2 genes which encode products that promote partitioning of the plasmid between cells at division. Therefore, the YEp plasmids containing the region encompassing only ori and REP3 must be propagated in cir+ hosts containing the native 2 μm plasmid

Most YEp plasmids are relatively unstable, being lost in approximately 10-2 or more cells after each generation. Even under conditions of selective growth, only 60% to 95% of the cells retain the YEp plasmid. The copy number of most YEp plasmids ranges from 10-40 per cell of cir+ hosts. However, the plasmids are not equally distributed among the cells, and there is a high variance in the copy number per cell in populations.

Several systems have been developed for producing very high copy-numbers of YEp plasmids per cell, including the use of the partially defective mutation leu2-d, whose expression is several orders of magnitude less than the wild-type LEU2+ allele. The copy number per cell of such YEp leu2-d vectors range from 200-300, and the high copy-number persists for many generations after growth in leucine-containing media without selective pressure. The YEp leu2-d vectors are useful in large-scale cultures with complete media where plasmid selection is not possible. The most common use for YEp plasmid vectors is to overproduce gene products in yeast.

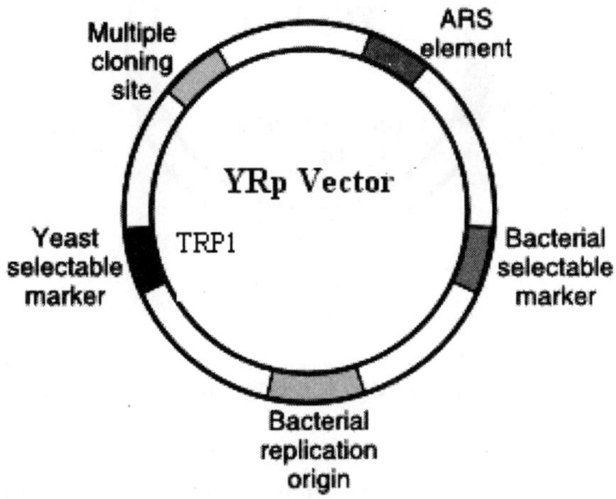

Fig. 18 : Showing Yeast Replicative plasmid

YRp Vectors:

Yeast Replicative Plasmids are able to replicate as independent entities/ plasmids as they carry a chromosomal DNA sequence that includes an origin of replication. The replication origins are located very close to the yeast genes. The yeast genes also include one or two selectable markers. A good example of YRp is YRp7 (Fig. 18). It is made up of pBR322 and the yeast gene TRP1.

This gene which is involved in tryptophan biosynthesis, is located near to the chromosomal origin of replication. The yeast DNA fragment present on YRp7 (Fig. 18) contains both TRP1 and the origin of replication.

YCp Vectors

The YCp yeast centromere plasmid vectors are autonomously replicating vectors containing centromere sequences, CEN, and autonomously replicating sequences, ARS (Fig. 19). The YCp vectors are typically present at very low copy numbers, from 1 to 3 per cell, and possibly more, and are lost in approximately 10-2 cells per generation without selective pressure. In many instances, the YCp vectors segregate to two of the four ascospores from an ascus, indicating that they mimic the behavior of chromosomes during meiosis, as well as during mitosis. The ARS sequences are believed to correspond to the natural replication origins of yeast chromosomes, and all of them contain a specific consensus sequence.

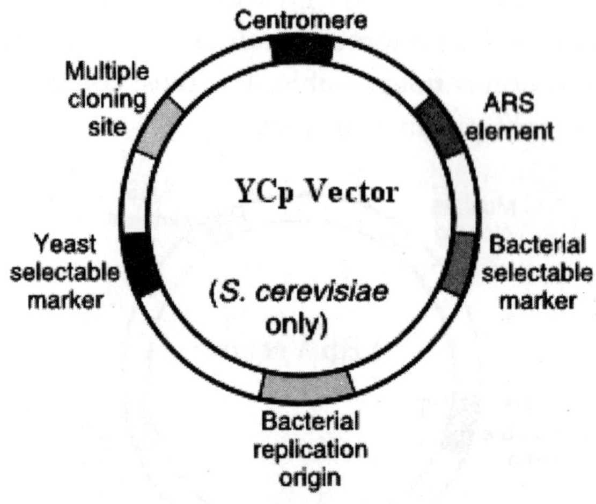

Fig. 19 : Showing Yeast Centromere plasmid vectors

The CEN function is dependent on three conserved domains, designated I, II, and III; all three of these elements are required for mitotic stabilization of YCp vectors. YRp vectors, containing ARS but lacking functional CEN elements, transform yeast at high frequencies, but are lost at too high a frequency, over 10% per generation, making them undesirable for general vectors.

The stability and low copy-number of YCp vectors make them the ideal choice for cloning vectors, for construction of yeast genomic DNA libraries, and for investigating the function of genes altered in vivo. ARS1, which is in close proximity to TRP1, is the most commonly used ARS element for YCp vectors, although others have been used.

CEN3, CEN4 and CEN11 are commonly used centromeres that can be conveniently manipulated. For example, the vector YCp50 contains CEN4 and ARS1.

Yeast Artificial Chromosomal Vectors:

The initial step in the molecular characterization of eukaryotic genomes generally requires cloning of large chromosomal fragments, which is usually carried out by digestion with restriction endonucleases and ligation to specially developed cloning vectors. Usually 200 to 800 kb fragments are cloned as Yeast Artificial Chromosomes (YACs), and 100-200 kb fragments are cloned as Bacterial Artificial Chromosomes (BACs or PACs). The importance of YAC technology has been heightened by the recently developed methods for transferring YACs to cultured cells and to the germline of experimental animals. The major components of the yeast artificial chromosomes are:

- The centromere, which distributes the chromosome to the daugthter cells correctly during cell division.
- Homologous or heterologous DNA sequences that function as telomeres (TEL) in vivo. The two telomeres at the ends of the chromosomes are needed for the ends to be replicated correctly. Besides, they also prevent the chromosome from being acted upon by exonucleases.
- amp^r for selective amplification and markers such as TRP1 and URA3 for identifying cells containing the YAC vector.
- Recognition sites of restriction enzymes (e.g., EcoRI and BamHI)
- The origin of replication, a position from where the replication starts.

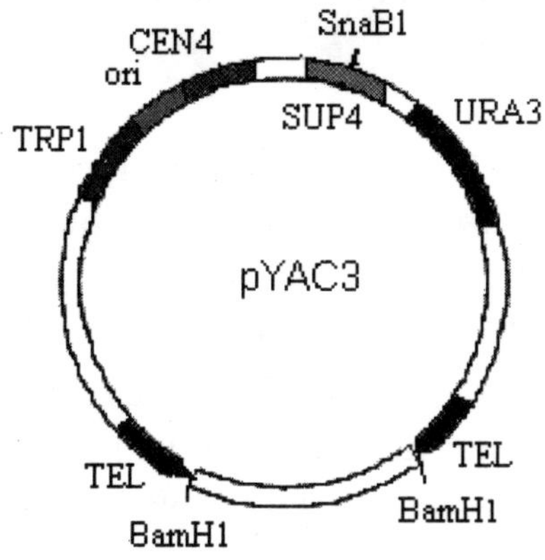

Fig. 20 : Showing a YAC3 Vector

One typical example of yeast artificial chromosome is pYAC3. It is typically obtained from pBR322 but also contains a number of yeast genes (Fig. 20). Out of these two genes act as selectable markers, URA3 and TRP1 for YIp5 and YRp7 respectively. In YRp7 the DNA fragment which caries TRP1 also contains an origin of replication, but in pYAC3 this fragment is further extended and includes a sequence called CEN4. It is the DNA from the centromere region of chromosome 4. Thus it is seen that TRP1-ori-CEN4 fragment contains three out of the five components of the artificial chromosome.

The another component, the telomeres is provided the two sequences called TEL. though they are not the complete telomere sequences, yet when they are inserted into the yeast nucleus they act as seeding sequences on which the telomere sequences will built. The part of YAC thus left unexplained is SUP4. It is the selectable marker into which the foreign DNA is inserted during the cloning experiment. (See Plate Fig. H)

The Cloning Strategy:

1. The target DNA is partially digested by EcoRI and the YAC vector is cleaved by EcoRI and BamHI.
2. Ligate the cleaved vector segments with a digested DNA fragment to form an artificial chromosome.
3. Transform yeast cells to make a large number of copies.

Protoplast transfusion is then used to introduce the artificial chromosome into *S. cerevisiae*. The yeast strain used is double auxotrophuic mutant, *trp1*$^-$ *ura3*$^-$, which is converted to *trp*$^+$ *ura3*$^+$ by the two markers on the artificial chromosomes.

Transformants thus obtained are selected by plating onto minimal medium. On this minimal medium only those cells will grow which contain a correctly constructed artificial chromosome. The cell transformed with an incorrect artificial chromosome, containing two left or two right arms rather than one of each, is not able to grow onto minimal medium.

Applications of YAC vectors:

YACs act as stable artificial chromosomes during propagation in the in yeast cells which raised the possibility of their use as vehicles for genes that are too long to be cloned as a single fragment in an *E.coli* vector. Thus most of the mammalian genes which are greater than 1000kb in length can be cloned in YAC systems.

YACs also have shown great importance in establishment of gene libraries. YAC vectors are routinely used to clone fragments upto 600kb size and certain special types are able to handle DNA upto 1400 kb in length. This brings the size of the DNA vectors to just 6500 clones.

Plant Viruses as Vectors:

Modified versions of M13 and λ phage are important as cloning vectors for *E. coli*. Since most plants are subject to viral infection, they can be used to clone genes in plants. In such a way they can prove to be much more convenient than any other vectors, because with many viruses transformation can be achieved just by rubbing the virus DNA onto the surface of a leaf. The virus then spreads throughout the plant.

One of the problems encountered with plant viruses was that most of them had their genome of RNA and not DNA. As a result only two classes of DNA viruses, namely Caulimoviruses and Geminiviruses classes, are known to infect higher plants and they can be used as cloning vectors since the manipulation with the RNA becomes difficult.

Caulimovirus:

They are a group of plant viruses which have double stranded DNA. These have a very narrow host range. One such virus is Cauliflower Mosaic Virus (CaMV). This virus infects plants of the family Cruciferae. The caulimoviruses are isometric particles and are 50 nm in diameter. The cells of the infected plants accumulate virus containing proteinaceous inclusion bodies. These infected regions are considered as sites of virus assembly.

There were two general difficulties encountered while using caulimoviruses as vectors. Firstly, the total length of DNA that can be inserted was small. This was due to the fact that genome of caulimovirus also contained the genes for packaging it into the protein coat. Even after the deletions of the non essential portions of the genome the total capacity of the caulimovirus genome to insert the foreign DNA was very limited. Later on this problem was overcome by the use of a Helper Phage.

Secondly, the caulmoviruses had an extremely narrow host range. This restricts the cloning experiments to a few plants.

The genome of CaMV is 8024 bp long. The DNA is a circular double stranded molecule. It has three discontinuities at specific sites on each strand. Two gaps are present on one strand whereas other strand has only one gap. These gaps have short regions of overlapping DNA produced by strand displacement.

The free genome of CaMV is infectious and it has no tissue restrictions. It has a total of 6 genes.

Gene I: responsible for the production of essential factors involved in cell to cell spread of virus

Gene II: produces Aphid transmission factor

Gene III: encoded a virion structural protein

Gene IV: production of Coat protein

Gene V: responsible for the production of enzyme reverse transcriptase

Gene VI: encoded for the inclusion body protein

CaMV as Vector:

The naked CaMV is capable of infecting host cells. Infections can simply be initiated by mechanical inoculation of the DNA extract of the host plant. CaMV has a single Sal

I site the Sal I linearized DNA is infectious *in vitro* even in the absence of religation. Recircularization occurs in the plant cell. The viral particles assembled in the host cells have the ability to spread from cell to cell as it is a systemic virus. Plant cells infected by the CaMV can be cultured *in vitro* to produce clones.

Geminivirus Vectors:

Geminiviruses are rapidly becoming major plant pathogens in tropical and subtropical countries affecting an increasing number of crops with a disastrous impact on productivity. The recent spread of geminivirus-associated diseases has been linked to the emergence of new virus strains and the increasing abundance of viruliferous whitefly vectors, possibly as a result of enhanced insecticide resistance and global warming.

Geminivirus are interesting because their natural host plant includes maize and wheat and they could therefore be potential vectors for these and other monocots. The most common problem encountered in using geminivirus as vectors s that during infection cycle the genomes of some geminiviruses undergo rearrangements and deletions, which would scramble up any foreign DNA inserted.

Geminivirus comprise a group of single stranded DNA plant viruses. They replicate via ds DNA forms and are subdivided into two groups, one infecting monocots and transmitted by leafhoppers and the other infecting dicots and transmitted by whitefly. The dicot geminivirus have two DNA segments in their genome, DNA A and DNA B. both of these are essential for infection by mechanical inoculation.

Virus Vectors for animal cells:

A number of animal viruses are also used as vectors. They can be used either for the delivery of the nucleic acid into cultured cells followed by integration with host genome or, for the amplification and high level expression of foreign genes. These cloned genes have a variety of application including gene therapy in mammalian cells and synthesis of important proteins by cloned genes in cell cultures.

SV 40 as Vectors: It is a spherical virus with circular, double stranded 5243 bp genome. It encodes 5 proteins in total: Small T, Large T (early proteins), VP1, VP2, and VP 3 (VP = virion protein). It has an origin of replication of about 80 bp and is complexed with histones to form chromatin. Large T is essential for viral replication whereas VP1, VP2, and VP 3 form the viral capsid. (See Plate Fig. I)

SV 40 genome is used to develop three major types of vectors: 1) Transducing vectors, 2) Plasmid vectors.

SV 40 Transducing Vectors: These vectors produce viral particles after infecting monkey cells. They must have these three features:

1. The SV40 origin including the surrounding region containing the transcriptional regulatory signals (i.e. regions at which splicing and polyadenylation occur),
2. Their size should be between 3900 and 5300 bp for packaging into virions,
3. They should contain the genes encoding for the large T, VP1, VP2 and VP3. This leaves very little space for the foreign DNA. This problem was solved by the incorporation these genes in another virus or within the host genome.

Late Replacement vectors: The region encoding VP1, VP2 and VP3 may be replaced in the vector by DNA insert; such a vector is called late replacement vector. A vector of this type is used for infection of host cells in conjugation with another virus, called helper virus, which has the VP1, VP2 and VP3 genes intact but has a defective large T gene. In such a case, only those host that are infected by both the vector and the helper virus will lyse and produce virions since cells infected by either the vector or the helper alone will not support packaging or replication (respectively) of the virus.

Early replacement vectors: In this kind of vectors the essential genes missing from the vector may be present within the genome of the host cells.

Eg COS (CV -1, Origin of SV- 40; CV -1 is a monkey cell line) cell line of African green monkey kidney cell cultures contains in its genome the gene for large T of SV 40. Therefore, a vector having the origin of replication and genes for VP1, VP2 and VP3 will replicate and produce virions in COS cell line cells.

SV 40 Plasmid Vectors: SV 40 genome with the insertion of a segment of a plasmid could produce an SV 40 plasmid vectors. eg pSV1GT5. It is derived from SVGT5. It contains a fragment of pBR322 DNA 4 kb in length. This plasmid segment codes for β lactamase and contains the origin of replication of pBR322 genome. Thus these vectors can be used in both monkey cells and *E.coli.* This SV 40 –plasmid has been used to express several genes.

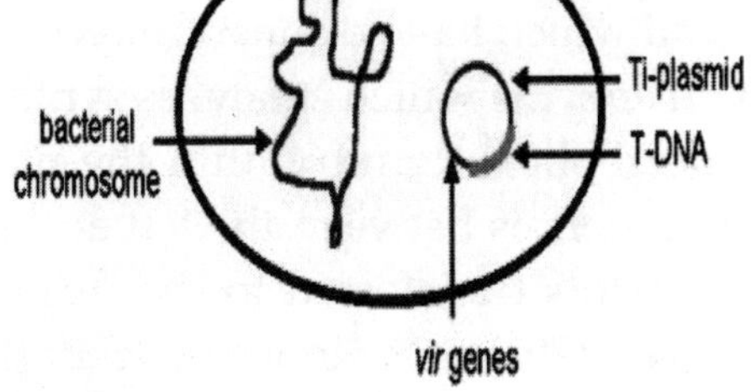

Fig. 21: Showing *Agrobacterium tumefaciens* cell with bacterial chromosome and Ti plasmid

Gene Cloning Strategies in Plants - Gene cloning strategies in plants are more or less basically similar to the strategies discussed earlier but with some variations. The most promising method of transforming plant cells makes use of a plasmid called the Ti plasmid, which is found, within the bacterium *Agrobacterium tumefaciens*. Fraley *et al.* (1983) and An *et al.* (1985) exploited the natural ability of *Agrobacterium tumefaciens* to transfer DNA into plant chromosomes.

Agrobacterium tumefaciens is a soil-borne phytopathogen that uses genetic engineering processes to subvert the host plant cell's metabolic machinery. This gram negative, rod shaped, motile bacterium lives in soil and invades many dicotyledonous plants and some gymnosperms when they are damaged at soil level. The bacterium enters the fresh wound and attaches itself to the wall of an intact cell, after which it transfers a relatively small part of its Ti plasmid into the nucleus of the cell (Fig. 21).

This plasmid integrates some of its DNA; (T-DNA) into the chromosome of its host plant cells. This is a unique morphogenetic phenomenon wherein a permanent incorporation of a portion of bacterial genome completely diverts the host cells from their predetermined path of development.

This infection results in a crown gall, i.e. a lump or callus of tumor tissue that grows in an undifferentiated way at the site of infection.(See Plate Fig. J) The cells of the crown gall acquire the properties of independent, unregulated growth. When crown gall cells are cultured, they grow to form a callus even in media devoid of the plant hormones that must be added to induce normal plant cells to grow in culture.

Ti plasmids have evolved solely for the benefit of the bacterium. The potential of the *Agrobacterium* Ti plasmid as a vector arises from the ability of the bacterium to somehow transfer and stably integrate a piece of the plasmid DNA into the plant nuclear genome (this is a case of a natural vector system). *Agrobacterium*-mediated transformation of plant tissue generally results in a low transgene copy number, minimal rearrangements, and higher transformation efficiency than direct DNA delivery techniques such as microparticle bombardment (Pawlowski & Somers 1996; Gelvin 1998).

The transferred DNA is known as T-DNA and carries several genes which are expressed within the plant and which have dramatic effects on its metabolism (See Plate Fig. K). One gene codes for an enzyme which catalyzes synthesis of an opine from amino acids and other common metabolites found within the plant cell. [Until 1995, it was generally assumed that the sequences between the left and right borders of the T-DNA were the only transgenic elements transferred to the recipient host. Ramanathan and Veluthambi (1995), Wenk *et al.* (1997) and Kononov *et al.* (1997) all demonstrated that plasmid backbone sequences beyond the borders of the T-DNA could also be integrated along with the genes of interest. Experiments by Kononov *et al.* (1997) demonstrated that plasmid backbone sequences could be integrated into the host genome coupled with either the right or left border sequences, or as an independent unit unlinked from the T-DNA. Matzke & Matzke (1998) state that backbone sequences that join T-DNA and host DNA appear to be especially deleterious for gene expression, an observation supported by the authors' finding that backbone fragments separated from T-DNA have been found associated with stably expressed transgenes].

Opines are never found in normal plants and cannot be metabolized by them, but they can and are used as sources of carbon and nitrogen and the ability both to induce and to metabolize opines is encoded by plasmids in the bacteria.

Two types of opines are produced, namely octopine and nopaline; accordingly there are two kinds of plasmids, namely octopine plasmid and nopaline plasmid. The two have very similar physical and genetic structures. They are large circles, measuring 140-235 kilobase pairs. The two plasmids have genes for transfer of the plasmid between bacteria, for opine and arginine catabolism, to ensure that there will not be two types of plasmids in the same cell and to exclude a certain bacteriophage.

They also have genes outside the T-DNA that are necessary for virulence. It is not known exactly what these control, but it might be attachment of bacteria to plant cell walls, transfer of the plasmid into a plant cell, or integration of the T-DNA into the plant genome. The two plasmids have the T-DNA itself.

The genes which determine oncogenicity (one) are distributed over wide segments of the plasmids and can also be found as chromosomal markers in the host bacterium. The genes responsible for the biosynthesis of opines (octopine, ocs: nopaline, nos) as well as genes specifying their degradation (occ: ncc) are found at well defined regions.

Another gene found in nopaline synthesizing plasmids but not in octopine synthesizing plasmids, directs the synthesis of another opine, agrocinopine.

The system for utilization of opines by the bacterial cell requires their active transport into the cell and subsequent cleavage to give the parent amino acid and a keto acid, both being plasmid-encoded functions.

The T-DNA represents a single contiguous region of the plasmid. This region is often moved into the plant genome as a single piece but in some tumors the T-DNA is split in two and integrated into the plant DNA in two separate places. There are at least four T-DNA genes that influence tumor morphology. The first two, called tms or Shi genes, apparently inhibit shoot formation since mutants of these genes induce tumors that form shoots. The third gene, called tmr or Roi, apparently inhibits root formation.

The internal genes of the T-DNA are not necessary for integration although they are necessary for tumor formation. The internal genes can be replaced with foreign genes, which are then integrated under control of the border regions. This provides a mechanism for transforming cells with a wide variety of genes.

Two properties of the T-DNA of Ti plasmids make them virtually ideal vectors for introducing foreign genes into plants:

- The host range of *Agrobacterium* is quite broad; they are capable of transforming cells of virtually all dicotyledonous plants. Whether the range can eventually be broadened to include monocotyledons remains to be seen.
- The integrated T-DNA is inherited in a Mendelian way and its genes have their own promoters to which foreign genes can be coupled and expressed.

The Cloning Strategy: Due to the above mentioned features of the Ti plasmid it was soon known that it can be used to transport new genes into plant cells if we are successful at inserting the new genes into the T- DNA and then the bacterium is able to insert them into the plant chromosomal DNA. The following properties of Ti plasmids did not allow their direct use as vectors:

(i) Large size,

(ii) Absence of unique restriction enzyme sites and

(iii) Tumor induction property.

Therefore, only the useful attributes of Ti plasmid have been used in designing several plant transformation vectors. Since tumor cells can not develop into normal shoots, disarming of Ti plasmids was an essential step in designing useful vectors. This was achieved by replacement of tumor inducing genes in T-DNA (T-DNA is found in both Ti and Ri plasmids) by selectable markers (like npt II), providing resistance against antibiotics like kanamycin.

Promoters and polyadenylation signal isolated from octopine or nopaline synthase genes were used for expression of selectable markers. Other powerful promoters included CaMV35S and CaMV19S isolated from cauliflower mosaic virus.

T-DNA (disarmed) and vir genes are two essential elements for designing transformation vectors. While T-DNA with border sequences allows manipulation of DNA sequences intended to be transferred, vir genes allow infection of host plant. Since vir genes can function even in trans configuration, it is not essential that vir genes be present on the same plasmid, which functions as a vector and carries DNA segment to be transferred.

Two main strategies are used in view of above:

The Binary Vector Strategy: This strategy is based upon the fact that T-DNA does not need to be attached to the rest of the Ti Plasmid. The two plasmids A and B complement with each other when present together in the same A. tumefaciens cell. The T- DNA carried by the B plasmid is transferred to the plant chromosomal DNA by proteins coded by the genes carried by plasmid A (Fig. 22).

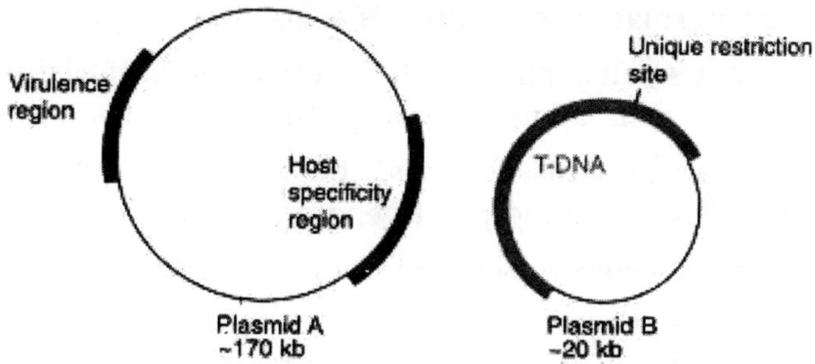

Fig. 22: The binary vector strategy showing two plasmids

T-DNA is found on a separate vector (binary vector) designed to replicate in both *E. coli* and *Agrobacterium* and capable of conjugal transfer between these two bacterial species. Many binary vectors have been developed, which differ in size and the source of 25bp repeat sequences, plant selection marker, bacterial selection marker and cloning sites. pBin19 which was designed in 1984, is still popular and is based on wide host range replicon pRK252.

This vector contains the following elements: kanamycin resistance gene (APH-l)-for selection in bacteria, T-DNA borders derived from pTiT37, a plant selectable transformation marker (npt-11) isolated from transposon, TnS (this marker is associated with promoter and polyadenylation signal derived from nopaline synthesis gene or nos), a multiple cloning site derived from pUC19 and housed within lac Z (β-galactosidase gene) region.

Bacterial colonies containing pBin19 are recognized by loss of blue colour on IPTG/ X-GAL plates. Unlike the cointegrative vectors, binary vectors need not have ai1Y homology with resident Ti plasmid and are capable of autonomous replication.

In the binary vector system, the two different plasmids employed are:

- A **wide-host-range small replicon**, which has an origin of replication (ori) that permits the maintenance of the plasmid in a wide range of bacteria including *E. coli* and *Agrobacterium*. This plasmid typically contains:

1. Foreign DNA in place of T-DNA,
2. The left and right T-DNA borders (or at least the right T-border),
3. Markers for selection and maintenance in both *E. coli* and *A. tumefaciens*,
4. A selectable marker for plants.

The plasmid is said to be "disarmed", since its tumor-inducing genes located in the T-DNA have been removed.

- A **helper Ti plasmid**, harbored in *A. tumefaciens*, which lacks the entire T-DNA region but contains an intact *vir* region. (Fig. 23)

In general, the transformation procedure is as follows:

- The recombinant small replicon is transferred via bacterial conjugation or direct transfer to *A. tumefaciens* harboring a helper Ti plasmid,
- The plant cells are co-cultivated with the *Agrobacterium*, to allow transfer of recombinant T-DNA into the plant genome, and
- Transformed plant cells are selected under appropriate conditions.

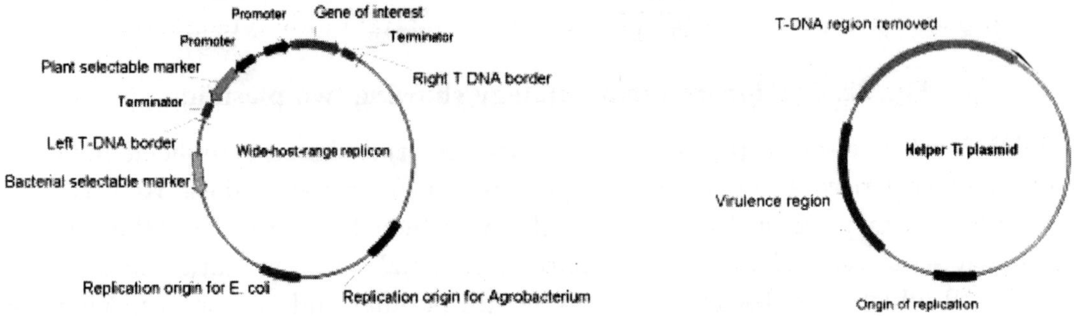

Fig. 23: Note the two types of plasmids employed in Binary vector system

Limitations

A possible disadvantage may ensue from the fact that the stability of wide host range replicons in *E. coli* and *Agrobacterium* varies considerably. Depending on the orientation, plasmids with two different origins of replication may be unstable in *E. coli* where both origins are active.

The Cointegration Strategy: These vectors are constructed by homologous recombination of a bacterial plasmid with the T-DNA region of an endogenous Ti plasmid in *Agrobacterium*. Integration of the two plasmids requires a region of homology present in both.

Three vectors are necessary in this system:

- Disarmed ***Agrobacterium*** Ti plasmids: In these Ti plasmids, the oncogenes located in the T-DNA region have been replaced by exogenous DNA. Examples of these vectors include:

1. ***SEV series:*** the right border of the T-DNA together with the phytohormone genes coding for cytokinin and auxin are removed and replaced by a bacterial kanamycin resistance gene while the left border and a small part of the left segment (TL) of the original T-DNA (referred to as Left Inside Homology (LIH)) are left intact.

2. ***pGV series:*** the phytohormone genes are excised and substituted by part of pBR322 vector sequence. The left and right border sequences as well as the nopaline synthase gene of the Ti plasmid are conserved.

• ***Intermediate vectors:*** These are small pBR322-based plasmids (*E. coli* vectors) containing a T-DNA region. They are used to overcome the problems derived from the large size of disarmed Ti plasmids and their lack of unique restriction sites. Intermediate vectors are replicated in *E.coli* and are transferred into *Agrobacterium* by conjugation. They cannot replicate in *A. tumefaciens* and therefore, carry DNA segments homologous to the disarmed T-DNA to permit recombination to form a co-integrated T-DNA structure.

• ***Helper vectors:*** These are small plasmids maintained in *E. coli* that contain transfer (*tra*) and mobilization (*mob*) genes, which allow the transfer of the conjugation-deficient intermediate vectors into *Agrobacterium*.

A resulting **co-integrated plasmid** assembled by *in vitro* manipulation normally contains:

1. The *vir* genes,
2. The left and right T-DNA borders,
3. An exogenous DNA sequence between the two T-DNA borders, and
4. Plant and bacterial selectable markers.

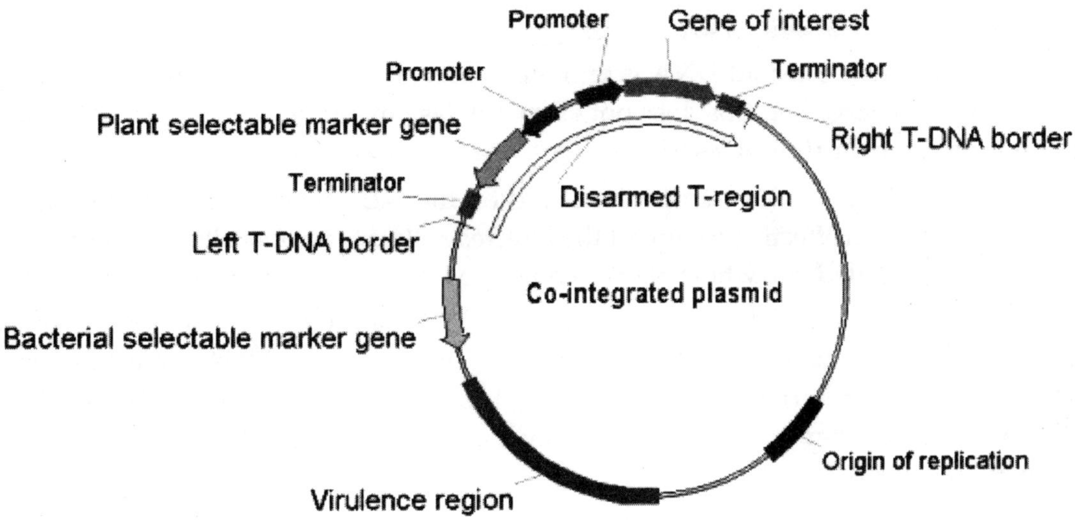

Fig. 24: Showing Cointegrative vector and its components

Some Drawbacks:

Although co-integrated vectors have been designed to allow site-specific recombination based on the recombination system of the phage P1 (e.g., wP1*lox*P-Cre series), co-integrated vectors in general are less popular due to:

- Long homologies required between the Ti plasmid and the *E. coli* plasmids making them difficult to engineer and use, and
- Relatively inefficient gene transfer compared to the binary vectors.

One of the first cointegrative vectors, pGV3850 was developed from a nopaline type Ti plasmid (C58), where almost all T-DNA has been deleted and replaced by pBR322, a common small *E. coli* cloning vector. The intermediate vector (e.g. pGVllO3) based on pBR322 is conjugated into pGV3850 at the region of pBR322 homology.

After cointegration, pGV3850 : : 1103 T-DNA contains the whole of intermediate vector including the unwanted DNA, which is also transferred to host plant along with the gene intended to be transferred. To overcome this difficulty the right border repeat or both borders are introduced to flank a cloning site in the intermediate vector, so that only a part of intermediate vector cointegrates. The utility of this vector system is however limited.

Advantages of Binary Vector over Cointegration Vector

Compared with co-integrated vectors, binary vectors present some advantages:

- No recombination process takes place between the molecules involved.
- Instead of a very large, recombinant, disarmed Ti plasmid, small vectors are used, which increases transfer efficiency from *E. coli* to *Agrobacterium*.

Limitations of Ti Plasmid DNA

Even though Ti plasmid DNA is an ideal vector, it poses several problems. Firstly, the plasmids are large and this does not allow for easy manipulation of their DNA; also, they have a large number of restriction sites which are not usefully distributed.

Secondly, the plasmids transfer functions on the T-DNA which specify production of substances which effectively convert the infected cells into tumor cells and these cannot be readily regenerated into whole plants. Thirdly, only dicotyledonous plants are infected.

In spite of these problems, the potential of Ti plasmid as a vector for the genetic manipulation of plants was quickly recognized. The genes in T-DNA are eukaryotic in nature, even though derived from a bacterial plasmid, and are transcribed by the plant's RNA polymerase; they contain introns, which are excised correctly during maturation of the mRNA.

REFERENCES

An G (1986) Development of plant promoter expression vectors and their use for analysis of differential activity of nopaline synthase promoter in transformed tobacco cells *Plant Physiol* **81**: 86-91

Becker A and H Murialdo (1990) Bacteriophage lambda DNA: the beginning of the end *J Bacteriol* **172**: 2819-2824

Becker A H Murialdo H Lucko and J Morell (1988) Bacteriophage lambda DNA packaging: the product of Fl gene promotes the incorporation of the prohead to the DNA terminase complex *J Mol Biol* **199**: 597-607

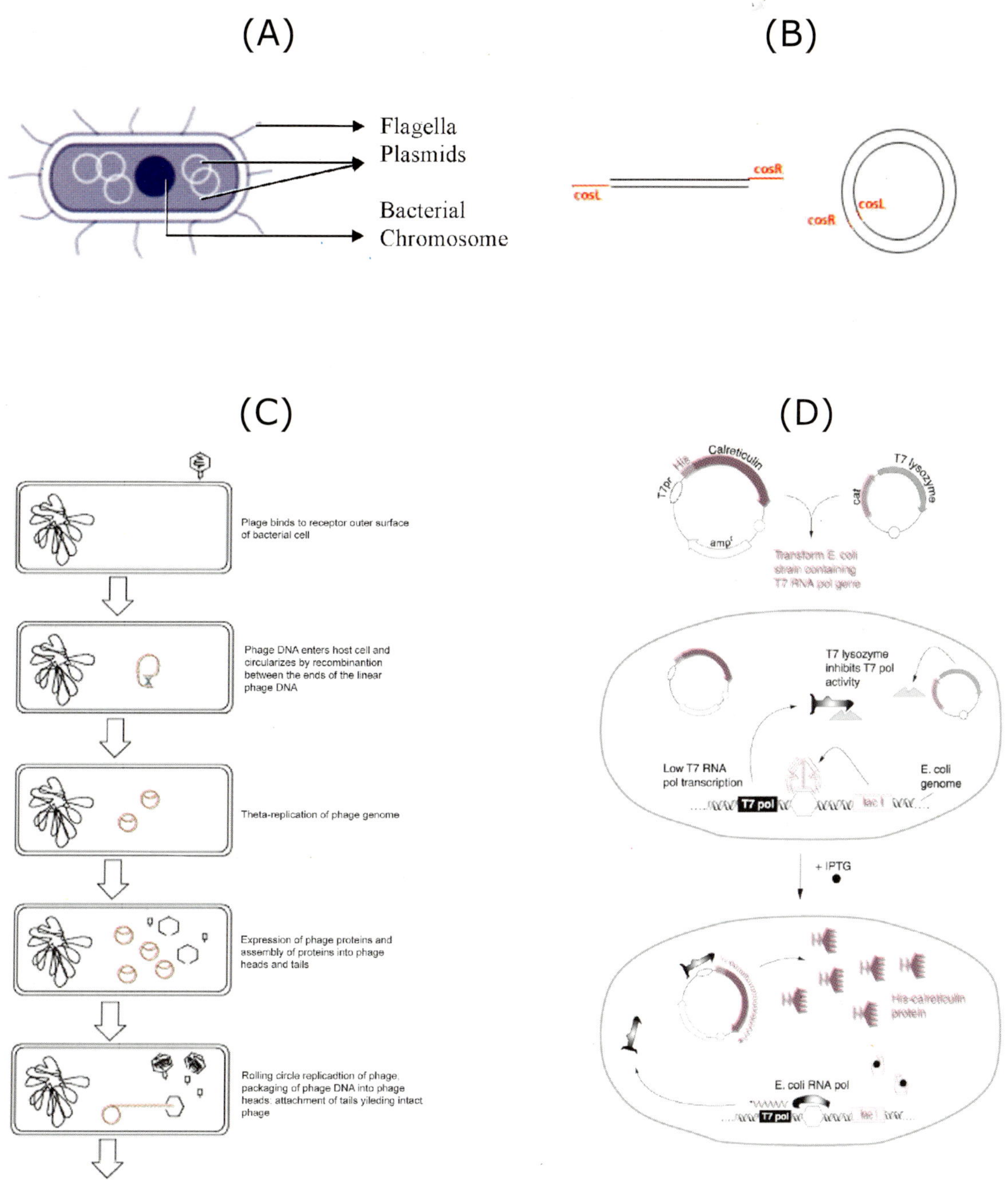

Fig: (A) Bacteria showing plasmid and genomic DNA, **(B)** The linear form of phage DNA converting to circular form, **(C)** The phage lysis cycle, **(D)** Inhibition of T7 RNA polymerase in presence of T7 lysozyme encoding plasmid.

(E) (G)

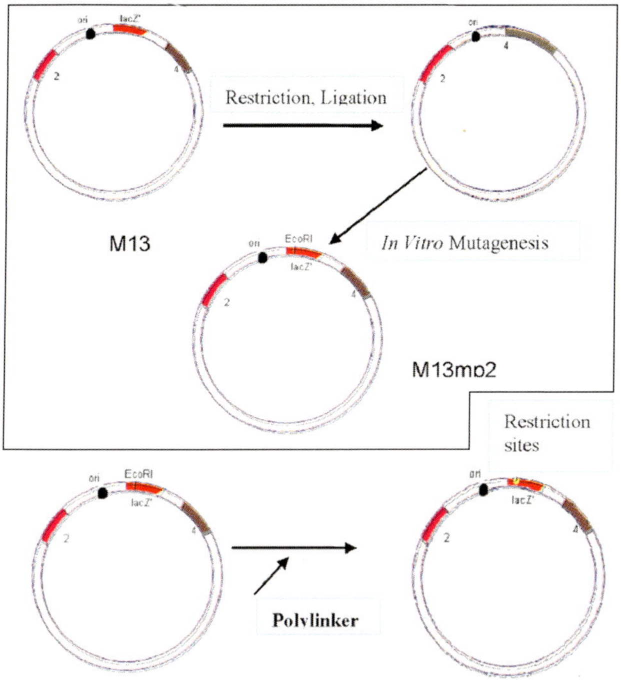

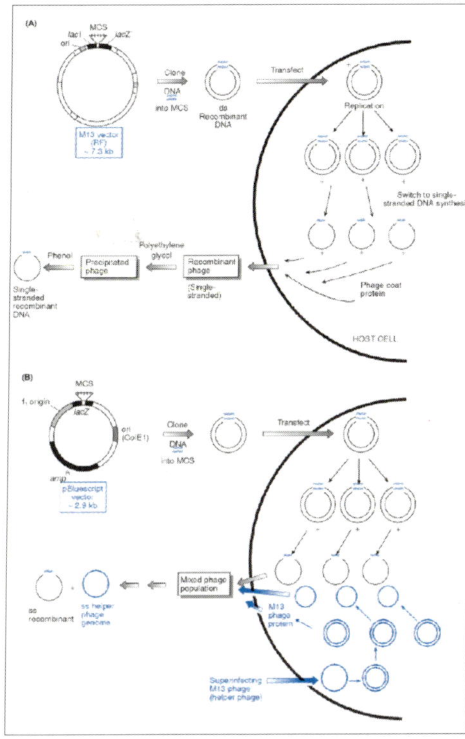

(F)

Fig: E) Construction of M13mp1 and M13mp2 from the wild type genome of M13, **F)** Construction of M13mp7 from M13mp2, **G)** Producing single-stranded recombinant DNA using M13 and phagemid vectors. **(G-A)** M13 vectors. M13 vectors are replicative (RF) forms of M13 derivatives containing a nonfunctional component of the lacZ ß-galactosidase system which can be complemented in function by the presence of a complementary lacZ component in the E. coli JM series. The double-stranded M13 recombinant DNA enters the normal cycle of DNA replication to generate numerous copies of the genome, prior to a switch to production of single-stranded DNA (+ strand only). Maturo recombinant phage exit from the cell without lysis. **(G-B)** Phagemid vectors. The pBluescript series of plasmid vectors contain two origins of replication: a normal one from ColE1 and a second from phage f1 which, in the presence of a filamentous phage genome, will specify production of single-stranded DNA. Abbreviation: MCS, multiple cloning sites.

(H)

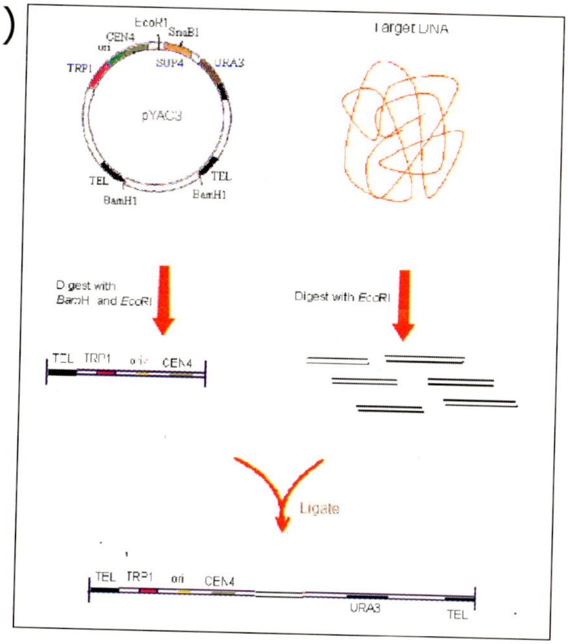

(I)

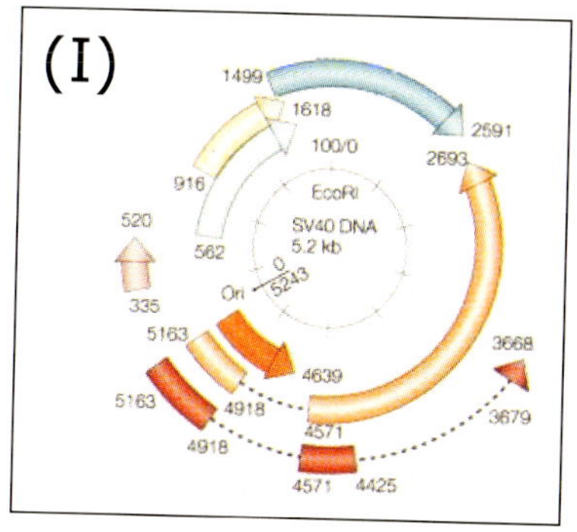

(J)

(K)

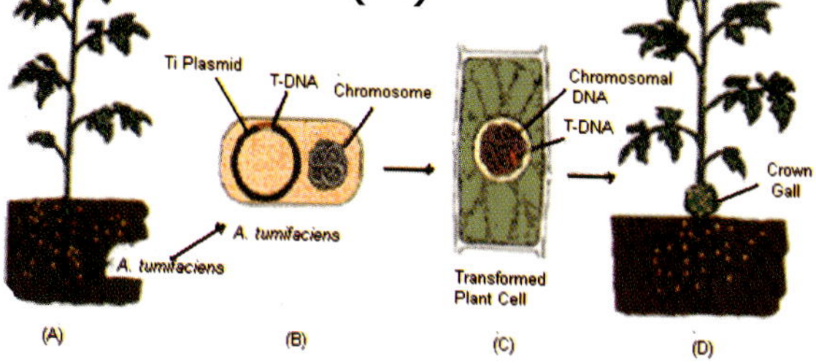

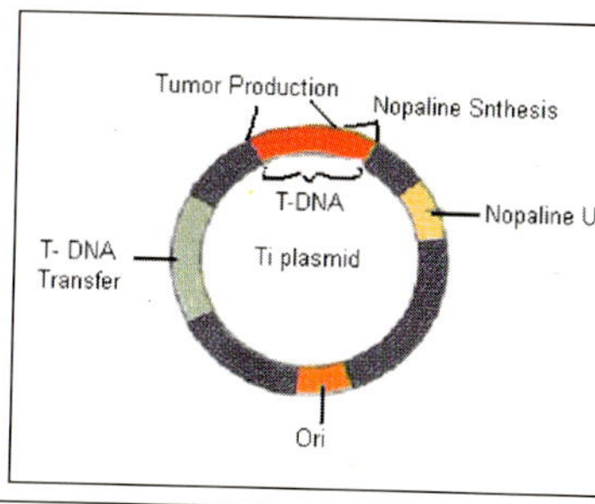

(L)

(M)

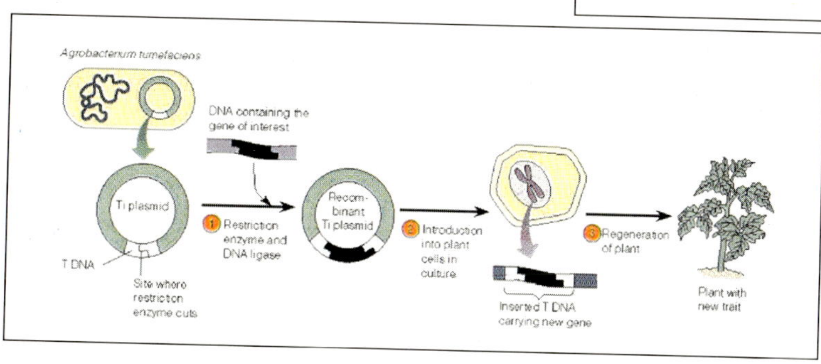

Fig: (H) The cloning strategy for the YAC vector, **(I)** The genome of SV40 virus, **(J)** Crown Gall formation in plant infected with A. tumifaciens, **(K)** Infection of the plant with Agrobacterium tumefaciens and development of crown gall in plant, **(L)** Ti Plasmid, **(M)** basic strategy for integration of the DNA of interest into plant cells.

Bevan M (1984) Binary Agrobacterium vectors for plant transformation *Nucleic Acids Res* **12**: 8711-8721

Bolivar F Rodriguez RL Greene PJ Betlach MC Heyneker HL Boyer HW (1977) Construction and characterization of new cloning vehicles II A multipurpose cloning system *Gene* **2(2)**: 95–113

Brown TA (2001) Gene cloning and DNA analysis : An introduction 4th ed *Blackwell Science Ltd* 105-129.

Chauthaiwale V M Therwath A Deshpande V V (1992) Bacteriophage Lambda as a cloning vector *American Society for Microbiology* **56**: 577-591

Christou P (1997) Rice transformation: bombardment *Plant Mol Biol* **35**: 197-203

Collins J and B Hohn (1978) Cosmids: a type of plasmid gene-cloning vector that is packageable *in vitro* in bacteriophage lambda heads *Proc Natl Acad Sci* USA **75**: 4242-4246

Colson C Glover SW Symonds N Stacey KA (1965) The Location of the Genes for Host-Controlled Modification and Restriction in ESCHERICHIA COLI K-12 *Genetics;* **52(5)**: 1043–1050

Dale J W Schantz M V (2002) From Genes to Genomes: Concepts and Applications of DNA Technlology 1st ed John Wiley & Sons Ltd

Davis R W and J S Parldnson 1971 Deletion mutants of lambda III Physical structure of att *O J Mol Biol* **56**: 403-423

De Block Marc ;Herrera-Estrella Luis ;Van Montagu Marc ; Schell Jeff ; Zambryski (1984) Patricia Expression of foreign genes in regenerated plants and in their progeny *EMBO J* **3(8)**: 1681–1689

De Greve H Decraemer H Seurinck J Van Montagu M Schell J (1981) The functional organization of the octopine *Agrobacterium tumefaciens* plasmid pTiB6s3 *Plasmid* **6(2)**: 235–248

Deblaere R Reynaerts A Hofte H Hernalsteens JP Leemans J and Van Montagu M (1987) Vectors for cloning in plant cells *Meth Enzymol* **153**: 277-292

Dotto G P and N D Zinder (1983) The morphogenetic signal of bacteriophage fl *Virology* **130**: 252-256

Dotto G P and N D Zinder (1984) Increased intracellular concentration of an initiator protein markedly reduces the minimal sequence required for initiation of DNA synthesis *Proc Natl Acad Sci* USA **81**: 1336-1340

Dotto G P K Horinuchi and N D Zinder (1984) The functional origin of bacteriophage fl DNA replication: its signal and domains *J Mol Biol* **172**: 507-521

Fraley RT Rogers SG Horsch RB Sanders PR Flick JS Adams SP Bittner ML Brand LA Fink CL Fry JS Galluppi GR Goldberg SB Hoffmann NL SC Woo (1983) Expression of bacterial genes in plant cells *Proc Natl Acad Sci U S A* **80(15)**: 4803–4807

Freifielder D (1987) Molecular biology p 577-616 Jones & Bartlett Publishers Inc Boston

Frischauf A (1987) Construction and characterization of genomic library in lambda *Methods Enzymol* **152**: 190-199

Frischauf A L Hans A Poustuka and N Murray (1983) Lambda replacement vectors carrying polylinker sequences *J Mol Biol* **170**: 827-842

Frischauf A N Murray and H Lehrach (1987) Phage vectors-EMBL series *Methods Enzymol* **153**: 103-115

Hajdukiewicz P Svab Z Maliga P (1994) The small versatile pPZP family of Agrobacterium binary vectors for plant transformation *Plant Mol Biol* **25**: 989-994

Han J H and W J Rutter (1987) Lambda gt22 an improved lambda vector for directional cloning of full length cDNA *Nucleic Acids Res* **15**: 6304

Hernalsteens J-P Thia-Toong L Schell J Van Montagu M (1984) An Agrobacterium-transformed cell culture from the monocot Asparagus officinalis *EMBO J* **3(13)**: 3039–3041

Herrera-Estrella L Block M De Messens E Hernalsteens J-P Montagu M Van Schell J (1983) Chimeric genes as dominant selectable markers in plant cells *EMBO J* **2(6)**: 987–995

Hiei Y Ohta S Komari T Kumashiro T (1994) Efficient transformation of rice *(Oryza sativa L)* mediated by Agrobacterium and sequence analysis of the boundaries of the T-DNA *Plant J* **6**: 271-282

Hoekema A Hirsch PR Hooykaas PJJ Schilperoort RA (1983) Binary vector strategy based on separation of vir- and T-region of the *Agrobacterium tumefaciens* Ti-plasmid *Nature* **303**: 179-180

Hohn B Collins J (1980) A small cosmid for efficient cloning of large DNA fragments *Gene* **11(3-4)**: 291–298

Hood EE Gelvin SB Melchers S Hoekema A (1993) New *Agrobacterium* helper plasmids for gene transfer to plants (EHA105) *Trans Res* **2**: 208-218

Hood EE Helmer GL Fraley RT Chilton MD (1986) The hypervirulence of *Agrobacterium tumefaciens* A281 is encoded in a region of pTiBo542 outside of T-DNA *J Bac* **168**: 1291-1301

Hooykaas P J J and Beijersbergen A G M (1994) The virulence system of *Agrobacterium tumefaciens Annu Rev Phytopathol* **32**: 157-179

Kado C I (1991) Molecular mechanisms of crown gall tumorigenesis *Critic Rev Plant Sci* **10**: 1-32

Leemans J Deblaere R Willmitzer L Greve H De Hernalsteens JP Montagu M Van Schell J (1982) Genetic Identification of functions of TL-DNA transcripts in octopine crown galls *EMBO J* **1(1)**: 147–152

Leemans J Shaw C Deblaere R De Greve H Hernalsteens JP Maes M Van Montagu M Schell J (1981) Site-specific mutagenesis of Agrobacterium Ti plasmids and transfer of genes to plant cells *J Mol Appl Genet* **1(2)**: 149–164

Mead D A E Szczesna-Skorupa and B Kemper (1985) Single stranded DNASP6 promoter plasmids for engineering mutant RNAs and proteins: synthesis of a "stretched" preproparathyroid hormone *Nucleic Acids Res* **13**: 1103-1118

Meyer T F K Geider C Kurz and H Schaller (1979) Cleavage site of bacteriophage fd gene II protein in the origin of viral strand replication *Nature* (London) **278**: 365-367

Murphy K C (1998) Use of Bacteriophage Recombination Functions To Promote Gene Replacement in *Escherichia coli American Society for Microbiology 180*: 2063-2071

Murray K and N E Murray (1975) Phage lambda receptor chromosomes for DNA fragments made with restriction endonuclease III of Haemophilus influenzae and restriction endonuclease I of *Escherichia coli J Mol Biol* **98**: 551-564

Ohta S Mita S Hattori T and Nakamura K (1990) Construction and expression in tobacco of a (β-glucuronidase (GUS) reporter gene containing an intron within the coding sequence *Plant Cell Physiol* **31**: 805-813

Prentki P Karch F Iida S Meyer J (1981) The plasmid cloning vector pBR325 contains a 482 base-pair-long inverted duplication *Gene* **14(4)**: 289–299

Primrose S B Twyman R M Old R W (1980) Principles of Gene Manipulation 6th ed *Blackwell Science* Ltd 43-84

Rao RN Rogers SG (1979) Plasmid pKC7: a vector containing ten restriction endonuclease sites suitable for cloning DNA segments *Gene* **7(1)**: 79–82

Sambrook J E F Fritsch and T Maniatis (1989) Molecular cloning: a laboratory manual 2nd ed Cold Spring Harbòr Laboratory Press Cold Spring Harbor N Y

Sambrook J Fritsch E F and Maniatis T (1989) Molecular Cloning: A Laboratory Manual 2nd Ed Cold Spring Harbor Laboratory New York

Seal S E vandenBosch F and Jeger M J (2006) Factors influencing begomovirus evolution and their increasing global significance: implications for sustainable control *Crit Rev Plant Sci* **25**: 23-46

Short J M J M Fernandez J A Orge and W D Huse (1988) Lambda ZAP A bacteriophage lambda expression vector with in vivo excision properties *Nucleic Acids Res* **16**: 7583-7599

Siemering KR Golbik R Sever R Haseloff J (1996) Mutations that suppress the thermosensitivity of green fluorescent protein *Curr Biol* **6**: 1653-1663

Simoens C Alliotte T Mendel R *et al* (1986) A binary vector for transferring genomic libraries to plants *Nucleic Acids Res* **14**: 8073-8090

Stansfield Wiliam D (1996) Schaum's Outline of Theory and Problems of Molecular and Cell Biology McGraw-Hill NY 183-200

Strachan T Read A P (1999) Human Molecular Genetics Garland Science Publishers

Tanaka A Mita S Ohta S Kyozuka J Shimamoto K Nakamura K (1990) Enhancement of foreign gene expression by a dicot intron in rice but not in tobacco is correlated with an increased level of mRNA and an efficient splicing of the intron *Nucl Acids Res* **18**: 6767-6770

Turner P C McLennan A G Bates A D and White M R H (1997) Instant notes in Molecular Biology School of Biological Sciences University of Liverpool Liverpool UK 91-109

Van Haute E Joos H Maes M Warren G Van Montagu M Schell J (1983) Intergeneric transfer and exchange recombination of restriction fragments cloned in pBR322: a novel strategy for the reversed genetics of the Ti plasmids of *Agrobacterium tumefaciens EMBO J* **2(3)**: 411–417

Van Larebeke N Engler G Holsters M Van den Elsacker S Zaenen I Schilperoort RA Schell J (1974) Large plasmid in *Agrobacterium tumefaciens* essential for crown gall-inducing ability *Nature*; **252**: 169–170

Wang K Herrera-Estrella L Van Montagu M Zambryski P (1984) Right 25 bp terminus sequence of the nopaline T-DNA is essential for and determines direction of DNA transfer from agrobacterium to the plant genome *Cell* **38(2)**: 455–462

Yanisch-Perron C J Vieira and J Messing (1985) Improved M13 phage cloning vectors and host strains: nucleotide sequences of the M13mpl8 and pUC19 vectors *Gene* **33**: 103-119

Zagursky R J and M L Berman (1984) Cloning vectors that yield high levels of single-stranded DNA for rapid DNA sequencing *Gene* **27**: 183-191

Zambryski P Joos H Genetello C Leemans J Montagu M Van Schell J (1983) Ti plasmid vector for the introduction of DNA into plant cells without alteration of their normal regeneration capacity *EMBO J* **2(12)**: 2143–2150

5

GENETICALLY MODIFIED CROPS IN INDIAN PERSPECTIVE

DAPINDAR K. BAKSHI AND J.K ARORA

INTRODUCTION

In India, vigorous efforts have been initiated for development of GM crops, in view of their potential and importance. As of now, Bt cotton is the only transgenic crop approved for commercial cultivation in the country. Since its introduction in 2002, there has been a phenomenal increase in the area under Bt cotton which presently stands at 3.5 million hectare. As of now, 111 hybrids and 4 events viz. Bollgard 1 (BG-1), Bollgard 2 (BG-2), Event 1 and GFM event of Bt cotton are approved for commercial cultivation in the country. The first GM food crop of India i.e. Bt brinjal is in pipeline for introduction in the market. In fact, with the release of Bt brinjal, India is likely to take global lead in introduction of first GM eggplant. Further, various other GM crops including rice, potato, tomato, mustard, groundnut and several other food crops are under different stages of development and trials in the country. The genetic modification is primarily targeted at introduction of crop protection traits which include: resistance to pests, disease and herbicides; abiotic stresses such as drought, heat, cold or salinity and quality traits for enhanced nutrition, shelf life, improved taste, colour, fragrance and increased productivity. The results of socio-economic impact assessment studies on Bt cotton are encouraging in terms of economic and environmental benefits.

The agriculture sector is vital to economic growth of India with approximately 43 % of land under active agriculture and more than 60 per cent of Indian population depending on agriculture for their livelihood. Modernization of agriculture via introduction of new hybrid varieties and increased inputs such as irrigation, fertilizers, plant protection chemicals and modern implements, in 1970's brought in the Green Revolution which resulted in very high productivity of wheat and rice and made the country self-sufficient in food. However, over the years, intensive cropping patterns and monoculture led not only to various ecological problems like ground water depletion, micronutrient deficiency in the soil and increased use of farm chemicals but also a plateau in the agricultural productivity (Sharma *et al*, 2003).

In this context, biotechnology is strongly emerging as a powerful solution having all the technical depth and breadth to provide second push to 'Green Revolution'. Apart from biotechnological interventions like biofertilizers, biopesticides, tissue culture, etc., genetic engineering of crops has revolutionized the whole concept of modern agriculture. With the advent of recombinant DNA technology, it has become possible to transfer one or more genes in a quick, controlled and reliable manner, not from within the crops species or closely related plants but also from other species that is not achievable through conventional methods. Thus transgenic / Genetically Modified (GM) crops with new and beneficial traits like resistance to pests; diseases and herbicides; tolerance to abiotic stresses such as drought, cold, heat, salinity and enhanced nutrition / quality of the crop have been generated.

STATUS OF GM CROPS IN INDIA

APPROVED GM CROP

Cotton

India is the major cotton-producing country ranking third in global cotton production after the United States and China (MacDonald, 2000). Out of 35 million hectare global area under cotton cultivation, India contributes approximately 9 million hectares representing about one quarter of the total global area under the crop (Sen, 2005). At national level, this fiber crop contributes around 30% gross domestic produce of Indian agriculture. Cotton is planted by 4 million small farmers and is predominantly cultivated in the northern region of the country, mainly in the states of Punjab, Haryana, and Rajasthan. During the last five decades, the production of cotton has increased from 30 lakh bales in 1950-51 to 244 lakh bales during 2005-06 (*Source* : Cotton Advisory Board). In spite of large area under cotton cultivation, the average yield of cotton is 440 kg/ha which is very less as compared to global yield of 677 kg/ha on an average (Dureha *et al*, 2007).

Main losses in cotton production are due to its susceptibility to ~ 160 species of pests and various diseases like Bacterial blight, Fusarium wilt, Alternaria leaf spot, Grey mildew, etc. which damage cotton at various stages of growth (Manjunath, 2004). Together these pests and diseases result in an estimated loss of 50 - 60% of potential cotton yield. The pests attack the cotton plant at its varied stages of growth e.g. sucking pests like aphids (*Aphis gossypii*) and jassids(*Amrasca bigutulla*) attack at the initial stage followed by white flies (*Bemisia tabaci*) and finally the cotton bollworms including American bollworm (*Helicoverpa armiger*), Spotted bollworm (*Earias vittella*) and Pink bollworm (*Pectinophora gossipiella*) (Datt, 2001, Sharma *et al*, 2001). The farmers have been using combinations of several expensive chemical pesticides in multiple sprays to control pest infestations. In India, an estimated Rs 28 billion worth of pesticides are being used in agriculture, of which Rs 16 billion are being used on cotton alone (Sharma *et al*, 2001; Manjunath, 2004). Excessive and indiscriminate use of pesticides have led to development of resistance among some of the key pests, including the American bollworm (Kranthi *et al*, 2002). For revival of cotton, various steps and strategies were worked out by the Indian Government, the major being introduction of Bt cotton hybrids in the country.

The Govt. of India approved commercial cultivation of first and only GM crop i.e. Bt Cotton in March 2002, though the efforts to harness genetic engineering technology for bollworm resistance in cotton began as early as in 1990s with formation of Institutional Bio-safety Committees (IBSCs) in 1994 and import of genetically modified cotton seed from Monsanto, USA in 1995. Bt hybrid cotton seed carried a gene (cry) sourced from the soil bacterium *Bacillus thuringiensis* (Bt) subspecies *kurstaki* which produced a protein toxic to bollworm, thus providing resistance to the cotton plant. The approval of Bt cotton in India was preceded by various field / multi-centric research / large scale trials in different agro climatic regions between 1998 to 2001. Research on biosafety studies and agronomic evaluation of Bt cotton including its toxicological and allergencity effects was initiated in various national laboratories (IRC, 2002; Barwale, 2004). Indian Bt varieties were developed by Monsanto India Ltd. and the Maharashtra Hybrids Seeds Co. (Mahyco). Subsequently, several other companies took sub licence from MAHYCO that resulted in development of more and more Bt hybrids. At present there are 19 major seed companies involved in the process. For more details reader may refer to a Review by Malik *et al.*, (in this volume).

Area and approved hybrids / events of Bt cotton

Bt cotton hybrids were released for the first time to farmers in India in 2002 and were planted on approximately 44,500 ha. The six states, including Gujarat, Madhya Pradesh, Maharashtra from central zone and Andhra Pradesh, Karnataka, Tamil Nadu from south zone were the first ones to adopt Bt cotton in 2002. Since then, the following five years witnessed phenomenal increase in plantation of Bt cotton, with area expanding to nearly 1,00,000 ha in 2003, half a million hectare in 2004, 1.3 million hectare in 2005 and 3.8 million hectare in 2006 which has also exceeded the total area of 3.5 million hectare under Bt cultivation in China (Clive, 2006). The North Zone including Punjab, Haryana and Rajasthan got the Govt. approval for Bt cotton cultivation only in the year 2005. At present Maharashtra is leading the other states with 1.840 million hectares of area under the Bt cotton contributing 48% of the total Bt cotton in India. It is closely followed by Andhra Pradesh with 22% contribution (830,000 hectares), Gujarat with 12 % (470,000 hectares), Madhya Pradesh with 8 % contribution (310,000 hectares) and Northern zone contributing 6% (215000 hectares) and the balance by Karnataka, Tamil Nadu and other states (Clive, 2006).

India is the only country which is growing all the four species of cultivated cotton *Gossypium arboreum, G. herbaceum, G.barbadense, G.hirsutum* (PIB, 2006). *G.hirsutum* represents 90% of hybrid cotton production in India and all the current Bt cotton hybrids are *G.hirsutum*. A total of 62 cotton Bt hybrids were approved for plantation in 2006 as compared to 20 in 2005 and 4 in 2004. Till 2005, only one event known as bollgard 1 (BG-I) featuring cry1Ac gene, developed by MAHYCO sourced from Monsanto was approved for sale. The number of events increased from 1 to 4 in 2006 and three new events approved are : Bollgard II (BG-II) with event MON 15985, developed by MAHYCO and sourced from Monsanto; featuring stacked genes cry1AC and cry2Ab; Event 1, developed by JK seeds featuring the cry1Ac gene, sourced from IIT Kharagpur, India and GFM event developed by Nath Seeds, sourced from China, featuring the fused genes cry1Ab and cry1Ac (Clive, 2006). Recently in May 2007, the Genetic Engineering Approval Committee (GEAC) of India approved 53 new varieties of Bt cotton hybrids for commercial release for the current year. With these new approvals, farmers now have a choice to select among 111 Bt cotton hybrids for current 2007 season.

GM CROPS UNDER DEVELOPMENT

Besides, Bt cotton, there are various other genetically modified crops under different stages of development and trials in India. These include:

Brinjal

According to USDA/ ERS (2006), of the total brinjal world production, 55 % is contribued by China followed by India contributing 28%. In India, 25 million farmers cultivate brinjal on over 5.1 lakh hectares with an annual production of about 8.2 million tonnes (Rao, 2005). It is mainly cultivated by resource-poor farmers in various parts of the country and occupies third position amongst vegetable crops.. There are many local varieties grown in India, in addition to improved varieties and hybrids. West Bengal is the largest producer of brinjal followed by Maharashtra and Bihar. The other main states growing this crop are Karnataka, Maharashtra, Gujarat, Andhra Pradesh, Assam and Madhya Pradesh (*Source*: Ministry of Agriculture, GOI). It is a staple vegetable crop which is extensively damaged by brinjal fruit and shoot borer *Leucinodes orbonalis.* Due to high reproductive potential of the pest, the crop losses range from 50 - 70 % as the affected fruit looses its market value besides considerable reduction in yield (Rao, 2005). The farmers use huge quantities of chemical pesticides with an average of 25-80 sprays for effective control of the insect (Verma and Hota, 2006).

In view of this, the research for development of Bt Brinjal, the first GM food crop is at an advanced stage in India. Bt brinjal - a genetically modified food crop has been developed by the seed company Mahyco in 2000 by inserting a gene cry 1Ac from *Bacillus thuringiensis* (Rao, 2005). The developed crop is resistant against major pests attacking brinjal through out its life cycle i.e. the Brinjal Fruit and Shoot Borer *(Leucinodes orbonalis)* and Fruit borer *(Helicoverpa armigera) (Source:* GM Watch 2006). For more than five years, rigorous scientific studies and biosafety trials were carried out by Mahyco and after their successful completion, the company sought permission for large scale field trials of four Bt brinjal hybrids i.e. MHB-4 Bt, MHB 9 Bt, MHB 80 Bt and MHBJ -99 Bt from Genetic Engineering Approval Committee (*Source* : GEAC 2006). Of the three GM food crops in India : brinjal, mustard and potato, awaiting permission for large-scale field trials and seed production, brinjal is the first one to get GEAC approval in July 2006. No genetically-modified Brinjal has so far been released for an advanced stage of field trials in open conditions anywhere in the world.

Work on Bt brinjal is also being carried by number of other public/private sector institutions. Mahyco transferred the technology and basic breeding material of Bt brinjal to public sector institutions like Tamil Nadu Agricultural University, Coimbatore (TNAU), the University of Agricultural Sciences, Dharwad (UASD), and the Indian Institute of Vegetable Research, Varanasi under the agreement where public sector institutes are allowed to develop, breed and distribute their own Bt Brinjal varieties on cost-to-cost basis (Rao, 2005; Verma and Hota, 2006). Recently, Sungro Seeds, New Delhi, also got engaged in developing transgenic hybrids of Bt Brinjal for lepidopteran resistance in collaboration with Mahyco. Indian Agricultural Research Institute (IARI) has also developed a Bt brinjal with Cry1Ab gene (Rao, 2005) and took up agronomic trials in controlled environment from 1998-2001. In 2003, IARI was permitted to conduct field trials at five locations - Delhi, Karnal, Pune, Tamil Nadu Agricultural University and the Indian Institute of Horticultural Research. IARI was also a part of Monitoring Committee of Review Committee on Genetic Modification (RCGM,GOI) in monitoring multi-location field trials during the development of Bt brinjal by Mahyco in the year 2000.

Rice

Among the rice growing countries in the world, India has the largest area under rice crop and ranks second in production next to China. Rice is the staple food of 65% of Indian population and occupies about 23.3% of gross cropped area of the country. Average rice productivity in India during 1999-2000 was 1986 kg/ha, which was about 23% below the world average productivity of 2563 kg/ha during the same year (Source: DRD, 2002). There has been considerable increase in productivity of rice (2,000 kg per hectare) in the country during the recent past and the area under this cereal has increased from 43,000 hectares to 14.06 lakh hectares, while Assam, Uttar Pradesh, West Bengal and Orissa have reported a rise in the acreage, Punjab has reported its downfall (The Bureau, 2006).

A major disease afflicting rice crop in most rice-growing areas is sheath blight which is caused by the fungus *Rhizoctonia solani* (Balasubramanian, 2005). Though the attempts have been made to control sheath blight by fungicides, but the practice is neither practical nor sustainable and has been reported to have deleterious effects on both human health and environment. Further, yellow stem borer (YSB) and Rice Leaf Folder (RLF) have been reported to cause 5 -10% damage to rice crop with local catastrophic outbreaks causing up to 60% damage (Pasalu *et al,* 2004). The YSB bores into the rice stem and hollows it out, while the RLF causes the young larvae to feed on tender leaves. Chemical control of these pests is ineffective because larvae remain on the outer surface of the rice plant for only a short time before penetrating inside the tissues of the crop.

Genetical engineering of rice for resistance to major pests appears as a promising approach. Currently, no GM rice variety has been commercially released in India but a lot of research is underway on this aspect. Bt Rice varieties have been developed by Mahyco, IARI and Punjab Agricultural University, Ludhiana by incorporating *Bt cry1A(b)* and *Cry1Ac* genes to generate plants resistant to yellow stem borer, the lepidopetran pest. The work on development on herbicide-tolerant rice plants by using *bar* transgene has been carried out by Centre for Cellular and Molecular Biology, Hyderabad. Similar work on *Chitinase, α-1, 3-glucanase* and *osmotin* genes has been done by Madurai Kamaraj University, Madurai to develop plants resistant to fungal infection (Sharma *et al* 2003). Further, following development of rice with single transgenic traits, transgenic varieties with multiple traits have also been created. A stacked combination of Bt toxins Cry1Ab and Cry1Ac along with tolerance to the herbicide glufosinate (*bar* gene) have been created in order to develop transgenic line resistant to both stem borer and herbicide glufosinate (Cummins, 2004). On similar grounds, various other transgenes viz. *xa21* for conferring resistance to bacterial blight; fungal *glucose oxidase* (*gox*) gene for broad-spectrum disease resistance; *S-adenosylmethionine decarboxylase, codA, cor47* for biotic and abiotic stress tolerance and vitamin-A and iron enriched Golden Rice, have been created by various institutes like Bose Institute of Calcutta, Delhi University, South Campus, New Delhi; Central Rice Research Institute, Cuttack; Centre for DNA Fingerprinting and Diagnostics, Hyderabad. Recently, an agreement has been signed by Indian Council of Agricultural Research (ICAR) with the Las Banos-based International Rice Research Institute (IRRI) for collaboration in research for genetic enhancement of rice in terms of yield and quality (SABP, 2007).

Potato

Potato is world's fourth important food crop after wheat, rice and maize owing to its great yield potential and high nutritive value. It accounts for annual global production of about 300 million tones and is an economically important staple crop in both developed and developing countries. India ranks 5th in potato production after China, Russian Federation, Poland and Ukraine. It is most widely grown vegetable crop in the country with a share of 25.7 per cent and 1.28 million ha under cultivation with total production of 22.49 MT. Uttar Pradesh is the leading potato growing state in the country with a production of 9.53 million tonnes followed by West Bengal and Bihar (Kumar and Kumar 2005).

The overall potato productivity in India is however merely 16-19 tonnes/ha as compared to other nations i.e. European countries and USA with 30-40 tonnes/ha (TarKeshwar, 2007). The major pests leading to damage of potato are Potato Virus Y (PVY) and mosaic and leaf curl virus which account for about 30-40% crop losses and even total crop failure sometimes. On the lines of Bt cotton, Bt potato resistant to various viruses, fungi and diseases is being developed and is likely to be available for cultivation within the next few years. Recently, significant success has been achieved in developing PVY resistant transgenic potato using coat-protein mediated protection. The potato variety has been developed jointly by Central Potato Research Institute, Shimla and National Research Center on Plant Biotechnology, New Delhi (BS Agriculture Editor 2002; Ghosh *et al* 2002). The efforts are also being made to enhance nutritional value of potato by inserting non-allergenic seed albumin gene *Amal* derived from plant *Amaranthus hypochondriacus* into plants (Chakraborty *et al*, 2000).

Tomato

Globally, India occupies second position after China with respect to area under tomato cultivation and ranks fifth in terms of production. Nearly, 7.1 million tonnes of tomato is produced annually in the country from an area of 5.4 lakh hectares. Tomato occupies second position amongst the vegetable crops in terms of production. Andhra Pradesh is the largest grower of tomato with a production of 2.05 MT. The other main tomato growing states are Bihar, Karnataka, Maharashtra and Orissa (*Source* : IPNI; TNAU/DAM)

Tomato producers in India today lose 20-30 % of their crop through post harvest spoilage due to pests especially the fruit borer *Helicoverpa armigera* (Kumar and Kumar, 2004). The research for development of transgenic tomato for various traits is underway in the country e.g. introduction of *Bt. cry1A(b) gene* for resistance to lepidopteron pests; *ACC synthase gene* for controlling fruit ripening, *antisense replicase* gene for development of resistance to tomato leaf curl virus by IARI; *chitinase and glucanase* genes for resistance to fungal diseases by Indian Institute of Horticultural Research and *OXDC* gene for generation of fungal resistance by Jawaharlal Nehru University (Sharma *et al* 2003; Kumar and Kumar, 2004).

Mustard

Mustard, the source of edible oil is the second most important oilseed crop in India after groundnut sharing 27.8 per cent in India's oilseed economy. Though, last few decades witnessed Yellow Revolution due to phenomenal increase in mustard production and productivity from 2.68 million tonnes and 650 kg/ha in 1985-86 to 6.96 million tonnes and 1022 kg/ha in 1996-97 (Kumar 1999). In spite of this, India still has to import 1/3rd of its edible oil requirement, costing Rs 12,000 crore annually (Sharma, 2002). In order to make the country self reliant in edible oils, the only way is to increase productivity of this crop. Hybrids, which have been used in other agricultural crops to increase yield, have not been commercially viable in mustard till date, because of the several biological limitations associated with the naturally occurring systems of hybridisation in mustard.

In this regard, GM mustard has a promise of increased productivity in edible oils. Various pubic and private institutes are involved,in research on this aspect. Proagro Seed Co Pvt Ltd is the predominant private sector organisation working on GM mustard in India and has developed genetically engineered hybridisation system for development of high yielding mustard hybrids. GM mustard has been extensively tested for its food, feed and environmental safety and field trials have been conducted at 50 locations in Gujarat, Haryana, Madhya Pradesh, Rajasthan, and Uttar Pradesh which are being coordinated by ICAR, (Rao, 2007). Delhi University has developed mustard by use of genes:*bar, barnase, barstar* to generate herbicide-tolerant plants, male-sterile and restorer lines for hybrid seed production while IARI has used *Arabidopsis annexin* and *choline dehydrogenase* gene to generate stress-tolerant mustard plants. Similarly, Tata Energy Research Institute, New Delhi has used *Ssu-maize Psy* and *Ssu-tpCrtI* gene to generate

Mustard plants containing high levels of *β-carotene* (Jagannath *et al*, 2002; Sharma *et al* 2003; Tuli and Bhatia, 2005)

Groundnut

Groundnut is the major oilseed and single largest source of edible oil in India. The global production of groundnut seed and oil is 32 - 34 million tonnes and 4 - 5.5 million tonnes respectively. Approximately 5.9 million hectare area in India is under groundnut cultivation with total produce of 6.7 million tonnes. Statewise, Gujarat (4.25 million tonnes) tops in production followed by Andhra Pradesh (1.5 million tonnes), Tamil Nadu (1.2 million tonnes), Karnataka (0.5 million tonnes) and Maharastra (0.5 million tonnes) which totally contribute about 78 per cent of the total groundnut production (*Source* : TNAU/ DAM).

Groundnut crop contracts the devastating Indian Peanut Clump Virus (IPCV) via the soil fungus *Polymyxa graminis* which leads to stunted growth and discolored leaves. In the past, efforts have concentrated on eradicating or inhibiting the fungus. The lack of success in this endeavour was mainly due to highly resistant characteristics and complexity of *P. graminis*. The genetically modified (GM) groundnut that can resist IPCV has been developed by ICRISAT through *Agrobacterium* mediated transfer of IPCV's coat protein and polymerase to groundnut (Fernandez, 2003; ICRISAT 2003). The work on a nucleo capsid protein transgene of PBNV as well as "golden" groundnut variety which manufactures extra vitamin A for a more nutritious crop is also underway at ICRISAT.

SOCIO ECONOMIC IMPACT OF GM CROPS - BT COTTON : A CASE STUDY

With 5 years gone by, its release in the country, preliminary figures based on numerous studies conducted to gauge the agronomic and economic performance of Bt cotton have begun to arrive. Though the true extent of popularity of Bt cotton is yet to be determined but the interesting facts are the increasing area being brought under its cultivation by more and more farmers. Apart from Mahyco, who after launch of Bt hybrids in 2002, carried studies that supported higher yield and reduced insecticide sprays leading to economic benefit to the farmers, various other studies also showed potential gains and good performance of Bt cotton in India (Naik, 2001; Qaim and Zilberman, 2003). Qaim and Zilberman (2003) analyzed trial data from seed companies testing Bt cotton and compared it with conventional varieties (non Bt) and reported the reduction in usage of insecticides by one third and yield gains up to 80%. However, to assess the economic impact of Bt cotton under real conditions, Bennett *et al* (2004) collected the data from farmers growing Bt cotton crop under market conditions, rather than from trials. The research compared the performance of more than 9,000 Bt and non-Bt cotton farm plots in Maharashtra over the 2002 and 2003 growing seasons and reported the Bt cotton varieties having significant positive impact on average yields and economic performance of cotton growers. The work showed significant increase in yield of Bt cotton i.e. by 63% in 2001 and 45% in 2002.

In a nationwide survey conducted by Neilsen and Marg in the year 2003, Bt cotton hybrids yields were found to be approximately 22% better than conventional varieties which produced nearly 1,898 kg/ha versus 1,473 kg/ha for the conventional varieties leading to increased net profit of 78% for the Bt cotton growers over that of conventional growers (Manjunath, 2004). In the year 2005, Morse *et al* (2005) compared the performance of official genetically modified cotton with unofficial varieties along with non-Bt hybrids by selecting 622 farmers from Gujarat. Results reported the official Bt varieties (MECH 12 and MECH 162) to significantly outperform the unofficial varieties in terms of gross margin. However, unofficial, locally produced Bt hybrids also performed significantly better than non-Bt hybrids, although second-generation (F_2) Bt seed appeared to have no yield advantage compared to non-Bt hybrids but saved on insecticide use. Further, a Front Line Demonstration (FLD) study on cotton for 2005-06 recently released by ICAR reconfirmed a net 33.7% increase in seed yield of Bt cotton hybrids over non-Bt hybrids and 73.8% increase over open-pollinated cotton varieties (OPV). Data in the study came from 1,200 demonstration and farmers plots in 11 cotton growing states in India. In the demonstration plots, the Bt cotton hybrids proved to be highly productive with an average yield of 2,329 kg/ha of seed cotton compared to the non-Bt cotton hybrids (1,742 kg/ha) and varieties (1,340 kg/ha). Similarly, the average yield of Bt cotton hybrids was higher in farmers' plots at 1,783 kg/ha compared to non-Bt cotton hybrids (1,362 kg/ha) and OPV in farmers' field (1,072kg/ha) (Crop Biotech Update 2007).

Cotton farmers have earned an additional income of Rs 7,039 crore in 2006 after a 50% increase in yield due to use of Bt cotton seed, a study conducted jointly by the Associated Chamber of Commerce and Industry and IMRB International (Source: Monsanto) has revealed. Introduction of two-stacked genes Bollgard II Bt cotton has benefited famers by making a saving on pesticide use to the tune of Rs 1,600 per acre. Bollgard II Bt cotton was allowed for commercial cultivation in central and western India in 2006 and according to Assocham-IMRB study, farmers growing conventional cotton spend Rs 2,900 per acre on pesticide use, while those growing Bt cotton (with one gene *cry 1 Ac*) spend Rs 2,000 per acre and farmers growing Bollgaurd II Bt cotton spend Rs 1,300 per acre. A recent study by Gandhi and Namboodhri (2006) also showed the yield gain of 31% with the total reduction in number of pesticide sprays by 39% and 88% increase in profits.

However, certain studies have shown the poor performace of Bt cotton and argued that Bt cotton adopters are actually worse off relative to non-Bt cotton growers (Shiva, 2004; Qayum and Sakkhari, 2005) and Bt cotton has not lived up to expectations, and instead has led to adverse socio-economic impacts for many small farmers in India.

Summing up the GM Crops scenario in India, the adoption of Bt cotton since its commercialization in 2002 has increased multifold in the country. The socio-economic and environmental impact assessment studies are indicating increased farm income and environmental benefits in terms of lesser usage of chemical insecticides and reduced levels of exposure of farmers and farm workers to insecticides, thus favouring the introduction of recombinant DNA technology in agriculture. Further, a lot of research on transgenic brinjal, ricè, potato, tomato, mustard, groundnut and several other food crops is being undertaken in public/private sector in India. However, there are challenges to ensure their safety, agronomic, environmental and other advantages before clearance for commercial application.

REFERENCES

Balasubramanian P (2005) Biotechnology in plant disease control. Foundation for Biotechnology Awareness and Education. http://www.fbae.org/channels/biotech_in_plant_disease_control/biotechnology_in_plant_disease_c.htm

Barwale RB, Gadwal VR, Zehr U, Zehr B (2004) Prospects for Bt cotton technology in India. *Agbioforum* **7(1&2)**: 23-26.

Bennett RM, Ismael Y, Kambhampati U, Morse S (2004) Economic Impact of Genetically Modified cotton in India. *AgBioForum* **7**: 3.

BS Agriculture Editor (2002) Transgenic potatoes developed. http://www.rediff.com/money/2002/may/10potato.htm

Chakraborty S Chakraborty N Datta A (2000) Increased nutritive value of transgenic potato by expressing a nonallergenic seed albumin gene from *Amaranthus hypochondriacus Proc Natl Acad Sci* USA **97**: 3724–3729

Clive J (2006) Global Status of Commercialised Biotech /GM crops:2006 ISAAA Brief No 35 ISAAA:Ithaca NY

Cotton Advisory Board Cotton balance sheet http://www ictextile com/cotton/cotton_balance_sheet html#a3

Crop Biotech Update (2007) Indian Council of Agricultural Research reconfirms higher yield of Bt cotton *Seed Quest* March 2007

Cummins J (2004) GM rice in India Institute of Science In Society (ISIS) Press Release http://www i-sis org uk/GMRII php

Datt K (2001) Bt cotton or better cotton? CCS working paper; 8 http://www ccsindia org/RP01_8 asp

DRD (2002) Rice Productivity Analysis in India Directorate of Rice Development Deptt of Agriculture and Cooperative Ministry of Agriculture GoI Patna

Dureha R Rao S Kulkarni N (2007) India beats China in Bt cotton cultivation *Biospectrum* (India) March 2007

Fernandez RA (2003) ICRISAT develops world's first GM peanut *Seed Quest* May 2003

Gandhi VP Namboodiri NV (2006) The adoption and economics of Bt-cotton in India W P No 2006-09-04 IIM Ahmedabad

GEAC (2006) Development of fruit and shoot borer tolerant brinjal *Seed Quest* June 2006

Ghosh SB Nagi LHS Ganapathi TR Khurana SNP Bapat VA (2002) Cloning and sequence of potato virus Y coat protein gene from in Indian isolate and development of transgenic tobacco for PVY resistance *Current Science* **82(7)**: 855 - 859

GM Watch (2006)

Briefing paper on Bt Brinjal in India http://www gmwatch org/archive2 asp?arcid=6591

ICRISAT (2003) Annual report

IPNI (International Plant Nutritional Institute) Potassium Improves Yield and Quality of Tomato in the Red and Lateritic Soils of West Bengal with Low Inherent Fertility http://www ipni net/ppiweb/eindia nsf/$webindex/F3AEF8AD0713C1D265257049002A248C

IRC (2002) Chronology of Bt cotton in India http://www indiaresource org/issues/agbiotech/2003/chronologyofbt html

Jagannath A Arumugam N Gupta V Pradhan A Burma PK Pental D (2002) Development of transgenic barstar lines and identification of a male sterile *(barnase)*/restorer *(barstar)* combination for heterosis breeding in Indian oilseed mustard (*Brassica juncea*) *Current Science* **82**: 46–52

Kranthi KR Jadhav DR Kranthi S Wanjari RR Ali SS Russell DA (2002) Insecticide resistance in five major insect pests of cotton in India *Crop Protection* **21(6)**: 449-460

Kumar A Kumar AV (2005) Potato biotechnology in India : present status and future prospects 6th World Potato Congress

Kumar H Kumar V (2004) Tomato expressing cry1a(b) insecticidal protein from *Bacillus thuringiensis* protected against tomato fruit borer *Helicoverpa armigera* (Hübner) (lepidoptera: noctuidae) damage in the laboratory greenhouse and field *Crop Protection* **23 (2)**: 135-139

Kumar PR (1999) Rapeseed - mustard research in India : 21st century strategies 10th international rapeseed Congress Canberra Austraila

MacDonald (2000) USDA Economic Research Service The new agricultural trade negotiations: background and issues for the U S Cotton sector cotton and wool Situation and outlook series CWS-17-23

Manjunath TM (2004) Bt - Cotton In India: The Technology Wins as the Controversy Wanes *Agbioworld* December 2004

Ministry of Agriculture GOI Current status: fruits and vegetables http://agricoop nic in/hort/hortrevo5 htm

Monsato News Releases Two Studies Show Benefits of BT Cotton in India http://monsanto mediaroom com/index php?s=43&item=508

Morse S Bennett RM Ismael Y (2005) Genetically modified insect resistance in cotton: Some economic impacts in India *Crop Protection* **24(5)**: 433-440

Naik G (2001) An analysis of socio-economic impact of Bt technology on Indian cotton farmers Centre for Management in Agriculture Indian Institute of Management Ahmedabad India p- 19

Pasalu IC Mishra B Krishnaiah NV Katti G (2004) Integrated pest management in rice in india: status and prospects In : Birthal PS Sharma OP editors Integrated pest management in indian agriculture proceedings 11 NCAP Publications p 25 - 50

PIB (2006) Cotton and Cotton Textiles Industry Ministry of Textiles Govt of India Annual Report http://pib nic in/release/rel_print_page1 asp?relid=23740

Qaim M Zilberman D (2003) Yield effects of genetically modified crops in developing countries *Science* **299**: 900-902

Qayum A Sakkhari K (2005) Bt Cotton *in Andhra Pradesh: A Three-year Assessment* Deccan Development Society Hyderabad

Rao CK (2005) Indian Bt Brinjal in Public-Private Partnership Foundation for biotechnology awareness and education Bangalore India (http://www fbae org/channels/views/indian_bt_brinjal_in_public htm)

Rao K (2007) GEAC torpedoes commercialization of GM mustard in India Foundation for biotechnology awareness and education http://www fbae org/channels/agri_biotech/mustard/geac_torpedoes_commeracialisation htm

SABP (2007) Seeking Better Biotech Yields *SABP Newsletter* **3(3)**

Sen A (2005) Cotton scenario in India http://www indiaonestop com/cotton/cotton htm

Sharma D (2001) The introduction of transgenic cotton in India *Biotechnology and Development Monitor* **44**: 10-13

Sharma D (2002) It's mustard this time India together http://www indiatogether org/agriculture/opinions/dsharma/mustardgm htm

Sharma M Charak K S Ramanaiah T V (2003) Agricultural biotechnology research in India; status and policies Current Science **84(3)**: 297-302

Shiva V (2004) The future of food:countering globalization and recolonisation of Indian agriculture *Futures* **36**: 715–732

Tarkeshwar (2007) Potato Commodity Online http://www commodityonline com/commodities/others/potato-tarkeshwar php

The Bureau Oilseeds cotton sowing lags behind Business line Chennai Jun 27 2006

TNAU / DAM (Tamil Nadu Agricultural University and Deptt of Agriculture Marketing) http://www tnagmark tn nic in/price_forecast htm)

Tuli R Bhatia CR (2005) Accelerating the commercialization of home-grown genetically engineered crops *Current Science* **88(5)**: 716-721

USDA / Economic Research Service (2006) Commodity highlight : Eggplant vegetables and melons outlook/vgs-318

Verma DD Hota M (2006) Development of Bt Brinjal *Biosafety News letter* **2(1)**: p-5

6

THE MECHANISMS OF COTTON FIBER DEVELOPMENT

C.P. MALIK, BHAVNEET KAUR, CHITRA WADHWANI AND AMAN VERMA

INTRODUCTION

Cotton, a worldwide cash crop, is the source of the premiere natural fiber that meets one of man's most basic needs of textiles and clothing, and number of other consumer products. Cotton is the most important textile crop because of its long cellulose-enriched mature fibers. The demand for cotton fiber is increasing hence the crop is getting added attention from the economies of both developed and developing countries.

Cotton has played an important role in the history of several countries and regions of the world, including India. The crop was grown and weaved into cloth in the Indus River Valley as early as 3,000 years B.C. The Egyptians were also weaving cotton fabrics around the same time. In Europe, cotton arrived through Arab merchants around 800 A.D. In fact Christopher Columbus encountered cotton in the Bahamas islands around 1492. By 1500, cotton was internationally known. Now the crop is grown in more than 100 countries occupying about 2.5% of the world's arable land, and more than 100 million families are engaged directly in its production.

Today, the world uses more cotton than any other fiber. The cotton belongs to genus *Gossypium* and has more than 50 species, yet spinnable (lint) fiber, which accounts for nearly 90% of cotton's value, is produced by four cultivated species *G. arboreum* L. and *G. herbaceum* L. (2n = 26), two Asiatic A-genome diploid species *G. hirsutum* L. and *G. barbadense* L(2n = 4x = 52), two allotetraploid AD-genome species,. These are also known as Upland and Pima cotton, respectively.

Cotton fibers are seed trichomes, single cells that differentiate from individual cells of the outer epidermis of the ovule. These fibers elongate rapidly during the course of their development and eventually attain the proportions of the longest cell types known through out the plant kingdom. The mature fibers is an elongated epidermal cell of the ovule with a thickened secondary cell wall comprising chiefly of cellulose (>94%). A developing fiber experiences four discrete yet overlapping phases: (1) initiation, (2) elongation, (3) secondary cell wall synthesis, and (4) maturation (Basra and Malik, 1984). Primordial fiber cells, destined to become lint fibers, arise as balloon-like outgrowths on the ovule surface at anthesis [0 days post-anthesis (dpa)], and soon after enter into the phase of rapid cell elongation; ~30% of the fiber primordia are initiated, producing ~20,000 fibers/ovule (Berlin, 1986). Other epidermal cells forming fuzz fibers arise 4 to 10 dpa, but never attain lengths greater than 10 mm. Typically, the fuzz fibers are 1.5 to 3.3 mm long and usually remain attached to the seed coat during ginning, thus giving the seed a fuzzy appearance.

Fuzz fibers fail to elongate as much as lint fibers due to lack of the initiation and rapid elongation which requires the expression of sucrose synthase (Sus) and, potentially, a transient closure of plasmodesmata. Suppression of Sus activity by 70% or more in the ovule epidermis causes a fiberless phenotype (Ruan *et al.*, 2001). The fiber initials in such ovules were few and shrunken or generally collapsed. Ruan *et al.* (2005) studied the patterns of Sus expression and plasmodesmata in fuzz-like short fibers of a cotton mutant that showed delayed initiation and much slower and reduced elongation of the fiber cells. Through immunolocalization studies, delayed expression of Sus in the mutant seed-coat epidermis was made out, which correlates temporally and spatially with the initiation of the fiber cells. These short fibers differed from the normal lint in having basal ends enlarged immediately after initiation, while majority of the normal lint on wild-type seed did not show such enlargement until the cessation of elongation. Ruan *et al.* (2005) made following conclusions: (i) the delayed initiation of fuzz-like short fibers from the mutant seed correlates with delayed or insufficient expression of Sus in a subset of seed-coat epidermal cells destined to become fibers and (ii) the much shortened elongation of the fibers from the mutant may be related to their inability to close plasmodesmata.

Cotton fibers are developmentally similar to *Arabidopsis* trichomes, though the molecular control of fiber initiation is not vivid. Suo *et al.* (2003) isolated 55 MYB domain-containing sequences expressed in ovules during fiber initiation. Among them, *GhMYB109*, encoding a R2R3 MYB transcription factor of 234 amino acids, and was found to be structurally related to *AtMYBGL1* and *AtWER* controlling the trichome initiation in *A. thaliana*. RNA expression analysis showed that it is specifically expressed in cotton fiber initial cells and also elongating fibers, suggesting that *GhMYB109* plays a direct role in the initiation and elongation of cotton fibers. Similarly, the studies by Wang *et al.* (2004) suggested that cotton and *Arabidopsis* use similar transcription factors for regulating trichome production, with *GaMYB2* being a key regulator of cotton fiber development.

Two cotton fiber cDNAs (*GhMyb7 and GhMyb9*) and their corresponding genes, encoding R2R3-MYB proteins, were recently isolated from the *G. hirsutum* L. cv. DES119 and characterized (Hsu *et al.*, 2005). Northern blot analysis showed that *GhMyb7/9* expression in fibers was developmentally regulated and very responsive to auxin application. Hsu *et al.* (2005) demonstrated that the *Escherichia coli*-expressed GhMYB7 fusion protein binds to the promoter of Ltp3, a cotton-fiber-specific lipid transfer gene, in an *in vitro* DNA-protein-binding assay, suggesting that GhMYB7/9 possibly plays a role in the transcriptional regulation of the Ltp3 gene during fiber development.

Cotton ovules with fibers besides being economically important are also a good experimental model for studying plant cell elongation and wall biogenesis due to their fast elongation and high cellulose content. Cotton fibers are also utilized for biochemical, molecular and genomics research because large quantities of the single cell type can be easily obtained. In addition, fibers can be sampled from field grown, growth chambers or from cultured ovules, providing diverse tools to meet different experimental needs (Feng and Brown, 2000; Kim and Triplett, 2001).

Though fibers originate and elongate as single epidermal cells *in vivo,* without any cell division, there is evidence to show that fully differentiated fiber cells are able to divide up to 4 dpa *in vitro* (Van't Hof and Saha, 1997; Basra and Saha, 1999). This tempts one to assume that genes controlling fiber cell division exist and can possibly be identified. Ultimately, their expression could be modified to alter fiber properties that are contributory to improved textile performance.

Fiber elongation occurs over a period of approximately 21 days, with peak growth rates (in excess of 2 mm per day) occurring at 12 dpa (Meinert and Delmer, 1977). The rapid elongation of cotton fibers is among the most remarkable cell elongation events. The fiber cells undergo enormous elongation during development, increasing their length 1000 to 3000 times the diameter of the cell, which is typically between 12 to 20 μm.

The rate of fiber elongation and ultimate length are regulated by both environmental and genetic factors. The type of growth mechanism (diffuse versus tip) associated with the rapid growth of elongating cotton fibers has been widely debated (Seagull, 1990; Wilkins and Jernstedt, 1999). However, structural and physiological data provide clear evidence in support of diffuse growth (Tiwari and Wilkins, 1995). Pertinently, no genes are known to be specific for tip-growth (e.g. as in pollen tubes and root hairs) have been identified in developing cotton fibers, supporting a diffuse-growth mechanism during rapid polar elongation (Wilkins and Arpat, 2005).

At ~15 dpa, developing fibers enter into a "transition" phase that signals a mechanism switching from primary cell wall to secondary cell wall deposition, and this continues for several weeks thereafter (Meinert and Delmer, 1977; Beasley, 1979). There is distinct overlap between the two phases varying from 5 to 10 days (Schubert *et al.*, 1973; Meinert and Delmer, 1977; Beasley, 1979), depending upon the cultivar and environmental conditions. In long-staple cultivars overlap tends to be greater than in short-staple cultivars (Naithani *et al.*, 1981).

Termination of fiber elongation is accompanied by cessation of growth, transcriptional activity (Kosmidou-Dimitripoulou, 1986), and protein complexity (Graves and Stewart, 1988). At this stage, cross-linking of cellulose microfibrils and non-cellulosic matrices presumably fixes the structure of the primary cell wall (Wilkins and Jernstedt, 1999), leading to the first significant increase in fiber strength (Hsieh, 1999). The fiber secondary cell wall, which consists of >94% cellulose, is deposited in massive amounts and in a way that enables the fiber to be spun into threads and yarns. The space originally occupied by a large central vacuole (the lumen) is progressively filled with secondary wall cellulose which is deposited inside the primary wall (Goynes *et al.*, 1995). Very little is known about the biology of later stages of fiber development (past 25 dpa) because biochemical and molecular studies are hindered by a lack of proper cellular and biochemical fractionation imposed by active deposition of cellulose.

Cotton fruits open around 50 to 60 dpa when the mature fibers desiccate into a flattened twisted ribbon. Optimally mature fiber has high single fiber strength, high absorption per unit volume, high dyeing intensity, and interfaces well with high-speed rotor spinning technology for the production of high-quality textiles (El Mogahzy, 1999). Adverse environmental conditions during fiber development can result in either immature fibers or excessively thick fibers (Ramey, 1986; Roberts *et al.*, 1992; Pettigrew 1994). Probably a single major gene or closely linked cluster of genes results in increased fiber strength (Meredith, 2005).

Fiber length and diameter are important attributes of fiber quality, as are the physical properties of cellulose. Secondary wall thickness is a major determinant of fiber yield and fiber quality parameters, including maturity, dye ability, water absorption and single fiber strength (Hsieh, 1999). To optimize fiber yield and quality, it would be desirable to control cellulose biosynthesis and alleviate its sensitivity to environmental stress (Kohel *et al.*, 1974; Gipson, 1986; Benedict *et al.*, 1994; Haigler *et al.*, 2005).

FACTORS CONTROLLING FIBER ELONGATION

Fiber elongation is subject to temporal and spatial control imposed both from within (its developmental program) and from the external environment. Plant hormones play an important role in mediating both the internal and external control of fiber growth. Fiber elongation is also sensitive to environmental factors, especially the temperature.

Temperature

Temperature is an important environmental factor influencing the final fiber length within a species or cultivar (Gipson and Joham, 1969; Gipson and Ray, 1969, 1970; Gipson, 1986). It is well documented that developing fibers exposed to cool temperatures (generally at night) have reduced rates of elongation and thickening (Hawkins, 1930; Gipson and Joham, 1969; Gipson and Ray, 1970; Thaker *et al.*, 1989), and growth rings in their secondary walls (Balls, 1919; Kerr, 1936; Haigler *et al.*, 1991; Xie *et al.*, 1993). Cell wall development slows during cool nights, giving rise to rings that appear similar to the growth rings of trees. These rings may be weak points in the fibers, as slippage at the rings has been frequently observed.

Even though the duration of the elongation period is increased at lower temperatures, the fibers attain the same final length as those grown under more favorable conditions (Hawkins, 1930; Gipson and Joham, 1969; Gipson and Ray, 1969; Thaker *et al.*, 1989). Fibers cultured *in vitro* under cycling temperatures of 34°C/22°C and 34°C/15°C experience delayed onset of substantial elongation and require prolonged periods to attain fiber length equal to controls (Haigler *et al.*, 1991 Malik *et al.*, 1995). After an initial delay of variable length, the maximum rates of elongation get similar under all conditions, suggesting that the prolonged elongation period under cool temperatures is due to the delay in its onset rather than to its reduced rate.

Secondary wall synthesis may be relatively more sensitive to cool temperatures than primary wall synthesis (Ryser, 1985), possibly due to differences in the diversity of proteins and noncellulosic carbohydrates in the primary wall (Buchala, 1999), compared with the almost exclusively cellulose composition of the secondary cell wall (Delmer, 1999). Cool temperatures significantly decrease the rate of cellulose synthesis (Gipson, 1986), and cause perturbations in fiber sugar metabolism (Roberts *et al.*, 1992; Malik *et al.*, 1995). Comparatively little is known about the regulation of fiber growth and metabolism by other environmental factors such as drought (Bradow *et al.*, 1997).

Plant Hormones

Hormonal Influence on Fiber Development

Cotton fiber differentiation and development involves several biochemical, physiological and morphological alterations. The phenotypic expression is determined by genotype, though the development of fiber is regulated by a combination of internal stimuli, hormonal contents and external influence (Basra and Malik, 1984; Kaur, 1988; Malik *et al.*, 1995). Most studies dealing with fibers have been concerned with the two phases of development, elongation and secondary wall deposition.

Some of the functions of hormones in mechanisms concerning cell differentiation and growth have been reviewed and major progress has been made in their biosynthesis and metabolism, even though their precise mode of action and interactions is still elusive (Malik *et al.*, 1993; Malik, 1994). Several researchers have contributed to understanding the hormonal influences on cotton fiber development (Beasley, 1977; Jasdanwala *et al.*, 1977; De Langhe, 1986; Kosmidou-Dimitropoulou, 1986; Basra and Malik, 1984; Nayyar *et al.*, 1990; Seagull, 1990). Much of the information has flowed from the works of Dhindsa *et al.*, (1976), Dhindsa (1978), Beasley and Ting (1974), Basra and Malik (1984), Nayyar *et al.*, (1989 a, b), Basra *et al.*, (1992a, 1993).

In general two approaches have so far been adopted in cotton regarding hormonal effects and these are: (a) isolation, identification and quantification of hormones from bolls, ovules and fibers (Naithani *et al.*, 1982; Bhardwaj and Verma, 1985; Nayyar *et al.*, 1989 a, b). Several workers have isolated gibberellins, auxins, cytokinins, abscisic acid and ethylene from cotton fruits and/or ovules having attached fibers at different stages of boll development (Kosmidou, 1986; Nayyar *et al.*, 1990). In most studies no attempt was made to correlate the level of hormone(s) with fiber development. Thus, very scanty information exists on hormonal evolution in fiber development only. Even then the available data provide some clue on the role(s) of hormones in fiber differentiation and elongation, fluctuations of hormones taking place in the ovule, embryo and also fibers. The second approach, involves external application of hormones *in vitro* as well as *in plantae*. Exogenously, divergent plant hormones have been sprayed on the cotton plants at different stages of development to influence boll growth and hence fiber quality. For this several techniques have been used and these are spraying the whole plant, soaking the ovaries, or injecting the desired hormones in the bolls (Nayyar *et al.*, 1990). Under field conditions cotton plants respond differently to growth regulators than to trials in greenhouse or growth chambers.

The practical uses of growth regulators in cotton production have been discussed (Nayyar *et al.*, 1990). In any case extending the period of fiber elongation or secondary wall thickening and stimulating the translocation of photosynthates to the fruits through growth regulators was sure to enhance cotton production and improve the quality of fiber. However, the whole plant is a complex system considering that other factors limit fiber improvement. The applicability of this technique has some limitations. It may be necessary to control processes like pollination, pollen tube growth, fertilization embryo and proper endosperm growth which affect the internal supply and balance of hormones, essential inputs for fiber growth. It must be internal hormonal balance which in combination with other factors was likely to affect fiber growth and quality. Cotton plants have indeterminate growth habit and consequently growth hormones must be made available at a suitable time to elicit favorable response. Very few studies on this aspect have been carried out. Due to concerned contributions of several research groups it has been possible to establish basic protocol for cotton ovule culture, aseptic fiber growth and understanding some of the environmental factors conducive for fiber growth, etc. (Beasley and Ting, 1974; Triplett *et al.* 1989; Nayyar *et al.* 1989 a, b). Most of the information about the hormonal influence on fiber growth available to date is due to researches done in this system.

Hormonal Influence on Fiber Differentiation

The available evidence suggests that gibberellins and auxins are important factors for fiber differentiation. Hormones or their inhibitors are added in the culture medium to establish their role. Recently SEM studies of the cultured ovules have been undertaken (Malik *et al.*, 1995; Malik, unpublished). The role of polyamines and phenolics has also been suggested (Malik, 1994; Malik unpublished). This is consistent with the earlier report that ovules with excessive fiber formation possess phenols whereas lintless ovules contain les phenolics (Kosmidou, 1986). The role of o-diphenol oxidase activity during fiber initiation and its role in shifting redox balance towards oxidation possibly play a significant role in fiber initiation (Malik *et al.*, 1995; Singh *et al.*, 1995). A study of IAA oxidase in developing fibers indicated high and low IAA catabolism during fiber elongation and the secondary wall thickening phase (Rama Rao *et al.*, 1982; Nayyar, 1988). The essentiality of IAA for fiber initiation and GA for the expansion of ovules has been demonstrated (Dhindsa, 1978; Malik *et al.*, 1995; Singh *et al.*, 1995) using metabolic inhibitors of two hormones through the estimation of IAA oxidase or peroxidase activities as well as through SEM studies. *In plantae*, pollination seems to trigger hormone induction whereas *in vitro* unfertilized ovules required the addition of the two hormones. Malik (1994), Malik *et al.* (1995) and Nayyar (1988) have proposed that levels of auxin or its rate of degradation possibly regulate the culmination of fiber elongation and the secondary wall synthesis. In brief, the significance of exogenously added hormones seems to be dependent on whether the ovules used in culturing were fertilized or not. Thus, unfertilized ovules require both IAA and GA for the initiation and production of fibers. In the fertilized ovules, only GA is needed to support ovules growth. Kinetin did not support fiber growth and induced callusing in the ovules. Thus, a precise balance of hormones is required to ensure proper fiber development. During fiber initiation and elongation, the nucleus and nucleolus undergo marked changes in size and structure pointing towards enhanced capacity for ribosome synthesis. Studies on effect of GA, IAA and ABA alone or in combination have shown that they affect the nucleolar evolution and concomitant fiber length (Malik *et al.*, 1995) especially in cultured, unfertilized ovules. Briefly, GA: IAA stimulated early nucleolar vacuolation and higher fiber elongation, whereas ABA blocked the synthesis of nucleolus material. The input of all the three hormones in a balanced fashion seems to be important for the regulation of synthesis and release of nucleolar material. No discernible effect of increased free IAA content on fiber length was observed in transgenic cotton transformed with two auxin biosynthetic genes, *iaaM* and *iaaH* from *Agrobacterium tumefaciens,* linked to the fiber-specific E6 promoter (John, 1999).

Exogenous cytokinins and ABA inhibit fiber growth in both fertilized and unfertilized ovules *in vitro* (Beasley, 1973; Beasley and Ting, 1973, 1974; Zhang, 1982; Chen *et al.*, 1988). Increased cytokinin content in the ovules of mutant cotton was suggested to be the main factor for inhibition of fiber elongation (Chen *et al.*, 1996). However, over expression of the cytokinin biosynthetic gene isopentenyl transferase into cotton did not influence either elongation or other fiber traits (John, 1999).

ABA reduced the capacity of the ovule to produce fibers in the presence of IAA or GA_3 *in vitro* (Beasley and Ting, 1974). Inhibition of ABA biosynthesis by addition of fluridone (FLU) to the culture medium caused a marked promotion of fiber production and elongation (Nayyar *et al.*, 1989a). Concomitant with a reduction in ABA levels, FLU treatment also caused increased IAA levels. Chen *et al.* (1996) reported a gradual decline in the ABA level of cotton ovules during the period of fiber elongation. An inverse correlation between fiber elongation and ABA content was also reported by Gokani *et al.* (1998). A low level of endogenous polyamines seems to promote fiber elongation, as reflected in the declining levels of polyamines observed during the active period of fiber elongation (Davidonis, 1995).

Recently, the role of brassinosteroids (BRs) in cotton fiber development has been highlighted by several studies. Direct evidence for the role of brassinosteroid (BR) - stimulated gene expression in cotton fiber development comes from transgenic cotton plants that overexpress a BR-responsive cotton *XTH* (Xyloglucan transferase/hydrolase) gene known as *KC22*. These plants produce fibers that are significantly longer than those from control plants (Allen *et al.*, 2000). Sun *et al.* (2004, 2005) demonstrated that treatment of cultured ovules with the BR biosynthesis inhibitor brassinazole (Sekimata *et al.*, 2001) strongly inhibited cotton fiber development. Simultaneous addition of brassinazole (Brz) and brassinolide (BL) to the cultured ovules partially restored fiber development, indicating that the inhibition of fibers by Brz is, at least for the most part, mediated via inhibition of BR biosynthesis. Furthermore, treatment of cotton floral buds with Brz resulted in the complete absence of fiber differentiation, indicating that BR was required for both fiber initiation as well as elongation. Expression of fiber genes associated with cell elongation increased in ovules treated with BL and was suppressed by Brz treatment, establishing a clear correlation between brassinosteroid-regulated gene expression and fiber elongation.

More recently, Yoshida *et al.* (2005) have demonstrated that addition of low concentrations of the BL promoted fiber elongation while Brz inhibited fiber development. In order to elucidate the role of BRs in cotton fiber development further, they performed a more detailed analysis of the effects of these chemicals on cultured cotton ovules. Their results confirm that exogenous BL promotes fiber elongation while treatment with Brz inhibits it. Furthermore, treatment of cotton floral buds with Brz resulted in the complete absence of fiber differentiation, indicating that BR is required for fiber initiation as well as elongation. Expression of fiber genes associated with cell elongation increased in ovules treated with BL and was suppressed by Brz treatment, establishing a correlation between BR-regulated gene expression and fiber elongation. These results establish a clear connection between BRs and fiber development and open the door for genetic analysis of cotton development through direct modification of the BR signal transduction pathway.

Though ethylene is an important hormone and potentially interacts with hormones studied above, its role in fiber development is not fully understood. This is particularly worthy of investigation in future studies.

Influence of Hormones on Indian Cultivars

In vitro studies with Indian cvs. have revealed the following data:

(i) Fertilized, isolated *hirsutum* or *arboreum* ovules are deficient in their capacity to synthesize optimal levels of GA, near optimal to synthesize auxin(s), and ABA was essential for the cessation of fiber elongation (Kaur 1988; Nayyar *et al.* 1989 a, b). Further, cvs. with longer fibers contained higher levels of auxins and GA than their counterparts with short staple (Nayyar *et al.* 1989 a, b). Interestingly, ABA content was considerably high at all phases of fiber development in short staple cvs. than long staple ones. Correlation coefficient and regression values indicated the development of ABA in the regulation of fiber elongation in short staple cottons (Nayyar *et al.* 1989 a, b; Malik, 1994).

(ii) Quantitatively the endogenous levels of ABA in cultured and field grown ovules did not vary, though IAA content decreased by 40% in the cultured ovules. Supplementing culture medium with FLU (inhibitor of ABA) enhanced the endogenous level of IAA and GA and suppressed ABA content. There was an increase in fiber length over the control, possibly due to reduced ABA level (Nayyar *et al.*, 1989 a, b).

(iii) Basra *et al.* (1992a) have reported a close correlation between the cessation of elongation growth and increase in wall peroxidases pointing to their significant role in the termination of the elongation phase.

(iv) ABA and calcium affected soluble and wall-bound activities or peroxidases and o-diphenol oxidase of fibers at primary and secondary wall development stages (Basra *et al.*, 1992a). ABA significantly reduced the activities of both the fractions whereas Ca^{2+} was promotory. The incorporation of EGTA, a calcium chelator, to the incubation medium decreased enzyme activity suggesting its role in enzyme synthesis. This suggests involvement of Ca^{2+} as second messenger in mediating the inhibitory effects of ABA.

(v) The high content of free ABA at the elongation stage was not the consequence of high activities of ABA-Glc splitting enzymes. During the secondary wall thickening stage, it is suggested that ABA conjugate was the final product of the ABA metabolism.

(vi) Hormonal partitioning of ^{14}C-sucrose at two growth stages of fibers has been studied by Basra *et al.*, (1992b). ABA enhanced incorporation into callose which might determine the onset of secondary wall deposition and extent of fiber elongation achievable in a species. ABA also reduced incorporation into amino acids and organic acid fractions but GA was promotory. IAA showed inconsistent effect. The *in vitro* significance of these findings is apparent.

(vii) ABA and Ca^{2+} also affect dark metabolism of CO_2 in *hirsutum* fibers. Both inhibited activity of PEP carboxylase and also the fixation of $^{14}CO_2$ into organic acids. ABA also inhibited incorporation of label into amino acids. It is inferred that ABA and calcium modulated the fiber capacity for dark fixation of CO_2 and hence the rate of fiber elongation.

(viii) Gill *et al.*, (1999) determined the role of ABA in inhibiting the fiber growth via polyamines. These workers added ABA, FLU and α-difluoromethylarginine, DFMA (an inhibitor of polyamine biosynthesis), to the cultured medium with ovules to see the relationship between the two. Both ABA and DFMA inhibited fiber growth *in vitro* along with a corresponding decrease in spermidine, putrescine and spermine levels. Contrarily, FLU greatly enhanced fiber growth and putrescine levels. When used in combination, FLU reduced the effect of ABA on fiber polyamine levels, especially spermidine and putrescine. Clearly, endogenous polyamines especially putrescine, are required for normal fiber growth and morphogenesis. ABA appears to act by decreasing putrescine.

(ix) Singh *et al.* (1999) presented combined evidence from *in vivo* and *in vitro* studies regarding essentiality of polyamines in the process of cotton fiber differentiation and initiation.

(x) Sharma and Malik (2005) investigated changes in endogenous levels of IAA, GA and ABA during *in vivo* fiber growth in *G. arboreum* cv. LD 327 and demonstrated overriding influence of ABA in limiting the rate of fiber elongation. FLU applications markedly decreased the ABA content and concomitant increase in IAA, GA in cultured ovules. The enzymic hydrolysis of α-glucopyranosyl ester of ABA by α-glucosidases and esterases. The activity of two hydrolases did not correlate with the ABA contents in the growing fibers. They concluded that high content of ABA at 15-DAA was not a consequence of the increased activity of the ABA-Glc splitting enzymes.

(xi) Malik (2004) investigated the influence of IAA, Brassinolide, ABA, calcium along with calcium inhibitor and calcium sequestering agent on total fiber units. He correlated their effect with the activities of soluble and wall bound activities of peroxidase. Cotton seeds 10 days after culturing (DAC) were supplemented with various chemicals; of these IAA, BR and calcium chloride markedly enhanced fiber elongation whereas ABA and different calcium inhibitors, calcium sequestering agents inhibited elongation of fibers. IAA, BR supplemented with calcium chloride promoted fiber elongation. On the contrary calcium inhibitors supplemented with IAA or BR were ineffective in inducing fiber growth. It was inferred that IAA and BR act via calcium and/or calmodulin protein, especially for two fractions of peroxidase. Calcium promoted peroxidase activity whereas CPZ or EGTA decreased peroxidase activity suggesting the role of calcium ions in the synthesis and / or secretion of the enzyme. On the contrary ABA appeared to inhibit peroxidase activity through calmodulin synthesis or its mobilization to the site of action.

(xii) Subsequent studies by Malik *et al.* (2006) demonstrated the role of BR in promoting fiber elongation while treatment with brassinazole 2001 (BRZ) inhibited fiber elongation. When the two chemicals were exogenously supplied to the just fertilized ovules, BRZ completely suppressed fiber differentiation though BR promoted the differentiation. The authors demonstrated the role of BR in fiber differentiation and elongation processes. It would be interesting to undertake genetic analysis of cotton fiber development through direct modification of the brassinosteroid signal transduction pathway.

(xiii) Malik *et al.* (2007) reported that cotton ovules collected during late September when night cool temperature (15°C) prevailed, were cultured at 32°C/15°C and 32°C/12°C. At low temperature, activities of acid invertase, α-galactosidase and α-glucosidase were adversely affected. i.e. cycling temperatures for 12 h each in Beasley and Ting medium had very few epidermal cells bulging. When NO or Brassinolide were added individually to the medium, the suppression of the protrusions could be restored. The two phytohormones stimulated fiber elongation. Compared with TFU in control, cool temperature reduced it but NO and Brassinolide mitigated the suppression effect. Activity of o-diphenol oxidase increased at 15 and 35 DAC compared with 2 DAA. With NO and BR the activities enhanced at both the stages. The enhanced activity during the elongation and wall thickening phases points toward the fact that phenols are oxidized and thus the redox balance shifts towards oxidation, causing an increase in S—S bonds which check the process of wall loosening. The activity of acid invertase increased in response to two phytohormones suggesting the role of the enzyme in promoting apoplastic hydrolysis, preventing sucrose accumulation during the elongation phase. Total sugars decreased in response to cool temperature though NO and BR restored their levels. α-galactosidase is believed to be concerned with metabolism of cell wall polysaccharides. Both the phytohormones promoted the enzyme activity indicating restoration of wall synthesis during elongation phase.

Dark Metabolism of CO_2

As cotton fibers grow in the dark interior of the cotton, fruit they experience elevated levels of CO_2. *In vitro* studies have shown inhibition of fiber growth in the absence of K^+ and CO_2 (Basra and Malik, 1983). Malate produced through dark metabolism of respired CO_2 serves as an indispensable osmolyte for fiber growth. The latter is shown to depend on the turgor pressure produced in the fibers by K^+ and malate which are osmotically active. In the elongating fibers, malate and K^+ levels fluctuate and correlate with fiber length (Basra and Malik, 1984). ABA inhibits malate synthesis and also inhibits fiber elongation. During the period of active fiber elongation the activities of malate dehydrogenase (MDH), malic enzyme and PEP carboxylase enhance (Basra and Malik, 1983). Kloth (1992) also observed variability of MDH among cotton cultivars with differing fiber traits. The significance of nonphotosynthetic CO_2 fixation in various developmental systems has been discussed by Basra and Malik (1985) and Malik *et al.* (1991).

Nayyar *et al.*, (1989) studied incorporation of labeled acetate at two stages of fiber growth and reported low incorporation during the secondary wall thickening phase. Kaur *et al.* (1990) have provided evidence for the involvement of ABA in suppressing the activities of enzymes of dark metabolism of CO_2. With FLU the inhibition was overcome and the activity of enzymes was stimulated, and concomitantly fiber length also enhanced.

Carbohydrate Metabolism

Fibers accumulate carbohydrate, mainly glucose, fructose and sucrose during development (Cooner *et al.*, 1972; Basra *et al.*, 1990). During secondary wall thickening rhamnose, arabinose, uronic acid enhance but non-cellulosic glucose and arabinose residues decrease. Among the nucleotide sugars UDP sugars were predominant whereas GDP sugars were not detectable in significant quantities. Up to 25 DAA UDP glucose level increased several fold (Capita and Delmer, 1981). UDP glucose is the precursor of secondary wall thickening. Several researchers have assayed the carbohydrate metabolizing enzymes at different stages of fiber development (Buchala *et al.*, 1990; Nayyar, 1988; Basra *et al.* 1990; Hendrix 1990; Wafler and Meir, 1994; Malik, 1994). Most enzymes assayed included invertases, sucrose synthase, ATP-PEK, PPi, UDP-glucose pyrophosphorylase, glycosidase, etc.

Basra and Malik (1983), Dhindsa *et al.* (1975), Nayyar *et al.* (1989), Kaur *et al.* (1992) have shown that growth of epidermal cells on the ovules outer integument was dependent on positive turgor pressure. More than half of the total osmotic potential necessary to produce a positive turgor pressure is accounted for by K^+ and malate. The concentration of these two osmolytes increased and decreased with the increase and decrease of the fiber growth rate (Basra and Malik, 1983b, 1984). Kloth (1992) examined the role of MDH in regulating the final step in the synthesis of malate and hence the possibility of its being a strong candidate for limiting growth of fiber. Using 24 inbred cultivars of *Gossypium* he attempted to correlate variability in MDH with eight quantitative traits of the cotton fiber. A significant correlation was observed especially with regard to traits of fiber wall, maturity, and micronaire and wall thickness.

Cotton Fiber development in response to cool temperature

Due to tropical habitat cotton is highly sensitive crop to cool temperature (Haigler *et al.*, 1991; Roberts *et al.*, 1992; Malik *et al.*, 1995). Temperature less than 16°C essentially inhibits fiber growth (Haigler *et al.*, 1991; Malik *et al.*, 1995). Field studies have revealed that cool nights (12°C) result in suboptimal cotton yield and reduced fiber maturity (Gipson, 1986).

Fiber initiation and elongation

Night temperature (10, 15, 20, 27°C) regimes affected elongation of fibers in field grown *hirsutum* cv. (Gipson and Johan, 1969). Fiber age and night temperature determined rate of fiber elongation, and temperature coefficient of elongation decreased with progressive fiber age and night temperature (Xie *et al.*, 1993; Malik *et al.*, 1995). The initial stages of fiber elongation are highly temperature sensitive, while the later phases (elongation and secondary wall formation) were temperature independent. Early delay in fiber initiation by cool temperature extended the elongation period (Malik *et al.*, 1995). On the basis of the response of fibers to cool temperature, three stages of fiber development have been recognized; fiber initiation, very early elongation and subsequent elongation. Of these, the first two stages are highly sensitive to cool temperature. Fibers of plants exposed to adverse diurnal temperature fluctuations are shown to elongate at reduced rate for an extended period (Haigler *et al.*, 1991) but attain the comparable final length as those grown under favourable conditions.

The adverse response of cotton fiber development to cool temperatures is an ovule/ fiber specific event. Low temperature altered fiber development even though sufficient carbon source was available (Roberts *et al.*, 1992). Secondary wall synthesis is insensitive to cool conditions compared with wall synthesis (Malik *et al.*, 1995).

Cellulose synthesis and respiration respond differently to cool temperature (Roberts *et al.*, 1992). A decrease in temperature from 28°C to 15°C reduces respiration and cellulose synthesis and the two processes increased roughly in a linear manner between 18°C and 28°C. The rate of cellulose synthesis plateaued up to about 37°C and dropped off at 40°C, suggesting that more energy was released through respiration above 28°C but was not necessarily used for increased cellulose synthesis. Therefore, methods can be optimized for utilization of increased energy at high temperature for increased rate of cellulose synthesis. After exposure to cool temperature, cellulose synthesis can be fast recovered. Haigler (1992) transferred genes that confer cool temperature tolerance for cellulose synthesis in existing cotton varieties to commercial cotton cultivars (Haigler 1992).

Effect of cool temperature on Initiation and Early Elongation of Cotton Fiber Development as Assessed by Scanning Electron Microscope (SEM)

The pattern of fiber initiation is affected by cool temperature (Malik *et al.*, 1995). Cotton ovules at – 1 DPA and 0 DAA culture *in vitro* (Xie *et al.*, 1993; Malik *et al.*, 1995) revealed some developmental link between fiber initiation and time of anthesis *in situ*. This could not be overcome by culture manipulation. By -2 DPA the ovules possibly acquire an internal signal that permitted initiation only on the normal day of anthesis. Initiation and early fiber elongation were found as two different physiological processes and were independently affected by cool temperature. Similar developmental response of *in vitro* cultured ovules to cool temperature has been shown. Ovules subjected to 32°C/15°C cycling temperature had delayed initiation of ballooning by 1 to 2 days followed by slow initiation and fiber elongation (Malik *et al.*, 1995). Once the fibers attained a length of nearly 0.4 mm their further elongation was independent of temperature effect. Furthermore, *in vitro* culturing of ovules with fibers about 0.4 mm long at 32°C/15°C cycling and 32°C constant had comparable rates of fiber elongation suggesting that -1 DPA to -2 DAA stages of fiber development were highly sensitive to cool temperature. Interestingly, the delay in fiber development induced by cool temperature could be overcome by the supplementation of culture medium with guainiolide derivative-I (S-3307). The compound caused preponderance of bulging of epidermal cells and was highly promotive. Furthermore, S-3307 (antigibberellin) completely suppressed the initiation phase of fiber development and no fiber formation was observed even after many days of culturing. Malik *et al.* (1995) have demonstrated that cool temperature affected biochemical/physiological processes also. Sucrose and glucose levels decreased in ovules cultured in cool temperature (32°C/22°C), sugar accumulation was reported up to the phase of fiber elongation, though the content was more at 32°C/22°C. The trend was increased during secondary wall thickening. Even at higher concentration glucose was not used efficiently to synthesize and deposit secondary wall around the fibers. Seemingly, low temperature does not affect availability of glucose during secondary wall thickening phase though primordial intiation was delayed by the limitation of glucose. Enzymatic studies have shown low secondary wall deposition process. Thus, cool temperature adversely affected the activities of α-glucosidase, acid invertase and o-diphenol oxidase (Malik *et al.* 1995) with marked decrease in the activities of all three enzymes at -1 DPA suggesting their role during initiation and early elongation of fibers. Cool temperature markedly decreased the activity of o-diphenol oxidase and this in turn caused irregular and delayed initiation of fibers. The role of o-diphenol oxidase, pertinent of auxin turnover and redox balance of developing cotton fibers, has been suggested by Basra *et al.* (1992). The decreased activity of acid invertase at -1 DPA showed its considerable influence on the immediate metabolism of sucrose entering the fiber primordia. Cool temperature reduced the activities of oxidases in -1 DPA ovules and their subsequent increase till 5 DAA is correlated with essentiality of auxin and fiber initiation (Singh *et al.* 1995). Decrease in IAA oxidase activity and increase in o-diphenol oxidase activity in 1DPA was reported with guaianolide derivative, possibly by regulating auxin degrading system during the entire period of fiber growth. The activity pattern of o-diphenol oxidase indicated a shift in redox balance from a reduced to an oxidative state in developing ovules (Basra *et al.*, 1992a; Sharma *et al.*, 1995). O-dihydroxyphenols are implicated in redox potential pertinent to cell division (reduction) and cell differentiation and senescence (oxidation). The low o-diphenol activity with guaianolide derivative compared with the level during the imitation phase suggested that dihydroxyphenols maintained-SH group in a reduced state hence aided initiation and early elongation of fibers. A growth retardant (E)-1 (4-chlorophenyl)-4,4-dimethyl-2 (1,2,4, triazol (-ly-1) pentene-3-ol (S-3307) is an antigibberellin and completely suppressed fibers on ovules suggesting role of GA in differentiation of initials (Sharma *et al.*, 1995). Considerable decline in the activity of acid invertase by S-3307 is indicative of the role of GA in regulating the activity of acid invertase and hence high apoplastic hydrolysis by preventing sucrose accumulation. The higher influence of the compound on wall bound acid invertase suggested that wall bound invertase possibly concerns with sucrose uptake from apoplast of the cotton fiber by preventing phloem-reloading. Relationship between acid invertase and the rate of fiber elongation has already been mentioned (Basra *et al.*, 1990). Enzymes hydrolyzing non-cellulosic polysaccharides like α-glucosidase and α-galactosidase also depicted reduced levels with S-3307 during all stages of cotton fiber development. Their role to provide precursor for cellulose biosynthesis has been postulated (Sharma *et al.*, 1995). S-3307 reduced the activity of o-diphenol oxidase at -1DPA but enhanced the secondary wall thickening. Its role in auxin catabolism and redox balance has already been documented.

In vitro studies have revealed that the developing fibers have growth characteristics comparable to systems exhibiting intercalary or tip growth. In fibers both the processes are operative, though to what extent these two forms of growth contribute to the final length of the fiber is not clear. It may be stated that the precise mechanisms of tip synthesis in cotton fiber is variable from that reported in pollen tubes, fungal hyphae and root hair. Cotton fibers exhibit tapered and obtuse tips, and their significance remains to be deciphered. Unlike the systems mentioned previously cotton fibers synthesize secondary cell wall thickening which limits intercalary growth and cause relationship between the secondary thickening and cessation of fiber elongation which remains to be worked out. It must be realized that several factors regulate the ultimate length of the cotton fiber, the extent of secondary wall thickening, and the transition of primary wall and secondary wall synthesis ultimately determine the quality of the fiber. In brief, developing cotton fibers raised *in vitro* provide a good tool for the study of several fundamental problems in plant cell biology.

BIOTECHNOLOGY OF COTTON

Cotton has been in the biotechnological forefront in several areas including ovule and cell culture and regeneration of plants from cells. These developments have enabled cotton to be genetically engineered. As discussed earlier, cell and organ culture have been used for basic studies on cotton physiology and genetics. Cotton biotechnology involves several programmes: identification of genes which confer desirable traits; identification of a system for the stable and optimal introduction of new genes; and recovery of transgenic plants via regeneration of meristem culture.

Organ Culture

Conditions for culture of pre- and post-anthesis ovules/seeds from cotton were established by Beasley and co-workers (Beasley, 1977; Beasley & Ting 1974). The cotton ovule culture system has been used extensively in studies pertaining to cellulose synthesis and cytoskeletal elements during fiber development (Seagull, 1991), activities of various enzymes (Kaur *et al.* 1990; Basra *et al.*, 1992a), Wafler and Meir 1994; calcium mediated changes in the activities of enzymes (Basra *et al.* 1992a), influence of hormones on ^{14}C-partitioning (Basra *et al.*, 1992b) and to cool temperature effects (Xie *et al.*, 1993; Malik *et al.*, 1995; Singh *et al.*, 1995; Sharma *et al.*, 1995). Their intense studies have helped in the understanding of the physiology of fiber development. Stewart (1986) has reviewed the tissue culture work on cotton and to date the ovule/seed culture system continues to be improved and constitutes an important aspect of current research on fiber physiology and molecular biology (Triplett *et al.*, 1989; Malik *et al.*, 1995). Fibers grown in *in vitro* cultured ovules/seeds have the same chemical composition as fibers grown in *in plantae* (Meinert & Delmer, 1977). However, *in vitro* fibers attain lengths nearly one half of that reached by *in plantae* fibers. Recently, cells released from cotton ovules in suspension cultures have been raised *in vitro* (Trolinder *et al.*, 1988). The single cells elongated continuously in culture medium and attained a length of 2-3 mm. Some workers have investigated the hormonal differences between the fibers of *hirsutum* and *arboreum* (Nayyar *et al.*, 1989) to seek answers to why the fibers of the former continue to elongate whereas that of the latter cease. Tight regulation of hormonal level seems to be a key factor though significant differences in auxin and ABA levels between the two species are also observed.

Through the use of culture system it has been demonstrated that cotton ovule tissues, like exposed tissues, possess defence mechanisms against fungal attack (Mellon & Triplett, 1989). Steware & Hsu (1977) adopted the ovule culture system to culture seed to maturity of the enclosed embryo; wide hybridization of diploid *Gossypium* species (Stewart, 1981) to overcome physiological barriers. Ovule culture is used to introgress the lintless seed/ linted plant trait from *G. sturtianum* into upland cotton (Altman *et al.*, 1990). In fact these workers have recommended *in vitro* techniques to recover germplasms material from its native habitat when seeds are not available during collection. However, precise protocols have to be developed. In the USA, introgression on exotic germplasm is in the cotton biotechnology programme at the University of Arkansas. Diploid Asiatic *Gossypium* species are being hybridized to introduce pest resistance and bridge species (D genome) is used to facilitate genetic recombination with upland cotton. Here ovule culture methods are routinely used. Additionally, several genetic nuclear traits are being transferred from wild species to cultivated ones, and the cytoplasm of six wild species has been placed in the *barbadense* nuclear background.

Meristem culture of cotton is being used for transformation with *Agrobacterium.* Meristem culture has advantage over cell culture since plants can be obtained from any genotype. Thus far cotton plants have not responded to this technique. Hence *Agrobacterium*-infected meristems have not been transformed in a repeatable fashion. Cotton team at Agracetus Inc. used meristem culture to secure transformed plants via particle acceleration. Using cultured mertistems, they are able to transform plants of any cotton genotype, rather than being limited to those genotypes with regenerative ability (Coker lines).

Cell Culture

Cells may be cultured as cell suspension or callus; and suspension cultures are widely preferred to investigate basic biochemical mechanisms and to raise somaclonal variation. Apostol *et al.* (1989) are studying primarily the mechanisms of elicitation of phytoalexin production in cultured cells of *Verticillium dahliae*. In cotton phytoalexins are sesquiterpenoid compounds. H_2O_2 is involved as a second messenger in the stimulation of phytoalexin production. Suspension cell culture is also used to study the interaction of

Xanthomonas malvacearum with blight resistant lines of cotton. Trolinder *et al.* (1988) have succeeded in development of single cells in this study using GA. Investigating fiber development free from ovule has a great utility. With added perfection cell cultures should constitute successfully raised *in vitro* fibers though they failed to elongate beyond 0.3 mm.

Callus has been derived from three cultivars including the embryogenic line Coker 310, and exposed to *Phymatotrichum ommivorum* toxins. Through repeated culturing and exposure to toxin, 60 plants were regenerated from the selected cell lines. Success has been achieved in obtaining stable regeneration of cotton from embryogenic suspension cultures.

Some workers are using somaclonal variations to select cell population tolerant to a variety of stresses. A line resistant to 10^{-4} M atrazine has been developed. Cells tolerant to low and high temperature have been selected and plants have been regenerated.

Regeneration

Davidonis (1995) 2 yr old culture of Coker 310

Ability to regenerate was genotype-specific and heritable as a quantitative trait. Most genotypes are recalcitrant though new regenerative stocks are being developed which will be regenerable.

GENETIC ENGINEERING OF COTTON

Most accomplishments in genetic engineering have been made by several biotech companies, due to sufficient modification of the triplet codes of Bt gene to make it more compatible with the genetic codes preferred by the plant protein synthesizing system. Strains of *Bacillus thuringiensis* have been identified which are active against the Colorado potato beetle and nematodes. Hence modified genes of the Bt toxin will be available to combat a range of insects. *Agrobacterium*-mediated transformation of cotton suspension cultures have been perfected with marker genes and then recover mature plants. Several groups have successfully introduced *B. thuringiensis* toxin genes for pest resistance and also various herbicide resistance genes have been manipulated in cotton. Cotton has been transformed via particle bombardment. The introduction of resistance genes may affect fiber quality and yield. This aspect is being carefully looked into by several research groups. More recently array of neuropeptides acting on the neural system have been identified. They have low molecular weight and are effective at minute quantifies.

Environmental Stress

Responses of plants to stress are complex and dependent on the expression of many genes. Two stresses, drought and temperature extremes, are being investigated indepth in cotton. Search for genes whose expression has altered in drought-stressed plants is on. Dr. N. Trolinder has obtained a cDNA library from drought-stressed cotton and is now screening for uniquely expressed mRNA species. The influence of over production of hydroxyl-proline as a mechanism for stress resistance is being examined. Cotton is being transformed with the bacterial "osmogene" under the regulation of a constitutive promoter. The role of superoxide dismutase in stress cDNA library made to combat anaerobic respiration during flooding is being studied.

Hybrid Cotton

Superior hybrid cultivars are being produced having high expression of heterosis to improve yield. Cotton is self-compatible and hence cytoplasmic male sterile lines are being raised to produce hybrid seeds. TA-29 specific promoter is being used for inducing male sterility in cotton.

Product Quality: Modifying Cotton Fibers

Fiber quality is determined by several measured physical and mechanical properties of mature cotton fiber. Basically, these properties are determined by fiber structure and biochemistry. In fact for every bale of cotton as many as seven quality measurements are considered. Fiber length, strength, fineness, non-fiber content, and length uniformity are measured by an electronic high volume instrument (HVT). Trained specialists evaluate color and grade visually and also measure with HVT colorimeter. Each of the physical characteristics is determined by complex, undetermined relationship between many biologically determined factors. For instance, fiber strength is correlated with the molecular weight distribution of cellulose polymers, the orientation of cellulose microfibrils, the frequency and distribution of reversals, and properties of fiber surfaces. Truly speaking, no systematic measurements of germplasm of these four strength related parameters have been undertaken anywhere in the globe. Perhaps multivariate analysis of these characteristics for a group of cotton cvs. should provide some information on general distribution of each factor contributing to the strength of fiber.

Very scanty information is available on how these strength-related parameters are biologically regulated. The precise knowledge on the initiation and cessation of cellulose biosynthesis giving rise to polymeric products with a fairly narrow molecular weight distribution is lacking. Clearly, the ultimate strength of a fiber is a composite expression of many genes. Fundamental questions relating to cellulose synthesis in plant cells may improve fiber quality, but the role of other processes in involving genes is also visualized. Reversal points are the spots where breakage occurs: why reversal occurs and what determines their optimal numbers are still unresolved questions! Inter-fiber interactions must be determining the bundle strength of the fibers and so information on fiber cuticle to identify a relationship between fatty acid chain length or degree of saturation or modification with fiber friction potential and fiber strength is needed. Similarly genetic background of cv. is crucial for determining the fiber length. Several processes linked with turgor driven cell expansion are involved and possibly regulate gene products including malate formation (Basra & Malik, 1983a; Kloth, 1992) and vacuolar proton ATPases (Wilkins, 1991). Wall hydrolases and peroxidases appear to be other enzymes concerned with the determination of fiber length (Basra *et al.*, 1992a, Seagull, 1990).

Generally, longer fibers tend to be fine fibers, and the longer fibers result from extended period of fiber elongation or high rate of fiber elongation. Precisely no information is available on the regulatory triggers which enforce cessation of fiber elongation and initiation of secondary wall thickening.

Two complementary approaches could be undertaken for the identification of new genes that will improve cotton fiber quality. The first, concerns the isolation and characterization of genes which are suspected to be involved in some aspect of fiber growth e.g. cellulose synthesis, the fiber cytoskeleton, wall-associated proteins, cuticle lipid synthesis, turgor generating metabolic processes, identifying genes that are regulated by phytohormones at the time of differentiation, expressed differentially during a cool temperature event; ABA biosynthesis and IAA synthesis identification of key components in the fiber during cellulose and secondary wall deposition notably modified by cool temperature to isolate the corresponding gene and fuse these with an appropriate promoter from another species, eg., pea which is more tolerant to cool temperature regimes. Most of these identifications are being attempted through quantitative comparisons of mRNA and/or protein product levels in cotton varieties differing in physical fiber traits; heterologous probes could be used for the identification of cellulose synthase genes in cotton. The second approach concerns identification of new fiber quality related genes through genome mapping techniques using RFLP or PCR methods. In the following a brief account of accomplishments is given:

Reference	*Application to cotton biotechnology*
Beasley (1977)	Ovule culture; established essentiality of auxin and GA, several minerals for fiber elongation.
Dhindsa (1978); Basra & Malik (1984)	Role of K^+ and dark metabolism of CO_2.
Delmer *et al.* (1977)	Synthesis of polysaccharides in the fiber wall; secondary wall synthesis.
Steward (1979); Basra & Malik (1984); Malik *et al.* (1995); Graves & Steward (1988)	Fiber cells differentiated on ovules 2-3 days before anthesis/ initiation of fiber primordia after anthesis. Detection of as many as 10 proteins exclusively expressed in fibers.
John & Steward (1992)	Isolated and characterized more than 12 genes that are fiber-specific, some genes coding for polypeptides containing high percentage of amino acids with –OH groups (serine, tyrosine, hydroxyproline).
Triplett, B., USAD, ARS, Southern Regional lab., Orleans (personal communication)	Biological determinants of fiber quality with emphasis on fiber strength and length (deposition of cellulose, formation of reversals in the gyre of the secondary cell wall under water and temperature stresses).
Mellon & Triplett (1989)	Characterization of peroxidase related to fiber formation; a putative clone of the gene isolated; using antibodies to tubulin and actin (cytoskeletal elements of fiber).
Davidonis (1989, 1990)	Effect of PGRs on fiber elongation in *arboreum* and fluridon overcoming it.
Xie *et al.* (1991, 1993)	Cool temperature affects fiber initiation; cellulose biosynthesis.
Malik (1994); Malik *et al.* (1995)	Cool temperature effects on fiber initiation; enzymatic activities affected by cool temperature at initiation, elongation, secondary wall thickening phases (SEM studies).
Sing *et al.* (1995)	Chemical to overcome the effect of cool temperature.
Haigler *et al.* (1991)	Identification of genes expressed differentially during a cool temperature event.
Wikins (1991)	Isolation, identification and characterization fiber-specific genes expressed during differentiation, especially vacuolar ATPase.
John & Crow (1992)	Several genes active in fiber cells similar to those in leaf, ovules, or root. 5 cDNA clones preferentially expressed in fiber. Eb promoter suggestively will be a valuable tool in the study of gene expression in fiber.

In our opinion added attempts should be made to delineate the role of plant hormones in modifying fiber initiation as well as fiber traits. Approaches specifically targeting modification of cotton fiber would be extremely useful in altering the qualities of dye uptake and wrinkle-resistance, pathogen free material product, etc. The cellulose synthase operon from *Acetobacter xylinum* has been cloned in cotton and wrinkle-resistance cotton has already hit the USA market. Similarly naturally coloured cottons are no more fantasies. Cotton with natural pink, cream and blue color is a reality. It is expected that among the genetically engineered crops, cotton will be one of the earliest crops to reach the market (See chapter on Bt Cotton in this volume). Of course, release with a single additional gene which confers insecticidal or herbicidal properties has been made. Concomitantly chemical companies are developing new, more eco-friendly herbicides that would be effective at low doses compared with those currently being used. The chief aim of exploiting new biotechnological techniques is to improve quality parameters etc. strength, length and micronaire in transgenic plants of cotton.

In summary, added information on physiology of fiber development and availability of biotechnological tools would provide opportunities to improve cotton fiber hitherto not visualized. Genomic mapping and exploitation of targeted approach of gene identification and transfer should help in achieving better cotton.

MOLECULAR BIOLOGY STUDIES

Little is known about the developmental pathway causing fiber to differentiate from ovular epidermal cells even though limits on the number of cells that differentiate into fiber will limit yield. Cotton fiber development is a fundamental biological phenomenon, yet the molecular basis of fiber cell initiation is poorly understood.

Identification and Characterization of Novel Genes Associated with Fiber Development

John (1995) reported two genes, each corresponding to fiber mRNA E6, were isolated from cotton cultivars Coker 312 (*Gossypium hirsutum* L.) and Sea Island (*G. barbadense* L.). E6 is one of the predominant fiber-specific mRNAs present during early fiber development. The distinguishing feature of the nucleotide-derived E6 protein is the presence of a motif where a dimer, Ser-Gly, is repeated several times. Two of the Sea Island genes contained a pentameric motif, Ser-Gly, while one of the Coker genes had one and the other had four motifs each. cDNA clones containing one or five Ser-Gly motifs were also identified. Thus, it appears that the strict conservation of this motif may not be critical to E6 protein function. Sequence characterizations of the genes and cDNAs showed that multiple members of the E6 family are transcribed in fiber and may result in proteins 238 to 246 amino acids long. The proteins 238 to 246 amino acids long. The 3[1] ends of the genes and cDNAs showed considerable heterology among them. Transgenic plants containing antisense genes were generated to decipher E6 function. Transgenic cotton with reduced E6 protein levels in the range of 60 to 98% were identified. However, no discernible phenotypic changes in fiber development or properties were apparent. This result leads to the conclusion that E6 is not critical to the normal development or structural integrity of cotton fibers.

Alcaligenes eutrophus genes encoding the enzymes, β-ketothiolase (phaA), acetoacetyl-CoA reductase (phaB), and polyhydroxyalkanoate synthase (phaC) catalyze the production of aliphatic polyester poly-D-(-)-3-hydroxybutyrate (PHB) from acetyl-CoA. PHB is a thermoplastic polymer that may modify fiber properties when synthesized in cotton. Endogenous β-ketothiolase activity is present in cotton fibers. Hence cotton was transformed with engineered phaB and phaC genes by particle bombardment, and transgenic plants were selected based on marker gene, β-glucuronidase (GUS), expression. Fibers of 10 transgenic plants expressed phaB gene, while eight plants expressed both phaB and phaC genes. Electron microscopy examination of fibers expressing both genes indicated the presence of electron-lucent granules in the cytoplasm. High pressure liquid chromatography, gas chromatography, and mass spectrometry evidence suggested that the new polymer produced in transgenic fibers is PHB. Sixty-six percent of the PHB in fibers is in the molecular mass range of 0.6 Da. The presence of PHB granules in transgenic fibers resulted in measurable changes of thermal properties. The fibers exhibited better insulating characteristics. The rate of heat uptake and cooling was slower in transgenic fibers, resulting in higher heat capacity. These data show that metabolic pathway engineering in cotton may enhance fiber properties by incorporating new traits from other genetic sources. This is an important step toward producing new generation fibers for the textile industry.

The number of genes that are active in fiber cells is similar to those in leaf, ovule, or root tissues. Through differential screening of a fiber cDNA library, John and Crow (1992) have isolated five cDNA clones that are preferentially expressed in fiber. One of the cDNA clones, pCKE6, corresponded to an abundant mRNA in fiber. Transcripts for E6 were detected throughout the development of the fiber. Immunoprecipitation of *in vitro* translation products and Western blot analysis of fiber proteins showed two polypeptides in the range of 30-32 kDa as the products of E6 mRNA. Sequence analysis and hybrid-selected RNA translation also suggested that E6 mRNAs encode two polypeptides. Concentrations of E6 mRNA and protein are highest during the late primary cell wall and early secondary cell wall synthesis stages. Sequence comparison of E6 with other known eukaryotic and prokaryotic genes reveals no significant homology (GenBank; December 1991). E6 or a homologous gene(s) is conserved in several members of Malvaceae as well as in one other fiber-producing plant, kapok, but is not found in several other plants examined or in *A. xylinum*. A genomic clone corresponding to pCKE6 was isolated, and the promoter element of the *E6* gene was shown to direct the expression of a carrot extensin mRNA in a tissue-specific and developmentally regulated fashion in transgenic cotton plants.

Taliercio and Boykin (2007) have developed a method to isolate RNA from fiber initials 1 dpa. Complementary DNA libraries representing 1 dpa fibers and other cotton tissues were sequenced and analyzed. Assembly of *G. hirsutum* Expressed Sequenced Tags (ESTs) identified over 11,000 sequences not previously represented in GenBank. New genes identified among these ESTs were represented on microarrays. The microarrays were used to identify genes enriched in fiber initials (1 dpa fibers) and elongating fibers. Analyses of Gene Ontologies (GO) of differentially expressed genes determined that terms associated with the membranes were statistically over represented among genes increased in expression in fiber initials and 10 dpa fibers. Staining ovules with a fluorescent dye confirmed an increase in Endoplasmic Reticulum (ER) occurred in fiber initials on the day of anthesis, persisted through 3 dpa and was absent in a fiberless mutant. Two genes similar to the CAPRICE/TRIPTYCHON (CPC) gene that inhibits differentiation of leaf trichomes in *Arabidopsis* were also characterized. Genes associated with novel regulation of BRs, GTP mediated signal transduction, cell cycle control and components of a Ca^{+2} mediated signaling pathway were identified. Staining of cellular Ca^{+2} indicated that fiber initials had more Ca^{+2} than other ovule cells supporting a role for Ca^{+2} in fiber development. Based on their studies, these authors concluded that analysis of genes expressed in fiber initials identified a unique stage in fiber development characterized by an increase in ER and Ca^{+2} levels that occurred between 0 and 1 dpa. The gene similar to CPC has a MYB domain but appeared to lack a transcription activating domain similar to the *Arabidopsis* gene. According to these workers the method used to stain the ER also could be used to count fiber initials and showed that fiber cells developed from adjacent cells unlike leaf trichomes.

Lee *et al.*, (2007) examined molecular and cellular events of fiber cell development in the naked seed mutant (N1N1) and its isogenic line of *G. hirsutum L.* (cv. Texas Marker-1, TM-1). The dominant mutation not only delayed the process of fiber cell formation and elongation but also reduced the total number of fiber cells, resulting in sparsely distributed short fibers. Gene expression changes in TM-1 and N1N1 mutant lines among four tissues were analyzed using spotted cotton oligo-gene microarrays. Using the *Arabidopsis* genes, they selected and designed approximately 1,334 70-mer oligos from a subset of cotton fiber ESTs. Statistical analysis of the microarray data indicated that the number of significantly differentially expressed genes was 856 in the leaves compared with the ovules (3 dpa), 632 in the petals relative to the ovules (3 dpa), and 91 in the ovules at 0 dpa compared to 3 dpa, all in TM-1. Moreover, 117 and 30 genes were expressed significantly differently in the ovules at 3 and 0 DPA, respectively, between TM-1 and N1N1. Quantitative RT-PCR analysis of 23 fiber-associated genes in seven tissues including ovules, fiber-bearing ovules, fibers, and non-fiber tissues in TM-1 and N1N1 indicated a mode of temporal regulation of the genes involved in transcriptional and translational regulation, signal transduction, and cell differentiation during early stages of fiber development. Suppression of the fiber-associated genes in the mutant possibly suggested that the N1N1 mutation disrupts temporal regulation of gene expression, leading to a defective process of fiber cell elongation and development.

For genetic characterization of rapid cell elongation in cotton fibers, Arpat *et al.*, (2004) assembled approximately 14,000 unique genes from 46,603 expressed sequence tags (ESTs) from developmentally staged fiber cDNAs of a cultivated diploid species (*Gossypium arboreum* L.). Conservatively, the fiber transcriptome represents 35-40% of the genes in the cotton genome. *In silico* expression analysis revealed that rapidly elongating fiber cells exhibit significant metabolic activity, with the bulk of gene transcripts, represented by three major functional groups - cell wall structure and biogenesis, the cytoskeleton and energy/carbohydrate metabolism. Oligonucleotide microarrays revealed dynamic changes in gene expression between primary and secondary cell wall biogenesis showing that fiber genes in the dbEST are highly stage-specific for cell expansion - a conclusion supported by the absence of known secondary cell wall-specific genes from fiber dbEST. During the developmental switch from primary to secondary cell wall syntheses, 2553 "expansion-associated" fiber genes appear to be significantly down regulated. Genes significantly up-regulated during secondary cell wall synthesis are involved in cell wall biogenesis and energy/carbohydrate metabolism, which is consistent with the stage of cellulose synthesis during secondary cell wall modification in developing fibers. This investigation provides the first in-depth view of the genetic complexity of the transcriptome of an expanding cell, and lays the groundwork for studying fundamental biological processes in plant biology with applications in agricultural biotechnology.

Zhang *et al.* (2007) have identified microRNAs (miRNAs) and their targets in cotton. In this study, they have found 30 potential cotton miRNAs using a comparative genomic approach based on genomic survey sequence analysis and miRNA secondary structure. These cotton miRNAs belong to 22 miRNA families. Expressed sequence tag (EST) analysis indicated that the predicted miRNAs were expressed in cotton plants. Based on the characteristic that miRNAs exhibit perfect or nearly perfect complementarity with their targeted mRNA sequences, a total of 139 potential miRNA targets were identified in cotton genome. A majority of these targets belong to transcriptional factors which regulate cotton growth and development, including leaf, root, stem, flower, and even fiber development. Those miRNAs may also be involved in other cellular and metabolic processes, such as stress response, signal transduction, and secondary wall synthesis and deposition. Some of the newly identified miRNA targets may be unique to cotton species. At least 3 miRNA families (miR 396, 414, and 782) target callous synthase, fiber protein Fb23, and fiber quinone-oxidoreductase, suggesting that miRNAs play an important role in cotton fiber differentiation and development.

Wu *et al.* (2006) have identified genes that possibly regulate lint fiber initiation, expression profiles of 0 days dpa whole ovules from six reduced fiber or fiberless mutants and compared these with wild-type linted cotton using cDNA microarrays. Numerous clones were differentially expressed, but only those genes that are normally expressed in the ovule outer integument (where fibers develop) were considered. Thirteen different cDNA clones were down-regulated in some or all of the mutants. These included: a Myb transcription factor (GhMyb25) similar to the *Antirrhinum* Myb AmMIXTA, a putative homeodomain protein (related to *Arabidopsis* ATML1), a cyclin D gene, some previously identified fiber-expressed structural and metabolic genes, such as lipid transfer protein, alpha-expansin and sucrose synthase, as well as some unknown genes. Laser capture microdissection and reverse transcriptase-PCR were used to show that both the *GhMyb25* and the homeodomain gene were predominantly ovule specific and were up-regulated on the day of anthesis in fiber initials relative to adjacent non-fiber ovule epidermal cells. Their spatial and temporal expression pattern therefore coincided with the time and location of fiber initiation. Constitutive over expression of *GhMyb25* in transgenic tobacco resulted in an increase in branched long-stalked leaf trichomes. The involvement of cell cycle genes prompted DNA content measurements that indicated that fiber initials, like leaf trichomes, undergo DNA endoreduplication. Cotton fiber initiation therefore has some parallels with leaf trichome development, although the detailed molecular mechanisms are clearly different.

As a main natural fiber source, cotton plays an important role in human life. Liu *et al.* (2006) attempted to identify genes preferentially expressed during early and late cotton fiber development, two fiber subtracted libraries were constructed on the basis of PCR-selected subtraction using a pool of nonfiber tissues as the same driver and 10 dpa and 20 dpa fiber cells as testers, respectively. Through differential screening, 292 clones in both libraries were identified as being preferentially expressed during fiber development. Sequence analysis showed that 31 unique sequences were found in the library of 10 dpa fibers and 48 unique sequences were obtained in the library of 20 dpa fibers. In addition, there were 13 unique clones in common in both libraries. Many previously reported cotton fiber-related genes were included in both libraries. Northern hybridization was performed to further confirm the differential expression and to determine the pattern of mRNA accumulation of selected clones. As a result, all selected cDNAs showed highly preferential expression in developing cotton fibers. Four genes-putative gibberellin-regulated protein, putative tonoplast intrinsic protein, putative plasma membrane intrinsic protein, and *G. hirsutum* putative membrane protein-were identified in 10 dpa fiber subtracted library, and they were found to be expressed a lot during early fiber development. On the other hand, those genes screened out of 20 dpa fiber subtracted library, like arabinogalactan protein and fiber glycosyl hydrolase family of 19 proteins, were found highly expressed in fibers with the maximal transcription level during developmental switch from elongation to cellulose deposition. By subtraction between fibers and five nonfiber tissues, two sets of genes were identified, and their fiber-specific or fiber-preferential expression indicated that they are involved in the network that controls cotton fiber development.

In view of the fact that 14-3-3 protein, a highly conserved protein in all eukaryotic cells, is an important regulatory protein and plays an important role in the growth, amplification, apoptosis, signal transduction, and other crucial life activities of cells, Shi *et al.* (2007) isolated a cDNA encoding a putative 14-3-3 protein from cotton fiber cDNA library. The cDNA, designated as Gh14-3-3L (*G. hirsutum* 14-3-3-like), is 1,029 bp in length (including a 762 bp long open reading frame and 5'?3'-untranslated regions) and deduced a protein with 253 amino acids. The Gh14-3-3L shares higher homology with the known plant 14-3-3 proteins, and possesses the basic structure of 14-3-3 proteins: one dimeric domain, one phosphoralated-serine rich motif, four CC domains, and one EF hand motif. Northern blotting analysis showed that *Gh14-3-3L* was predominantly expressed during early fiber development, and reached the peak of expression in 10 dpa fiber cells, suggesting that the gene may be involved in regulating fiber elongation. The gene is also expressed at higher level in both ovule and petal, but displays lower or undetectable level of activity in other tissues of cotton.

Li *et al.* (2005) have investigated the role of actin cytoskeleton during fiber development. Fifteen *G. hirsutum* ACTIN (GhACT) cDNA clones were characterized. RNA gel blot and real-time PCR analysis revealed that *GhACT* genes are differentially expressed in different tissues and can be classified into four groups. One group, represented by *GhACT1*, is expressed predominantly in fiber cells and was studied in detail. A 0.8-kb GhACT1 promoter sufficient to confirm its fiber-specific expression was identified. RNA interference of *GhACT1* caused significant reduction of its mRNA and protein levels and disrupted the actin cytoskeleton network in fibers. No defined actin network was observed in these fibers and, consequently, fiber elongation was inhibited.

The data suggested that GhACT1 played an important role in fiber elongation but not fiber initiation.

Considering that a large number of genes are required for fiber differentiation and development, Li *et al.* (2002) investigated gene expression patterns in fiber, a cDNA, *GhTUB1*, encoding beta-tubulin was isolated from a cotton fiber cDNA library. The analyses of RNA northern-blot hybridization and reverse transcriptase-polymerase chain reaction demonstrated that GhTUB1 transcripts preferentially accumulated at high levels in fiber, at low levels in ovules at the early stage of cotton boll development, and at very low levels in other tissues of cotton. The corresponding *GhTUB1* gene including the promoter region was isolated by screening a cotton genomic DNA library. To demonstrate the specificity of the GhTUB1 promoter, the 5'-flanking region including the promoter and 5'-untranslated region was fused with the beta-glucuronidase reporter gene. The expression of the reporter chimera was examined in a large number of transgenic cotton plants. Histochemical assays demonstrated that GhTUB1: α-glucuronidase fusion genes were expressed preferentially at high levels in fiber and primary root tip of 1- to 3-d-old seedlings and at low levels in other tissues such as ovule, pollen, seedling cotyledon, and root basal portion. Their results suggested that the *GhTUB1* gene may play a distinct and required role in fiber development. In addition, the GhTUB1 promoter possibly has great potential for cotton improvement by genetic engineering.

The α-galactosidases (EC 3.2.1.23) constitute a widespread family of glycosyl hydrolases in plants and are thought to be involved in metabolism of cell wall polysaccharides. A cDNA of the cotton (*G. hirsutum*) α-galactosidase gene, designated *GhGal1*, has previously been identified and its transcripts are highly abundant at the elongation stage of the cotton fiber. To examine the temporal and spatial control of GhGal1 expression, a transcriptional fusion of the GhGal1 promoter region (1770 bp) with the α-glucuronidase (*GUS*) reporter gene was introduced into tobacco plants by the *Agrobacterium* infection method. The resulting transgenic plants showed higher GUS activity of fruit in the transgenic plants than that in the negative and positive controls. Histochemical localization of GUS activity demonstrated that the expression of the *GUS* gene could be found in the meristem zones of roots, cotyledons, vascular tissues, fruit and trichomes in transgenic tobacco plants.

Additionally, sequence analysis of the regulatory region also revealed several conserved motifs among which some were shared with previously reported fruit/seed-specific elements and the others were related with trichome expression (Wu and Liu, 2006). These results indicate the temporal and spatial expression characterization of the GhGal1 promoter in transgenic tobacco plants and provide an important insight into the roles of GhGal1 in cotton fiber development.

Reversibly glycosylated polypeptides (RGPs) are thought to be involved in polysaccharide metabolism. A cDNA of the cotton (*G. hirsutum*) RGP gene, designated *GhRGP1*, is preferentially expressed in fiber cells (Wu and Liu, 2006). Its temporal and spatial control was investigated by isolating a 624 bp fragment upstream of the *GhRGP1* coding sequence using PCR-based genomic walking method. The 624 bp promoter sequence was fused to the α-glucuronidase (GUS) gene, and gene expression in tobacco was analyzed. In 4-week-old transgenic tobacco plants, the highest expression level was observed in roots, and the GUS activity was 1.13 and 6.65 fold higher than that in stems and leaves, respectively. In the reproductive growth stage, the *GUS* expression level was highest in the pistils and the GUS activity in the stigmas and styles were 17.6 fold higher than that in the ovaries. High GUS activity was also detected in the anthers. In addition, histochemical staining for GUS activity on transgenic tobacco plants further indicated a higher expression in the trichomes, seeds and vascular tissues of stems. Abiotic stress treatments on transgenic tobacco plants showed that wounding and dehydration induced GUS expression. These results demonstrated the spatial and temporal regulation of a cotton RGP promoter in a model plant, and provided an important insight into the factors that control the fiber development and stress responses of the gene.

Globulins are the most abundant seed storage proteins in cotton and, therefore, their regulatory sequences could potentially provide a good source of seed-specific promoters. Sunilkumar *et al.* (2002) isolated the putative promoter region of cotton á-globulin B gene by gene walking using the primers designed from cotton staged embryo cDNA clone. PCR amplified fragment of 1108 bp upstream sequences was fused to gusA gene in the binary vector pBIN101.3 to create the test construct. This was used to study the expression pattern of the putative promoter region in transgenic cotton, *Arabidopsis* and tobacco. Histochemical GUS analysis revealed that the promoter began to express during the torpedo stage of seed development in tobacco and *Arabidopsis*, and during cotyledon expansion stage in cotton. The activity quickly increased until embryo maturation in all three species. Fluorometric GUS analysis showed that the promoter expression started at 12 and 15 dpa in tobacco and cotton, respectively, and increased through seed maturation. The strength of the promoter expression, as reflected by average GUS activity in the seeds from primary transgenic plants, was vastly different amongst the three species tested. In *Arabidopsis*, the activity was 16.7% and in tobacco it was less than 1% of the levels detected in cotton seeds. In germinating seedlings of tobacco and *Arabidopsis*, GUS activity diminished until it was completely absent 10 days post imbibition. In addition, absence of detectable level of GUS expression in stem, leaf, root, pollen, and floral bud of transgenic cotton confirmed that the promoter is highly seed-specific. Analysis of GUS activity at individual seed level in cotton showed a gene dose effect reflecting their homozygous or hemizygous status. The results show that this promoter is highly tissue-specific and it can be used to control transgene expression in dicot seeds.

Two homologous cotton (*G. hirsutum* L.) genes, GhCTL1 and GhCTL2, encode members of a new group of chitinase-like proteins (called the GhCTL group) that includes other proteins from two cotton species, *Arabidopsis*, rice, and pea. Members of the GhCTL group are assigned to family GH19 glycoside hydrolases along with numerous authentic chitinases, but the proteins have novel consensus sequences in two regions that are essential for chitinase activity and that were previously thought to be conserved. Maximum parsimony phylogenetic analyses, as well as Neighbor-Joining distance analyses, of numerous chitinases confirmed that the GhCTL group is distinct. A molecular model of GhCTL2 (based on the three-dimensional structure of a barley chitinase) had changes in the catalytic site that are likely to abolish catalytic activity while retaining potential to bind chitin oligosaccharides. Studies of Zhang *et al.* (2004) on RNA blot analysis showed that members of the GhCTL group had preferential expression during secondary wall deposition in cotton lint fiber. Cotton transformed with a fusion of the GhCTL2 promoter to the α-d-glucuronidase gene showed preferential reporter gene activity in numerous cells during secondary wall deposition. Together with evidence from other researchers that mutants in an *Arabidopsis* gene within the GhCTL group are cellulose-deficient with phenotypes indicative of altered primary cell walls, these data suggest that members of the GhCTL group of chitinase-like proteins are essential for cellulose synthesis in primary and secondary cell walls. However, the mechanism by which they act is more likely to involve binding of cotton oligosaccharides than catalysis.

Suppression of BR biosynthesis in cotton ovules by treatment with brz inhibits fiber formation, indicating that BR plays an important role in cotton fiber development. Plant responses to BR are mediated through a plasma membrane-bound leucine-rich repeat (LRR) receptor-like protein kinase known as BRI1. Mutations in the BRI1 genes of several species result in dwarfed plants with reduced sensitivity to BR. A single expressed sequence tag (EST) from cotton with strong sequence similarity to *Arabidopsis* BRI1 (AtBRI1) was identified in a search of publicly available databases (Sun *et al.*, 2004). With this EST as a starting point, full-length cDNAs and genomic coding sequences from upland cotton (*G. hirsutum*) BRI1 (GhBRI1) have been characterized. Ectopic expression of this coding sequence in BR-insensitive *Arabidopsis* plants resulted in recovery of normal growth indicating that GhBRI1 is a functional homologue of AtBRI1. *G. hirsutum* is an allotetraploid (AADD) derived from diploid ancestors. Analysis of several GhBRI1 cDNAs showed two distinct sequences indicating the presence of two GhBRI1 genes, denoted GhBRI1-1 and GhBRI1-2. Sequence comparisons between these GhBRI1 coding sequences and those from related A and D genome diploid *Gossypium* species (*G. arboreum* and *G. thurberi*) indicated that GhBRI1-1 is likely to the A sub-genome orthologue while GhBRI1-2 is from the D sub-genome.

BRs promote the elongation of cotton fibers and may be a factor in determining their final length. To begin to understand the role of BR-mediated responses in the development of cotton fibers Sun and Allen (2005) have characterized the *BIN 2* genes of cotton. BIN 2 is a member of the shaggy-like protein kinase family that has been identified as a negative regulator of BR signaling in *Arabidopsis*. Sequence analyses indicate that the tetraploid cotton genome includes four genes with strong sequence similarity to BIN 2. These genes fall into two distinct subclasses based on sequence and expression patterns. Sequence comparisons with corresponding genes from cotton species that have the diploid A and D genomes, respectively, shows that each pair of genes comprises homologs derived from the A and D sub-genomes. Transgenic *Arabidopsis* plants that express cotton BIN 2 cDNAs show reduced growth and similar phenotypes to the semi-dominant bin 2 mutant plants. These results indicate that the cotton BIN 2 genes encode functional BIN 2 isoforms that can inhibit BR signaling.

Cotton fibers are single, highly elongated cells derived from the outer epidermis of ovules, and are developmentally similar to the trichomes of *Arabidopsis thaliana*. To identify genes involved in the molecular control of cotton fiber initiation, Humphre *et al.* (2005) isolated four putative homologues of the *Arabidopsis* trichome-associated gene TRANSPARENT TESTA GLABRA1 (*TTG1*). All four WD-repeat genes are derived from the ancestral D diploid genome of tetraploid cotton and are expressed in many tissues throughout the plant, including ovules and growing fibers. Two of the cotton genes were able to restore trichome formation in ttg1 mutant *Arabidopsis* plants. Both these genes also complemented the anthocyanin defect in a white-flowered *Matthiola incana* ttg1 mutant. These results demonstrate parallels in differentiation between trichomes in cotton and *Arabidopsis*, and indicate that these cotton genes may be functional homologues of AtTTG1.

MECHANISMS OF FIBER ELONGATION

Fiber elongation is the net result of a complex interplay between cell turgor and cell wall extensibility (Wilkins and Jernstedt, 1999). In conjunction with increased water uptake through aquaporin activity (Ferguson *et al.*, 1997), the sequestration of osmotically active solutes (osmolytes) by the expanding vacuole is an important aspect of turgor-driven wall extension. In elongating fibers, K^+ and malate are the predominant osmolytes, and their levels fluctuate in correlation with the growth rate (Dhindsa *et al.*, 1975; Basra and Malik, 1983a). Plant K^+ channels reside in both the plasma membrane and the tonoplast (Cheryl *et al.*, 1996), as do the two membrane-bound electrogenic proton pumps that hydrolyze ATP to pump H^+ ions against a concentration gradient, thereby generating both an electrochemical and a pH gradient across the membrane (Wilkins and Jernstedt, 1999). The electrochemical potential provides the energy to activate osmolyte transport. Joshi *et al.* (1985) demonstrated a positive relationship of ATPase activity to the process of fiber elongation. No ATP-specific enzyme activity was seen in non-elongating epidermal cells. As the fibers started elongating, the enzyme activity gradually increased in the tonoplast of enlarging vacuoles. The developmental regulation of the vacuolar H^+-ATPase, with a key role in osmoregulation and therefore in turgor-driven cell expansion, often serves as a convenient stage-specific marker for fiber expansion and elongation (Smart *et al.*, 1998; Wilkins and Jernstedt, 1999; Rataiczak and Wilkins, 2000).

Malate is derived by dark metabolism of CO_2 via PEP carboxylase (Dhindsa *et al.*, 1975; Basra and Malik, 1983a). Fiber growth is inhibited *in vitro* in the absence of CO_2 and K^+ in the culture medium (Dhindsa *et al.*, 1975). CO_2 concentration in the cotton bolls is high enough to support high rates of carbon fixation. Besides malate production, dark fixation of CO_2 is intimately involved in supporting other biosynthetic reactions via anaplerotic metabolism (Basra and Malik, 1985).

ABA, which inhibits fiber elongation, also inhibits the enzyme activities associated with dark metabolism of CO_2 (Dhindsa, 1978b). Conversely, cotton ovules cultured in the presence of FLU, an inhibitor of ABA biosynthesis, respond with a marked increase in these enzyme activities (Kaur *et al.*, 1990). Calcium also modulates the fiber capacity for CO_2 fixation and is possibly involved in ABA action (Basra *et al.*, 1993). Recent work has indicated that nitric oxide (NO) and its synthesis are important elements of signal cascades in ABA responses. NO regulates inward-rectifying K^+ channels of *Vicia faba* guard cells through its action on Ca^{2+} release from intercellular Ca^{2+} stores, but alternative pathways are indicated for its action on the outward-rectifying K^+ channels which are Ca^{2+} insensitive (Sokolovski and Blatt, 2004). NO either directly modifies the K^+ channel or a closely associated regulatory protein, probably by nitrosylation of cysteine sulfhydryl groups. The role of NO in cotton fiber development has not been investigated.

Apart from K^+ and malate, the contribution of other osmolytes, such as sugars, amino acids, and their derivatives, is equally important, which needs to be precisely quantified. There exist separate "storage" and "metabolic" pools of sugars in the fiber showing mutual exchange (Carpita and Delmer, 1981). Since the fibers are highly vacuolated, the storage pool is probably the vacuole. This is supported by the fact that only about 11 per cent of the total of reducing sugars is released by treatment of the fibers with 7.5 per cent dimethylsulfoxide (Carpita and Delmer, 1981). This treatment alters permeability of the plasma membrane, while having much less effect on the vaculoar membrane.

In addition to the generation of the fiber turgor, the primary cell wall must also be able to undergo plastic extension and incorporate new cell wall material. A thin primary cell wall (0.2-0.4 μm) deposited during fiber elongation contains <30% cellulose, whereas the thick (8-10 μm) secondary cell wall is composed of >94% cellulose, no lignin, and few proteins (Meinert and Delmer, 1977). Fiber elongation is initially achieved largely by cell wall loosening and finally terminated by increased wall rigidity and loss of higher turgor. Ruan *et al.* (2001) demonstrated that the gating of plasmodesmata in fibers is developmentally reversible and is coordinated with the expression of sugar and K^+ transporters, and the cell wall-loosening gene. Studies have provided further evidence that the role of plasmodesmatal regulation in cotton fiber elongation is mediated by callose turnover (Ruan *et al.*, 2004). In all cases, the plasmodesmata closure occurred at the rapid phase of elongation. Aniline blue staining and immunolocalization studies showed that callose deposition and degradation at the fiber base correlates with the timing of plasmodesmata closure and reopening, respectively. Northern analyses showed that the expression of a fiber-specific α-1, 3-glucanase gene, GhGluc1, was high in the short fiber genotype and weak in the intermediate and long fiber genotypes (Ruan *et al.*, 2004). They have suggested that the expression of GhGluc1 could play a role in degradation of callose, leading to opening of the plasmodesmata.

Sustained rapid and prolonged growth of developing fibers requires the extensive involvement of numerous metabolic and biosynthetic pathways (Basra and Malik, 1982, 1983 a, b, c; Kumar *et al.*, 1987; Nayyar *et al.*, 1988; 1990; Basra *et al.*, 1990; 1992; Kaur *et al.*, 1992 a,b; Qin *et al.*, 2005).

To identify genes involved in cotton fiber elongation, Ji *et al.* (2003) performed subtractive PCR using cDNA prepared from 10 dpa wild-type cotton fiber as tester and cDNA from a fuzzless-lintless mutant as the driver. cDNA macroarrays showed that 172 genes were significantly up-regulated in elongating cotton fibers. Twenty-nine genes, encoding a putative vacuolar (H^+)-ATPase catalytic subunit, a kinesin-like calmodulin binding protein, several arabinogalactan proteins, and key enzymes involved in long chain fatty acid biosynthesis, accumulated to >50-fold in 10 dpa fiber cells as compared to 0 dpa ovules. Various upstream pathways, such as auxin signal transduction, the MAPK pathway, and profilin- and expansin-induced cell wall loosening, were simultaneously activated during the period of rapid elongation.

Using the suppression subtractive hybridization method, Feng *et al.* (2004) isolated five gene families, including proline-rich proteins (PRPs), arabinogalactan proteins (AGPs), expansins, tubulins and lipid transfer proteins (LTPs), from fast elongating cotton fiber cells. Expression profile analysis using cDNA array technology showed that many transcripts accumulated over 50-fold in 10 dpa fiber cells than in 0 dpa samples.

Conclusively, it can be stated that a concerted mechanism involving multiple cellular pathways is responsible for cotton fiber elongation (Li *et al.*, 2002; Ji *et al.*, 2003; Feng *et al.*, 2004).

In cotton fibers, cortical microtubules undergo dramatic reorganization during fiber development (Preuss *et al.*, 2003). To dissect the role of the actin cytoskeleton, Li *et al.* (2005) characterized 15 *G. hirsutum* ACTIN (GhACT) cDNA clones by RNA gel blot and real-time RT-PCR analysis. They found that one group of GhACT genes, represented by GhACT1, was expressed predominantly in elongating fibers. Its RNA interference disrupted the actin cytoskeleton network as well as elongation of fibers, suggesting that GhACT1 plays an important role in fiber elongation but not fiber initiation (Li *et al.*, 2005). Over expression of a α-tubulin-like cDNA (Gh-BTubL) expressed specifically in elongating cotton fibers, in fission yeast induced longitudinal growth of the host cells by 1.74-fold, with no apparent effect on other aspects of the host cells (Ji *et al.*, 2002). Whittaker and Triplett (2002) noticed two discrete patterns of α-tubulin transcript levels in elongating cotton fibers. Whereas transcripts of α-tubulin genes GhTua2/3 and GhTua4 increased in abundance from 10 to 20 dpa, GhTua1 and GhTua5 transcripts were abundant only through to 14 dpa, and dropped significantly at 16 DPA with the onset of secondary wall synthesis. Collectively, these results suggest that elucidation of the control mechanisms for expression of tubulin-like proteins may help in modifying fiber quality and productivity.

The high level of metabolic activity in developing fibers during turgor-driven cell expansion and elongation (Smart *et al.*, 1998) is reflected in the genetic complexity of the fiber transcriptome. Of the approximately 14,000 unique gene sequences identified in elongating cotton fibers, almost two-thirds of annotated genes assigned to functional categories based on gene ontology fall into three major categories (energy/ carbohydrate metabolism, cellular structure/ organization/ biogenesis, and protein metabolism), a profile that is certainly consistent with metabolically active cell types (Arpat *et al.*, 2004).

Increase or decrease in the susceptibility of walls to turgor-driven stretching can be described in general terms as wall "loosening" and "tightening," respectively. The underlying mechanisms causing these physical changes in the fiber cell walls are not well understood. Orford and Timmis (1997) isolated five fiber-specific cDNA clones by differential screening of a cDNA library. One cDNA encoded a lipid transfer protein (LTP), and a second encoded a member of a group of well-characterized proline-rich proteins (PRP) from plants. The presence of signal peptide-encoding sequences indicated that both the LTP and the PRP are targeted to the extracellular matrix of the fiber, possibly playing a role in cell elongation. Shimizu *et al.* (1997) studied changes in the levels of mRNAs for cell wall-related enzymes in growing cotton fibers by RT-PCR analysis. Both endo-1, 4-α-glucanase and expansin mRNA levels were high during cell elongation, but decreased when cell elongation ceased and xyloglucan decreased. The endo-1, 4- α-glucanase may be responsible for solubilization of xyloglucan in the primary wall, which may be associated with the auxin-induced cell elongation in plants (Hayashi, 1989). Xyloglucan breakdown during cotton fiber development has been demonstrated (Tokumoto *et al.*, 2003), and the decrease was most prominent in cell wall preparations obtained from the rapidly elongating fibers. Zhang and Liu (2005) isolated a full-length cDNA encoding a typical α-galactosidase protein, designated GhGal1, and its very high abundance at the elongation stage suggested that the GhGall gene could be involved in metabolism of the primary cell wall.

The period between 16 to 21 dpa represents a switch in emphasis from primary to secondary cell wall synthesis during fiber development, exemplified by the transition in cell wall metabolism (Shimizu *et al.*, 1997). A dramatic shift in the IAA/ABA ratio coincident with the decrease in the rate of fiber elongation characterizes this transition (Nayyar *et al.*, 1989 a, b). A precise understanding of this regulation is crucial for devising strategies for fiber modification that could affect ultimate fiber properties.

During the fiber elongation phase, the ionically bound cell wall peroxidase activity remained low, but it increased significantly during the secondary thickening phase (Rama Rao *et al.*, 1982). The role of wall peroxidase in cessation of elongation growth has been postulated. MacAdam *et al.* (1992) also established that ionically bound wall peroxidase plays a key role in determining when cell elongation ceases in growing leaves, both temporally and spatially. Few reports in the literature have suggested the involvement of ABA in control of cell wall-associated peroxidase activity, possibly through controlling the cellular redox state (Lee and Lin, 1996; Takahama, 1994). Evidence suggests a relationship of ABA and Ca^{2+} to peroxidase activity in cotton fibers (Basra *et al.*, 1992). H_2O_2 and redox regulation of plant metabolism and gene expression is extensively documented in the literature (Foyer *et al.*, 1997; Foyer and Noctor, 2005). Interestingly, H_2O_2 has been proposed as a signal molecule that is initiated exactly coincident with the onset of secondary wall formation and the termination of fiber elongation (Potikha *et al.*, 1999). Furthermore, H_2O_2 causes an increase in cytosolic Ca^{2+}, indicating a role for calcium in the signal transduction process (Price *et al.*, 1994).

Li *et al.* (2005) have reported the cloning and characterization of a receptor-like kinase gene (designated GhRLK1), expression of which is induced during the period of active secondary wall synthesis in the fiber cells. GhRLK1 is located in the plasma membrane and shows dual specificity as both a serine/threonine kinase and a tyrosine kinase, suggesting its possible role in the signal transduction pathway that is involved in the induction and maintenance of active secondary wall formation during fiber development.

Dynamic changes in the fiber transcriptome occur in response to developmental signals that control fiber elongation and the transition to secondary cell wall synthesis (Arpat *et al.*, 2004). Conservatively, the fiber transcriptome represents 35-40% of the genes in the cotton genome. During the developmental switch from primary to secondary cell wall synthesis, 2553 "expansion-associated" fiber genes are significantly down regulated. Many cell wall and carbohydrate metabolism-related genes (81) are temporally regulated during secondary cell wall synthesis in a manner typified by the developmental regulation of *GhCesA1* and *GhCesA2* genes encoding the catalytic subunit of cellulose synthase (Arpat *et al.*, 2004). Expression of *GhCesA* genes dramatically increases coincident with the rate of cellulose synthesis to reach peak levels at approximately 24 dpa (Meinert *et al.*, 1977; Pear *et al.*, 1996).

About 80% of available carbon is used to form cellulose in secondary wall stage of cotton fibers (Haigler *et al.*, 2001). Sucrose phosphate synthase (SPS), which regulates the synthesis of sucrose from fructose-6-phosphate and UDP-glucose in leaves, is upregulated during secondary wall deposition (Babb and Haigler, 2001). Understanding the mechanism of cellulose biosynthesis is one of the major challenges in plant biology (Haigler *et al.*, 2005; Saxena and Brown, 2005). Assembly of cellulose microfibrils is a multi-step process involving a number of proteins with the key protein being the cellulose synthase catalytic sub-unit (Saxena and Brown, 2005). Recent advances towards understanding and manipulating cotton fiber secondary wall deposition have been reviewed in detail (Haigler *et al.*, 2001; Delmer and Haigler, 2002; Haigler *et al.*, 2005).

CONCLUDING REMARKS

The study of cotton fiber biology is important for describing, defining and refining fundamental questions in the fields of cell and developmental biology, with applications in textile industry, and agriculture. For understanding the regulation of the development and functioning of cotton fibers, and ultimately perhaps to manipulate fiber properties, it is imperative to seek more fundamental information on the developmental and environmental signals and mechanisms associated with signal transduction, leading to changes in fiber gene expression. Characterization of the cotton fiber transcriptome has immediate applications in agricultural biotechnology and molecular breeding programs geared toward the genetic improvement of yield and fiber quality (Arpat *et al.*, 2004; Wilkins and Arpat, 2005). The wealth of genomic tools and resources now available to the cotton community are paving the way to manipulating the fiber transcriptome using molecular approaches to develop enhanced germplasm. Genetic control of fiber size and shape, and hence agronomic properties, is governed by the rate and duration of fiber elongation, which in turn, positively correlates to fiber length. Breeding for long staple genotypes has been successful, resulting in lines that possess improved yield and fiber quality but without unduly delaying crop maturity (Braden and Smith, 2004). Improved fiber yield and maturity would result from the ability to control secondary wall cellulose deposition quantitatively, including making the process less sensitive to environmental stress (Haigler *et al.*, 2005). Both genetic engineering and marker-assisted breeding are possible avenues for the control of fiber morphogenesis once key regulatory genes and gene products have been identified. In addition, future cotton fiber biotechnology will require strong fiber-specific promoters that drive targeted gene expression in order to avoid pleiotropic effects on plant development.

REFERENCES

Allen RD Burns TM Light G Fokar M (2000) Investigating the role of xyloglucan endotransglycosylase in cotton fiber quality In: *Genetic Control of Cotton Fiber and Seed Quality* Eds C Benedict G Jividen 166-174 Cotton Incorporated Cary NC USA signaling *Physiologia Plantarum* **100**: 241-254

Arpat AB Waugh M Sullivan JP Gonzales M Frisch D Main D Wood T Leslie A Wing RA Wilkins TA (2004) Functional genomics of cell elongation in developing cotton fibers *Plant Molecular Biology* **54**: 911-929

Babb VM Haigler CH (2001) Sucrose phosphate synthase activity rises in correlation with high-rate cellulose synthesis in three heterotrophic systems *Plant Physiology* **127**: 1234-1242

Baert T DeLanghe E Waterkeyn L (1975) *In vitro* culture of cotton ovules III Influence of hormones upon fiber development *La Cellule* **71**: 55-63

Balls WL (1919) The existence of daily growth rings in the cell wall of cotton hairs *Proceedings of the Royal Society of London B Biological Sciences* **90**: 542-555

Basra AS Malik CP (1982) Adenine nucleotides and energy charge during fiber elongation of two cottons differing in their lint lengths *Proceedings of Indian National Science Academy* **46**: 229-233

Basra AS Malik CP (1983a) Dark metabolism of CO_2 during elongation of two cottons differing in fiber lengths *Journal of Experimental Botany* **34**: 1-9

Basra AS Malik CP (1983b) Glucose oxidation of elongating fibers of two species of cotton differing in fiber lengths *Tropical Plant Science Research* **1**: 329-333

Basra AS Malik CP (1983c) A comparison of lipid synthesizing capacity in elongating fibers of two cottons differing in fiber lengths *Current Science* **52**: 30-31

Basra AS Malik CP (1984) Development of the cotton fiber *International Review of Cytology* **89**: 65-113

Basra AS Malik CP (1985) Non-photosynthetic fixation of carbon dioxide and possible biological roles in higher plants *Biological Reviews* **60**: 357-401

Basra AS Dhillon-Grewal R Sarlach RS Malik CP (1993) Influence of ABA and calcium on dark fixation of carbon dioxide in cotton fibers *Phyton* **33**: 1-6

Basra AS Saha SS (1999) Growth regulation of cotton fibers In: *Cotton Fibers: Developmental Biology Quality Improvement and Textile Processing* Ed A S Basra New York: Haworth Press 47-63

Basra AS Sarlach RS Dhillon-Grewal R Malik CP (1992a) Calcium-mediated changes in peroxidase and o-diphenol oxidase activities of cotton fibers (*Gossypium* spp) and its possible relationship to ABA *Plant Growth Regulation* **11**: 159-164

Basra AS Sarlach RS Nayyar H Malik CP (1990) Sucrose hydrolysis in relation to development of cotton (*Gossypium* spp) fibers *Indian Journal of Experimental Biology* **28**: 985-988

Basra AS Sarlach RS Nayyar H Malik CP (1992b) Hormonal effects on partitioning of ^{14}C-sucrose in cotton fibers *Acta Physiologiae Plantarum* **14**: 137-142

Beasley CA (1979) Cellulose content in fibers of cotton which differ in their lint lengths and extent of fuzz *Physiologia Plantarum* **45**: 77-82

Beasley CA (1973) Hormonal regulation of growth in unfertilized cotton ovules *Science* **179**: 1003- 1005

Beasley CA Ting IP (1973) The effects of plant growth substances on *in vitro* development from fertilized cotton ovules *American Journal of Botany* **60**: 130-139

Beasley CA Ting IP (1974) Effects of plant growth substances on *in vitro* development from unfertilized cotton ovules *American Journal of Botany* **61**: 188-194

Benedict CR Pan S Madden SL Kohel RJ Jividen GM (1994) A cellulose cotton fiber mutant: effect of fiber strength In: *Biochemistry of Cotton* Ed JG Benedict Cotton Incorporated Cary NC 115-119

Braden CA Smith CW (2004) Phenology measurements and fiber associations of near-long staple upland cotton *Crop Science* **44**: 2032-2037

Bradow JM Bauer PJ Hinojosa I Sassenrath-Cole G (1997) Quantitation of cotton fiber quality variations arising from boll and plant growth environments *European Journal of Agronomy* **6**: 191-204

Buchala AJ (1999) Noncellulosic carbohydrates in cotton fibers In: *Cotton Fibers: Developmental Biology Quality Improvement and Textile Processing* Ed AS Basra New York: Haworth Press 113-136

Carpita NC Delmer DP (1981) Concentration and metabolic turnover of UDP glucose in developing cotton fibers *Journal of Biological Chemistry* **256**: 308-315

Chen JG Du XM Zhao HY Zhou X (1996) Fluctuation in levels of endogenous plant hormones in ovules of normal and mutant cotton during flowering and their relation to fiber development *Journal of Plant Growth Regulation* **15** 173-177

Chen YN Shen CY Zhang ZL Van JQ 1988 Study of the fiber development of cotton ovules *Acta Biol Exp Sinica* **21**: 417-421

Cheryl I Daram P Gaymard F Horeau C Thibaud B Sentenac H (1996) Plant K$^+$ channels: Structure activity and function *Biochemical Society Transactions* **24**: 961-971

Claussen M Luthen H Blatt M Bottger M (1997) Auxin-induced growth and its linkage to potassium channels *Planta* **201**: 227-234

Davidonis G (1995) Changes in polyamine distribution during cotton fiber and seed development *Journal of Plant Physiology* **145**: 105-112

DeLanghe EAL (1986) Lint development In: *Cotton Physiology* Eds J R Mauney J M Stewart Memphis TN: The Cotton Foundation 325-350

Delmer DP (1999) Cellulose biosynthesis in developing cotton fibers In: *Cotton Fibers: Developmental Biology Quality Improvement and Textile Processing* Ed AS Basra New York: Haworth Press 85-112

Delmer DP Haigler CH (2002) The regulation of metabolic flux to cellulose a major sink for carbon in plants *Metabolic Engineering* **4**: 22-28

Dhindsa RS (1978a) Hormonal regulation of cotton ovule and fiber growth: Effects of bromodeoxyuridine AM0-1618 and p-chlorophenoxyisobutyric acid *Planta* **141**: 269- 273

Dhindsa RS (1978b) Hormonal regulation of enzymes of non-autotrophic CO_2 fixation in unfertilized cotton ovules *Zeitschrift für Pflanzenphysiologie* **89**: 355-362

Dhindsa RS Beasley CA Ting IP 1975 Osmoregulation in cotton fiber: Accumulation of potassium and malate during growth *Plant Physiology* **56**: 394-398

El Mogahzy YE (1999) Fiber-to-fabric engineering: optimization of cotton fiber quality In: *Cotton Fibers: Developmental Biology Quality Improvement and Textile Processing* ed A S Basra New York: Haworth Press 339-376

Feng JX Ji SJ Shi YH Xu Y Wei G Zhu YX (2004) Analysis of five differentially expresses gene families in fast elongating cotton fiber *Acta Biochimica et Biophysica Sinica* **36**: 51-56

Feng R Brown RM (2000) A novel cotton ovule culture: Induction growth and characterization of submerged cotton fibers (*Gossypium hirsutum* L) *In Vitro Cellular and Developmental Biology-Plant* **36**: 293-299

Ferguson DL Turley RB Kloth RH (1997) Identification of a TIP cDNA clone and determination of related A and D genome subfamilies in *Gossypium* species *Plant Molecular Biology* **34**: 111-118

Foyer CH Noctor G (2005) Redox homeostasis and antioxidant signaling: A metabolic interface between stress perception and physiological responses *Plant Cell* **17**: 1866-1875

Foyer CH Lopez-Delgado H Oat JF Scott IM (1997) Hydrogen peroxide- and glutathione-associated mechanisms of acclimatory stress tolerance and signaling *Physiologia Plantarum* **100**: 241-254

Gipson JR (1986) Temperature effects on growth development and fiber properties In: *Cotton Physiology* eds JR Mauney JM Stewart Memphis TN: The Cotton Foundation 47-56

Gipson JR Joham HE 1969 Influence of night temperature on growth and development of cotton *(Gossypium hirsutum* L) III Fiber elongation *Crop Science* **9**: 127-129

Gipson JR Ray LL (1969) Fiber elongation rates in five varieties of cotton *(Gossypium hirsutum* L) as influenced by night temperature *Crop Science* **9**: 339-341

Gipson JR Ray LL (1970) Temperature-variety interrelationships in cotton Boll and fiber development *Cotton Growers Review* **47**: 257-271

Gokani SJ Kumar R Thaker VS (1998) Potential role of abscisic acid in cotton fiber and ovule development *Journal of Plant Growth Regulation* **17**: 1-5

Goynes WR Ingber BF Triplett BA (1995) Cotton fiber secondary wall development time vs thickness *Textile Research Journal* **65**: 400-408

Graves DA Stewart JM (1988) Chronology of the differentiation of cotton *(Gossypium hirsutum* L) fiber cells *Planta* **175**: 254-258

Haigler CH Ivanova-Datcheva M Hogan PS Salnikov VV Hwang S Martin K Delmer DP (2001) Carbon partitioning to cellulose synthesis *Plant Molecular Biology* **47**: 29-51

Haigler CH Ramo Rao N Roberts EM Huang JY Upchurch DR Trolinder N (1991) Cultured ovules as models for cotton fiber development under low temperatures *Plant Physiology* **95**: 88-96

Haigler CH Zhang D Wilkerson CG (2005) Biotechnological improvement of cotton fiber maturity *Physiologia Plantarum* **124**: 285-294

Hawkins RS (1930) Development of cotton fibers in Pima and Acala varieties *Journal of Agricultural Research* **40**: 1017-1029

Hayashi T (1989) Xyloglucans in the primary cell wall *Annual Reviews of Plant Physiology and Plant Molecular Biology* **40**: 139-168

Hsieh YL (1999) Structural development of cotton fibers and linkages to fiber quality In: *Cotton Fibers: Developmental Biology Quality Improvement and Textile Processing* ed A S Basra New York: Haworth Press 137-144

Hsu CY Jenkins JN Saha S Ma DP (2005) Transcriptional regulation of the lipid transfer protein gene LTP3 in cotton fibers by a novel MYB protein *Plant Science* **168**: 167-181

Humphries JA Walker AR Timmis JN Orford SJ (2005) Two WD-repeat genes from cotton are functional homologues of the Arabidopsis thaliana TRANSPARENT TESTA GLABRA1 (TTG1) gene *Plant Molecular Biology* **57(1)**: 67-81

Jasdanwala RT Singh YD Chinoy JJ (1977) Auxin metabolism in developing cotton hairs *Journal of Experimental Botany* **28**: 1111-1116

Jasdanwala RT Singh YD Chinoy JJ (1980) Changes in components related to auxin turnover during cotton fiber development *Beitrage Biologie Planzenphysiologie* **55**: 23-36

Ji SJ Lu YC Feng JX Wei G Li J Shi YH Fu Q Liu D Luo JC Zhu YX (2003) Isolation and analyses of genes preferentially expressed during early cotton fiber development by subtractive PCR and cDNA array *Nucleic Acids Research* **31**: 2534-2543

Ji SJ Lu YC Li J Wei G Liang XJ Zhu YX (2002) A beta-tubulin-like cDNA expressed specifically in elongating cotton fibers induces longitudinal growth of fission yeast *Biochemical and Biophysical Research Communications* **296**: 1245-1250

John M E Keller G (1996) Metabolic Pathway Engineering in Cotton: Biosynthesis of Polyhydroxybutyrate in Fiber Cells *Proceedings of the National Academy of Sciences* USA **93**: 12768-12773

John ME Crow LJ (1992) Gene expression in cotton (Gossypium hirsutum L) fiber: cloning of the RNAs *Proceedings of the National Academy of Sciences* **89(13)**: 5769-5773

John ME (1999) Genetic engineering strategies for cotton fiber modification In: *Cotton Fibers: Developmental Biology Quality Improvement and Textile Processing* ed A S Basra New York: Haworth Press 271-292

Joshi PC Stewart JM Graham ET (1985) Localization of α-glycerophosphatase activity in cotton fiber differentiation *Protoplasma* **143**: 1-10

Kaur K Nayyar H Basra AS Malik CP (1992a) Cytochemical activity of dehydrogenases during early ovule and fiber development in cotton *Journal of Plant Science Research* **8**: 17-20

Kaur K Nayyar H Basra AS Malik CP (1992b) Activity profiles of some transaminases during early ovule and fiber development in cotton *Journal of Plant Science Research* **8**: 56-58

Kaur K Nayyar H Basra AS Malik CP (1990) Stimulation of enzymes of non photosynthetic C_4 metabolism in cultured cotton ovules by fluridone *Acta Physiologiae Plantarum* **12**: 3-6

Kerr T (1936) The structure of the growth rings in the secondary wall of the cotton hair *Protoplasma* **27**: 229-243

Kim HJ Triplett BA (2001) Cotton fiber growth *in planta* and *in vitro* Models for plant cell elongation and cell wall biogenesis *Plant Physiology* **127**: 1361-1366

Kohel RJ Quisenberry JE Benedict CR 1974 Fiber elongation and dry weight changes in mutant lines of cotton *Crop Science* **14**: 471-474

Kosmidou-Dimitripoulou K (1986) Hormonal influences on fiber development In: Mauney JR Stewart J Mc D (eds) *Cotton Physiology* eds JR Mauney JM Stewart Cotton Foundation Memphis TN 361-373

Kumar H Kaur K Basra AS Malik CP (1987) The relationship of hydrolases to fiber development in *Gossypium hirsutum* L *Proceedings of Indian National Science Academy* **53**: 259-261

Lee JJ Hassan OS Gao W Wei NE Kohel RJ Chen XY Payton P Sze SH Stelly DM Chen ZJ (2006) Developmental and gene expression analyses of a cotton naked seed mutant *Planta* **223(3)**: 418-432

Lee M Lin YH (1996) Peroxidase activity in ethylene-treated ABA-treated or Meja-treated rice *(Oryza sativa)* roots *Botanical Bulletin of Academia Sinica* **37**: 201-207

Li CH Zhu YQ Meng YL Wang JW Xu KX Zhang TZ Chen XY (2002) Isolation of genes preferentially expressed in cotton fibers by cDNA filter arrays and RT PCR *Plant Science* **163**: 113-120

Li XB Cai L Cheng NH Liu JW (2002) Molecular characterization of the cotton GhTUB1 gene that is preferentially expressed in fiber *Plant physiology* **130(2)**: 666-674

Li XB Fan XP Wang CL Cai L Yang WC (2005) The cotton ACTIN1 gene is functionally expressed in fibers and participates in fiber elongation *Plant Cell* **17**: 859-875

Li YL Sun J Xia GX (2005) Cloning and characterization of a gene for an LRR receptor like protein kinase associated with cotton fiber development *Molecular Genetics and Genomics* **273**: 217-224

Liu D Zhang X Tu L Zhu L Guo X (2006) Isolation by suppression-subtractive hybridization of genes preferentially expressed during early and late fiber development stages in cotton *Molecular Biology (Mosk)* **40(5)**: 825-834

Mac Adam JW Nelson CJ Sharp RE (1992) Peroxidase activity in the leaf elongation zone of tall fescue I Spatial distribution of ionically bound peroxidase activity in genotypes differing in length of the elongation zone *Plant Physiology* **99**: 872-878

Malik C P (2004) The role of calcium in auxin and brassinolide induced fiber elongation in *Gossypium arboreum Jour Plant Sci* Res **20**: 32-36

Malik C P Kaur Bhavneet and Wadhwani Chitra (2006) Involvement of Brassinolide and Ethylene in cotton fiber elongation *in vitro Jour Pl Sci Res* **22**: 181-183

Malik C P Kaur Bhavneet Verma Aman and Wadhwani Chitra (2007) The role of Nitric Oxide and Brassinolide in amelioration of low temperature effect on fiber elongation in *Gossypium arboreum* L *Jour Pl Sci Res* **23**: (in press)

Malik C P Singh B Sharma P and Basra A S (1995) Fiber initiation and early elongation *in vitro* of cotton ovules as influenced by low temperature: Scanning electron microscopic comparison and biochemical changes *Physiology and Molecular Biology of Plants* **1**: 129-134

Malik C P and Thind S K (2000) Achievements and perspectives in cotton improvement through Biotechnology In *Advances in Biotechnology* Ed J S Singh; A K Rai BHU Varanasi 213-240

Meinert MC Delmer DP (1977) Changes in biochemical composition of the cell wall of the cotton fiber during development *Plant Physiology* **59**: 1088-1097

Meredith WR (2005) Minimum number of genes controlling cotton fiber strength in a backcross population *Crop Science* **45**: 1114-1119

Naithani SC Rama Rao N Singh YD 1981 Physiological and biochemical changes associated with cotton fiber development I Growth kinetics and auxin content *Physiologia Plantarum* **54**: 225-229

Nayyar H Basra AS Malik CP 1988 $H^{14}CO_3$ fixation into lipids of elongating cotton fibers in two *Gossypium* species *Journal of Nuclear Agriculture and Biology* **17**: 243-245

Nayyar H Kaur K Basra AS Malik CP 1990 Activity profiles of pyruvate orthophosphate dikinase and phosphoenolpyruvate carboxylase in developing cotton fibers *Beitrage Biologie Pflanzenphyiologie* **65**: 319-324

Nayyar H Kaur K Basra AS Malik CP 1989a Hormonal regulation of cotton fiber elongation in *Gossypium arboreum* L *in vitro* and *in vivo Biochemie und Physiologie der Pflanzen* **185**: 415-421

Nayyar H Kaur K Basra AS Malik CP 1989b Regulation of differential fiber development in cotton by endogenous plant growth regulators *Proceedings of Indian National Science Academy* **55**: 463-468

Orford SJ Timmis JN (1997) Abundant mRNAs specific to the developing cotton fiber *Theoretical and Applied Genetics* **94**: 909-918

Pear JR Kawagoe Y Schreckengost WE Delmer DP Stalker DM (1996) Higher plants contain homologs of the bacterial *celA* gene encoding the catalytic subunit of cellulose synthase *Proceedings of National Academy of Sciences* **93**: 12637-12642

Pettigrew WT (1994) Source-to-sink manipulation effects on cotton fiber quality *Agronomy Journal* **87**: 947-952

Potikha T Collins C Johnson DI Delmer DP Levine A (1999) The involvement of hydrogen peroxide in the differentiation of secondary walls in cotton fibers *Plant Physiology* **119**: 849-858

Preuss ML Delmer DP Liu B (2003) The cotton kinesin-like calmodulin-binding protein associates with cortical microtubules in cotton fibers *Plant Physiology* **132**: 154-160

Price AH Taylor A Ripley SJ Griffiths A Trewavas AJ Knight MR (1994) Oxidative signals in tobacco increase cytosolic calcium *Plant Cell* **6**: 1301-1310

Qin YM Pujol FM A Shi YH Feng JX Liu YM Kastaniotis AJ Hiltunen JK Zhu YX (2005) Cloning and functional characterization of two cDNAs encoding NADPH dependent 3ketoacyl CoA reductases from developing cotton fibers *Cell Research* **15**: 465-473

Rama Rao N Naithani SC Singh YD (1982) Physiological and biochemical changes associated with cotton fiber development II Auxin oxidizing system *Physiologia Plantarum* **55**: 204-208

Ramey HH Jr 1986 Stress influences on fiber development In: *Cotton Physiology* eds JR Mauney JM Stewart The Cotton Foundation Memphis TN 351-360

Ratajczak R Wilkins TA (2000) Energizing the tonoplast In: *Vacuolar Compartments* eds DG Robinson JC Rogers Sheffield Academic Press Ltd England 133-173

Roberts EM Nunna RR Huang JY Trolinder NL Haigler CH (1992) Effects of cycling temperatures on fiber metabolism in cultured cotton ovules *Plant Physiology* **10**: 979-986

Ruan YL Llewellyn DJ Furbank RT (2001) The control of single-celled cotton fiber elongation by developmentally reversible gating of plasmodesmata and coordinated expression of sucrose and K^+ transporters and expansin *Plant Cell* **13**: 47-63

Ruan YL Llewellyn DJ Furbank RT (2003) Suppression of sucrose synthase gene expression represses cotton fiber cell initiation elongation and seed development *Plant Cell* **15**: 952-964

Ruan YL Llewellyn DJ Furbank RT Chourey PS (2005) The delayed initiation and slow elongation of fuzz-like short fiber cells in relation to altered patterns of sucrose synthase expression and plasmodesmata gating in a lintless mutant of cotton *Journal of Experimental Botany* **56**: 977-984

Ruan YL Xu SM White R Furbank RT (2004) Genotypic and developmental evidence for the role of plasmodesmatal regulation in cotton fiber elongation mediated by callose turnover *Plant Physiology* **136**: 4104-4113

Ryser U (1985) Cell wall biosynthesis in differentiating cotton fibers *European Journal of Cell Biology* **39**: 236-256

Ryser U (1999) Cotton fiber initiation and histodifferentiation In: *Cotton Fibers: Developmental Biology Quality Improvement and Textile Processing* ed AS Basra New York: Haworth Press 1-45

Saxena IM Brown RM Jr (2005) Cellulose biosynthesis: Current views and evolving concepts *Annuals of Botany* **96**: 9-21

Schubert AM Benedict CR Berlin JD Kohel RJ (1973) Cotton fiber development Kinetics of cell elongation and secondary wall thickening *Crop Science* **13**: 704-709

Seagull RW (1990) Tip growth and transition to secondary wall synthesis in developing cotton hairs In: *Tip Growth in Plant and Fungal Cells* ed IS Heath San Diego CA: Academic Press 261-284

Seagull RW Giavalis S (2004) Pre- and post-anthesis application of exogenous hormones alters fiber production in *Gossypium hirsutum* L cultivar Maxxa GTO *Journal of Cotton Science* **8**: 105-111

Sekimata K Kimura T Kaneko I Nakano T Yoneyama K Takeuchi Y Yoshida S Asami T (2001) A specific brassinosteroid biosynthesis inhibitor Brz evaluation of its effects on *Arabidopsis* cress tobacco and rice *Planta* **213**: 716-721

Sharma P and Malik CP (2007) Effect of ABA Calcium chloride on fiber elongation and enzyme activities in *Gossypium arboreum The Icfai Jour of Life Sciences* **1**: 31-38

Sharma P Malik CP (2005a) Effect of inhibitors on enzyme activities in developing cotton ovules *Jour Plant Sci Res* **21**: 170-173

Sharma P Malik CP (2005c) Influence of some chemicals on partitioning of ^{14}C-sucrose in *Gossypium arboreum Jour Plant Sci Res* **21**: 246-248

Sharma P Malik CP (2005b) Incorporation of ^{14}C into various biochemical constituents and effect of guaianolide derivative in developing cotton fibres *Jour Pl Sci Res* **21**: 59-62

Shen TY Chang SC Yhn CC (1978) The growth of fibers on excised cotton ovules and the formation of seedlings *Acta Phytophysiol Sinica* **4**: 183-187

Shi H Wang X Li D Tang W Wang H Xu W Li X (2007) Molecular characterization of cotton 14 3-3L gene preferentially expressed during fiber elongation *Journal of Genetics and Genomics* **15**: 1-9

Shimizu Y Aotsuka S Hasegawa O Kawada T Sakuno T Sakai R Hayashi T (1997) Changes in levels of mRNAs for cell wall-related enzymes in growing cotton fiber cells *Plant and Cell Physiology* **38**: 375-378

Singh B Gill R Malik CP Basra AS (1999) Are polyamines essential for cotton fibre growth? *Jour Plant Sci Res* **15**: 3-10

Singh B Malik CP (1997) Interior biochemical environment and cotton fibre development : An enzymatic approach *Environment and Development* (IS Grover and AK Thukral (Eds) 283-292

Singh B Sharma Malik CP Kalsi PS Sanju (1995) Effects of Guaianolide derivative on fibre initiation and elongation in *Gossypium arboreum* L in relation to adverse effects of cool temperature *Plant Growth Regulation* **17**: 101-107

Smart LB Vojdani F Maeshima M Wilkins TA (1998) Genes involved in osmoregulation during turgor-driven cell expansion of developing cotton fibers are differentially regulated *Plant Physiology* **116**: 1539-1549

Sokolovski S Blatt MR (2004) Nitric oxide block of outward-rectifying K+ channels indicates direct control by protein nitrosylation in guard cells *Plant Physiology* **136**: 4275-4284

Sun Y Allen RD (2005) Functional analysis of the BIN 2 genes of cotton *Molecular Genetics and Genomics* **274(1)**: 5-19

Sun Y Fokar M Asami T Yoshida S Allen RD (2004) Characterization of the brassinosteroid insensitive 1 genes of cotton *Plant Mol Biology* **54(2)**: 221-232

Sun Y Veerabomma S Haggag A Abdel-Mageed HA Fokar M Asami T Yoshida S Allen RD (2005) Brassinosteroid regulates fiber development on cultured cotton ovules *Plant and Cell Physiology* **46**: 1384-1391

Sunilkumar G Connell JP Smith CW Reddy AS Rathore KS (2002) Cotton alpha-globulin promoter: isolation and functional characterization in transgenic cotton *Arabidopsis* and tobacco *Transgenic Research* (2002) **11(4)**: 347-59

Suo JF Liang XO Pu L Zhang YS Xue YB (2003) Identification of GhMYB109 encoding a R2R3 MYB transcription factor that expressed specifically in fiber initials and elongating fibers of cotton (*Gossypium hirsutum* L) *Biochimica et Biophysica Acta* **1630**: 25-34

Takahama U (1994) Changes Induced by abscisic acid and light in the redox state of ascorbate in the apoplast of epicotyls of *Vigna argularis Plant and Cell Physiology* **35**: 975-978

Taliercio EW Boykin D (2007) Analysis of gene expression in cotton fiber initials *Plant Biology* **7(1)** 16-22

Thaker VS Saroop S Vaishnav PP Singh YD (1989) Genotypic variations and influence of diurnal temperature on cotton fiber development *Field Crops Research* **22**: 1-13

Tiwari SC Wilkins TA (1995) Cotton *(Gossypium hirsutum* L) seed trichomes expand via a diffuse growing mechanism *Canadian Journal of Botany* **73**: 746-757

Tokumoto H Wakabayashi K Kamisaka S Hoson T (2003) Xyloglucan breakdown during cotton fiber development *Journal of Plant Physiology* **160**: 1411-1414

Van't Hoff J Saha S (1997) Cotton fibers can undergo cell division *American Journal of Botany* **84**: 1231-1235

Wang S Shen ZY Zhang ZL (1985) A study of elongation of the cotton fiber cell *Acta Phytophysiol Sinica* **11**: 409-417

Wang S Wang J W Yu N Li CH Luo B Gou JY Wang LJ Chen XY (2004) Control of plant trichome development by a cotton fiber MYB gene *Plant Cell* **16**: 2323-2334

Whittaker DJ Triplett BA (1999) Gene-specific changes in alpha-tubulin transcript accumulation in developing cotton fibers *Plant Physiology* **121**: 181-188

Wilkins TA Arpat AB (2004) The cotton fiber transcriptome *Physiologia Plantarum* **124**: 295-300

Wilkins TA Jernstedt JA (1999) Molecular genetics of developing cotton fibers In: *Cotton Fibers: Developmental Biology Quality Improvement and Textile Processing* ed AS Basra New York Haworth Press 231-267

Wu A Liu J (2006) Isolation of the promoter of a cotton beta-galactosidase gene (GhGal1) and its expression in transgenic tobacco plants *Sci China C Life Sci* **49(2)**: 105-114

Wu AM Ling C Liu JY (2006) Isolation of a cotton reversibly glycosylated polypeptide (GhRGP1) promoter and its expression activity in transgenic tobacco *Journal of Plant Physiology* **163(4)**: 426-435

Wu Y Machado AC White RG Llewellyn DJ Dennis ES (2006) Expression profiling identifies genes expressed early during lint fiber initiation in cotton *Plant and Cell Physiology* **47(1)**: 107-127

Xie W Trolinder NL Haigler CH (1993) Cool temperature effects on cotton fiber initiation and elongation clarified using an *in vitro* culture *Crop Science* **33**: 1258-1264

Yoshida S Sun Y Asami T Fokar M Allen RD Veerabomma S Abdel-Mageed HA (2005) Brassinosteroid regulates fiber development on cultured cotton ovules *Plant and Cell Physiology* **46(8)**: 1384-1391

Zhang B Wang Q Wang K Pan X Liu F Guo T Cobb GP Anderson TA (2007) Identification of cotton microRNAs and their targets *Gene* (in press)

Zhang D Hrmova M Wan CH Wu C Balzen J Cai W Wang J Densmore LD Fincher GB Zhang H Haigler CH (2004) Members of a new group of chitinase-like genes are expressed preferentially in cotton cells with secondary walls *Plant Molecular Biology* **54(3)**: 353-372

Zhang HM Liu JY (2005) Molecular cloning and characterization of a beta-galactosidase gene expressed preferentially in cotton fibers *Journal of Integrative Plant Biology* **47**: 223-232

Zhang SZ (1982) Advances in cotton fiber cell development *Plant Physiology Communications* **6**: 1-7

Zheng ZR Xu DW (1982) The role of plant hormones in the reproductive growth of cotton plant *Sci Agric Sinica* **5**: 40-47

7

TRANSGENIC *Bt* COTTON AND ITS COMMERCIAL USE

C.P. MALIK, AMAN VERMA, CHITRA WADHWANI AND BHAVNEET KAUR

Introduction

In cotton production pests are a serious problem. To control them, more than 50% of pesticides are consumed on cotton. This causes many environmental, ecological and health hazards. Breeding of transgenic *Bt* cotton is the effective and environment-ecological-and health-friendly strategy for controlling pests and enhancing cotton yield. In 1990 the first *Bt* cotton was produced, and subsequently great progress has been made in this field. The lab studies and field practices demonstrated that *Bt* cotton is highly resistant to many prevalent cotton pests, such as bollworm, budworm and armyworm. Thus, cotton farmers have saved 70-80% pesticides, labor and costs on planting *Bt* cotton compared with traditional non-*Bt* cotton. Yield in *Bt* cotton has increased by 0-50% depending on its location. *Bt* cotton was first commercialized in China in 1994 and in 1996 in USA. In the past 10 years, *Bt* cotton has had a dramatic effect on cotton production, as the area under its cultivation has increased more than 6.7 fold i.e. from 0.8 million hectares in 1996, to 5.38 million hectares in 2003. Right now, up to 5 million farmers in 9 countries commercially grow *Bt* cotton. Approximately 25,000 large farmers benefited from the technology in the industrial countries of the USA and Australia. The vast majority of the 4 to 5 million *Bt* cotton farmers in the developing countries may be benefited in future by planting *Bt* cotton. Ten years of commercial *Bt* cotton use demonstrates that *Bt* cotton has achieved the goal of providing an effective tool for controlling cotton pests especially lepidopteran, and it is safer to human beings and environment than other strategies.

Cotton is one of the most important economic and fiber crops, grown in 70 tropical/ subtropical and temperate countries around the world. The cultivation of cotton has affected the economic development of several countries since it was first cultivated, 5,000 to 10,000 years ago (Stephens and Mosley, 1974). Over 180 million people are associated with the fiber industry that annually produces 20 to 30 billion dollars worth of raw cotton (International Advisory Committee, 1996). Trade in cotton related products presents almost 50% of the total $115 billion trade in textiles and the $133 billion trade in clothing (Rabobank, 1996). About 20 million cotton farmers are growing about 33.5 million hectares of cotton. Of these, about 70% of cotton is planted in developing countries by 19.3 million cotton farmers (James, 2002). In most cases, the major income of these farmers comes from cotton.

Cotton pests and pest control

Pests are a major problem in cotton production. Although the infestation levels of specific pests vary enormously from year to year, from country to country, and from region to region (Benedict and Altman, 2001), pests significantly damage yield by 15% to 50% and also the fiber quality. Up to 1326 species of insects have been reported on cotton worldwide (Matthews, 1994), but only few pests cause in yield dramatic decrease. Most of these pests belong to the caterpillar species (Lepidoptera), such as bollworm and budworm. In the absence of any specific control measure, about 35-41% of potential yield losses are reported due to these pests on global basis (Table 1). Thus, it is very important to adopt suitable measures to control cotton pests.

Table 1: Range of actual and potential losses from cotton insect pests for different global regions (Oerke, 2002).

Global regions	*Actual loss % with controls*	*Potentials loss % without controls*
Africa		
Eastern	24	35
Western	23	34
Southern	21	37
Northern	9	41
Asia		
South East	18	33
South	17	36
East	9	37
North East	9	38

America		
South	13	39
Andean	13	39
North	11	38
Central	7	37
CIS	11	37
Europe	9	35
Oceania	7	38

During 1999, global consumption of pesticide was estimated around US$ 8000 million, of which nearly 22% was spent on cotton. In India, pesticide consumption is worth Rs. 2,800 crores in agriculture of which Rs. 1,600 crores is spent on cotton alone and of these Rs. 1100 crores is spent for controlling bollworms. So nearly 45% of the pesticides are being used for the control of bollworms. Amongst them, *Helicoverpa armigera* is a polyphagous pest feeding apart from cotton, on pigeon pea, chick pea, tomato, okra, sunflower, safflower, sorghum, maize and many other species and its control today is becoming very difficult. It appears to have developed resistance to particular group of pyrethyroid. Indiscriminate pesticide application and use of substandard brands is worsening the situation still further. The cumulative effect of all this is increasing the financial burden of the cotton cultivations, reducing the profit margins and causing pollution hazards. Almost every year new molecules are being tried for controlling bollworms which remain effective for a very short period and the quest for new molecules then goes on. This is a never ending race and farmers are the victims of this situation.

The damage caused by various bugs on different plants parts in cotton is listed in Table 2.

Table 2: Cotton plant stage, plant parts fed upon and symptoms of plant bug and stink bug damage.

Plant stage	*Plant part*	*Bug type*	*Damage symptoms*
Pre-bloom	Terminals	Plant Bug	With heavy feeding, terminals may be deformed or killed, resulting in a loss of apical dominance (crazy cotton)
	Small Square		Squares, yellowing turning brown, then black, and finally aborting, leaving a scar

Blooming	Various squares		Small squares, turning brown, then black, and finally aborting, larger square, suffer internal damage to pollen anthers.
	White blooms		Darkened pollen anthers (dirty blooms), small petal deformations
	Bolls	Plant bugs & stink bugs	Aborted small bolls, external spotting, internal feeding "stings", wart like growth, stained lint, often in seed area, damaged seed

Many different strategies are employed to control cotton pests; of these, biological control, host-plant resistant varieties and certain cultural practices (such as timing of planting, crop rotation, and destruction of cotton residues) are less expensive and the most biorational. However, when the pest number and injury to cotton begins to exceed acceptable levels, it suffers serious damage in growth. Finally, the cotton farmers suffer serious economic losses. The common and suitable method for control of cotton pest outbreak is foliar spraying with conventional synthetic insecticide since pesticides are relatively inexpensive, fast action, effective, and easy to apply. However, control of pests with pesticides results in many environmental, ecological and health issues, such as environmental pollution, health risk to human and animals, and pest resistance to pesticides. Thus, scientists all over the world planting cotton try to figure out new strategies to control cotton pests. Of all the pest control methods, adoption of pest resistant varieties is the most ecological and biorational methods (Zhang and Feng, 2000). Thus cotton breeders have been attempting to breed pest resistant varieties for many years.

Traditional breeding has done much to improve the resistance of cotton to insect pests, and continues to provide new varieties that require less chemical intervention than old varieties. New characters such as glabrous ness (absence of hairs on the foliage), forego bract (outward bending, thin bracts around the cotton boll), nectar less (absence of nectar-forming nectary glands on the leaves), and high gland content are now being assessed for their capacity to reduce the attractiveness of the cotton plant to insect pest (Zhang and Feng, 1998). There are, however, limits to improvement in natural resistance to insects provided by alterations in plant shape and structure. Up to now, none of the cotton varieties developed have shown more than a moderate level of resistance. Thus, it is not enough to reduce the damage by pests by such efforts. Therefore, control of insect pests in cotton cultivation depends mainly on the use of chemical insecticides that are under serious public debate for the human safety and environmental pollution. Thus, scientists have been looking for new strategies to control cotton insect pests. Of them, an attractive alternative is the production of proteins with insecticidal activity by cotton plant itself.

Table 3: Size, shape and activity of Crystal proteins

Gene	*Crystal shape*	*Protein size (KDa)*	*Insect activity*
Cry I (Several subgroups) A(a), A(b), A(c), B, C, D, E, F, G)	Bipyramidal	130-138	Lepidoptera larvae
Cry II (Subgroups A, B, C)	Cubodial	69-71	Lepidoptera and diptera
Cry III (Subgroups A, B, C)	Flat/irregular	73-74	Coleoptera
Cry IV (Subgroups A, B, C, D)	Bipyramidal	73-134	Diptera
Cry V-IX	Various	35-129	Various

Genetic engineering should enhance the capacity to produce more tolerant plants by accessing a much wider gene pool for novel insect resistance characters not present in any of the *Gossypium* species or their close relatives. Numerous labs and field tests confirm that the most efficient and cheapest method for protecting cotton from pests is utilization of transgenic cotton as high pest-resistant cotton varieties. *Bt* is the popular abbreviation for *Bacillus thuringiensis*, a bacterium commonly found in soil, with ubiquitous distribution. Hence it is popularly called a 'soil bacterium', but it is also common in other habitats like dead insects, water, dead plants, grain dust, etc. The insecticidal property of *Bt* was first discovered in 1901 in Japan by Ishiwata and later in 1911 in Germany by Berliner who described this species and gave the present name. More than 80 varieties or subspecies of *Bt* have been described so far. A unique feature of *B. thuringiensis* is that it is a gram-positive, endospore-forming bacterium characterized by the presence of Cry (acronym for 'crystal' protein as the insecticidal protein aggregate to form insoluble crystals in the bacterium) protein within the cytoplasm of the sporulating cells.

Different varieties or sub-species of *Bt* produce different insecticidal proteins. Currently about 250 Cry proteins, sourced from about 80 sub-species of *Bt*, have been characterized (Table 3). Each of these proteins affects only a narrow range of insects belonging to a particular group. Thus, there are *Bt* proteins that can selectively kill certain larvae of only lepidoptera (moths and butterflies), coleoptera (beetles), hemiptera (bugs), diptera (flies and mosquitoes) and so on. A particular *Bt* protein active on one group of insects may not affect other insects or other organisms. Such host specificity, is also a unique feature of *Bt*-proteins which require certain specific conditions for them to be active against the insect. Firstly, the concerned *Bt* protein has to be ingested by the susceptible insects since it has no contact effects. In *Bt*-plants, this happens when the larvae feed on plant tissues. The protein requires an alkaline gut with a suitable pH (9.5 and above) for its activation. Specific receptors, in the insect mid-gut epithelial cells are a must for protein-binding before it can kill the insect. All these conditions are available only in the susceptible insects and, therefore, they succumb when they feed on *Bt*-plants.

In the commercialized transgenic *Bt*-cotton plants, the expression of *Bt* protein is constitutive i.e., the protein is expressed in all parts of the plant. When the larvae feed on *Bt* plants, they ingest *Bt* protein along with the plant tissues. If it is a susceptible insect like bollworms, the *Bt* protein gets activated in the mid-gut and the activated molecules bind themselves to certain receptors present on the gut membrane, very much like a specific key fitting into a lock. Such a specific interaction between the activated *Bt* protein and receptor results in 'holes' being formed in the insect intestine, causing destruction of the gut lining. The haemolymph (insect blood) carrying ions and vital nutrients leak into the intestine. This leads to paralysis of the insect gut as a result of this the insect stops feeding. This sequence of events can take place within a few hours. The affected larvae may die after a day or two, since it stops feeding, and any further damage to plants is prevented. *Bt* proteins can affect only those insect species possessing the specific receptors and conditions for toxin activation and, therefore, pose no threat to higher order organisms as they lack such specific conditions in their gut. This includes warm blooded animals. Ingestion (while feeding plant tissues), solubilization (alkalinity), activation (pH >9.5), binding (specific receptor), insertion (damage gut wall), pore formation, cell lysis and then death.

Bt proteins have served as the chief active ingredient of a number of commercial bacterial insecticides and these have been used as spray formulations (a mixture of endospores and insecticidal crystals) for control of various insect pests since the mid-1950s in several countries. The pests targeted included various caterpillars (lepidoptera) and beetles (coleoptera) attacking several crops as also mosquito and black fly (diptera) vectors of human and animal diseases.

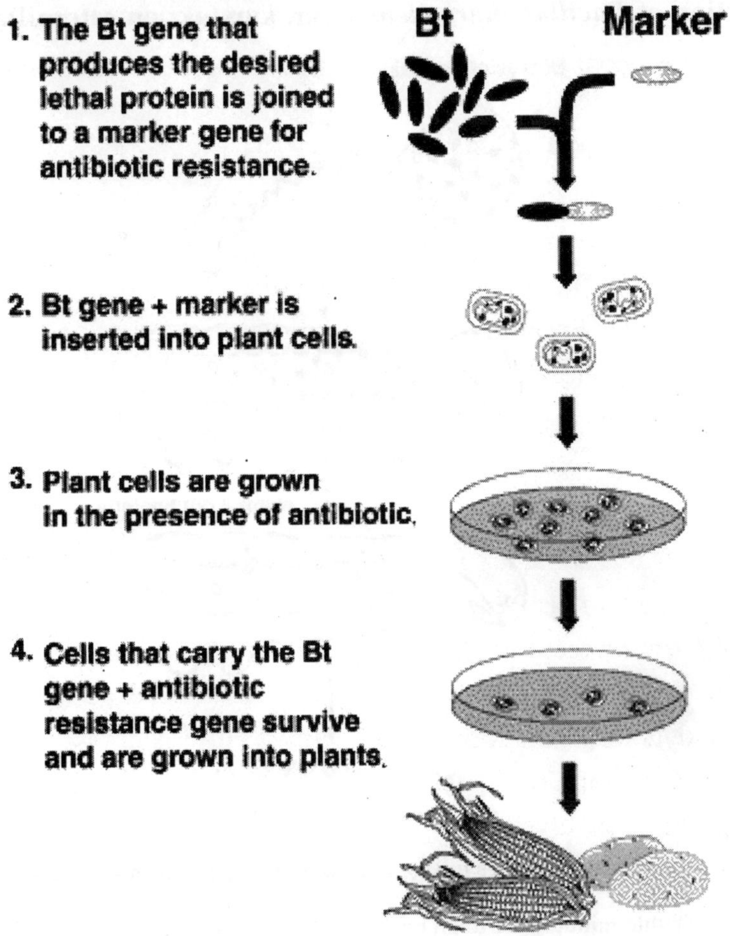

Fig 8: Procedure of introduction of *Bt* gene into plant cells for production of insect resistant cell lines.

The concept of genetic engineering of *Bt* strain

The insecticidal property of *B. thuringiensis* toxins from different sources is found to reside in the central region of each protein molecule. This segment of the toxin molecule has been called the hypervariable region, because it is distinct for each of the many insecticidal toxins that were examined. The remainder of the protein molecules in each case is largely invariant. With the identification of the portion of the insecticidal toxin that is responsible for insect specificity, the development of specific hybrid toxin molecules can now proceed in a more systematic manner (Fig. 8). Cotton plants have been protected by using a strong promoter to enhance the expression of *Bt* gene and altering its coding sequence, so that it can be translated into plants. The entire process of *Bt* gene introduction into cotton plant involve major steps like cleavage of amino terminal protein (Fig. 8).

Action of *Bacillus thuringiensis* var. *kurstaki* on caterpillars

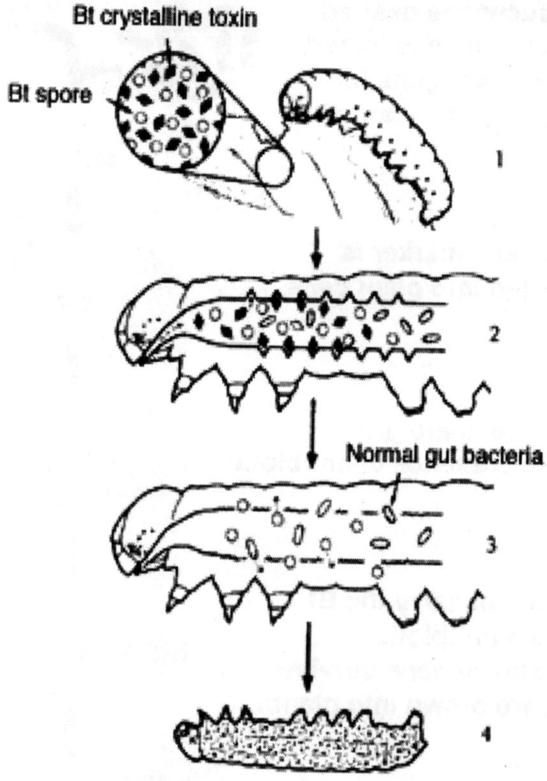

1) Caterpillar consumes foliage treated with *Bt* (spores and crystalline toxin).
2) Within minutes, the toxin binds to specific receptors in the gut wall, and the caterpillars slops feeding.
3) Within hours, the gut wall breaks down, allowing spores and normal gut bacteria to enter the body cavity, the toxin dissolves.
4) In 1-2 days, the caterpillar dies from septicemia as spores and gut bacteria proliliterate in its blood.

Fig 9: Methodology followed by *Bacillus thurigiensis* to attack caterpillars harmful for the crop.

The roots of many plants are attacked by insects that, because of their underground location, would normally be protected from *B. thruingiensis* – based insecticides. However, it is possible to introduce toxin gene from a *B. thruingiensis* strain into a bacterial species that colonizes the region adjacent to plant roots (the rhizosphere). The engineered bacteria could be introduced into the soil, where they would synthesize and release the insecticidal toxin into the area immediately surrounding the plant roots, thereby conferring protection against root-attacking insects. As long as the engineered bacteria were able to synthesize the insecticidal toxin, the need for repeated sprayings of either biological or chemical insecticides was obviated.

Other examples of genetic manipulation and genetic engineering with *Bt* include the following:

- In the agricultural product foil, the gene for a toxin with activity against beetles was transferred through conjugation (sexual reproduction in bacteria) to a *Bt*k cell that only affected butterflies and moths. The resulting cell showed insecticidal properties against beetles, butterflies, and moths. Since "During the 1992 Asian gypsy moth spray program in Oregon, a woman who was exposed to Foray 48B had a pre-existing allergy to a carbohydrate that was present as an inert ingredient. Within 45 minutes of exposure, the woman suffered from joint pain and neurological symptoms." EPA considers the organisms resulting from conjugation to be genetically manipulated rather than genetically engineered. Foil was registered for use in the US in 1990.
- *Pseudomonas fluorescens* cells can be engineered to produce the *Bt* delta-endotoxins without production of a spore. The Crystal protein remains inside the *P. fluorescens* cell wall. In the product MVP and M-Trak, the *P. fluorescens* cell is killed after it produces the Crystal protein. When the product is applied, the delta-endotoxin remains protected within the now dead cell wall. In this way, the *Bt* delta-endotoxin retains its effectiveness for two to three times longer than other *Bt* formulations. MVP and M-Trak were the first genetically engineered products to be registered by The United States Environmental Protection Agency (EPA).
- *(B. thuringiensis var israelensis)* used to control mosquito and blackfly larvae that live on the water surface begins to sink, away from the target larvae, within 24 hours. Bacteria that naturally live on the water surface (in the same environment as mosquito or blackfly larvae) have been engineered to produce the *Bt*i Crystal proteins.
- Over thirty different crops have been engineered to produce the *Bt* Crystal protein throughout their plant structure. Any pest that feeds on any part of these plants will be exposed to the *Bt* delta-endotoxin, and those susceptible to the toxin will be killed.

Clearly, the possibilities for the genetic engineering of *Bt* delta-endotoxin seem endless. However, researcher knows so little about the ecology and genetic stability of *Bt*, that the potential ecological effects of these transgenic organisms are impossible to predict with certainty. Monsanto Inc., a US multinational developed a modified *Cry* 1 Ac gene in order to improve its properties and for better expression in the plant system. Using this modified gene they carried out genetic manipulation of cotton plants. The transgenic plants so developed contain *Cry*1 Ac gene in its genome and such plants themselves have started producing the Crystal protein. This *Bt* cotton has certain properties.

1. It is effective against specific lepidopteran pests.
2. It offers no control on egg laying of the lepidopteran pests.
3. *Bt* cotton containing *Cry* 1 Ac offers no protection against the sucking pests like aphids, jassids, thrips, whiteflies etc. It is also safe to the beneficial insects and natural predators like honey bees, green lacewings and lady bird beetles.
4. In a transgenic *Bt* cotton plant, the Crystal protein is produced in all the plants parts and when an insect larva feeds on any of the plant parts, it consumes the toxin and dies within hours.

Monsanto company has developed a second *Bt* cotton variety, "Bollgard II" which produces two different toxins. In 2004, Dow Agro-sciences introduced "Widestrike", another *Bt* cotton producing two toxins, while Syngenta has introduced its *Bt* cotton, "VIP Cotton". *Bt* cotton is a critical technology in the biotech industry i.e. to push GM agriculture around the world.

Though "Bollgard I" provides excellent, season-long control to "Lashkari Sundi" and pink bollworm, and a high level of suppression of the cotton bollworm. "Bollgard II" also provides fantastic control of most of the sundies and pests like "Lashkari Sundi", "American Sundi", pink bollworm, fall armyworm, beet armyworm, cabbage and soybean loopers, and other second day leaf-or fruit-feeding caterpillars. *Bacillus thuringiensis* products available in the United States are comprised of one of five varieties of *Bt* (Table 4): *Bt* var. kurstaki and var. morrisoni cause disease in moth and butterfly capterpillars; *Bt* var. israelensis causes disease in mosquito and blackfly larvae; *Bt* var. aizawai causes disease in wax moth caterpillars); and *Bt* var. tenebrionis, also called var. san diego, which causes disease in beetle larvae. Other strains of *Bt* have been discovered that exhibit pesticidal activities against enmatodes, mites, flatworms, and protozoa (Feitelson *et al.*, 1992).

Table 4 : Popular *Bt* strains used (De Barjac Frachon, 1990)

Bt strain	***Effective against***
Bt kurstaki (*Bt*k)	Types of lepidopterous insects life gypsy moth, cabbage looper
Bt aizawai (*Bt*a)	Wax moth larvae in honey combs
Bt isaraelensis (*Bt*i)	Mosquitoes, black flies, midges
Bt San diego	Certain beetle species, boll weevil

Bt products are used to control moth pests in fruits, vegetables, and beehives; blackfly and mosquito pests in ponds and lakes; and several beetles pests in vegetables and shade trees. Common brand names include Dipel, Foray, Thuricide (all *Bt* kurstaki), Vectobac, Mosquito Attack (all *Bt* israelensis), and M-Trak (*Bt* tenebrionis).

Factors affecting selectivity

One of *Bt*'s most desirable characteristics is its selectivity, only certain insects are susceptible to the delta-endotoxin. Scientists have identified at least 29 different Crystal proteins and delta-endotoxin. Each is effective against specific insects. Each variety of *Bt* can produce one or more of these toxins. Alkaline solutions activate the delta-endotoxin, and different varieties may require different pH (Gill *et al.*, 1992). Certain enzymes must also be present in the insect's gut to break the crystal into its toxic elements. In addition, certain cell characteristic in the insect gut encourage binding of the endotoxin and subsequent pore formation. The age of the insect is also a factor, the young larvae being more susceptible.

Since *Bt* is a live microbial, testing for the possible hazards of *Bt* is conducted differently chat for conventional pesticides. Microbial toxicity is described using pathogenicity (the ability of the microbe to cause disease) and infectivity (the ability of an organism to reproduce within the body), EPA requires no testing of *Bt* for carcinogenicity, mutagenicity, or chronic toxicity.

Laboratory tests of acute toxicity

Each of the more than 800 strains of *Bacillus thuringiensis* may exhibit different toxicity to insects, rodents and humans. This fact complicates any discussion about the toxicity of *Bt*. The following are summaries of the acute toxicity data available for two commonly used commercial varieties of *Bt*.

***Bacillus thuringiensis* var. kurstaki (*Bt*k)**: *Bt*k and commercial products containing *Bt* generally have low oral acute toxicity to rats. In tests with laboratory animals, researchers did not observe any adverse effects after feeding large doses.

Others types of exposures have some acute effects. Rats who breathed air containing *Bt* spores, experienced respiratory depression, and injection of *Bt* spores into rats veins aggravated pre-existing disease. Both *Bt*k and Foray 48B are irritating to rabbit skin, and Foray 48B is moderately irritating to rabbits eyes.

***Bacillus thuringiensis var. israelensis* (*Bt*i)**: In studies assessing *Bti*'s acute toxicity to mammals, morality only occurred when *Bt*i was injected into the abdomen or the brain. In one study conducted on rats, 79 per cent mortality occurred after a single injection into the brain. Effects other than mortality can also occur. For example, in mice injected with a *Bt*i suspension, spleens became enlarged.

Bti is irritating to both eyes and skin. Injection of both viable and inactivated *Bti* spores under the skin resulted in abscesses in mice. Rabbit's eyes are irritated by *Bti* to eyes depends on the physical characteristics of the formulation; a dry, dusty formulation with smaller particles is less irritating and cleared from the eye more quickly than a clumped formulation with larger particles.

In a purified form, *Bti*'s endotoxin is barely toxic to mammals. When the delta-endotoxin from *Bti* was injected intravenously into mice, they exhibited rapid paralysis, followed by death within 12 hours. When the same dosage was injected under the skin of suckling mice, death occurred in 2-3 hours. The delta-endotoxin also caused destruction of rat, mouse sheep, horse and human red blood cells. When a small protein isolated from the endotoxin was administered to mice at sub lethal levels, mice suffered from severe hypothermia and their heart beat slowed.

Acute toxicity in humans

***Bacillus thuringiensis* var. kurstaki**: There have been few experimental studies assessing the toxicity of *Btk* to humans. Most information comes from occupational exposures, or from exposures occurring during large-scale *Btk* programs.

One case of *Btk* infection resulted from a farmer splashing a *Btk* formulation, Dipel, in his eye. The man developed an ulcer on his cornea from which positive *Btk* culture were taken. Another man working on a spray program splashed *Btk* on his face and eyes. He then developed skin irritation, burning, swelling, and redness. *Btk* was cultured from a sample taken from his eye. Ground-spray applicators using Foray 48B reported symptoms of eye, nose, throat, and respiratory irritation. The frequency of their complaints was found to be related to the degree of exposure. Workers with similar pre-existing health problems were more likely to report adverse effects from the ground spray.

A woman exposed to a *Btk* formulation as a result of drift went to the hospital due to burning, itching and swelling of her face and upper chest. She later exhibited fever, altered consciousness, and suffered seizures. No *Bt* was cultured from tissue samples, but her doctor believed that *Bt* was the cause of the clinical symptoms.

Monitoring studies following large-scale *Bt* spray programs have shown that exposed people carry *Bt* in their tissues. For example, more than 11 per cent of nasal swab samples taken from patients surveyed by doctors in Vancouver (Canada) following a gypsy moth spray program were found to contain *Btk*. *Bt* was also found in cultures taken from patients in Lane County, Oregon following a gypsy moth spray program there. Monitoring studies also show that exposed people report a variety of health problems that they believe to be associated with *Bt* exposure. For example, during the Vancouver spray program, almost 250 people reported health problems, mostly allergy like or flu like symptoms. During a Washington gypsy moth spray program, over 250 people reported health problems and 6 were treated in emergency rooms for allergy or asthma problems. Physicians have so far been unable to definitively link *Bt* exposure to these health problems (Drobniewski, 1994).

Bacillus thuringiensis* var. *israelensis: There has only been one case of documented adverse effects of *Bti* on humans. This case involved a researcher who accidentally injected himself with a mixture of *Bti* and another kind of bacteria commonly found on human skin. He suffered from a toxic reaction and irritated lymph vessels. When two bacteria were later injected into rodents the combination was consistently lethal, but each bacteria injected separately caused only slight inflammation.

Special concerns about *Bt* toxicity

Exotoxins: The earliest test done regarding *Bt*'s toxicity were conducted using *Bt* var. thuringiensis, a *Bt* strain known to contain a second toxin called beta-exotoxin. The beta-exotoxin is toxic to vertebrates, with an LD50 (median lethal dose; the dose that kills 50 per cent of a population of test animals) of 13-18 milligrams per kilogram of body weight (mg per kg) in mice when injected into the abdomen. An oral dose of 200 mg per kg per day killed mice after eight days. Beta-exotoxin also causes genetic damage to human blood cells. *Bt* formulations containing beta-exotoxin have not been used in most countries although attempts are currently being made to register beta-exotoxin as an insecticide in the United States. Another toxin produced by *Bt* is the alpha-exotoxin that is highly acutely toxic to mice. Current *Bt* production methods are such that alpha-exotoxin is not a "significant component" of *Bt* formulations.

Related Bacteria: *Bt* belongs to a small group of closely related *Bacillus* species, including *B. cereus,* a bacterium that is an agent of food poisoning, and *B. anthracis,* the pathogen of the virulent animal disease, anthrax. These three bacteria are so similar it has been theorized that they are all varieties of the same species. If *B. cereus* is cultured with *Bt* cells, genetic material is transferred to the *B. cereus* cells that allows *B. cereus* to produce *Bt*'s crystal proteins. Transfers of genetic material between *B. anthracis* and *Bt* have also occurred.

A toxin produced by *B. cereus* that causes diarrhoea in monkeys is also produced by certain strains of *Bt,* although this toxin is not likely to be present in *Bt* spore formulation (Drobnie, 1994). Human volunteers suffered from nausea, vomiting, diarrhoea, colic-like pains, and fever after eating food contaminated with one *Bt* strain, *Bt* var. galleriae. These examples indicate the close relationship between *Bt* and disease-causing pathogens.

Increased susceptibility and people with compromised immune systems or pre-existing allergies may be particularly susceptible to the effect of *Bt.* In mice with reduced immune function, the dose required to kill more than 50 per cent of the mice when injection was of several orders of magnitude smaller than the highest dose tested in normal mice. Mice with impaired immune functions, also showed higher mortality than regular mice when one dose of *Bt* was injected into the abdominal cavity.

A memo from Novo Nordisk, the manufacturer of Foray 48B, states that the amount of the spray a person would be exposed to would be too small to develop new allergies. However, "It is possible that someone who already has developed an allergy to one of the components of Foray 48B or has asthma could be affected by exposure to small quantities of Foray 48B. The 1991 Material Safety Data Sheet for Foray 48B states "Repeated exposure via inhalation can result in sensitization and allergic response in hypersensitive individuals."

Contaminants: In the mid 1980s, several *Bt* products were contaminated with other bacteria, including *Streptococcus faecium* and *S. faecalis*. While *Bt* products are routinely monitored for bacterial contaminants, the risk of contamination with disease-causing bacteria is always present.

Inert ingredients: All *Bt* products contain ingredients other than *Bt*. These are identified only as "inert" ingredients and are called trade secrets by the manufactures of the products. The "inert" ingredients are potentially the most toxic components of the formulations. For example, during the 1992 Asian gypsy moth spray program in Oregon, a woman who was exposed to Foray 48B had a pre-existing allergy to a carbohydrate that was present as an inert ingredient. Within 45 minutes of exposure, the woman suffered from joint pain and neurological symptoms.

Because "inerts" are called trade secrets, there is little public information about their identity, but the information that is available indicates they could cause health problems. Foray 48B contains sodium hydroxide, sulfuric acid, phosphoric acid, methyl paraben, and potassium phosphate, as "inerts". While these ingredients make up less than 10 per cent of Foray 48B, they pose hazards. Sodium hydroxide, more commonly known as lye, cause "severe corrosive damage to the eyes, skin, mucous membranes and digestive system. Breathing sodium hydroxide dust of mist leads in milk cases to irritation of the mucous membranes of the nose and in severe cases, to damage of the upper respiratory tract." Sulfuric acid and phosphoric acid are both corrosive. Sulfuric acid can cause severe deep skin burns and permanent loss of vision. When inhaled as a mist, sulfuric acid may cause severe bronchial constrictions, and bronchitis. Phosphoric acid is an irritant to skin and mucous membranes, and its vapors may cause coughing and throat irritation. Both methyl paraben and potassium phosphate were once registered by EPA as pesticide active ingredients (U.S. EPA. Prevention, Pesticide and Toxic Substances, 1994).

Sodium sulfite has been identified as inert ingredients of the *Btk* formulation, Dipel 8AF. Up to 10% of asthmatics (about one million people in the United States) may react to sulfites, particularly those people who are treated with steroids. Symptoms of exposure in those sensitive to sulfites usually involve the respiratory system, and can also include nausea, diarrhoea, lowered blood pressure, hives, shock, and loss of consciousness.

Environmental fate: Very little is known about the natural ecology of *Bt*. It occurs naturally in many soils. In one study, *Bt* was isolated from 70 per cent of soil samples taken from around the world, and was most abundant in samples taken in Asia. More than half of these isolates were undescribed varieties of *Bt*. *Bt* has also been isolated from insect bodies, tree leaves and aquatic environments. It has even been recovered from paper.

Soil: *Bt* generally persists only a short time in soil. T½ of the insecticidal activity (the time in which half of the insecticidal activity is lost) of the crystal is about 9 days. However, small amounts can be quite persistent. In one experiment, *Bt* spore numbers declined by one order of magnitude after 2 weeks, but then remained constant for 8 months following application.

Bt does not appear to move readily in soil. In one study, two varieties of *Bt* were applied in adjacent plots, but did not become cross-contaminated, indicating that *Bt* does not move laterally in soil. It has also been found that *Bt* was not recovered past a depth of 6 centimeters after irrigation, and that movement beyond the application plot was less than 10 yards.

Foliage: *Bt* deposited on the upper side of leaves (exposed to the sun), may remain effective for only 1-2 days, but *Bt* on the underside of leaves (*i.e.* protected from the sun) may remain active for 7-10 days. It is possible for it to be significantly more persistent, however, viable spores of *Bt*k were recovered from white spruce foliage one year after application. In one experiment conducted in Japan, *Bt* persisted for two years in a citrus orchard and remained toxic to caterpillars.

Water: *Bt*k has been recovered from rivers and public water distribution systems after aerial applications of Thuricide 16B. Standard water treatment processes are not adequate to destroy *Bt*k spores.

Bt i spores and crystals bind readily to sediments in the water column, which reduces their efficacy by making them inaccessible to mosquito and black fly larvae.

In one test, *Bt*i was applied to water and then allowed to contact mud particles. Over 99 per cent of the *Bt*i spores were found in the mud, rather than in the water, after 45 minutes. The *Bt*i retained viability and toxicity for at least 22 days, killing 90 per cent of the mosquito larvae when the mud was stirred and reintroduced to the water column

In another experiment, viable cells were recovered from the water for up to 200 days and in the sediment for up to 270 days after application.

Air: *Btk* has been found to drift over 3,000 meters downwind during an aerial application. The distance *Btk* is capable of drifting depends upon the amount and method of application, as well as the climatic conditions. *B. thuringiensis* was measured in air for up to 17 days following an application.

Insect is also a factor, the younger larvae being more susceptible than older larvae.

Bt's Ecological Impacts

Some of the most serious concerns about widespread use of *Bt* as a pest control technique come from the effects it can have on animals other than the pest targeted for control. All *Bt* products can kill organisms other than their intended target. In turn, the animals that depend on these organisms for food are also impacted.

Beneficial insects: Many insects are not pests, and any pest management technique needs to be especially concerned about those that are called beneficial, the insects that feed or prey on pest species. *Bt* has impact on a number of beneficial species as well. For example, studies of a wasp that is a parasite on the meal moth (*Plodia interpunctella*) found that treatment with *Bt* reduced the number of eggs produced by the parasitic wasp, and the percentage of those eggs that hatched. Production and hatchability of eggs of a predatory bug were also decreased. On collards, aphid-eating flies in the family Syrphidae were reduced by Dipel treatment. Both *Bt* tenerbrionis and Dipel have caused mortality of predatory spider mites. Dipel also has caused mortality of the cinnabar moth, used for the biological control of the weed tansy ragwort (James 2002). Finally, *Bti* has caused mortality of a moth (*Synclita obliteralis*) that helps control aquatic weeds in Florida.

Other insects: Many insects that do not have a direct beneficial importance to agriculture are important in the function and structure of ecosystems. A variety of studies have shown that *Bt* applications can disturb insect communities. Research following large-scale *Bt* applications to kill gypsy moth larvae in Lane Country, Oregon, found that the number of oak-feeding caterpillar species was reduced for three years following spraying, and the number of caterpillars was reduced for two years. Similar results were found in a study of caterpillars feeding on tobacco brush following *Bti* application to control spruce budworm in Oregon. In untreated areas, the number of species was 30 per cent higher, and the number of caterpillars 5 times greater, than in *Bti*-treated areas two weeks after treatment. The number of caterpillars was still reduced in treated areas the following summer. In Washington, *Bt* applications in King and Pierse countries to kill gypsy moths reduced spring moth populations by almost 90 per cent. In addition, one rare species appeared to have been eradicated from the treatment zone, and moth population were "heavily impacted in an area more than double that which was actually sprayed" as moths moved into the treatment zone from surrounding area. In West Virginia, application of Foray 48B reduced the number of caterpillar species and the number of caterpillars. The year following application, the number of moth species and the number of moths were reduced. A (1994) study in four different Oregon plane communities found that total weight of caterpillars was reduced between 90 and 95 per cent by *Bt* treatment; the number of caterpillars was reduced by 80 per cent and the number of caterpillar species was reduced by over 60 per cent.

Aquatic insects are also affected by *Bt* treatments. Canadian studies found that certain steam insects (*Simulium vittatum* and *Taeniopteryx nivalis*) were killed by applications of Thuricide and Dipel respectively. Midges (chironomids) have repeatedly been shown to be killed by *Bt*.

Birds: Because many birds feed on the caterpillars and other insects affected by *Bt* applications, it is not surprising that impacts of *Bt* spraying on birds have been well documented. In Lane Country, Oregon studies of chickadees following a gypsy moth spray program found that birds nesting in *Bt*-treated areas brought fewer caterpillars to their nests than did birds nesting in untreated areas. The birds were able to-find other foods, so that nesting success was not significantly impacted. In new Hampshire, when *Bt*-treatment reduced caterpillar abundance, black-throated blue warblers made fewer nesting attempts and also brought fewer caterpillars to their nestlings. A Canadian study found that numbers of caterpillars, followed by numbers of two species of warblers and a thrush, were reduced by *Bt* treatment. In addition, there were fewer spruce grouse chicks in *Bt* treated areas, and the chicks in those areas grew more slowly than chicks in untreated area.

Other animals: Because shrews often feed on caterpillars, impacts from *Bt* treatments are likely. A study in northern Ontario (Canada) found that treatment with Dipel changed the structure of the shrew population. Adult males emigrated, so that the promotion of juveniles increased. The juveniles and adult females who did not emigrate shifted from a diet of caterpillars to alternative prey.

Foray 48B at high concentration (about 3 per cent) is acutely toxic to rainbow trout, probably because the product is highly acidic. *Bt* treatments can also affect other animals. Low concentration of *Bt* endotoxins decrease the weight of tadpoles and delay their metamorphosis. The *Bt* formulation Vectobac is acutely toxic to fathead minnows, probably because "inerts" in the product deplete the dissolved oxygen in water (Snarski, 1990). The *Bti* formulation Teknar was acutely toxic to brook trout fry, probably because of xylene used as an "inert" in the product.

Comparison with synthetic insecticides: Where comparative studies have been done, the ecological impacts of a *Bt* treatment are almost always less than those of synthetic insecticides. For example, *Bt* treatment of collards caused less of an increase in aphid numbers than did treatment with carbaryl, which killed many aphid predators. Vectobac was much less acutely toxic to an estuary fish than other mosquito insecticides including tempephos, fenoxycarb, diflubenzuron; and methoprene.

DEVELOPMENT OF *Bt*-COTTON

Depending upon the type of pests to be controlled-whether it is lepidoptera, coleoptera etc.-the relevant genes from the soil bacterium, *Bacillus thuringiensis* (*Bt*), are isolated, studied, suitably modified and introduced into the desired plant species through genetic engineering. The new *Bt*-gene gets stably integrated in the host genome and becomes an inheritable trait. Such transgenic plants containing the *Bt*-gene(s) are popularly called '*Bt*-plants.' For example, *Bt*-cotton is incorporated with the lepidopteran specific gene(s) as it is designed to control bollworms which belong to this insect order. Similarly we have *Bt*-corn, *Bt*-potato, *Bt*-brinjal (eggplant), *Bt*-rice, etc. with their encoded proteins providing insect control.

The first genes encoding the insecticidal *Bt* proteins were cloned in the early 1980s. This paved the way for constructing recombinant bacterial insecticides containing novel combinations of these proteins and to the development of *Bt*-plants. The first *Bt*-cotton plants were developed by the Chinese Academy of Agricultural Sciences (CAAS) in China and by Monsanto Company in the USA in the early 1990s. However, regulatory approval and large scale commercial cultivation of *Bt*-crops which included *Bt*-cotton along with *Bt*-corn and *Bt* potato developed by Monsanto, took place in the USA in 1996. Thus, 1996 marked the beginning of commercialization of transgenic crops.

It can be accomplished by introducing the required *Bt* gene into the desired cotton cultivars by genetic engineering. For example, the cry1Ac gene encoding the insecticidal protein Cry1Ac is first isolated from the soil bacterium, *Bacillus thuringiensis*. The gene is then modified to be more similar to a plant gene and combined with the 35-S promoter (derived from cauliflower mosaic virus) to achieve the expression of Cry1Ac protein in all parts of the plant (i.e., constitutive expression). In order to identify plant cells that contain this new gene, a selectable marker the npt II gene which encodes the enzyme neomycin phosphotransferase II (NPT II) is used. Plant cells containing these new genes are isolated on a medium containing the antibiotic kanamycin. The AAD gene is a bacterial selectable marker 3"(9)-0-aminoglycoside adenyltransferase (AAD) which allows selection of bacteria containing the Cry1Ac plasmid on a medium containing spectinomycin or streptomycin. Cotton transformation is achieved through tissue culture (starting with the American cotton variety Coker 312) technique that allows the soil bacterium, *Agrobacterium tumefaciens*, to transfer the DNA contained on the plasmid with the above sequences. The transformed cotton lines were screened to identity those with these genes inserted into the cotton genome in the most favorable position to achieve the desired insect control and agronomic performance.

Monsanto Co., provided the *Bt*-gene, cry1Ac (Bollgard®), incorporated in the seeds of the American cotton variety called Coker 312, to its Indian licensee Maharashtra Hybrid Seed Company (Mahyco) along with the technology. Through routine back-crossing with the parental lines of Mahyco's proprietary hybrids, the gene was transferred from Çoker 312 into them (American cotton).

It took 6 to 7 years for Mahyco to complete all the regulatory trials related to biosafety and agronomic impact and got the official approval of the first batch of three *Bt*-cotton hybrids in India in March 2002. The chronology of events is summarized in the following:

Chronology of Development and Approval of *Bt*-Cotton in India

March 10, 1995: Department of Biotechnology (*DBT*), Government of India, permits import of 100 g of transgenic Cocker-312 variety containing the *Cry*1Ac gene from the bacterium *Bacillus thuringiensis*.

April 1998: Monsanto-Mahyco tie up. Monsanto gave permission for small trials of *Bt* cotton 100g per trial by Department of Biotechnology (*DBT*).

July 13, 1998: LOI by *DBT* without involving GEAC, 15th MAHYCO agrees to LOI.

July 27, 1998: Impunged permission by *DBT* for trials at 15 locations, in August another 15 locations approved.

November 16, 1998: News broke of trials being carried out by Monsanto on *Bt* cotton without permission of regional governments and without consent of local communities.

November 28, 1998: Thousands of farmers occupied and burned down *Bt* cotton trial fields in Karnataka as part of Operation Cremation Monsanto.

December 2, 1998: about 200 farmers of Andhra Pradesh set fire to the experimental field of the "Bollgard" cotton crop raised by Monsanto in Urgonda village of Warangal district. The Andhara Pradesh government has asked the Mahyco Monsanto Biotech (I) Private Limited to stop forthwith the field trials of Bollgard gene seeds of cotton going on in seven districts in the State.

December 16, 1998: "Monsanto, Quit India" Campaign by various groups. Monsanto used the Andhra Pradesh Rytha Sangha (Federation of all the Farmers Organizations of Andhra Pradesh) through its President Shri Malla Reddy.

January 6, 1999: Vandana Shiva's Research Foundation for Science, Technology and Ecology went to Supreme Court challenging the "illegality" of the field trials authorized by the *DBT*.

January 8, 1999: RCGM expressed satisfaction over the trial results at 40 locations and on April 12th directed MAHYCO to submit applications for trials at 10 locations before MEC.

May 25, 1999: Revised proposal top RCGM submitted to MAHYCO.

June-November, 1999: Permission granted for different field trials. Oct.-Nov. Shiva's group undertake field visits.

May, 2000: Mahyco wrote to GEAC seeking approval for 'release of large scale commercial field trials and hybrid seed production of indigenously developed *Bt* cotton hybrids'.

July, 2000: GEAC cleared for large scale field trial on 85 hectares and seed production on 150 hectares. Mahyco allowed the conduction of large-scale field trials including seed production at 40 sites in six states. The permission was granted based on the "totally confidential" data from the small trials that allowed regulators to infer that *Bt* cotton was "safe". The D*BT* set up Committee to "independently" monitor and evaluate large scale field trials.

October, 2000: RFSTE filed application to amend petition challenging fresh clearance. Geeta Bharathan wrote in Current Science on the controversy that evoked responses from scientists and Mahyco in subsequent issues.

January 5, 2001: A 10 member delegation comprising US judges and Supreme Court Chief Justice A.S. Anand to educate him - and other members of the judiciary - on Biotechnology.

June 18, 2001: An open dialogue held between Monsanto and Greenpeace to discuss *Bt* cotton with scientists, Ministry of Environment representatives and farmers. No data on field trials was presented, though farmers vociferously demanded *Bt* cotton be commercialized. Technical questions and concerns raised by Greenpeace remained unanswered.

June, 2001: PAN AP and others took urgent action at the request of activists both within India and abroad in protesting Mahyco/Monsanto's aggressive attempt to fast-track government approval for *Bt* cotton commercialization. Some 300 emails were received from all over the world by Indian government.

June 21, 2001: Genetic Engineering Approval Committee (GEAC) extended field trials of *Bt* cotton by another year. Mahyco conducted large-scale trials on 100 hectares in seven states.

October, 2001: Mahyco discovered commercial *Bt* cotton farming over thousand hectares in Gujarat. Soure of the cotton is traced back to Navbharat Seeds Pvt. Ltd.

October, 2001: GEAC ordered *Bt* cotton fields to be burnt in Gujarat. No action was taken after farmer's protest order.

November 20, 2001: Gene Campaign filed a case in the Delhi High Court charging the Government with negligence in allowing large-scale field trials to be conducted without appropriate monitoring, regulation and safety precautions.

December 12, 2001: Dr. Manju Sharma, Secretary, DBT wrote an open letter to PM warning of the potentially devastating impact GE cotton seeds could have on farmers in India and the scientific fraud involved in the way the research trials had been conducted and monitored "the biggest scientific scam" to have ever hit India.

January 23, 2002: Dr. Manju Sharma, declared that the latest round of *Bt* cotton trials were satisfactory and that it was up to the GEAC and the Ministry of Environment to decide on a date of commercial release.

February 20, 2002: The Indian Council of Agricultural Research (ICAR) submitted a positive report to the Ministry of Environment on the field trials of *Bt* cotton. It was then expected that the Genetic Engineering and Approval Committee (GEAC) of the environment ministry would approve commercial use of *Bt* cotton within a month.

March 25, 2002: Approval was given to *Bt* cotton by GEAC.

May 2002: Piyush Patel of Navbharat seeds was finally arrested. Monstanto was ready to release *Bt*-gene *Cry*2x also known as *Cry*2Ab for pest resistance in cotton. Clearance in India was expected by 2005. Rasi Seeds got nod to conduct *Bt* cotton trials.

June 2002: Chinese environmental protection agency - Greenpeace report showed fall in diversity index in *Bt* cotton fields, other pests increased. Several groups in India asked for ban on *Bt* cotton as a result through press release in Banglore.

August 8, 2002: Suman Sahai of gene campaign demanded release of *Bt* cotton data.

August 9, 2002: Government banned sale of *Bt* cotton seeds temporarily.

August 24, 2002: There were reports from Khargone of 100% failure of *Bt* crops.

Prospects of *Bt* Cotton in India

Bt cotton in India took six years of experimentation, during which time agronomic, environmental, and biosafety data was generated and reported. The trials conducted prior to commercialization clearly established the superior performance of *Bt* cotton, as demonstrated by increased yields and reduction in application of pesticides (Table 5). Transgenic technology is suitable fro the India farmer despite small farm holdings. The area under *Bt* cotton is projected to increase rapidly in the coming years.

India grows all the four species of cultivated cotton. Barring the state of Punjab, Haryana, Rajasthan, cotton is predominantly cultivated under dry land conditions throughout the country. Today almost 67% area in cotton is plasticizing dry land cultivation and as far as Maharashtra is concerned, area under irrigated cotton is less than 5% of the total cotton cultivated area; nearly 35-40% area grows hybrid cotton. State wise area and production in cotton in the past five years is given in Table 6.

Table 5: Timeline summary for regulatory processes leading to commercial release of *Bt* cotton in India

Years	*Studies undertaken*	*Government of India oversight committees*
1995-96	Application and permit for importation of *Bt* cotton seed containing the *Cry1Ac* gene	D*BT*
1996-2000	Greenhouse breeding for integration of the *Cry1Ac* gene into Indian germplasm, seed purification and stock increase	D*BT*
1996-2000	Limited field studies for potential of pollen escape, aggressiveness, and persistence	RCGM (D*BT*)
1998-2000	Biochemical and toxicology studies	RCGM (D*BT*) GEAC
1998-2000	Multilocation field trials : agronomic and entomology performance of first generation *Bt* cotton hybrids, Agriculture Universities	RCGM (D*BT*), MEC
2000-2001	Soil rhizosphere evaluations and protein expression analyses from multilocation field trials	RCGM (D*BT*), GEAC
2001	Advanced stage multilocation field performance trials of first generation *Bt* cotton hybrids, conducted by ICAR	GEAC, ICAR, D*BT*, MEC
2002	Submission of final biosafety, environmental safety, gene efficacy and performance documentation to GEAC; commercial release of first generation *Bt* cotton hybrids by GEAC	GEAC
2002-ongoing	Continued field performance trials of second generation *Bt* cotton hybrids for regulator approval	RCGM (D*BT*), GEAC, ICAR, MEC

[a]D*BT*=Department of Biotechnology; GEAC=Genetic Engineering Approval Committee; RCGM=Review Committee for Genetic Modification (constituted by D*BT*); ICAR=Indian Council of Agricultural Research; MEC=Monitoring & Evaluation Committee (constituted by GEAC and RCGM)

Table 6: State wise area and production of cotton during 1997-2001

State	*1997-98*		*1998-99*		*199-00*		*200-01*		*2000-01*	
	A	*P*	*A*	*P*	*A*	*P*	*A*	*P*	*A*	*P*
Punjab	7.02	7.50	5.47	5.50	4.75	9.50	5.5	11.5	6.0	9.0
Haryana	6.55	9.00	5.87	7.50	5.10	11.0	5.8	11.5	6.1	6.5
Rajasthan	5.80	11.5	6.38	12.0	4.64	12.5	4.82	11.5	3.47	6.5
Gujarat	14.6	42.0	16.97	45.0	15.16	35.0	15.78	27.0	16.87	34.0
Maharashtra	31.0	20.5	31.99	25.0	32.53	38.0	27.93	24.0	29.80	34.0
Madhya Pr.	5.48	23.0	5.32	20.0	5.41	15.0	5.54	17.5	6.23	15.0
Andhra Pr.	8.50	24.75	10.03	25.0	9.09	23.0	8.87	26.0	10.02	27.0
Karnataka	6.00	7.50	6.09	8.50	5.29	8.00	5.32	9.00	5.11	8.00
Tamilnadu	2.65	5.50	2.20	5.50	2.50	5.50	1.33	5.50	1.60	5.50
Others	0.50	1.00	0.80	1.25	0.81	1.50	0.53	1.50	0.73	1.50

A: Area in lakh hectares
Source: Project Coordinator's Report
P = Production in lakh bales
All India Coordinated Cotton Improvement

Bt cotton-Approval for Commercial Cultivation

The Genetic Engineering Approval Committee (GEAC), in its 32nd meeting, held on March 26, 2002, made the landmark decision of approving cultivation of *Bt* cotton in India; three hybrids (MECH 162 *Bt*, MECH 184 *Bt*, and MECH 12 *Bt*) were approved for cultivation with the stipulation that certain conditions be met (Table 7). These hybrids are high yielding and produce medium-long-to long-staple fiber. The GEAC also approved production of seed; subsequently, seed for cultivating 100,000 acres (40,485 ha) was available for planting by Indian farmers. The seed was distributed in the state of Maharashtra, Madhya Pradesh, Karnataka, Andhra Pradesh, Gujarat, and Tamil Nadu in the kharif 2002 planting season. The area planted to *Bt* cotton was 14746, 2220, 3908, 5608, 8854 and 2702 hectares, respectively in the states mentioned above. Rasi Seeds had been allowed to conduct trials of RCH 118 *Bt* and RCH 559 *Bt* in Central India, RCH 368 *Bt* in South India and RCH 317 *Bt* in North India. Ankur Seeds also conducted trials of Ankur 651 *Bt* and Ankur 2534 *Bt* in North India and Ankur 651 *Bt* and Ankur 09 *Bt* in Central India. Mahyco went ahead for trials of MRC 6301 *Bt* and MRC 6160 *Bt* in Central India and MRC 6301 *Bt* and MRC 6322 *Bt* in South India.

Table 7: Commercial cultivation of *Bt* cotton hybrids in India, 2002 (hectares)

State	*MECH-12*	*MECH-162*	*MECH-184*	*Total*
Maharashtra	112	9300	5334	14746
Madhya Pradesh	60	404	1756	2220
Karnataka	–	3828	80	3908
Andhra Pradesh	44	5564	–	5608
Guajrat	76	4136	4642	8854
Tamil Nadu	–	2042	660	2702
Total	**292**	**25274**	**12472**	**38038**

Bt spray versus *Bt* plants

Bt spray is not preferred over *Bt* plants. The major reasons are as follows:

- To spray, *Bt* formulation has to be purchased separately whereas in transgenic *Bt*-plants, *Bt* is incorporated in the seed itself. During spray, *Bt* is applied externally on the plants whereas in transgenic plants, *Bt* protein is produced within the plant.
- When sprayed, it is difficult to get uniform coverage of the entire plant/crop and also there is wastage of spray fluid whereas in transgenic plants, *Bt* protein is present in all the plants, in all the parts and all the time.
- When sprayed, *Bt* may get rapidly degraded if exposed directly to solar radiation or washed off by rain whereas in transgenic plants, as *Bt* is present within the plant, it is not greatly affected by external conditions.
- One needs to monitor the pests closely so as to manage effective sprays at appropriate time, whereas in transgenic plants, since the control measure is in-built, the insecticidal activity is always present, providing protection day and night. Spray formulations can be applied, as and when required, on any crop which may be considered as an advantage whereas in the case of transgenic, the advantage is restricted only to those crops that have been transformed into *Bt*-plants.

Bt-cotton plants as such were not introduced in India. It should be clearly understood that *Bt* is an insecticidal trait, not a cultivar. Mahyco has obtained license from Monsanto to use their Bollgard® *Bt*-gene to develop *Bt*-cotton in India. This gene has been deployed by Mahyco in the cotton hybrids that have been developed by them. In fact, these hybrids which bear the Indian germplasm, are already under commercial cultivation in India even before the incorporation of *Bt*-gene in them. The *Bt*-gene has fortified such hybrids to express a particular protein that can kill bollworm larvae and derive protection. The gene can be introduced into any desired cotton cultivars as has been accomplished by Mahyco and subsequently by Rasi, Ankur, Nuziveedu and several other seed companies. The *Bt*-gene has empowered such hybrids to express a particular protein that can kill bollworm larvae and derive protection. The gene can be introduced into any desired cotton cultivars as has been accomplished by Mahyco and subsequently by Rasi, Ankur, Nuziveedu and several other seed companies. It is alleged that the 'American Bollworm' was introduced India from America and now the *Bt* cotton technology, also from America, is offered as a remedy against the same pest. This accusation defies facts.

The so-called 'American Bollworm', *Helicoverpa armigera,* does not occur in America at all! The distribution of this species is limited only to the Old World (India, Pakistan, China, Australia, Africa, etc). Therefore, it is more appropriate to call *H. armigera* the 'Old World. *H. armigera* was primarily a pest of tomato (it was popularly called the 'Tomato Fruit Borer') in India before it switched to cotton, chickpea, pigeon pea and a large number of other crops (more than 80 plant species). Two closely related species, *Helicoverpa zea* (cotton bollworm) and *Heliothis virescens* (tobacco budworm) have been serious pests of cotton in the USA for a long time. When *H. armigera* was first noticed on cotton in India in the early 1970s, perhaps it was mistaken as an accidental introduction of the American species and hence the misnomer 'American Bollworm.' The fact that India had earlier introduced American cotton *(Gossypium hirsutum)* varieties to enlarge the genetic base of cotton in our country further added to this suspicion. However, this confusion was cleared long ago and it is a well-established fact that *H. armigera* is a native species. Therefore, the above allegation is not valid. It is important to note that *Bt*-cotton is effective against the bollworm species found in the USA as well as those found in India. Thus, this technology is useful to us.

Bt-cotton has several advantages. Some of these are:

- *Bt*-technology for control of bollworms is made available in the seed itself. Farmers have to just sow the *Bt*-cotton seeds as they do with conventional seeds. The resulting plants have the in-built ability to produce *Bt*-protein within their body and defend themselves from bollworms. No extra efforts or equipment are needed to utilize this technology.
- *Bt* protein shows constitutive expression and is expressed in all parts of the plant providing bollworm control day and night, almost throughout the plant life. There is no need to monitor the bollworms to initiate control measures.
- The newly hatched larvae feeding on any part of the plant will ingest *Bt*-protein and die within one or two days, thereby preventing any potential serious damage to the crop.

- *Bt*-proteins, being lepidopteran specific, affect only the bollworms and are safe to biological control agents and other non-target beneficial organisms, higher animals and plants.
- *Bt*-cotton is compatible with other control measures such as biological control, pheromones, botanical insecticides and also chemicals that are recommended for Integrated Pest Management. In fact, *Bt*-cotton can serve as a major component of IPM in cotton crops.
- *Bt*-cotton helps to avoid or minimize chemical sprays, thus contributing to cleaner environment and conservation of biological control agents and biodiversity.
- *Bt*-cotton offers protection from bollworms right from the early days of the crop, leading to a healthy crop, better boll retention, greater harvest and more profit.
- The *Bt*-farmers experience a far lower tension and are certainly better off than the earlier scenario of "spray & pray."

It is important to know that *Bt*-cotton offers protection only against bollworms, not sucking pests and other non-lepidopteran pests. Therefore, separate control measures have to be taken against such pests as and when required. Another factor is that the expression of *Bt* protein in cotton plants may decline after 90-100 days, calling for supplementary control measures on rare occasions if warranted. It is also a fact that the expenditure on the *Bt*-trait cannot be de-linked from the seeds even if there is no serious bollworm infestation. *Bt* should be always treated as an insurance against bollworms. Further, enforcing 'refuge' crop on farmers poses several practical challenges. It is always necessary to understand clearly the scope of a technology for its proper utilization.

Besides bollworms, *Bt*-cotton is effective against secondary lepidopteran pests like the semilooper (*Anomis flava*) and leaf roller (*Sylepta derogata*). However, it is not designed to offer protection against sucking pests (whiteflies, aphids, thrips, jassids) and other non-lepidopteran pests, or against diseases and adverse environmental conditions like drought, cold, salinity etc.

However, by introducing multiple genes with different traits, scientists have already made some progress in solving more than one problem in a single crop. For example, the same cotton cultivar resistant to bollworms as well as certain herbicide has already been developed. Similarly, corn plants with combined resistance to European corn borer, rootworm and herbicide have also been successfully developed. These are already under commercial cultivation in other countries. More research is underway. However, it may not be possible to answer all the problems with any single technology. We should try to integrate and take advantage of several technologies. Science should not discriminate between one technology and the other as long as it is safe and beneficial.

There need not be such a fear whether the non-*Bt* farmers in the neighborhood of *Bt*-cotton will be affected? Will their crops suffer more pest damage? *Bt*-cotton and non-*Bt* cotton plants look alike with very little phenotypic differences between them. The bollworm moths lay their eggs on both the cottons without discrimination. However, the subsequent results would be different. The tiny larvae hatching from the eggs laid on *Bt*-cotton will perish within one or two days as they ingest *Bt*-protein while feeding on the plant tissues. On the other hand, the larvae hatching on non-*Bt*-cotton continue to feed and grow and cause serious damage if appropriate control measures are not taken. Merely the presence of *Bt*-cotton in the vicinity, therefore, does not influence the bollworm populations in the non-*Bt* crops. However, if *Bt*-cotton becomes very broadly planted in India, it is possible that it will lead to a decrease in bollworm populations, thereby reducing damage to non-*Bt* crops as well.

It may be stated that *Bt*-cotton is most effective against young larvae that are in the first or second instar. They get killed within one or two days after ingesting the *Bt* protein while feeding on the plant. The older larvae may not die, but they suffer a setback in their overall health and vigor. Such sick larvae feed far less. Since *Bt* protein is constitutively expressed in *Bt*-cotton and the neonates ingest the protein as soon as they take the first bite of the plant, the chances of larvae escaping and surviving beyond first or second instar are rare.

The chances of older larvae migrating from the adjacent non-*Bt* fields and damaging the *Bt*-plants have also been examined. The larvae of *Helicoverpa armigera* generally have a tendency to feed on the same or a few adjacent plants. Plot-to-plot en mass larval movement generally does not occur. Only the adult moths are highly migrant and known to travel long distances. In the case of Spotted Bollworm (*Earias vittella*) and Spiny Bollworm (*E. insulana*) and more so with Pink Bollworm (*Pectinophora gossypiella*), the movement is even more limited. Therefore, one need not be unduly worried about the presence of non-*Bt* crops in the adjacent fields.

The expression of *Bt* protein is more in tender leaves as compared to squares, bolls, flowers and pollen. The optimum expression in leaves is most critical as a greater number of bollworm eggs are laid on leaves and the newly hatched larvae, while feeding on chlorophyll in the leaves, ingest *Bt* protein and perish. High protein expression in fruiting parts will add to further efficacy. The extreme temperature and other environmental factors may also play a role. The protein expression remains more or less consistent up to about 100 days, but gradually declines as the plants age. Keeping such factors in mind, an 'Optimum Dose' strategy is deployed in *Bt*-plants so that they express far greater quantity of proteins (>25 times) than actually required for causing larval mortality. Thus, in the later stages of crop growth, even if the protein level has dropped, the remaining protein is adequate to bring about larval mortality, thereby ensuring satisfactory bollworm control throughout the season. Another objective of this strategy is to ensure that all susceptible larvae and even those which are heterozygous for resistance (i.e., likely to develop resistance with constant exposure to *Bt*-crops) are killed so that there is a very low probability of insects gaining resistance to *Bt* protein. The bollworm populations from different geographical regions as well as from the same region vary considerably in their susceptibility to *Bt* proteins and occasionally, especially later in the growing season, some larvae may survive, although at a greatly reduced level. Necessary alternative control measures have to be taken on such occasions if warranted. *Bt*-cotton should not be treated as a silver bullet.

Bt protein expression varies with different cultivars depending upon their genetic background. Therefore, it is necessary to choose the right hybrids to derive optimum benefit from this technology.

The mode of feeding by sucking pests and bollworms is different. Sucking pests like aphids, whiteflies, thrips and jassids feed by sucking the sap whereas bollworms feed by chewing the plant parts. They do not compete for the same source of food within the same plant. Therefore, control of bollworms need not necessarily result in encouraging sucking pests, but certain environmental factors like dry spells may do so. On the other hand, due to very limited or no application of chemical pesticides for bollworm control in *Bt*-cotton, the natural enemies of various pests are conserved and these contribute to keeping them in check. However, to get the full benefit of *Bt*-technology, it is always better to be watchful and take appropriate control measures to protect the crop from sucking and other pests as also from diseases and environmental factors that are not controlled by *Bt*-cotton. The ideal approach is integrated pest management (IPM) with *Bt*-cotton as the major thrust.

There is unnecessary fear that the health of humans and animals exposed to *Bt* cry proteins may be affected in some way. DNA (is present in all living organisms such as plants, animals and microorganisms and is eaten by human beings and animals in one form or the other with every meal. Most of it is broken down into more basic molecules during the digestive process while the remaining amount is either absorbed into the blood stream or excreted. Nevertheless, tests are carried out with each newly introduced DNA into the plant. This applies to cry proteins as well. A large number of experiments, as prescribed by the regulatory authorities in each country, have been carried out to examine the safety of cry proteins. Experimental animals like mice, rats, rabbits and sheep fed with unusually high doses (500, 1,000 and 4,300 mg/kg body weight) of cry protein showed no acute toxic effect on their health. These animals were found to be equivalent to those not fed with cry proteins with respect to body weight, food consumption and other respects. The U.S. Environmental Protection Agency (EPA) has concluded "toxicity and infectivity risks of cry proteins to non-target organisms like avian, freshwater fish, freshwater aquatic invertebrates, estuarine and marine animals, arthropod predators/parasitoids, honey bees, annelids and mammalian wildlife will be minimal to non-existent at the recommended rates of registered *B. thuringiensis* active ingredients." This provides confidence that cry proteins produced in *Bt*-crops would pose little risk to non-target organisms.

Biosafety and risk assessment studies have been carried out on *Bt*-cotton in India in the laboratories and fields designated by RCGM/GEAC as a part of the regulatory requirements, prior to its approval. Feed-safety studies using *Bt*-cotton seed-meal were conducted on goats, buffaloes, cows, rabbits, birds and fish by competent scientists at various institutions like the Industrial Toxicological Research Centre, Lucknow; National Dairy Research Institute, Karnal; Central Institute of Fisheries Education, Mumbai; Central Avian Research Institute, Izatnagar; National Institute of Nutrition, Hyderabad; and G. B. Pant University of Agriculture and Technology, Pantnagar. The results revealed that the animals fed with *Bt*-cotton seed-meal showed no ill-effects and was comparable to control animals in the various tests. In other words, *Bt*-cotton seed meal was substantially equivalent to its non-*Bt* counterpart. Based on such scientific data, the regulatory authorities considered *Bt*-cotton as safe.

Studies have been conducted to determine the amount of *Bt*-protein leached by roots as also from other plant parts incorporated in the soil and its effect on soil rhizosphere and non-rhizosphere microflora, soil Collembola, earthworms, etc. It was found that there was no adverse effect. In fact, there was no difference between the soils obtained from the *Bt* and non-*Bt* plots in this respect. It was also found that *Bt* insecticidal proteins are readily susceptible to metabolic, microbial and abiotic degradation once they are ingested or excreted into soil. The half-life of the Cry1Ac protein in plant tissues has been found to be to be a maximum of 41 days. Therefore, it cannot bio-accumulate causing delayed effects.

Except for the presence of *Bt*-gene, which has been introduced for the specific purpose of controlling the target pests, *Bt*-plants are substantially equivalent to their non-*Bt* counterparts in terms of protein, carbohydrate, ash and moisture content as well as other acid contents in the foliage. In other words, the impact of *Bt*-plants on other plants or non-target organisms is no different from that of their traditional counterparts. On the other hand, due to reduced application of chemical pesticides, the populations of non-target beneficial insects like ladybird beetles, green lacewings and other biological control agents as well as honey bees have been reported to be greater in *Bt*-fields thereby sustaining or enhancing biodiversity in the crop ecosystem.

The monarch butterfly and silkworm are both lepidopteran insects and it will not be a surprise if their young larvae die if they are made to ingest *Bt*-protein, as has been shown by certain laboratory experiments. However, the field realities are different. The habitats and food plants of the monarch butterfly (it feeds on milkweed) and silkworm (it feeds on mulberry) are different and there is very little chance for them to come in contact with adequate quantity of *Bt* protein from the pollen or other parts of *Bt*-corn or *Bt*-cotton in nature. Thus the threat to monarch butterfly, perceived on the basis of reported mortality of caterpillars as a result of force-feeding them on the milkweed leaves artificially coated with *Bt*-corn pollen in the laboratory at Cornell University, USA, in 1999, was highly exaggerated in the media and became one of the most controversial issues with the activists in the USA demanding a moratorium on further planting of *Bt*-corn. The USDA, on its part, has constituted several teams of scientists to investigate various aspects and reassess the potential risks to monarch butterfly by *Bt*-corn. Based on the results of their comprehensive studies which were reviewed by EPA (Environmental Protection Agency), it was concluded that *Bt*-corn is not a threat to monarch butterfly populations. A number of other independent studies also supported this view. Perhaps the results would be on the same lines with silkworm in India. The monarch story highlighted the need for drawing realistic conclusions based on detailed scientific investigations rather than jumping to false conclusions and resorting to agitations based on preliminary laboratory studies.

The possibility of pollen being transferred to weeds, resulting in 'Super Weeds'/ or 'Gene Pollution' or 'Gene Contamination' has also been examined. It may be stated that cotton has only one close weed relative in India. It is *Gossypium stocksii* and is found in northern Gujarat where incidentally cotton is not cultivated. Besides, there is no record of bollworms feeding on this weed and also there is no other major lepidopteran common to both cotton and this weed. Further, the cotton pollen is heavy and cannot move beyond a few metres away from cotton fields. Therefore, the possibility of gene transfer and the development of 'Super Weed' is a remote possibility. Even in other countries and with other *Bt*-crops, there is no evidence that 'Super Weeds' have ever enveloped over the past decade. In India, we have two types of cotton: the American cotton (*Gossypium hirsutum* and *G. barbadense*) and the 'Desi' (local) cotton (*Gossypium herbaceum* and *G. arboreum*). All the *Bt* cotton hybrids developed in India are *G. hirsutum*. The American cotton is tetraploid in genetic makeup whereas the 'Desi' cotton is diploid. There is no reproductive compatibility between the two. Even if cross pollination occurs between the tetraploid and diploid cotton plants, the zygotic embryo will not develop. The terms such as 'Gene Pollution' and 'Gene Contamination' are mere jargons in this case.

A number of short-term and long-term potential consequences related to bio-safety and environmental safety, as perceived by the experts, are considered and investigated through scientific studies before any biotech product is approved by the regulatory authorities. Any genuine new concerns will also be handled thus. A technology is generally developed to overcome the difficult problems faced currently, for example cotton bollworms, while keeping the safety and benefits for future as well. It is not possible to foresee everything beyond a certain time frame. We should follow the precautionary principle and minimize the risks as far as possible. No technology is absolutely free from risk. For fear of speculated risks, if we close the doors for new technologies, we cannot make progress and solve our problems.

Bt-technology has already lasted for a decade in several countries with no negative consequences whatsoever and is becoming more refined and popular. Furthermore, the use of this technology has led to significant reduction in chemical sprays with the concomitant benefits to the environment and farmer health as well as income. It is the responsibility of the scientists and regulators to sustain these benefits. It is imperative that planting non-*Bt* cotton as border rows to *Bt*-cotton is currently recommended as a step towards delaying or preventing the development of resistance *to Bt* protein by bollworms. It is called 'Refuge.' This is practiced in other countries like the USA and Australia and has also been recommended by Govt. of India. The 'refuge' crop could be 20% of the *Bt*-cotton area with the intervention of plant protection measures when required or 5% of the area without providing any chemical protection. However, studies are underway to determine whether alternative host crops of bollworms are common enough in Indian cotton-growing areas to fill this 'Refuge' function without the need for a structured non-*Bt* cotton 'Refuge'.

The refuge strategy is designed to ensure the availability of the *Bt* susceptible pest population so as to mate with the *Bt*-resistant population, should they arise. It is known that a great majority of larvae that feed on 'refuge' (i.e., non-*Bt*) crop, in the absence of adequate control measure, would complete their development and emerge as moths. Their number will be large and these remain as susceptible strain (SS). On the other hand, almost the entire population of newly hatched larvae that feed on *Bt*-plants would perish with only a very few developing into moths. These are the moths carrying the resistance gene (RS). Because of the overwhelming population of the SS moths in the vicinity, coming from the refuge (either from structured or natural) crops, and a very scant population of RS moths, the possibility of SS mating with RS is far greater than RS with RS. Resistance in this case being recessive, the pairing of SS and RS would result in a susceptible progeny. Thus, refuge helps in maintaining susceptible population.

Theoretically, the refuge does not have to be cotton but could be any plant that supports *Bt*-susceptible bollworms. Insect resistance management (IRM) is very important to conserve the *Bt*-technology for long-term benefits and, therefore, refuge cannot be considered as a burden or waste. However, in India, *Helicoverpa armigera*, the most destructive bollworm, has a large number of alternative host crops (more than 80 plant species). Of these, crops like pigeonpea, chickpea, sunflower, corn, chili, tomato and okra (bhendi) are heavily infested and support very high populations of eggs and larvae of *H. armigera*. Some of these crops are cultivated in close proximity to each other and around the same time as cotton in several parts of central and southern India. In fact, studies have revealed that pigeon pea and chickpea are more preferred than cotton by *H. armigera*. These crops can act as natural refuge. Further, in India, the area presently occupied by *Bt*-cotton constitutes only around 42% of the total cotton area. Although it is expected to increase in the coming years, a large area of non-*Bt* cotton crop would still be available which can also serve as natural refuge. Under the circumstances, there appears to no need for structured refuge crop in our country. In fact, in China, in view of the availability of alternative host plants, growing refuge is not mandatory. Perhaps this needs to be reviewed by the regulators in our country also, but until such time, the prevailing recommendation should be followed.

It is known that *Bt* causes quick mortality of only the early instars, not the grown up larvae. Therefore, it is advisable not to use the non-*Bt* cotton for gap filling as the larvae may complete their early days on such plants and then move over to the adjacent *Bt*-plants and damage them. The cost of *Bt* seed is small as compared to the benefits.

Besides 'refuge', several other proactive insect resistance management (IRM) strategies have been developed and practiced. These include the 'Optimum Dose' strategy wherein the plant is empowered to express *Bt* protein at a much higher dosage (>25 times) than normally required for larval mortality. The objective is to bring about maximum mortality of pest larvae, giving minimum chance for the potential resistant larvae to survive. Another IRM strategy is 'Gene Stacking' or 'Gene Pyramiding.' In this approach, more than one gene, each having a different mode of action and binding to a different receptor in the insect gut, is incorporated into the same plant against the same pests. The idea is that even if the pest develops resistance to one gene, it would succumb to the other as the possibility of developing resistance to both the genes simultaneously is very remote. Monsanto has deployed 'gene stacking' in Bollgard® II (BG II) which has two *Bt* genes, namely cry1Ac and cry2Ab2, in the same cotton plant for control of bollworms. BG II has been approved for commercial cultivation in Australia and USA in 2002. MAHYCO has completed the regulatory trials with BG II in India and it has been approved by GEAC in May 2006. Similarly, the *Bt*-cotton developed by Nath Seeds has two genes, namely cry1Ab and cry1Ac with GFM event and that by J.K Seeds has only one gene, cry1Ac, but with an event (Event 1) different from Monsanto's. These were also approved in May 2006. Such diversity in genes or events also helps in resistant management. Various IRM strategies may be deployed so as to prevent or delay the development of pest resistance.

Since its first commercialization in the USA in 1996, *Bt*-cotton has been cultivated on millions of hectares in 9 countries. During the last 11 years, although the bollworms might have passed through more than 100 generations (bollworms can complete a generation in 3-4 weeks under normal conditions), there has not been any scientific evidence indicating field resistance to *Bt*-cotton by any of the bollworm species in any country. This shows that it is not easy for the pest to develop resistance to *Bt* protein expressed *in planta*. It also suggests that the prevailing IRM strategies are working well. Some of the predictions about the resistance have come from laboratory 'selection' experiments where the bollworms were continuously fed sub-lethal doses of *Bt* proteins in order to induce resistance. Such populations were used for investigating the mechanisms of resistance. Although the results did not match with the field realities these studies are helpful so that one can be always on the alert as pest resistance is one of the genuine and important concerns.

One cannot expect any technology to last for ever. We should try to use scientific means to extend its durability as far as possible. One should also be on the look out for new insecticidal proteins, from *Bt* and other sources, and try to rotate these crops once in a few years. Fearing resistance, we cannot give up any useful technology on hand.

The duly approved *Bt*-cotton trials were conducted by the applicants as per the guidelines and with the full knowledge of the government regulators and the data were also submitted to them. These were not expected to be publicized. Regarding involving other agencies, be it ICAR, IARI, Universities, State Dept of Agriculture or any other, it is the prerogative of the regulatory authorities. In fact, senior scientists from these and also several other organizations were/are included as experts in several regulatory committees such as IBSC, RCGM and GEAC and they were/are involved in one way or the other at some stage during the laboratory and field trials as well as during the approval process of *Bt*-cotton.

Field trials conducted by public institutions like ICAR and by private seed companies like Mahyco, Rasi, Ankur, Nuziveedu and several others, before as well as after regulatory approvals, have clearly indicated that *Bt*-cotton hybrids provided satisfactory control of bollworms in all the locations and seasons, bringing profit to farmers. However, some of the NGOs who made parallel observations, claimed that *Bt*-cotton has failed to control bollworms and did not bring any benefits to farmers. Therefore, the final decision is best left to farmers as they will not adopt a technology if it is not beneficial to them. The fact remains that there is an ever-increasing demand for *Bt* cotton seeds in India.

Bt-cotton has been developed for the specific purpose of controlling bollworms. Therefore, it can be rightfully blamed if it has failed to provide protection against bollworms and the reason should be investigated. However, in some parts of Andhra Pradesh, Karnataka etc., cotton crops, both *Bt* and non-*Bt*, have sometimes suffered due to drought, para-wilt or some other biotic or abiotic stresses. Such failures have often been attributed to *Bt*-cotton which is not justified. The various hybrids that have been developed by different seed companies as well as by the same company vary in their attributes, with or without *Bt*, and are generally meant for different agro climatic regions. *Bt*-trait has no direct influence on the agronomic performance of a cultivar. It brings additional value to a cultivar by providing bollworm control. The yield potential of various hybrids, their suitability to different agro-climatic regions and *Bt*-trait are separate entities. If a right combination is made, they compliment each other. The unsatisfactory performances are often due to wrong choice of hybrids.

Farmers' response has been overwhelming. The area under *Bt*-cotton in 2002, the first year of introduction, was about 29,000 ha (72,000 acres). It increased significantly from year to year to reach 3.8 m ha (9.4 m or 94 lakh acres) in 9 cotton growing states in 2006 – a remarkable growth rate in a short period of five years. Similarly, the number of farmers who adopted this technology also increased from a few thousand in 2002 to about 2.3 m (23 lakhs) in 2006 (see table 8). Such growths clearly reflect the farmers' confidence in this technology. Similar trends have been recorded in other countries also. *Bt*-cotton is presently cultivated in 9 states, and the area has consistently increased from year to year in almost every state. The figures for 2005 and 2006 are given in the table to exemplify this.

Table 8: Area under *Bt*-cotton in India, 2002-2006 (In 6 to 9 states)

Year	*Area in Hectares*	*Area in Acres*	*No. of Bt farmers*
2002	29,000	72,000	-
2003	86,000	2,13,000	75,000
2004	5,53,000	13,66,000	3,50,000
2005	12,67,000	31,31,000	10,00,000
2006*	38,00,000	94,00,000	23,00,000

Table 9: State-wise adoption of *Bt*-cotton in India, 2005 & 2006

State	*2005**	*2006***
Maharastra	6,23,000	18,40,000
Gujarat	145,000	4,70,000
Madhya Pradesh	1,43,000	3,10,000
Andhra Pradesh	2,27,000	8,30,000
Karnataka	29,000	85,000
Tamilnadu	19,000	45,000
Punjab, Haryana, Rajasthan	81,000	2,20,000
TOTAL, hectares	**12,67,000**	**38,00,000**

Experience in other parts of the world

Bt cotton has already been commercialized in six countries: The United State (1996), Australia (1997), South Africa (1997), Argentina (1998), Mexico (1996), China (1998) and Indonesia (2000). *Bt* cotton is being extensively field tested under permit in Brazil, Colombia, Thailand, and Zambia. Globally, *Bt* cotton was planted on one million acres (405,000 ha) in 2000. Due to its excellent performance, the benefits to the farmer of *Bt* cotton have been consistently superior for the six years of commercialization.

Bt cotton seeds were first tested in India for germination, vigor and insect efficacy. Other experiments were conducted to confirm the environmental safety of *Bt* cotton, including tests of gene flow, persistence of the transformed plants, weediness characteristics, crossability of the transgenic pollen with the non transgenic relative and near relatives, effect of the pollen on insects and non target organisms, and changes in the soil microbial flora (Table 5). These studies were conducted under the unique environmental conditions of India and with the *Bt* trait in Indian germplasms. Studies of the molecular characterization and stability of the Cry1Ac gene were also carried out, as well as feeding studies and tests of food and feed safety, toxicity, and allergenicity.

As of 2006, genetically modified cotton was cultivated by 9 countries on 13.4 million hectares. Of this, 8.0 m ha had only *Bt* gene(s) (including 3.8 m ha in India), 4.0 m ha had *Bt* stacked with herbicide tolerance while another 1.4 m ha had only herbicide tolerance. The USA, India, China, Argentina, Australia, Mexico, South Africa, Colombia and Brazil were the countries that grew *Bt*-cotton. The area occupied by the genetically modified (GM) cotton in these countries from 2004 to 2006 is indicated in the table 10. The reduction in area in 2006 over 2005 in Mexico was due to seed import constraints and in Australia due to reduction of total plantings of cotton owing to drought. In all other countries there was a significant increase in area with India recording an unprecedented 192% growth over the previous years.

Table 10: Global area under GM-Cotton*, 2004 to 2006 (In million hectares)

Country / year of introduction	*2004*	*2005*	*2006*
United States (1996)	4.2	4.6	5.3
India (2002)	0.5	1.3	3.8
China (1997)	3.7	3.3	3.5
Argentina (1998)	0.02	0.07	0.36
Australia (1996)	0.2	0.3	0.18
Brazil (2005)	-	-	0.02
Mexico (1996)	0.07	0.12	0.06
Columbia (2002)	<0.02	<0.1	0.03
South Africa 1998)	0.02	0.03	0.12
Total, million ha.	**9.0**	**9.8**	**13.4**

More countries and farmers are expected to adopt this technology in the coming years.

Bt cotton – Agronomic Benefits

Trials conducted in several locations in 1998-99, 1999/2000, 2000/01 and 2001/02 demonstrated the following agronomic benefits of *Bt* cotton:

- good control of bollworm species in different growing areas;
- significantly higher yield and boll retention (compared to control or non-*Bt* cotton);
- reduction in expense of insecticide application;
- additional revenue (Rs. 2,500-4,000 per acre) in farm income (compared to non-*Bt* cotton; and
- no adverse effect on non-target insets or adjacent non-*Bt* cotton crops.

The survey was conducted in selected locations in Maharashtra and Andhra Pradesh, which are two of the six states that have been granted permission to commercially cultivated *Bt* cotton. The survey included a total of 100 farming families selected by random sampling from those who had chosen to grow *Bt* cotton simultaneously. Of the total of 100 families surveyed, 25 were from Maharashtra and 75 from Andhra Pradesh. Scientist from the Agricultural University in Hyderabad accompanied the Gene campaign researchers.

At Ukkunda village in Ranebennur taluk in the traditionally male-bastioned cotton belt Haveri district, a group of twenty-odd cotton farmers, faces crusted with the vicissitudes of working their lands. Well aware of the controversies surrounding it, 37-year-old B.G. Belekere and 48-year-old V.S. Patil, both '*Bt* farmers' had managed 10-12 quintals per acre from 2002-2005, where previously local hybrids such as DCH 32 gave them 5-6 quintals per acre. Seed for *Bt* were expensive too in comparison to non-*Bt*. The non-*Bt* variety is long-stapled, commanding good demand and price in the market, compared to the *Bt* varieties, which are short-stapled. The long-stapled variety is said to be tougher and weaves better. Farmer Belekere in Ukkunda village feels that *Bt* cotton does well with larger farmers who have more resources and more lands for other crops.

Indeed, Beleker's astute observation is a telling comment on the trial-and-error method of *Bt* cotton being played out in the fields and lives of poor farmers across India. With *Bt* varieties being heavily dependent on intensive use of water, smallholder farmers using their lands for growing *Bt*, spurious or authentic, have no alternative resource if their crops fail, as has happened in AP. Whereas earlier 'mixed' *Bt* seeds (*Bt* seeds mixed with regular, non-*Bt* hybrids) gave good bollworm-resistant results in Gujarat, agriculturalists point out that the mixing up of *Bt* and non-*Bt* or other forms of spurious seeds produce poor yields, as Vidarbha's farmers have found (Maharashtra).

Downs and ups up to 2006 in India

A 2002 study of *Bt* cotton in the Warangal district of Andhra Pradesh found a 35% reduction in the total yield of *Bt* cotton with a net loss of Rs. 1,295, compared to a net profit of Rs. 5,368 for non-*Bt* cotton. *Bt* cotton yield was 50% lower than that promised by Monsanto. Bollworms were predominant on both *Bt* and non-*Bt* crops, showing that *Bt* cotton was ineffective against its target pest.

In 2003, there was 30% more rainfall than in 2002, and a new *Bt* hybrid compared favourably with the previous year; however it was still 9% less profitable than the non-*Bt* hybrids.

In 2004, farmers in the state of Andhra Pradesh grew *Bt* cotton on 10% of the cotton acreage. Half of the farmers growing *Bt* cotton bought licensed seed from Monsanto at Rs. 1,500 per 400 g packet, while the other half both unauthorized hybrid *Bt* seed at between Rs. 800 and Rs. 1,200 per packet. Non-*Bt* hybrid seed cost farmers about Rs. 400.

Farmers found that, with fluctuating weather as in 2002, much of the crop showed signs of wilt, and although some *Bt* cotton recovered from severe moisture stress, the yield was very poor compared to non-*Bt* types; also the yield from the unlicensed *Bt* cotton was better than Monsanto's seed because drought-tolerant female variety had been chosen for crossing to produce the hybrid. Monsanto is now demanding royalties of 70% from these seed producers.

So, during 2000-05, there were numerous reports of cotton-crop failures and farmer-suicides in Andhra Pradesh, Maharashtra and Tamilnadu. Running alongside these, were reports of cotton-crop success and increased revenue in some area of Maharashtra and Gujarat, according to an A.C. Nielsen survey in 2003-04.

The AC Nielsen study however which claimed a premium for bollgard cotton in the market, attracted criticism from NGOs and 'anti GM' activists for being conducted cursorily. Its credibility also became suspected due to its commissioning by seed corporate Monsanto-Mahyco Biotech Ltd. (MMB).

In 2003, a CICR-Nagpur (Central Institute of Cotton Research) study of eight commercial *Bt* cotton hybrids grown in India found that the expression of *Cry*1Ac gene inserted to be a repellent to Indian cotton's biggest threat, the bollworm pest, was far below levels needed to actually repel the bollworm, resulting in continuing bollworm attack. The study also revealed that incorporation of the *Cry*1Ac gene into local hybrids was far less effective than incorporation into original varieties being used in Australia, US and elsewhere. MMB has since gone into production of a second patented *Cry*2Ab gene called Bollgard-II which expects to be more effective in targeting both bollworm and sucking aphids.

Amongst reports of success, a 2002-2003 Maharashtra survey by Bennet, Ismael *et al* in *AgBioForum*, had reviewed two crop seasons from Khandesh, Marathawada and Vidarbha, this last district also the scene of rampant cotton distress. The researchers reported in *The Hindu* that "since its commercial release in 2002, *Bt* cotton has had a significant positive impact on yields and on the economic performance of cotton growers in Maharashtra.

The reduced pesticides costs were neutralized by high seed costs and that overall costs for *Bt* farmers were higher. The researchers concluded that the real benefit of growing *Bt* cotton was in higher output. "When costs are taken into account (gross margin=revenue-variable costs), the result is a much higher gross margin for *Bt* growers compared to growers of non-*Bt* varieties. It is worth nothing that the average gross margin gap between *Bt* adopters and non-adopters was larger in 2003 (74%) than in 2002 (49%), the study said.

Nevertheless, *Bt* cotton crop failures were reported in 2004 and 2005, most notably from Andhra Pradesh and Maharastra, where thousands of farmers reported crop losses, driving large numbers of them to suicide. The local government in AP's Warangal district demanded compensation from Monsanto Biotech Ltd., for farmers who lost their crop. The GEAC then banned the use of MMB's MECH-184, MECH 162 and MECH 12 in AP in 2005.

Seed cost went up in 2004-05 to Rs. 1800 per 450 g packet with the AP government challenging the company under the Monopolies and Restrictive Trade Practices Commission for hugely overcharging farmers for its seed. The AP government appealed for prices to be fixed at Rs. 750 per packet, making it comparable to rates internationally. The Central government has backed the State's appeal and ordered the company to charge Rs. 750 per packet. But conversations with cotton farmers in Karnataka in June 2006 reveal that the seed continues to be sold at Rs. 1250-1400 per 450 g packet.

Seed costs apart, central to the issue of crop behaviour, higher yields and yet less revenue, is the variety of *Bt* cotton seed that has been used by farmers till 2005. MMB's short-stapled MECH 162, MECH 184 in Karnataka had given average results in irrigated areas, with a very poor rate of success in non-irrigated areas in AP, giving small farmers in AP, as well as in Maharashtra, no butter at the failure.

In Karnataka, farmers reported middling yields and lesser revenue earnings from *Bt* than from non-*Bt* varieties existing within non-*Bt* varieties. But in 2005, at least one *Bt* long-staple variety, MECH 6918 was introduced which has reported good results in Karnataka, making the economics of long-stapled *Bt* comparable to long-stapled non-*Bt*.

It is the economics of short-stapled varieties like MECH 162 and similar others that have been criticized by academics, NGOs and agricultural consultants. In 2003-04, an FAO report from Karnataka showed that nearly 45% of all *Bt*-growing farmers had bought the seeds at an expense of Rs. 1600 for a 450 g packet on the expectation of high yields promised by seed agents and were then disappointed at the low market price.

Farmers in Karnataka in 2006, however, say they still make a savings of Rs. 1500 per acre. This, despite irrigation costs for farmers growing only *Bt* cotton being significantly higher than for farmers using both *Bt* and non *Bt* varieties.

The FAO had also shown, as a 'major conclusion' that irrigation for higher productivity seemed to be more important than *Bt* or non *Bt* cotton varieties. Under rainfed conditions, *Bt* cotton showed no advantage at all. The FAO study has asked for a carefully consideration of agro-climatic factors before using *Bt* seeds. "Economic performance of a cotton crop is not only determined by its genetic makeup, but also by the agro-ecological conditions under which it is grown," states the study.

Indeed, in the absence of any further definitive research and debate, *Bt* cotton's mixed results could well have more to do with the FAO study's concern on agro-climatic variables, than what meets the eye at present. Professor Ronald Herring of Cornell University points to the MECH 184 variety as an example, which wilts if not watered early; a point that not many farmers either knew of or understood until they learnt it the hard way (Herring, 2006). If this be true, then India's small-holder farmer (61% of total farmers, at last count by the government) has unfortunately borne the burnt of these poor and inadequate varieties.

Coming back to the present, *Bt* cotton did well in traditional and now-extended cotton-growing areas in Karnataka in 2005-06, adding even more weight to the need to discuss this unpredictability. But unpredictability and confusion not withstanding, the GEAC approved six new *Bt* cotton hybrids for commercial cultivation in northern India with more varieties shortly being commissioned.

New varieties might be what are needed, as per Herring's argument, it is the silence on these poor results by the seed corporates themselves, and the GEAC's lack of acknowledgement of *Bt* cotton's erratic performance that is disturbing. The existing varieties that have had such disastrous results in the southern Indian region however are still allowed in Madhya Pradesh, Gujarat and Maharastra till 2007.

This is not peculiar to *Bt*-cotton alone, but is applicable to hybrids of any crop. Hybrids of many crops are preferred over varieties by farmers because they give much higher yields than varieties. It is a well-known fact that if the hybrid seeds saved from one generation are used to raise the next crop, the hybrid vigor will decline and the resulting harvest will gradually becomes poorer from generation to generation. The cost of such yield losses will be far greater than the cost of new seeds. In India, *Bt* gene has been incorporated into cotton hybrids and, therefore, it is better that farmers buy fresh seeds that are specially produced under controlled conditions so that they can enjoy a better harvest and profit. In case the saved seeds are used, they do germinate and also express *Bt* protein, but yield will be poor because of the inherent weakness as explained above.

In India, cotton hybrids, being high yielding, are more preferred by farmers with about 70% (6.3 m ha) of the nearly 9.0 million hectares having been already occupied by them. A number of seed companies are coming out with newer and better hybrids almost every year and their area is bound to further increase. Historically, hybrids are more prone to pest infestation, including bollworms, than varieties and require more stringent pest control measures. Hence hybrids are preferred for incorporating *Bt* genes. Additionally, the companies also have a vested interest. Since they spend huge sums of money on research for several years for developing new technologies/products, they look forward to recovering their investment and making profit. Hybrid seeds offer a value capture mechanism through repeated sale of seeds. This keeps them in business. Perhaps, public institutions may think of introducing *Bt* genes into varieties as they generally do not look for return on investment.

The cost of *Bt*-cotton seeds should not be compared with that of conventional seeds without considering the benefits associated with it. When farmers use the conventional seeds, they spend anywhere from Rs.3,500 to Rs.10,000 or more for an acre on spraying chemical insecticides from 5 to 20 times during a crop season in an effort to control bollworms. If this cost is added, the conventional seeds are far more costly than *Bt*-seeds. Even then, they are unable to get satisfactory control due to several reasons, one of them being that a pest like *Helicoverpa armigera* has developed resistance to most of the recommended insecticides. Hence they often become frustrated. In the case of *Bt*-cotton, bollworm control is in-built in the seeds and, therefore, there is no need for farmers to spend so much extra money on insecticides. The amount of money saved on insecticides coupled with yield benefits accruing out of bollworm control far outweigh the cost of *Bt*-cotton seeds. The genuine cotton farmers are aware of such benefits; nevertheless they would certainly welcome a reduction in the seed price.

The companies which take the leadership role and risk huge investments in developing new technologies/products would be anxious to get back their investment. They may enjoy the benefits of the so-called monopoly only in the initial periods, that too when the sales are generally limited & the product is yet to be widely accepted. Therefore, the price of a product, be it *Bt*-seeds or any other, will be initially high. As the volume of sales increases, the prices generally get reduced. For example, MMB (Mahyco-Monsanto Biotech Company) has voluntarily reduced its technology (or trait) fee from the initial Rs.1,250 to Rs.900 for a pack containing 450 g of *Bt* seeds (plus150 g of non-*Bt* seeds of the same hybrid to be planted as refuge) meant for one acre. With further growth in sales and more competitors entering the market, the prices are likely to further come down. This is a market reality. It is important to ensure that in the name of competition, inferior or spurious products do not enter the market.

The Andhra Pradesh government has challenged the price of *Bt* cotton seeds as exorbitant under MRTPC (Monopolies and Restrictive Trade Practices) and arbitrarily fixed Rs.750 as the maximum selling price of 450 g of *Bt*-seeds. Several other states have also joined hands with A.P. On the other hand, MMB has counter-challenged this unilateral decision in the Supreme Court. Their contention is that the prices are market driven and farmers are getting attractive returns for their money. Perhaps it is wiser for the State Govt. and the seed companies to discuss the various dimensions and implications of this issue and find an amicable solution as legal challenges are complicated, expensive and time consuming.

Both pre-and post-commercialization studies conducted by several public institutions and private seed companies (under the monitoring of RCGM) have indicated that *Bt*-cotton has increased farmers' income. For example, the multi-location field trials conducted by ICAR in 2001 with Mahyco's three *Bt*-cotton hybrids, as a part of the regulatory requirements, revealed that these hybrids yielded 60 to 92% more than the local and national checks and fetched a net profit between Rs.4,633 and Rs.10,205/ha which was about 67% higher. Post-release, the nationwide surveys conducted by AC Nielsen-ORG-MARG in 2003 and by the International Market Research.

Bureau (IMRB) in 2004, revealed that on an average owing to effective bollworm control ranged from 29 to 58% (4.25 to7.4 quintals/ha), pesticide reduction from 60 to 72% (savings of Rs.2,800 to 3,200/ha) and increase in net profit to farmers from 60 to and 42% of the non-users expressed their intention to purchase *Bt*-cotton seeds in the next season. Another survey conducted by the Gokhle Institute of Politics and Economics, Pune, in 2003 in certain parts of Maharashtra has also indicated that *Bt*-cotton was profitable to farmers. Another report indicated that the net economic benefits to Indian farmers from *Bt*-cotton was, on an average, \$139/ha in 2002, \$324/ha in 2003 and \$260/ha in 2004, with a four-year average of \$225/ha.Other studies also reported results in the same range, acknowledging that the benefits will vary from year to year and also from place to place due to varying levels of bollworm infestation, agronomic conditions and cultivation practices. In a more recent (2006) study conducted in Maharashtra, Gujarat,Andhra Pradesh and Tamil Nadu by the Indian Institute of Management (IIM), Ahmedabad, the profit from *Bt*-cotton was found to be higher in all the states, both under irrigated and non-irrigated conditions. It reported an yield gain of 31%, reduction in the numberof pesticide sprays by 39% and an 88% increase in profit (Rs.11,250 or US\$250/ha) The farmers found advantage in pest incidence, pesticide cost, cotton quality, yield and profit. Almost all farmers indicated that they plan to plant *Bt*-cotton in the future. Similar results have been reported from several other countries also. However, the results of alternative experiments and surveys carried out independently by Gene Campaign; Centre for Sustainable Agriculture; Research Foundation for Science, Technology and Ecology; Greenpeace and a few other NGOs in India who have always been opposed to this technology, found no such benefits. According to them, *Bt*-cotton suffered more bollworm damage, required more pesticide sprays, yielded less and produced poor quality cotton than the non-*Bt* cotton.

The final judgment is best left to farmers. They have the option to choose *Bt*-cotton or non-*Bt* cotton whichever is beneficial to them. The fact remains that the number of *Bt*-farmers is increasing from year to year. *Bt* technology does not distinguish between small farmers and big farmers. It controls bollworms no matter who is growing the crop. In India, in 2006 about 2.3 m small cotton farmers were able to derive attractive economic benefits from *Bt*-cotton. Similarly thousands of mall farmers in China, South Africa, Argentina and other developing countries have been amply benefited by this *Bt* cotton-Approval for Commercial Cultivation.

The Genetic Engineering Approval Committee (GEAC), in its 32nd meeting, held on March 26, 2002, made the landmark decision of approving cultivation of *Bt* cotton in India; three hybrids (MECH 162 *Bt*, MECH 184 *Bt*, and MECH 12 *Bt*) were approved for cultivation with the stipulation that certain conditions be met (Figure 13; Table 11). These hybrids are high yielding and produce medium-long-to long-staple fiber. The GEAC also approved production of seed; subsequently, seed for cultivating 100,000 acres (40,485 ha) was available for planting by Indian farmers. The seed was distributed in the state of Maharashtra, Madhya Pradesh, Karnataka, Andhra Pradesh, Gujarat, and Tamil Nadu in the kharif 2002 planting season. The area planted to *Bt* cotton were 14746, 2220 3908 5608 8854 and 2702 hectares spread over the three hybrids.

On April 22, the GEAC, which functions under the Ministry of Environmental and Forests, approved large-scale field trials and production of seeds of 12 varieties of *Bt* cotton hybrids developed by Rassi Seed, Ankur Seeds (Another sub-licesee of Monstanto, and Mahyco). These *Bt* cotton hybrids contain the *Bt*Cry1Ac gene developed by Monsanto.

Rassi Seeds has been allowed to conduct trials of RCH 118 *Bt* and RCH 559 *Bt* in Central India, RCH 368 *Bt* in South India and RCH 317 *Bt* in North India. Ankur Seeds has been allowed to conduct trials of Ankur 651 *Bt* and Ankur 2534 *Bt* in North India and Ankur 651 *Bt* and Ankur 09 *Bt* in Central India. Mahyco has been given to go-ahead for trials of MRC 6301 *Bt* and MRC 6160 *Bt* in Central India and MRC 6301 *Bt* and MRC 6322 *Bt* in South India.

Table 11: Commercial cultivation of *Bt* cotton hybrids in India, 2002 (hectares)

State	*MECH-12*	*MECH-162*	*MECH-184*	*Total*
Maharashtra	112	9300	5334	14746
Madhya Pradesh	60	404	1756	2220
Karnataka	–	3828	80	3908
Andhra Pradesh	44	5564	–	5608
Gujarat	76	4136	4642	8854
Tamil Nadu	–	2042	660	2702
Total	**292**	**25274**	**12472**	**38038**

The cotton-growing season of 2002, in India was affected by high rainfall in some areas and a long dry spell followed by heavy downpours (resulting in unfavorable conditions for cotton cultivation) in other areas.

The overall pest pressure for the year, bollworm complex was also low. However, across the cotton-growing areas, control of the bollworm complex was observed in *Bt* cotton; the amount of spraying required, if any, was significantly lower for *Bt* than for non-*Bt* cotton hybrids. *Bt* cotton yields were significantly higher than those of non-*Bt* cotton and the average increase in yield was about 30% over non-*Bt* hybrids in similar conditions. The quality of *Bt* cotton was cleaner with better color, and *Bt* cotton provided rates commensurate with the quality of the fiber based on the hybrids available.

The data in Table 12 show that in addition to the substantial increase in yield, there is a significant decrease in the number of insecticide sprays associated with the use of *Bt* cotton-the overall average indicates a yield increase of 8.1 quintals of cotton and a reduction of 1.93 sprays. These two factors add to the total economic benefit. There is also an average additional income of more than Rs. 18,000 per ha for *Bt* compared to non-*Bt* cotton.

Table 12: ***Bt*** **cotton results from kharif 2002 season, June-December (yield in quintals)**

State	*Non-Bt yield*	*Bt yield with Bt*	*Yield increase*	*Non-Bt sprays*	*Bt sprays*	*Spray reduction with Bt*	*Economic benefit per hectare*[c]
Andhra Pradesh	14.42 (5-25)	20.52 (12.5-32.5)	6.10	4.81 (1-8)	2.08 (0-4)	2.73	Rs. 16747
Gujarat	19.80 (3.7-37.5)	28.35 (10-44)	8.55	3.42 (1-7)	2.09 (0-5)	1.33	Rs. 18430
Karnataka	10.50 (1.3-30)	17.82 (7.5-40)	7.32	2.53 (0.6)	1.00 (0-3)	1.53	Rs. 16170
Madhya Pradesh	15.00 (10-50)	25.82 (35-62.5)	10.82	3.29 (1-9)	0.93 (0-3)	2.36	Rs. 24000
Maharashtra	14.47 (2.5-45)	20.82 (2.5-62.5)	6.35	2.78 (0-7)	0.99 (0-4)	1.79	Rs. 14490
Tamil Nadu[d]	–	–	–	–	–	–	–
Total	**13.25**	**21.35**	**8.10**	**3.10**	**1.17**	**1.93**	**Rs. 18130**

Note: All figures given in the table are based on a survey conducted by Mahyco in the six states where *Bt* cotton seed cotton was sold in the kharif 2002 season[a]. The total sample size was 1069 farmers. Averages are on weighted average basis. Figures in parentheses represent the range for yield (quintals per hectare) and number of sprays

[a] Kharif refers to a crop that is harvested at the beginning of winter

[b] 1 quintal = 100 kg

[c] Economic benefit per hectare was calculated on the basis of an average cotton rate of Rs. 2000 per q and an average cost of each bollworm complex spray of Rs. 1000/-ha.

[d] Cotton picking still in progress in Tamil Nadu at date of writing.

Nothing much has been changed in its performance except for the hype by Mahyco Monstanto, since *Bt* cotton cultivation was permitted in Andhra Pradesh.

As a matter of fact, the industry continuously makes several tall claims to promote *Bt* cotton. Three of the most important claims are:

- Cultivation of *Bt* cotton will reduce pesticide use considerably.
- Cultivation costs will come down significantly and
- Profits for farmers will increase.

On all three counts *Bt* cotton failed in Andhra Pradesh for the second consecutive time. Extended study (2003-2004)taken up on behalf of the AP Coalition in Defence of Diversity of Mr. Abuld Qayum and Mr. Kiran Sakkhari, two agricultural scientists.

The study conducted in three districts of AP viz. Adilabad, WArangal an Kurnool had a sizeable sample of nearly 164 farmers. They were studied almost on a day to day basis throughout the cotton cultivation period, from the date of sowing to the date of harvesting. In Warangal Distinct the sample size was nearly 10% of all the farmers who cultivated *Bt* cotton in the district. Therefore the study bears a great significance to the entire *Bt* cotton phenomenon.

The number of very interesting results point out to the fact that *Bt* cotton has once again let down small and marginal farmers in the dry land districts of Warangal, Adilabad and Kurnool in Andhra Pradesh. The facts were

1. In comparison to non-*Bt* hybrids, pesticide use did not drastically come down in *Bt* cultivation.
2. Costs of cultivating *Bt* cotton was higher compared to non *Bt* cultivation.
3. The *Bt* cotton yields were barely 2% higher compared to non-*Bt* cotton.
4. Having paid higher price for *Bt* seeds and investing more money on its cultivation, farmers in fact earned more profits through cultivation of non *Bt* cotton than through *Bt* cotton.

For AP Coalition in Defence of Diversity, a matter of great concern is the hype built up around *Bt* cotton by the industry which directly contrasts with the findings of this painstaking year round study.

Monstanto AC Nielsen study the making rounds all over the world, without being challenged painted an extremely rosy picture for *Bt* cotton in India, where the field realities are completely different. Two or three findings of the Monsanto study were taken and compared with APCIDD study (Table 13).

Table 13: Comparison between Monsanto and APCIDD studies in state Andhra Pradesh

Findings	*Monsanto study*	*APCIDD study*
Bollworm pesticide reduction	58% 1856 Rs.	14% 321 Rs.
Yield increase	24% 1.98 qu/ac	2% 0.09 qu/ac
Increase in net profit	92% 5138 Rs/ac	(-) 9% (-) 750

APCIDD-Andhra Pradesh Coalition in Defence of Diversity

This clearly pointed out how far from truth are the claims made by the study conducted by a marketing agency, which contacted farmers through questionnaires just once after their crop period. The APCIDD study which worked with farmers continuously, contacting them every 15 days and therefore always close to the realities of the situation, brought out the truth which counters the hollow claims of the industry.

The industry has claimed four times more than the actual reduction in pesticide use, 12 times more yield and 100 times more profit than the actual. If left unchallenged, the industry which is losing ground in the rest of the world (including China) will completely overrun our agriculture and decimate farming as we know it.

Last when the results of APCIDD study for 2002-2003 cam out and unmasked the disaster that *Bt* cotton had brought upon the farmers of Warangal, there was a great outcry by the media which made the government sit up and take notice of the tragedy it had let lose on the farmers. It made the government institute is own survey, clearly came out with the finding that cost of cultivation for *Bt* was more and net r were too low in comparison with not-*Bt*.

Later the then minister for Agriculture Mr. Shobhanadrishwara Rao made a public statement asking farmers to stay away from cultivation of *Bt* cotton. But within a month or so the government went back on its statement and without an iota of remorse invited Monsanto to sell its *Bt* cotton seeds from the government outlets.

The results was that farmers who paid 3.5 times higher price for *Bt* seeds today suffer a loss of 9% compared to the farmers who cultivated non *Bt* cotton.

Over the last four years, farmers merrily mixed *Bt* and non *Bt* cotton when they took them to the market. Thus cotton seeds which are the source for edible oil and cattle feed have been contaminated with *Bt*.

The profit hungry industry is not happy, even with this sleeping authority. The latest moves from the powerful industrial lobby, in India has been instrumental in a process that might completely dismantle the Genetic Engineering Approval Committee of the Ministry of Environment and Forests and hand over the control to an industry dominated committee in the name of a fast track approval. As a consequence we are completely handing over the control over our agriculture, health, environment and our well being itself into the hands of an irresponsible industry which knows very little beyond its balance sheet.

In Kharif 2002, around 1200 farmers cultivated *Bt* cotton in Warangal district alone. More than 90% of them cultivated *Bt* MECH-162 cotton hybrid, which was marketed by Mahyco-Monstanto. A season long study conducted by AP Coalition in Defense of Diversity in 2002-03 sowed that the cotton hybrid MECH-162 failed miserably on the farmer's fields in Warangal district. This study and other studies which endorsed these results forced Mahyco-Monsanto to replace MECH-162 with MECH-12.

As a sequel to its earlier research in 2002, the AP Coalition in Defense of Diversity decided to enlarge its study for Kharif 2003 to three cotton-growing district viz., Warangal, Adilabad and Kurnool covering 28 villages with a sample size of 164 farmers. The season long study systematically collected field data from farmers using structured interview schedules which recorded each and every farmer's income and expenditure patterns with regard to cultivation of cotton, both *Bt* and non-*Bt*. This was done at fortnightly intervals fright from sowing of the cotton crop till it was harvested. The data collection was helped by 11 NGOs working in these districts. All the data was collected and analysed to arrive at the following results.

1. *Bt* seeds cost 23% more than non *Bt* hybrids.
2. Total investments for *Bt* was 8% higher than for cultivation of non-*Bt* cotton.
3. The reduction in pesticide consumption by *Bt* farmers was just 12%.
4. Net profits from *Bt* was 9% less compared to profits from non *Bt* hybrids.
5. The benefit cost ratio was in favor of non *Bt* hybrids.
6. For small and medium category of farmers, the yield difference between *Bt* and non-*Bt* was marginal.

The study clearly showed that, even though the over all yields were marginally more for *Bt* cotton, the overall benefit cost ratio is still in favor of non *Bt* hybrids. This was caused by the higher investments incurred for the cultivation of *Bt* cotton hybrids which was 8% more than the investment for non *Bt* hybrids. In addition, the results clearly show that, for small and medium farmers *Bt* is not a viable proposition as the net profit from *Bt* was 9% less than non *Bt* hybrids.

The study underlines the argument that Genetically Altered crops need more investment per unit area than their non GM counter parts while the net profits are higher for non GM crops. This is a worldwide phenomenon. Therefore, in a country like India, where majority of the farmers are small and medium, while looking or GM options, one need to explore a policy that takes a long term perspective on the sustainability of different options available.

Over and above all this, there is a definite impending danger of major *Lepidopteran* pests-such as American Boll work, spotted Bollworm and Pink bollworm developing faster resistance to the delta-endotoxin, which will be catastrophic to the other alternate host crops like redgram, sorghum, maize, sunflower, groundnut and beans besides cotton. This will be highly disastrous for farmers who grow crops like pigeonpea and above-mentioned crops for their food and other needs.

APCIDD study examined if any pre cultivation training was given to the farmers by any of the government agencies had provided any training. The seed company had made available pamphlets showing that spraying had to be done on *Bt* cotton if number of pests exceeded a certain level. The major efforts of the company were directed towards broadcasting taped messages extolling the virtues of *Bt* cotton rather than any tips of cultivation.

No technical assistance was provided to the farmers during cultivation either by any governmental agency or by Mahyco-Monsanto. In the absence of any extension help, farmers had no one to assist them when they faced problems during the cultivation and pest attacks.

One of the most stocking revelations of this investigation was the fact that neither State Level nor District Level Committee had been set up in either Maharashtra or Andhra Pradesh where *Bt* cotton was being commercially grown. This is a breach of law and a direct violation of the prescribed rules for the manufacture, use import, export and storage of hazardous microorganisms and genetically engineered organisms and cells, under the Environment Protection Act, 1989.

Enquiries made during the survey also revealed that no one had come to that area to collect field data for monitoring the insect attacks and crop performance so at least in this area, no baseline were collected to evaluate the impact of *Bt* cotton on the environment, on beneficial insects, on other cotton crops and on the ecosystem. This amounts to gross negligence.

A number of factors have probably contributed to the failure of Mahyco-Monsanto's *Bt* cotton.

1. The first is the poor quality of the variety itself. It is well known that Mahyco-Monsanto's cotton varieties, MeCH 162 and MECH 184, which were transformed to *Bt* 162 and *Bt* 184, are poor to modest performers, giving modest yields. A better performing cotton would give a better *Bt* cotton so the GEAC must answer why it approved this *Bt* cotton when better quality *Bt* cotton hybrids belonging to Indian companies are in the pipeline.

With the substantially higher cost of seeds, the economics of the *Bt* crop is not favourable for the farmers. In addition to the high cost of the seed is the modest saving in pesticides, which does not make up for the large expense incurred on seed. Tilting the balance further is the fact that *Bt* cotton must be grown with a refuge, necessary for resistance management. This is recommended as 20% of the cultivated area by the GEAC. "Wasting" 20% of the land on managing resistance makes the *Bt* cotton even more nonviable, especially for small farmers.

The GEAC has to be held accountable for the failure of *Bt* cotton as much as the company providing the seed. Why did it keep the field trail data of *Bt* cotton secret when there were so many demands to examine this data? It requires to be known from the GEAC how approval for commercial cultivation in Andhra Pradesh and Maharashtra was granted when no State and District level committee was set up. The GEAC must also fix responsibility for the failed cotton crop and make Mahyco-Monsanto compensate those farmers who have suffered losses. This is required under the Indian law, the Protection of Plant Variety and Farmers Rights Act 2001. Section 39.2, of the law state clearly that:

"Where any propagating material of a variety registered under this Act has been sold to a farmer or a group of farmers or any organization of farmers, the breeder of such variety shall disclose to the farmer or the group of farmers or the organization of farmers as the case may be, the expected performance under given conditions, and if such propagating material fails to provide such performance under such given conditions, the farmer or the group of farmers or the organization of farmers, as the case may be, may claim compensation ...."

Genetic Engineering Approval Committee (GEAC), India's regulatory authority for products, has approved yet another transgenic *Bt* cotton, rCH 2, for commercial cultivation in the central and southern parts of the country (Krishna Kumar, 2004).

COTTON FIASCOS AROUND THE WORLD

Table 14: *Bt* cotton trials around the world

ArgentinaMonsanto's Bt cotton approved in 2001. It is sown on around 13% of the total area planted to cotton in 2005/2006 (approximately 40,000 ha out of 309,000 ha).

Australia	*Bt* cotton introduced in 1996. Reports are mixed. Initially pesticide use declined dramatically then increased year after year. Farmers do not seed economic benefits, due to high cost of technology fee ($225 per ha). By 2002/2003, 30% of total cotton crop is *Bt* cotton and this increases to 80% in 2004 with the release of Monsanto's Bollgard II variety, which involves less stringent insect resistance management plans and is introduced for the time being, at the same price as the old variety.
Brazil	In March 2005, following the adoption of a new biosafety law strengthening its powers, the pro-GM National Technical Biosafety Committee approves limited experimental field trials of Monsanto's *Bt* cotton. However, *Bt* cotton seeds are sold in neighbouring Argentina and Paraguay, and illegal cultivation is widespread in Brazil. Monsanto claims that 5% of the 1.3 million tons of cotton produced in Brazil for the 2005-2006 season are "black market" varieties of its *Bt* cotton and the company says that it is partnering with Brazil's roughly 200 gins to establish a scheme to test farmer's cotton for its patented genes and collect royalties
Burkina	In 2003, Monsanto, Syngenta and Burkina Faso's Institut National de I'Environment et la recherché Agronomique (INERA) begin field tests of two *Bt* cotton varieties without the involvement or consent of the national biosafety committee which is tasked with developing a national regulatory regimes for GMOs. In October 2006, local media report that the government will approved the commercial release of *Bt* cotton in 2007.
China	*Bt* cotton released in 1997. Currently *Bt* cotton is planted on over half of the national cotton area. *Bt* cotton seed costs around 50-60% more than regular seed but there is a high level of unauthorized use. Reports are that initial decrease in pesticide use are now completely offset by pesticide; use against pests that were formerly of secondary importance.
Colombia	Monsanto imports *Bt* cotton in 2002, without an environmental clearance. Popular legal action results in the suspension of the authorisation.

Costa Rica	Monsanto began field trials without regulatory oversight in 1992. In 2004, 638 ha of *Bt* cotton were planted, mainly for the export of seeds.
Egypt	Monsanto and Egypt's Agriculture Genetic Engineering Research Institute currently collaborating in field trials of *Bt* cotton. They claim commercial introduction could take place as early as 2006.
Guatemala	Monsanto and Algodones Mayas S.A. have conducted a field trial of GM cotton tolerant to glufosinate (not a *Bt* cotton) and Monsanto is pursuing field trials of GM cotton with another local seed company, Semillas S.A.
India	In 1998 Monsanto's first field trials of *Bt* cotton disrupted by farmer protests. Commercial introduction of *Bt* cotton occurs in 2002, although sales of an unapproved *Bt* variety began as far back as 1998. In 2005, *Bt* cotton hybrids planted on over 500,000 ha despite widespread reports of *Bt* cotton crop failures. In May 2005, the Genetic Engineering Approval Committee rejects Monsanto's application to renew its temporary authorization for the sale of its three *Bt* cotton varieties in Andhra Pradesh. In 2006-2007 season Monsanto begins sales of Bollgard II cotton in the central zone of Madhya Pradesh, Gujarat and Maharashtra at nearly twice the seed price.
Indonesia	Monsanto's *Bt* cotton commercialized in South Sulawesi province in 2001. However two years later it is withdrawn after its failure to perform triggers farmers protests. Due to poor harvests, some 70% of the 4,438 farmers growing *Bt* cotton were unable to repay their credit after the first year of planting.
Kenya	Monsanto imports *Bt* cotton into Kenya in 2004 for field trials. The field trials begin in early 2006 at the Kenyan Agricultural Research Institute's research station in Mewea Central Kenya.
Mexico	*Bt* cotton introduced in 1996. Government subsidises purchase of *Bt* cotton seeds. In 2002-03, 25% of the national cotton area planted to *Bt* cotton, slightly less than the percentage in 2000.
Pakistan	In May 2005, the Pakistan Atomic Energy Commission provides 40,000 Kg of *Bt* cotton seed (varieties IR-FH-901, IR-NIBGE-2, IR-CIM-448 and IR-CIM-443) to farmers in the Punjab to grow on over 8,000 acres of land during the 2005-06 season.
Paraguay	In July 2005, The Minister of Agriculture announces that it will approve Monsanto's GM cotton as part of a joint project with the company.
Philippines	In January 2005, the Cotton Development Authority signs a memorandum of agreement with the Philippine Rice Research Institute of begin field trials of *Bt* cotton.

Senegal	National cotton company (SODEFITEX) and Monsanto undertake field trials in the Senegal River Valley without notifying regulatory agencies or informing the local population. SODEFITEX backs away from project after early results show no reduction in pesticide use.
South Africa	*Bt* cotton approved for commercial planting in 1997. Adoption was very rapid and by 2002/03 an estimated 75% of national cotton area planted to GM cotton. In 2003-04 only 35,700 ha of cotton was planted, an 80% reduction since 2000, ascribed to low world prices and droughts. In 2004-05 the area planted was 21,700 ha, an extraordinary 40% drop in area RR cotton. Small-scale farmers. 90% of whom adopted *Bt* cotton, are in debt with the total debt amongst small-scale cotton farmers in Northern KwaZulu Natal estimated at over US$3 million in 2004.
Thailand	Monsanto imports *Bt* cotton seeds in 1995 and begins field-testing in 1997. In 1999 farmer's groups plantings of cotton find samples taken from locations outside Monsanto's approve sites testing positive for the presence of the *Bt* gene. It is estimated that 8,000 hectares of *Bt* cotton are being grown illegally. An alliance of 35 farmer groups and NGOs threaten to stage a mass rally unless the government responds to their calls for a stop to the testing and commercial release of genetically engineered crops. The government reacts by setting up such a ban and terminating field trials of Monsanto's *Bt* cotton.
USA	Around 40% of the cotton area in the US is *Bt* cotton. Studies show reduction in pesticide use since *Bt* cotton introduced in 1996, but now secondary pests are becoming an increasing problem.
Vietnam	Although IPM techniques have dramatically reduced the use of pesticides on cotton in Vietnam over the last two decades, Vietnam Cotton Company is pursuing *Bt* cotton in partnership with the Chinese company Biocentury in an effort to expand dry-season irrigated cotton production. Field trials of *Bt* cotton have taken place.
Zimbabwe	Monsanto planted a *Bt* cotton in 1998 without official permission. Crop was burnt before flowering when uncovered by authorities.

Conclusion and future prospects:

Bt cotton has a successful story of genetically modified crops. Adopting *Bt* cotton is an efficient strategy to control cotton pest, especially Lepidoptera pests. It is quickly adopted by cotton farmers around the world. Ten years of commercialization of *Bt* cotton has shown that *Bt* cotton is safe to human and environmental health as well as bring huge environmental, economic and social benefits. To high efficiency use of *Bt* cotton, the second generation of *Bt* cotton with ever better pest resistance has been developed. In 2002, this new *Bt* cotton was approved for commercial release by Australia government, and became available to cotton growers for planting from 2003. In the 2003-2004 growing season, there are four second generation varieties of *Bt* cotton were planted in Australia, they are Sicot 11B, Sicot 12B, Sicot 13B and Sicot 14B (CRISO Plant Industry, 2003). In short future, this new kind of *Bt* cotton will be adopted in USA, China, India and other countries, and cotton farmers will get more benefits from the new technology.

Table 15: Methods for controlling cotton pests

Methods	*Control method*	*Example*	*Disadvantage*	*Advantage*
Biological	Reduction of pests by the introduction and permanent establishment or repeated release of lying organism.	Parasitic, predaceous, and pathogenic arthropods, nematodes, and microorganism including biopesticides.	Unsteady, only can control pest infection at moderate level.	Environment and health friendly.
Chemical	Reduction of pests, or preventing of injury by the use of chemicals to kill them, attract them away or repel them from the cotton.	Insecticide, insect pheromones, insect hormones and repellents.	Environment and health risk; pest resistant easy occurred. High cost.	High effective, easy to operate.
Natural	Reduction of pests by the conservation and enhancement of naturally occurring living organisms.	Parasitic, predaceous and pathogenic arthropods, nematodes, and microorganisms.	Unsteady, only can control pest infection at moderate level.	Environment and health friendly.
Cultural	Reduction of pests by the use of agricultural production practices.	Change planting dates, plowing, harvesting, irrigation, crop rotation, crop residue disposal, trap crops, and insect pest-resistant varieties.	Unsteady, only can control pest infection at moderate level.	Environment and health friendly.
Physical or mechanical	Reduction of pests by means of specially designed mechanical devices	Screens, traps, barriers, heating, cooling, attracting, repelling, sound	Unsteady, only can control pest infection at moderate level.	Environment and health friendly.

	that affect their physiology or biology directly, or that alter their physical environment making it less suitable.	waves, light and X-rays.		
Reproductive and genetic	Reduction of pests by treating them with substances or introducing genetic traits that after sexual behavior or otherwise disrupt pest reproduction and survival.	Radiation, chemosterilents, genetic manipulation, and artificial sex pheromones.	Hard to operate, high cost	Environment and health friendly.
Legal	Reduction or avoidance of pest problems by lawful regulation of pests and their plant hosts.	Plant quarantines, eradiation programs, inspection and certification of movement of potentially infested materials.	Hard to operate health friendly	Environment and
Transgenic	Using genetic engineering to obtaining high pest-resistant cotton varieties.	*Bt* cotton	High cost	Environment and health friendly. Save labor and money.

Table 16 : Global adoption of *Bt* cotton (hectares)

Year	*USA*	*China*	*Australia*	*India*	*Mexico*	*South Africa*	*Argentina*	*Indonesia*	*Total*
1994	-	600	-	-	-	-	-	-	600
1995	-	3,000	-	-	-	-	-	-	3,000
1996	730,000	45,000	39,842	-	900	-	-	-	815,742
1997	1,050,000	80,000	58,235	-	15,000	-	-	-	1,203,235
1998	1,170,000	100,000	99,561	-	37,000	12,000	5,500	-	1,424,061
1999	1,580,000	400,000	133,850	-	17,000	-	12,000	-	1,782,850
2000	2,100,000	900,000	138,570	-	26,000	30,000	22,000	-	3,216,570
2001	2,080,000	1,500,000	153,323	-	28,000	N.A.	9,000	-	3,774,723
2002	2,170,000	2,100,000	121,200	72,682	N.A.	N.A.	N.A.	N.A.	4,463,882
2003	2,310,000	2,800,000	66,150	216,000	N.A.	N.A.	N.A.	N.A.	5,382,150

All around the world, farmers and governments have had disappointing and unpleasant experiences with the cultivation of genetically modified cotton. Cotton is responsible for more than 10% of world pesticide use including some of the most hazardous, and 25% of all insecticide use. As weeds and insects become resistant, more and more pesticides are needed in a vicious circle that's a recipe fro socio-economic, health and environmental disaster.

GM cotton seed prices include a 'technology fee' that can go up every year, and is calculated on supposed savings from reduced pesticide use with the *Bt* variety in a particular location. All farmers Monsanto's *Bt* cotton sign a contract, called a Technology Use Agreement, that is strictly applied. It stipulates that:

i. farmers cannot save seed for replanting

ii. farmers are prohibited from supplying seed to anyone else

iii. farmers must pay 120 times the technology fee, plus the legal fees of Monsanto, if they violate the contract

The Indonesian experience: A cautionary tale

Indonesia was the first country in South-East Asia to permit commercial GM farming against the warnings of scientists and activists on the environmental and socio-economic impacts. Fortunately, permission was granted only on a year-by-year basis, and the Government reviewed the impact of the failed *Bt* crop. The review was scathing. This 'Gene Revolution', it said, seemed to be 'a modern tool for cementing farmers' dependence on seeds and transnational agrochemical corporations appearing in developing countries in different guises'. The evidence from Indonesia is that 'GM crops are nothing more than a profit-motivated deployment of scientific power dedicated to sucking the blood of farmer's.

Monsanto promised *Bt* cotton would return 3-4 tones of cotton per hectare while requiring less pesticide and fertilizer than Kanesia, the local cotton variety. The seed was given to farmers with pesticides, herbicide (including roundup) and fertilizer as part of a credit scheme costing 16 times more than non-*Bt* cotton. In fact, the average yield was 1.1 tones per hectare and 74% of the are planted to *Bt* cotton produced less than 1 tones per hectare. About 522 hectares experienced total crop failure. Despite that, the government extended approval for *Bt* cotton for another year, and the results were no better.

In 2001 farmers signed contracts, but in 2002 the seed price rose and the cotton price slumped. Farmers had no choice but to shoulder the debt and sell at the company's rate; as a result, 76% of farmers who joined the credit scheme couldn't repay their debt and many burned their cotton in protest against the government and the company.

In 2003, Monsanto halted operations saying that the Indonesian Government's decision to authorize *Bt* cotton production on a year-by-year basis had been a big obstacle to business investment. PT Monagro Kimia, a Monsanto subsidiary, was under investigation by the US Department of Justice and the Indonesian Corruption Eradication Commission on Suspicions that a payment of US$50,000 was made to Indonesian officials in 2002. In January 2005, Monsanto was found guilty of authorizing the bribe and fined $1.5 million.

_Bt_ cotton in China

Monsanto received a permit in 1997 for commercial production of *Bt* cotton in China and has since shared the *Bt* cotton market with domestically developed varieties that have expanded quickly over the country's cotton-growing area. China has been held up as the success story in GM cotton, and is the key to statistics claiming for small farmers from GM. However, earlier warnings of major problems have now been confirmed by a Chinese researcher who reports that the technology will not only be useless within six to seven years, but 'could cause a disaster'.

Liu Xiaofeng, researcher from Henan, China's second largest cotton-producing province, told Reuters that the cotton bollworm is indeed developing resistance and will not be susceptible to *Bt* cotton after 20-30 generations, or in six to seven years. Moreover, *Bt* cotton does not effectively control secondary pests such as Lygus bug.

The early warnings appeared in a study published in June 2002 based on the work of scientists at a research institute funded by China's Environmental Protection Agency. It found that although *Bt* cotton was effective in bollworm control, it had. The study also found the diversity indices of the insect community in *Bt* cotton fields to be lower than in conventional cotton fields, and that the cotton bollworm could develop resistance to *Bt* cotton. Liu's work received further collaboration by another study published in October 2004, which found that *Bt* cotton did not reduce the total numbers of insecticide sprays because additional sprays were required against sucking pests.

Field trials in Africa

South Africa, already the sixth biggest producer of GM crops in the world, grows *Bt* cotton on large and small commercial scales, extolling the benefits to small farmers in spite of the serious debts incurred. Although there is a glut of cotton in the world market and depressed prices caused by US subsidies to their own growers worth $3.7 billion per annum, the US government and the world's biggest agrochemical companies are putting pressure on West African countries to introduce *Bt* cotton, the 'trojan horse' for other GM crops waiting in the wings.

In West Africa there are wild relatives of cotton that may be contaminated, but in the US, GM cotton is prohibited in Florida where wild relatives grow.

In November 2003, USAID, with the official support of the International Institute of Tropical Agriculture, declared that it wants to 'GM-ise' Africa.

- Mali's National Agricultural Research Institute has been negotiating with Monsanto and Syngenta for field trials of *Bt* cotton. There is a plan to convert the country's crop to GM varieties over the next five years; local farmers and the public are unaware of this intention. West African farmers, already unable to sell enough natural cotton because of subsidies, are locked into a cycle of poverty with credit against next year's harvest.
- Brukino Faso has been field-testing *Bt* cotton since July 2003 in collaboration with Monsanto. But Francois Traore, president of the National Union of Cotton Procuers, says, 'If we already have the means to reduce pesticide use, why look for things that are going to complicate life?
- Benin has had a moratorium on GM products since March 2002, but is under constant pressure to introduce *Bt* cotton.
- Senegal ran an unofficial trail of Monsanto's *Bt* cotton, but efforts were abandoned after the cotton failed to perform.
- Egypt has a prog-GM policy with field trials underway for *Bt* cotton and many other crops.
- Kenya has many research institute pushing GM crops, and research on GM cotton is under way.
- Uganda has just published its first biosafety policy bill, which has yet to be made law by parliament. However, it is expected to take up *Bt* cotton soon.
- Zimbabwe's government destroyed some unsupervised field trials of *Bt* cotton conducted by Monsanto some years ago.
- Co-existence between conventional and GM cotton is not possible. If *Bt* cotton is introduced in the region, the contamination of non-*Bt* cotton is inevitable. As there is no way to easily distinguish between *Bt* cotton and non-*Bt* cotton will easily and up being mixed into the conventional cotton supply when farmers drop off their harvests, when the cotton is transported, or when seeds are cleaned and distributed. Contamination will also take place in the fields through

cross-pollination, either by way of wind or, more likely, by way of insect pollinators.

Few constraints to GM crops are –

1. First, genetic modification is a new technique that is far from fully understood and the impacts on the environment and human health could take years to appear, But, once GM crops are introduced it is difficult, if not impossible, to take them out of the environment, especially for poor countries with few resources, like those of West Africa.
2. Second, Monsanto and other transnational corporation hold patent rights over the transgenic genes in GM crops and they claim that their rights extend to any plants that contain those transgenes, even if these plants incorporated the trangenes through contamination. In other words, if a farmer's crop is contaminated with Monsanto's *Bt* cotton, then that crop becomes Monsanto's property and the farmers will have to pay Monsanto royalties on it. This is precisely what happened to Canadian farmers Percy Schmeiser. Monsanto that makes plants resistant to Monsanto's Roundup herbicide. Even though Schmeiser says that his fields were probably contaminated by neighbouring GM canola fields, Monsanto accused Schmesier of acquiring it GM canola illegally and used him for loss of royalties. While the court could nto determined if Schmeiser had deliberately acquired Monsanto's GM technology or benefited from it, the court ruled in favour of Monsanto, forcing Schmeiser to pay Monsanto over $200,000 in royalties and legal fees.
3. Third, *Bt* cotton can contaminate local cotton varities and their relatives, leading to consequences for the local ecology and breeding programmes. For these reasons, *Bt* cotton is prohibited in the US states of Hawaii and Florida where there are local cotton verieties and wild species of cotton. In West Africa, there are important varieties of local cotton (*G. arboreum* and *G. herbaceum*) and several wild species of cotton that exist throughout the region. Moreover, cotton is part of the Malvaceae family, which includes many plants and trees common to West Africa, such as hibiscus, baobab and cola.
4. Finally, *Bt* cotton contamination could destroy the emerging organic cotton production sector in the region. Organic cotton farmers received higher prices for their cotton, but their practices have to conform to stringent certification requirements. In general, these requirements prohibit genetically modified organisms (GMOs). Under the "Basic Standards for Organic Production and

Process" of the International Federation of Organic Agricultural Movements (IFOAM): "The use of genetically engineered organisms or their derivatives is prohibited. Organic processed products shall not use ingredients, additives or processing aids derived from GMOs." West African organic cotton farmers, therefore, could lose the organic status of their cotton if their fields are contaminated by neighbouring *Bt* cotton fields or if their seeds (which are currently provided by the national cotton companies) are contaminated.

The America

In the US, home of Monsanto's Bollgard first planted in 1996, there have been problems with erratic and disappointing yield, especially in southeast Arkansas, where costs were significantly higher on *Bt* acreage. In 2002, despite the use of supplementary pesticides, 7.5% of the *Bt* crop was destroyed by bollworm and 1.4% destroyed by Spdoptera and Pseudoplusia includens caterpillars. The total insecticide use has remained relatively stable due to the increasing importance of secondary pests; it is lower in dry states such as Texas, but increasing in the Mississippi delta. Research on Bollgard cotton adopted in north Carolina, conducted between 1996 and 2003 by Jack Bacheler, North Carolina State University Extension entomologist, found changes in insect communities, and that while damage from bollworms decreased, stink bug problems have increased.

In 2004, *Bt* cotton was grown in nine states and comprised more than 75% of all cotton grown. Most varieties are Roundup Ready (RR) or RR and *Bt* combined. The proposed 'technology fee' for Bollgard II was US$99 per hectare in 2004, this is to be added to the seed price. *Bt* cotton is also grown in Braziil, Argentina, Mexico and Columbia. In Columbia the vice-president of the biosafety council works of Monsanto and was thus able to both apply for a grant permission for release of a *Bt* crop in an area that is a centre of origin for some wild cotton species. Moreover, the pest responsible for 70% of pesticide use on cotton is the picudo, which is not targeted by Monsanto's cotton. The small farmers will once again lose out due to this folly.

In 1995, attack from these pests reduced UK cotton yields by over 4%, or by over a quarter billion dollars worth of cotton. In some states such as Alabama, budworm damage reduced Alabama yields by 29% despite growers applying the highest applications of insecticide in US.

Reduction in Pesticide Control Costs:

It is estimated that the average cost reduction in pesticides applied on *Bt* cotton fields from 1996 to 1998 is between $25 and $26 per acre.

Increased yield:

Bt cotton planted from 1996 to 1998 is estimated to have yielded 5% more on average than if traditional cotton would have been planted.

Bt cotton: The Myth and the Reality

United state was the first country in the world to promote commercialization of transgenic cotton crop. Today besides them Argentina, P.R. China, Australia, South Africa, Indonesia have started commercial cultivation of *Bt* cotton. Our neighbouring country P.R. China has invested in a large way in transgenic cotton technology in China both Chinese Academy of Agricultural Science (CAAS) and Monsanto are marketing their *Bt* cotton products. Commercial cultivation of *Bt* cotton has considerably reduced the usage of insecticides. Reports indicates that there was reduction in use of insecticides by 900 metric ton (active ingredient) in the year 1998 in U.S. and the 1200 metric ton in 1999. P.R. China has reported a reduction of 15000 metric tons of formulated insecticide in 1999 on cotton.

Advantages of *Bt* cotton:

Major advantages of using *Bt* cotton are as follows:

1. Reduction in use of insecticides by almost 50% on cotton.
2. Reduction in use of insecticides in turn prevents pollution of soil, water and the atmosphere.
3. In a normal year, with a moderate attach of bollworms farmers can expect yield benefit of 10-15% while in the reasons with heavy bollworms attach yield increase of 25-30% can be expected in comparison to non-*Bt* version of the same hybrid or variety.
4. Quality of Kapas improves because of less kawadi Kapas aı d because of better quality, one can expect better premium to its produce.
5. Reduction in use of insecticides favors built up of population of beneficial insect pests thus aiding to improved plant protection.
6. Lesser residues of insecticides in the fibers will reduce the changes of allergic reactions. There is no doubt that introduction of *Bt* cotton will be a boon rather than bane.

Myth about *Bt* cotton:

Some of the commonly heard arguments against *Bt* cotton can be listed as under. For the sake of the society it is necessary to clarify these doubts with scientific back up.

1. ***Bt cotton has terminator gene:*** Terminator technology is a technology wherein the second generation of the crop is rendered sterile without any economic yield. Govt. of India has opposed this technology from day one and whatever *Bt* cotton seed that will be brought in the market will not have this technology. As far as the purchase of seed every year is concerned, it is not new thing for Indian farmers. With the onset of Green Revolution, Indian farmers have understood the importance of quality seeds. Seed replacement ratio has been gradually increasing in the country. Even in the crops like Soybean, Wheat, Paddy etc. where bulk planting material is required, farmers have willingness to buy certified seeds every year from the market. In Cotton, 1969 saw the 4". Every farmer knows that hybrid seed is to be purchased from the market every year because he realizes that in a hybrid good characters have been assembled from two diverse parents. *Bt* cotton technology will be available to Indian farmers in hybrid cotton as hybrids are occupying nearly 45% of the cultivated cotton area of the country. Hence like any other cotton hybrid, farmers will required to purchase *Bt* cotton seed every year.
2. ***India will have to rely on Multinationals for Every Year Seed Requirement:*** These kinds of statements do not have any meaning. Govt. of India has laid clear norms for procurement of the *Bt* technology. The varieties grown in the America are not suitable for Indian Cultivation. The companies that will be bringing *Bt* technology will be transferring the *Bt* gene in the Indian varieties which are well adopted to our conditions. These companies will be producing, packing and marketing the seed in India itself. The farmers will not have to rely on foreign companies for seed material.
3. ***Bt will destroy native crop of the country:*** This too is a baseless claim. The studies carried out on pollen dispersal in *Bt* cotton has indicated out crossing of non-transgenic plants from 0.42 to 2.1% up to a distance of 2 meters. The results in India are consistent with those from USA. Considering this fact there is hardly any doubt that *Bt* cotton will be safe to native non-Bt crop. Secondary *Bt* cotton technology is coming in *G. hirsutum* species which is not compatible with *G. arboreium* and *G. herbaceum*. Thirdly no wild relatives of cotton have their origin in India & hence drifting of the gene to wild relatives of the crop does not arise.
4. ***Bt may enter animal & human food chain:*** Food safety and evaluation of seeds, oil and cake of *Bt* cotton have already been done at Industrial Toxicology Research Center (ITRC) Lucknow. Their four year trial reports have clearly

indicated that *Bt* cotton is absolutely safe to mammals. Data from United States about safety of *Bt* cotton to mammals, birds and fish confirm the same fact.

5. ***Bt cotton will spoil the environment:*** *Bt* cotton is absolutely safe to the environment. In fact by using *Bt* cotton are can prevent soil, water and air pollution by reducing the use of insecticides to a large extent.

6. ***Sprays will be required on Bt cotton:*** Yes, a farmer will be required to take sprays even on *Bt* cotton. But these sprays will be required for the control of other pests mainly sucking pets like Aphids, Jassids of Thrips which can be easily controlled by insecticides in a couple of sprays. If we selected proper hybrids for sowing containing *Bt* gene, this problem will seldom arise. Major expenditure on insecticides is for controlling bollworms. *Bt* will take care of them and one may be required couple of sprays depending on pest attack. Population build up of white flies that develops after excessive use of pyrethroids may not occur. Since feeding habits of sucking pets and lepidopteron pests are altogether different a farmer will have spray on *Bt* cotton and sucking pests.

7. ***Farmer will not get any benefit from Bt cotton:*** This statement is also made with a prejudice. There may also be partial truth in this as one can not predict the prevailing market prices and the productivity of the crop. It is thumb rule of commerce that market prices go down with ample availability of the product. But one can not undermine the fact that farmers will get some saving in *Bt* cotton cultivation. He will be saving his advantage from which he can easily bear high cost at the seeds. If one think of the society, then with reduce use of pesticides there will be less pollution hazards, poisoning etc. So saying that farmer will not be absolutely benefited is also untrue. The consumer will have more quality of cotton bolls are available on long term credit from foreign countries. For this one has to become more quality conscious and make up the deficit in support prices by increasing the productivity of the crop and its better management.

8. ***Bt cotton will be costly:*** This is true and this is because the world rights of using *Bt* Gene are with Monsanto Inc., USA and who-so-ever is using their gene will have to pay the company by way of royalty of the gene. Secondly monitoring, transferring the gene in Indian lines needs high-tech laboratories and the chemicals which are use have to be imported. This all is increasing cost of the seed. But as the time passes and total indigenous research competent with this technology comes in the market the price will go down.

Potential risk to using *Bt*:

- ***Invasiveness:*** Genetic modifications, through traditional breeding or by genetic engineering can potentially change the organism to become invasive. Few introduced organisms become invasive, yet it's a concern for the users.
- ***Resistance to Bt:*** The biggest potential risk to using *Bt* crops is resistance. Farmers have taken many steps to help prevent resistance.
- ***Cross-contamination of gene:*** Although unproven, genes from GM crops can potential introduce the new genes to native species.

Resistance to *Bacillus thuringiensis* Toxins: A major problem

Insecticide Resistance:

White *Bt* is very unlike other insecticides in its origin, mode of action, and use, it still shares some of the problems of any insecticides. One major problem with insect control via insecticides is the evolution in insects of resistance to those insecticides. The first reported cases of insecticides resistance to early synthetic insecticides occurred over 50 years ago. About thirty years later, in 1979, the United Nations Environmental Programme declared pesticide resistance one of the world's most serious environmental problems. Its seriousness to the environment stems from problems of human nutrition due to crop loss, spread of disease by resistant insects, addition to the environment of new and potentially dangerous insecticides after resistance has developed, and application of greater and greater amounts of chemicals to which pests have already gained resistance (Pimentel and Burgess, 1985).

How Resistance Develops

Resistance to insecticides is, in fact, one of the many reasons Bacillus thuringiensis has come into common use today. Insecticide resistance develops due to genetic variation in large insect populations (Fig. 17). A few individuals in the original insect population are unaffected by a given insecticides. Generally unaffected (resistant) individuals differ from affected (susceptible) individuals either in the nature of the insecticide's target molecules in the insect, or in the method the insect uses to break down toxin molecules (Michaud, 1997). When the insecticide is applied, individuals who are unaffected by it are those who survive to pass their genes on to following generations. Over time, greater and greater proportion of the insect population is unaffected by the insecticide. In addition to the damage done by the increasingly large number of surviving, resistant insects, attempts to control insecticide resistance can indirectly cause secondary pests outbreaks that do yet more crop damage (Hoy, 1998).

Bt-cotton has faced opposition from a few NGOs and certain individuals, right from 1998 when it was still undergoing the prescribed regulatory trials in India. Their tirade is continuing even now, five years after its approval and adoption on about 3.8 m ha by 2.3 m farmers. The issues raised by them were all encompassing scientific, social, economical, ethical, emotional and legal grounds. Most of the allegations made by them were/are based on perceptions without rationale, primarily aimed at sensationalism. No scientific data seem to satisfy them and it appears to be an unending debate.

The activists of the Karnataka Rajya Raitha Sangha (KRRS) launched a campaign against *Bt*-cotton and burnt a few RCGM approved experimental plots in certain parts of Bellary and Raichur (Karnataka) in November 1998 alleging that they contained 'Terminator Technology' and that the terminator gene would escape and cause gene pollution and sterility in surrounding plants. They regularly issued scaring statements, held demonstrations, threatened the *Bt*-farmers and also ransacked the Monsanto laboratories in Bangalore during 1998-1999. A few other NGOs had also joined them in linking terminator technology with *Bt*-cotton. The truth is that *Bt*-cotton does not contain the so-called 'Terminator Gene.' The colloquial name 'Terminator Technology' (TT) was coined by Rural Advancement Foundation International (RAFI), an NGO headquartered in Canada. This name became very popular with the media. The technology was originally designated as 'Technology Protection System' (TPS) and was jointly patented by USDA (United States Dept of Agriculture) and Delta & Pine Land (a leading cotton seed company in the USA) in March 1998. The special feature of TT or TPS is that the seeds derived from the parent plants cannot be utilized for sowing as these would not germinate as such, but are otherwise fit for all other purposes. Such seeds need to be treated with a specific activator chemical compound to induce germination. The TPS was conceived with an idea to protect the proprietary traits from being pirated. This technology has not been commercialized in any crop as of 2006, nor has it been successfully advanced beyond some preliminary experiments. However, the pros and cons of TT became a subject of great controversy and hot debate in several countries. On its part, Monsanto has clarified and reiterated that its *Bt*-cotton does not contain the terminator gene. Following the allegations by the NGOs and as per the direction of the regulatory authorities, the Dept of Genetics, University of Delhi (South Campus), Delhi, carried out molecular detection tests on Mahyco's *Bt*-cotton hybrids to ascertain the presence or absence of the 'Terminator Gene.' The PCR analysis of DNA isolated from individual seedlings derived from *Bt*-cotton hybrids revealed that these lines were positive only for cry1Ac gene and did not contain cre recombinant gene which is an integral component of TT. This conclusively demonstrated the absence of 'Terminator Gene' in *Bt* cotton hybrids. Another 'common sense' test was carried out by a progressive farmer in Haveri and a seeds of the F1 generation and demonstrated that these germinated like normal seeds. It was unfortunate that the activists resorted to such extreme steps based on false assumptions without verifying the scientific facts.

A Public Interest Litigation (PIL) was filed in the Supreme Court by the NGO, Research Foundation for Science, Technology and Ecology (RFSTE), New Delhi, against Govt. of India (i.e. GEAC & MoEF), Monsanto and Mahyco, charging violation of the biosafety procedures during the approval of *Bt*-cotton filed trials in India. The National Environment Appellate Authority who examined this petition, dismissed it as baseless in October 2003. Similarly, several other litigations filed against *Bt*-cotton by other NGOs also did not find favour in the court of law. Even so, it has not stopped them from filing petitions again and again. Safety concerns regarding *Bt*-cotton were addressed scientifically under the direction and supervision of the regulatory authorities in India as in the USA and several other countries before the crop was approved for commercial cultivation. The comprehensive studies carried out in India covered issues such as allegenicity, toxicity and effect on non-target organisms (goats, buffaloes, cows, rabbits, birds, fish, honey bees, ladybird beetles, earthworms and other soil organisms). The experiments were conducted by the concerned experts, mostly in various public-funded institutions, and the data were submitted to the regulators (also see Section V). It is only after satisfying that the *Bt*-proteins are safe and beneficial, regulatory approval was granted. Further, *Bt*-cotton has been cultivated on millions of hectares in several countries since 1996 (and in India since 2002) and scientific investigations have not shown any harmful effect. The detractors must realize that safety assessment related to environment and human/animal health has been given the top-most priority by the regulators as well as product developers and that they are as much concerned about these issues as they are!

The seed-saving issue is not peculiar to *Bt*-cotton alone, but is applicable to all hybrids of any crop. Hybrids of many crops are preferred over varieties by farmers because they give much higher yields than varieties. It is a well-known fact that if the hybrid seeds saved from one generation are used to raise the next crop, the hybrid vigour declines coupled with segregation of traits and the resulting harvest becomes poorer from generation to generation. The cost of such yield losses will be far greater than the cost of new hybrid seeds. Hence, it is more profitable for farmers to buy fresh hybrid seeds that are specially produced under controlled conditions of pollination and undergo quality control checks. They must always buy it from reliable resources. In fact, our farmers have been cultivating hybrids of several crops, including cotton, corn, vegetables etc., for many years and most of them are already aware of the basic differences between the seeds of traditional varieties and hybrids. In India, *Bt* gene has been incorporated into cotton hybrids and, therefore it is advisable for farmers to buy new seeds so that they can enjoy a better harvest and profit. In case they use the saved seeds, they will germinate but express *Bt* protein in a segregated pattern (i.e., a portion of the plant population will not contain the *Bt* gene). In addition, the yield will be poor because of the loss in hybrid vigour.

During its 68th meeting held on May 2005, the GEAC deliberated at length on the representation received from the Centre for Sustainable Agriculture (CSA) regarding the alleged sheep mortality after feeding on the stubble of *Bt*-cotton crop in Warangal. After reviewing the case and the available data, it was the general opinion of GEAC that the report was highly exaggerated and was based more on hearsay than on scientific facts. They asserted that *Bt*-cotton, prior to its approval for commercial cultivation, has undergone animal feeding studies at the Industrial Toxicological Research Institute, Lucknow; National Dairy Research Institute, Karnal and at the G.B.Pant University, Izatnagar. No toxic effect was found in any animals even when they were fed with high doses of *Bt* protein through *Bt* cotton seed meal. It should be realized that the *Bt* proteins produced in *Bt*-cotton are insecticidal, that too specific to lepidopteran insects. The intestine of higher animals like sheep does not possess the necessary pH conditions to activate the *Bt* protein and also the specific receptors for protein-binding. These two are essential steps for the *Bt* protein to have toxic effect. Further, the amount of protein produced in *Bt* cotton plants is so small that it is adequate to kill only the first and second instar larvae of bollworms. The sheep mortality in Andhra Pradesh must have been due to factors other than *Bt* and these should have been thoroughly investigated before blaming *Bt*-cotton.

It has already been demonstrated that *Bt*-cotton is effective against the Indian bollworms such as the False American Bollworm (*Helicoverpa armigera),* Pink Bollworm (*Pectinophora gossy*piella), Spotted Bollworm (*Earias vittella*) and Spiny Bollworm (*E. insulana*) under different agro-climatic conditions. The more recently (in 2006) approved second generation *Bt*-cotton technology stacked with two *Bt* genes, not only controls these cotton bollworms more effectively, but also the Tobacco Caterpillar (*Spodoptera litura*). The same *Bt*-cotton technology has also been successfully utilized in other countries like China, South Africa and Australia where the bollworm complex is almost the same as in India.

It is a very cruel allegation that has no empirical basis. On the contrary, *Bt*-cotton has come as a big relief to farmers who have been haunted by the bollworms for more than three decades. It has saved their crops and enabled them to reap a better harvest and profit. A national survey conducted by IMRB (International Market Research Bureau) in 2004 indicated that for every Rupee spent by the farmers, they received Rs.5.80 in value for reduced insecticide cost and increased yield over conventional cotton. In other words, it has fetched more income and improved their living standard. This is reflected by the fact that number of farmers who have adopted this technology which was only a few thousand in 2002, the first year, has steadily increased from year to year to reach about 2.3 m (23 Lakhs) in 2006-more than a hundredfold jump! Several independent committees set up by the state governments to investigate the cause for farmers' suicides have mostly cited several debt-related social and economic issues as being responsible for this tragedy, but definitely not *Bt*-cotton. Further, *Bt* cotton cultivation has started in India only since 2002 whereas the farmers' suicide has been an issue since decades.

Any technology introduced into India after being cleared by the union government as safe and beneficial-be it in agriculture, medicine or industry and no matter who has developed it-should be welcome. We are living in a globalize society today. While modern technologies from overseas have been readily absorbed in other sectors, for example healthcare, there is no reason why our farmers alone be denied of such benefits if these can contribute to their betterment. With regard to *Bt*-cotton, only the technology has been adopted as it was not available locally, but the *Bt*-trait has been incorporated in the local cotton hybrids developed by the Indian seed companies and made available to the farmers. Thus, the technology has been indigenized. The number of farmers who have adopted *Bt*-cotton in India has increased from year to year in the last five years and reached 2.3 m (23 lakhs) in 2006, clearly reflecting that they have accepted this technology and realized its benefits. Our farmers are wise and they know what is good for them. They cannot be exploited. Our scientists are also trying to develop indigenous technologies, but even most of these are based on foreign know-how. Several public and private organizations in India have been trying to develop their own *Bt*-cotton and other transgenic for the last several years and it is hoped that they will come out with some good products. This would offer more choice to our farmers to choose whatever is more beneficial to them. Healthy criticism is always helpful in identifying gaps and refining a technology. Some of the issues raised by certain scientists and NGOs such as those related to the methods adopted for field trials, implementation of regulatory policies, transparency of scientific data, importance of cultivars with superior agronomic genotypes etc have made positive contributions towards fine-tuning certain aspects. The technology developers and regulators have learnt their lessons – it is not just enough to have a good technology, but it is equally important to create adequate awareness and knowledge in the stakeholders about the new products and also address the perceptions created by misinformation. The regulators, on their part, must have realized that if the official approval of new products, especially those with proven merits elsewhere, is unduly delayed, some miscreants will introduce them in a clandestine way as exemplified by the illegal *Bt*-cotton. The message to the scientists and scientific bodies is that they should speak up and write unambiguously as otherwise it will be misinterpreted by the opponents and highlighted in the media leading to their own embarrassment.

Bt-cotton is a thoroughly researched product of biotechnology. Unfortunately, most of the people who are involved in organizing protests etc. are non-scientists, pseudo-farmers or professional agitators. Their destructive activities will only hurt our farmers who need modern technologies and proper guidance for their progress. Public/ farmer awareness and education on the safety and benefits of biotechnology should be given top priority. The biotech industry and scientific community should make united efforts in this endeavor, especially to mitigate the unsubstantiated anti-biotech stories being hoisted aggressively by the anti-biotech lobbies to create doubts and fear and mislead the public. Media can play a very helpful role in highlighting the positive contributions of biotechnology. They need to verify the scientific truth before reporting sensational stories.

Government should take strict action against vandalism and illegal seed producers.

Starting with Mahyco's 3 *Bt*-cotton hybrids (MECH 12 *Bt*, MECH 162 *Bt* and MECH 184 *Bt*) in March 2002, another 20 hybrids from four companies were duly approved by GEAC and commercially planted by 2005. In 2006, yet another 39 hybrids were approved which will be ready for planting in the 2006-07 cotton season. Thus, altogether 62 hybrids have been approved in India between 2002 and 2006. Of these, 55 (48 with a single gene and 7 with two genes) belonged to Mahyco-Monsanto and its sub-licensees, 4 to J. K. Seeds with a single gene and 3 to Nath Seeds with two genes. Altogether 15 companies were involved (see table).

The following organizations are trying to bring out transgenic cotton with insecticide genes sourced from *Bt* or plant, using a different event or as sub-licensee of other seed companies (see Table). Their experiments are at various stages of regulatory approvals. Besides, some of the above mentioned seed companies are continuing their efforts to introduce *Bt*-genes into newer hybrids.

Starting with Mahyco's 3 *Bt*-cotton hybrids (MECH 12 *Bt*, MECH 162 *Bt* and MECH 184 *Bt*) in March 2002, another 20 hybrids from four companies were duly approved by GEAC and commercially planted by 2005. In 2006, yet another 39 hybrids were approved which will be ready for planting in the 2006-07 cotton season. Thus, altogether 62 hybrids have been approved in India between 2002 and 2006. Of these, 55 (48 with a single gene and 7 with two genes) belonged to Mahyco-Monsanto and its sub-licensees, 4 to J. K.

Seeds with a single gene and 3 to Nath Seeds with two genes. Altogether 15 companies were involved.

The following organizations are trying to bring out transgenic cotton with insecticide genes sourced from *Bt* or plant, using a different event or as sub-licensee of other seed companies (see Table). Their experiments are at various stages of regulatory approvals. Besides, some of the above mentioned seed companies are continuing their efforts to introduce *Bt*-genes into newer hybrids.

Every organization tries to develop new and improved hybrids to suit different agro-climatic regions. Also, the hybrids developed by different companies possess different genetic backgrounds and their agronomic performances vary. It offers a wide choice to farmers. *Bt* gene has been introduced into some of these hybrids. For example, of the 62 *Bt* hybrids approved in India so far, 36 hybrids from 15 companies have been recommended for cultivation in Central Zone (Gujarat, Madhya Pradesh, Maharashtra), 14 from 6 companies for North Zone (Haryana, Punjab, Rajasthan) and 31 from 13 companies for South Zone (Andhra Pradesh, Karnataka, Tamil Nadu) with a few hybrids being recommended for more than one zone. *Bt*, being a trait, can be incorporated into any desired cultivar to confer resistance to bollworms. Several new hybrids developed by various companies are undergoing regulatory trials and awaiting approval. It is an ongoing process. Realizing the market potential of *Bt*-cotton in India, certain agencies are exploiting the situation through development and sales of unapproved *Bt*-cotton. In fact, illegal *Bt*-cotton crops were first detected in certain parts of Gujarat in 2000/2001 even prior to the regulatory approval of Mahyco's *Bt*-cotton in March 2002. Navbharat Seeds Pvt. Ltd., Ahmedabad, was identified as the offender. The company was marketing the seeds of its hybrid, Navbharat 151, claiming that it was naturally boll worm resistant. On suspicion, GEAC got such seeds tested by the Central Institute of Cotton Research, Nagpur. PCR and ELISA tests revealed the presence of cry1Ac gene in all the samples. Besides, the samples also contained seeds of F1 and F2 generations. Since this product had not gone through the mandatory biosafety screening and approval, it is a serious violation of regulatory procedure. Later, such illegal or spurious seeds were also found in Maharashtra, Madhya Pradesh, Andhra Pradesh, Karnataka and elsewhere where they continue to occupy several thousand hectares. According to a recent estimation, the illegal *Bt*-cotton occupied at least 30% of the total *Bt*-cotton grown in our country. It is really shocking.

Illegal *Bt*-cotton negatively impacts in several ways:

- Sale of illegal seeds is a blatant violation of bio-safety norms and business ethics.
- Spurious producers are not accountable for purity, performance and safety.

- They may spoil the credibility of the product and technology.
- They can afford to sell their products at a much lower price as their investment on research and product development is meager.
- It will affect the confidence and enthusiasm of the law-abiding technology developers.
- Tax revenue is lost at the government levels.
- Farmers will be misled and confused.

It is a serious challenge and urgent action is needed to curb this malpractice.

The Agriculture Ministry had ordered destruction of the illegal *Bt* cotton and some stock was also burned as warning, but the farmers and state government demanded compensation. This of course was not forthcoming. A case was also registered against the managing director of Navbharat Seeds Pvt. Ltd. Several state governments have also conducted a few raids and threatened to take penal action against those involved in such activities. Subsequently, the issue was pushed aside as it was taking a political turn. To curb this malpractice, the government should strictly implement the Environment Protection Act (EPA) and take punitive action against the guilty. The state authorities have a greater role to play in solving this problem by identifying illegal seed production plots, ginning mills and retailer warehouses. They should strengthen the hands of the State Biotech Coordination Committee (SBCC) and the District Level Committees (DLCs). Introducing a system for mandatory registration of cotton seed production plots similar to the mandatory registration system already in place for public hybrids will also help. Moreover, farmers and dealers should also cooperate with the governments in discouraging illegal seeds.

Several public and private institutions in India have introduced *Bt* genes into brinjal (egg plant), cabbage, cauliflower, corn, okra, pigeonpea and rice to control various lepidopteran pests. Work is also in progress on various other beneficial traits to overcome biotic (pests, diseases, weeds) and abiotic (drought, salinity, cold etc) stresses as well as to enhance the quality of food (improved nutrition, enhanced shelf-life, edible vaccines etc). Altogether about 17 crops are undergoing such experiments in various public and private laboratories in India. Globally, about 63 countries are working on about 57 crop species for various beneficial traits.

The Government of India has adopted a policy of precautionary principles, on case by case basis, for careful evaluation of the risks and benefits of biotech crops and other GM products, at various stages of their development, before they are approved for commercialization. Such rules were framed and guidelines notified in 1989 under the

Environment Protection Act 1986 (EPA). The guidelines were amended in 1990, 1994 and 1998 to keep pace with the progress made in the GMO research. Two nodal agencies, Ministry of Environment and Forests (MoEF) and Dept of Biotechnology (*DBT*) under the Ministry of Science and Technology (MoST) are responsible for implementation of the biotech policies. They have constituted four major committees to handle various issues: Recombinant DNA Advisory Committee (RDAC), Institutional Bio-Safety Committee (IBSC), Review Committee on Genetic Manipulation (RCGM) and Genetic Engineering Approval Committee (GEAC).

These committees are further assisted by State Biotechnology Coordination Committee (SBCC), District Level Committee (DLC) and Monitoring and Evaluation Committee (MEC). All these committees have specific responsibilities (see figure).

The major committees are comprised of eminent experts drawn from various fields from various organizations across the country. They view the safety of a product from various perspectives and at various levels. The safety assessments are made through experiments based on scientific principles under their expert guidance. It is unfair to say that the committees are not represented by experts. The biosafety regulations in India are as stringent as anywhere else in the world. However, there is scope of implementing these more strictly. If any one has any specific suggestions for improvement, these can be communicated to the regulatory authorities for their consideration. We should always strive for excellence.

The laws are clear, but to be effective their enforcement should also be very stringent. Any infringement of the prescribed regulations is a punishable act. The government did try to take some action against the offenders, but did not punish them adequately to serve as a warning to others. The presence of illegal seeds is not a healthy trend and calls for strict and urgent action.

The Indian agri-biotech policies and protocols have evolved and become more streamlined with *Bt*-cotton which was the first product that was scrutenized for biosafety. Monsanto, being the pioneer, has paid the price. It has been the focus of all the debates, opposition and disputes. Monsanto-Mahyco had to patiently pass through all the hurdles before it got the approval of *Bt*-cotton after struggling for 6-7 years. It had also cost huge sums of money. Undoubtedly, this has smoothened the way for future applicants. Today, the protocols are more clear, the applicants know before hand what exactly are to be done and the approvals are relatively faster. This is reflected by the fact that about 42 new *Bt*-cotton hybrids were approved in 2006 alone as compared to 3 hybrids in 2002, 1 in 2004 and 16 in 2005. The approval process has also been simplified from 'hybrid-based' approval to 'event-based' approval. If an honest effort is made, everything improves with time.

India made its long-awaited entry into agricultural biotechnology in March 2002 with the regulatory approval of three *Bt*-cotton hybrids developed by Mayco-Monsanto for control of bollworms. This approval was preceded by a large number of laboratory studies and about 500 field trials carried out from 1996 to 2001 to demonstrate the safety and benefits of *Bt*-cotton as per regulatory requirements. This technology came as a big relief to cotton farmers who have been haunted by bollworms for more than three decades. Realizing the potential of *Bt*-cotton, reputed seed companies like Rasi, Ankur, Nuziveedu and several others, numbering more than 20, have already become sub-licensees of MMB (Mahyco-Monsanto Biotech Company) for this technology while two other companies, J.K.Seeds and Nath Seeds, have come out with newer versions of *Bt*-cotton. Efforts by several other organizations, both public and private, are also in progress. The public institutions intend to introduce *Bt*-gene(s) into several public cotton varieties/hybrids that have been developed by them for different agro-climatic regions. By 2006, altogether 62 *Bt*-hybrids from 15 companies have been officially approved by GEAC for commercial cultivation. These included 52 hybrids with single *Bt*-gene (48 from MMB and its sublicensees and 4 from J.K.Seeds) and 10 hybrids with two genes stacked (7 from MMB and its sub-licensees and 3 from Nath Seeds). Altogether, three genes (cry1Ac, cry2Ab2, cry1Ab) and four events (MON 531, MON 15985, Event I and GFM event) have been involved. Such diversity in genes and genotypes offers wide choice to our farmers and also helps in preventing or delaying resistance development in bollworms. All the *Bt* hybrids commercialized so far are *Gossypium hirsutum*. The *Bt* technology could extend to *G. arboreum* and the premium long-staple cotton, *G. barbadense*. The response to *Bt*-cotton by Indian farmers and its performance have been very encouraging. The area occupied by *Bt*-cotton in 2002, the first year, was only about 29,000 hectares in six states which significantly increased from year to year to reach 3.8 million hectares (9.4 m acres) in nine states in 2006. Similarly, the number of *Bt* farmers also increased from a few thousand in the first year to about 2.3 million in 2006, clearly indicating their confidence in this technology.

The major benefits from *Bt*-cotton included higher yields owing to effective control of bollworms, drastic reduction in the application of chemical pesticides and greater profit to farmers. Coincidental with its steep increased adoption from 2002, the average yield of cotton in India, which had one of the lowest yields in the world, increased from 308 kg per hectare in 2001-02 to 450 kg per hectare in 2005-06, with most of the increase in yield of up to 50%, or more, attributed to *Bt* cotton. At national level, this is a major

1

2

3
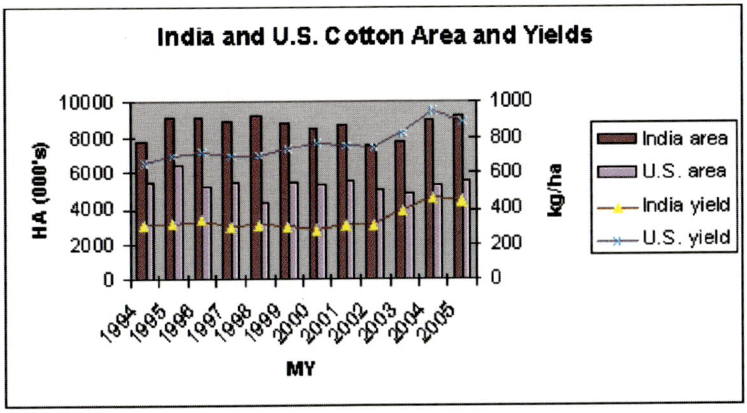

4
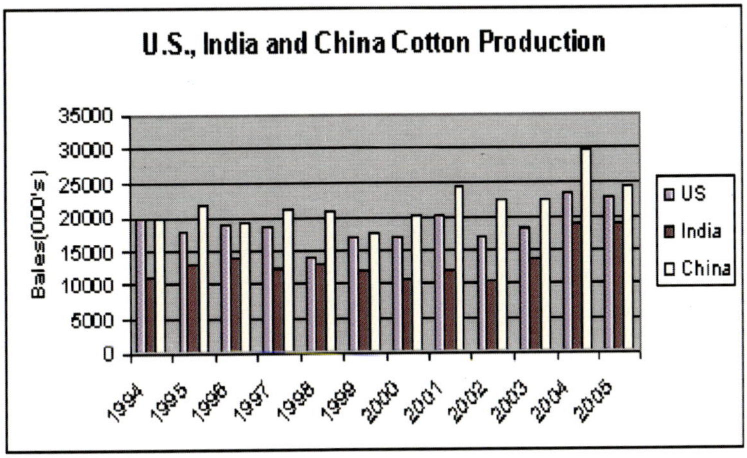

Fig: 1) Cotton cultivation across millions of hectares of land, **2)** Healthy cotton fruit ready to pluck, **3)** Total area under cotton cultivation in India and USA: Yield is also compared with the two countries, **4)** Comparison of the total cotton production between India, USA and China

5

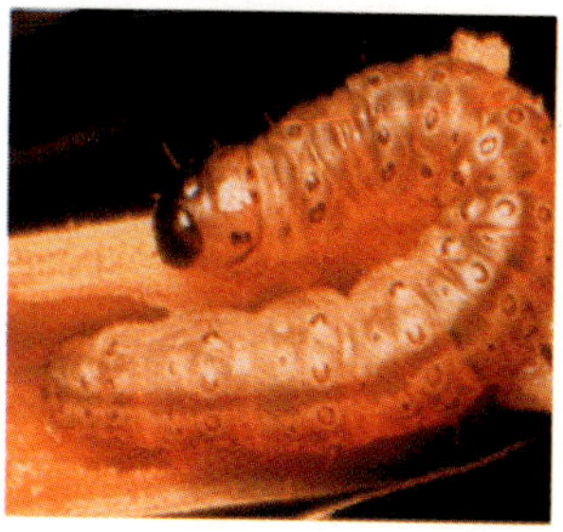

6

7

10

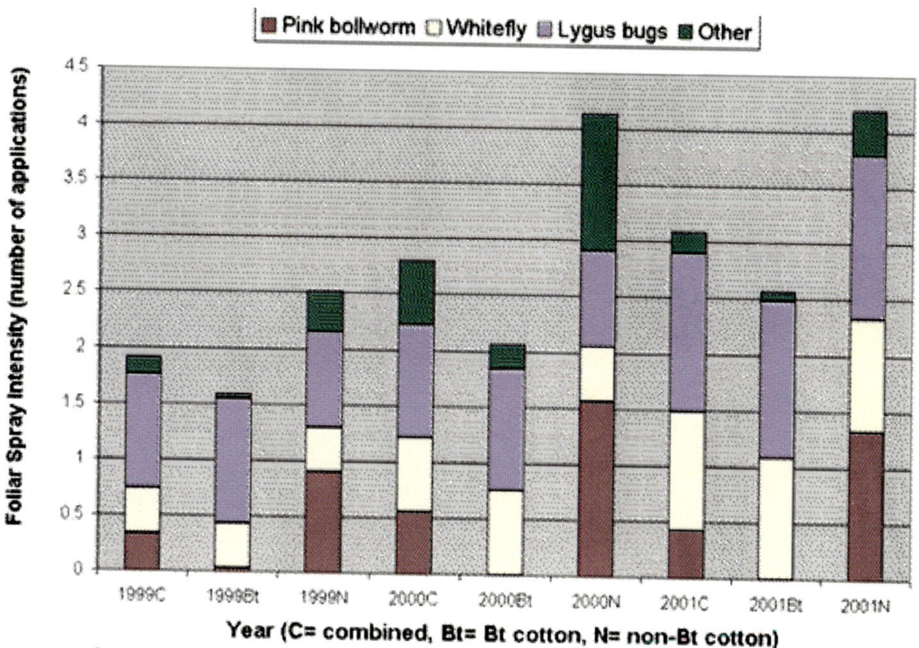

Source: P. Ellsworth, U. of Arizona

Fig: 5) A cotton Bollworm, **6)** Attack of cotton bollworm on a cotton bud, **7)** Bollworm rot as a result of attack on cotton bud, **10)** The foliar spray intensity of insecticides found to be least for Bt cotton.

*** 8-9)** Included within text

factor in higher cotton production increasing from 15.8 million bales in 2001-02 to 24.4 million bales in 2005-06, which is a record cotton crop for India (source: minutes of the third meeting of the Cotton Advisory Board for 2005-06; James, 2006). This is what a good technology can do to improve our productivity and economy.

India has a robust, multi-tier regulatory system comparable to any other country in the world. The major regulatory committees like RCGM (Review Committee on Genetic Modification) and GEAC (Genetic Engineering Approval Committee), constituted by Govt. of India, are composed of a number of eminent experts drawn from various reputed institutions across the country. They use their collective wisdom in scrutinizing the scientific data from various perspectives before approving any product as safe. In fact, safety is accorded the highest priority in biotechnology. In the last five years of its commercial cultivation in India and for eleven continuous years on thousands of hectares in several other countries, *Bt*-cotton has not caused any negative impact related to safety of human/ animal/environment nor has there been any crop contamination or pest resistance anywhere in the world. It has an impeccable global safety record as are the other *Bt*-crops. The detractors of biotechnology do not seem to take cognizance of these facts.

A few NGOs and certain individuals have opposed *Bt*-cotton as unsafe and pronounced it as a failure even before it completed the biosafety and agronomic trials as stipulated by the regulators and continue to do the same even years after its proven safety and benefits. They must realize that the regulators as well as product developers are as responsible and concerned about the safety of humans and environment as they are! Healthy criticism is welcome, but blind opposition and creating suspicion and fear through unsubstantiated allegations have no place in science. Fear is the worst enemy of progress. The current global scenario of agricultural biotechnology is quite encouraging. The year 2006 marked the completion of a decade of large scale commercial cultivation of transgenic crops, including *Bt* cotton, in multiple countries. It has been a tough journey punctuated with speculated risks, calculated opposition and mischievous controversies, as with some of the new technologies in the past, but finally leading to greater support, acceptance and success of the products following their proven merits, safety and benefits. This is reflected by the fact that in 2006, biotech crops were grown on 102.0 million hectares (or 252 million acres) in 22 countries (11 developing, 11 industrial), including India, on six continents, by 10.3 million farmers, marking a 60-fold increase in the area since their first commercialization on 1.7 m ha in 6 countries in 1996. Such a fast rate of adoption of a new technology is unprecedented in agriculture. According to ISAAA (James, 2006), the outlook for the next decade of commercialization, 2006 to 2015, points to continued growth in the global hectare of biotech crops, up to 200 million hectares, with at least 20 million farmers growing biotech crops in up to 40 countries or more by 2015. Most of this growth is expected to take place in the developing countries of Asia, led by China and India. The Indian Government is very supportive of biotechnology. Research on more exciting transgenic traits such as nutritional enhancement and tolerance to abiotic stresses like drought and salinity as well as biotic stresses like pests and diseases are making rapid progress in several public and private organizations.

These products would add value to sustainable agriculture and contribute towards meeting the food demand of the burgeoning population. With its vast resources, India has the potential to emerge as a supreme power in agriculture if modern technologies are appropriately reviewed and adopted. Farmers are the greatest beneficiaries of agricultural biotechnology.

REFERENCES

Adamczyk J J Holloway J W Church G E Leonard B R and Graves J B (1998) Larval survival and development of the fall armyworm (Lepidoptera: Noctuidae) on normal and transgenic cotton expressing the *Bacillus thuringiensis* CryIA(c) v delta-endotoxin *J Econ Entomol* **91**: 539-545

Akhurst R J James W Bird L and Beard C (2003) Resistance to the Cry1Ac-endotoxin of *Bacillus thuringiensis* in the cotton bollworm *Helicoverpa armigera* (Lepidoptera: Noctuidae) *J Econ Entomol* **96**: 1290-1299

Beever D E Glenn K And Phipps R H (2003) (A safety evalution of gentiaclly modified feedstuffs for livestock production: the fate of transgenic DNA and proteins *Asian-Australian J Anim Sci* **16**: 764-772

Benedict J and Altman D W (2001) Commercialization of transgenic cotton expression insecticidal crystal protein In: *Genetic Improvement of cotton: Emerging Technologies* (Eds Jenkins J and Saha S) Science Publication Enfiled New Hampshire USA 137-201

Benedict J H Altman D W Umbeck P F and Ring D R (1992) Behaviour growth survival and plant injury *Heliothis virescens* (F) (Lepidoptera: Nocutidae) on transgenic *Bt* cotton *J Econ Entomol* **85**: 589-593

Benedict J H Sachs E S Altman D W Deaton W R Kohel R J Ring D R and Berberich S A (1996) Field performance of cottons expressing transgenic CryIA insecticidal proteins for resistance to *Heliothis virescens* and *Helicoverpa zea* (Lepidoptera: Noctuidae) on transgenic *Bt* cotton *J Econ Entomol* **89**: 230-238

Betz F Hammond B and Fuchs R (2000) Safety and advantages of *Bacillus thuringiensis*-protected plants to control insect pests *Regul Toxicol Pharmacol* **32**: 156-173

Bharathan G (2000) *Bt*-cotton in India: Anatomy of a controversy *Curr Sci* **79**: 1067-1075

Caprio M A (2001) Source-sink dynamics between transgenic and nontransgenic habitats and their role in the evolution of resistance *J Econ Entomol* **94**: 698-705

Chesson A and Flaschowsky G (2003) Transgenic plants in poultry nutrition *World's Poultry Sci J* **59**: 201-207

Cousins Y J Lyon B R and Liewellyn D N (1991) Transformation of an Australian cotton cultivar: prospects for cotton improvement through genetic engineering *Aust J Plant Physiol* **18**: 481-494

CRISO Plant Industry (2003) Bollgard® II-the new generation of GM cotton http://www csiro au/proprietaryDocuments/PI_info_bollgard pdf

De Barjac and Frachon E 1990 Classification of *Bacillus thuringienesis strains Entomophaga* **35**: 233-240

Dong H z Li W J Tang W I and Zhang D M (2004) Development of hybrid *Bt* cotton in China-A successful integration of transgenic technology and conventional techniques *Curr Sci* **86**: 778-782

Drobniewski FA 1994 A Review: The safety of *Bacillus* species an insect vectors control agents *Bacteriol* **76**: 101-109

Eastick R (2003) Potential weediness of *Bt* cotton *Cotton Frontiers* **4**: 1-3

Eddleman B R Darmont D He Q and McCarl B A (1995) Potential economic benefits to society rom *Bt* cotton In *Proc Belwide Cotton Conf* p 393-396 San Antonio TX 4-7 Jan 1995

Edge J M Benedict J H Carroll J P and Reding H K (2001) Bollgard cotton: an assessment of global economic environmental and social benefits *J Cotton Sci* **5**: 121-136

Elzen G W and Hardee D D (2003) United States Department of Agriculture – Agricultural Research Service Research on managing insect resistance to insecticides *Pest Manage Sci* **59**: 770-776

Falck-Zepeda J B Traxler G And Nelson R G (1999) Rent creation and distribution from the first three years of planting *Bt* cotton *ISAAA Brief 14 ISAAA* Ithaca NY

Fan X Zhao J Z Fan Y and Shi X (2000) Inhibition of transgenic *Bt* plants to the growth of cotton bollworm *Plant Prot* **26**: 3-5

Feitelson J S Payne J Kim L (1992) *Bacillus thuriengenisis* Insects and beyond *Biotechnology* **10**: 271-275

Frisvold G B Tronstand R and Mortensen J (2000) Adoption of *Bt* cotton ; Regional differences in producers costs and returns In: *Proc Beltwide Cotton Conf* p 337-340 San Antonio TX 4-8 Jan 2000 Natl Cotton Counc Am Memphis TN

Fuchs R L Berberich S A and Serdy F S (1993) Safety evaluation of genetically engineered plants and plant products: insect-resistant cotton In: *Biotechnology and Safety Assessment* (Eds Thomas J A and Meyers L A) New York NY: Raven Press 199-212

Gore J Leonard B R Chruch G E Russell J S and Hall T S (2000) Cotton boll abscission and yield losses associated with first install bollworm (Lepidoptera: Noctuidae) injury to nontrasgenic and transgenic *Bt* cotton *J Econ Entomol* **93**: 690-696

Gould F (1998) Sustainability of transgenic insecticidal cultivars: Integrating pest genetics and ecology *Annu Rev Entomol* **43**: 701-726

Gould F (2000) Testing *Bt* refuge strategies in the field *Nature Biotech* **18**: 266-267

Gould F Anderson A Reynolds A Bumgarner L and Moar W (1995) Selection andgentic analysis of *Heliothis virescens* (Lepidoptera: Noctuidae) strain with high levels of resistance to *Bacillus thuringiensis* toxins *J Econ Entomol* **88**: 1545-1559

Halcomb J L Benedict J H Cook B Ring D R and Correa J C (2000) Feeding behavior of bollworm and tobacco budworm (Lepidoptera: Noctuidae) larvae in mixed stands of nontransgenic and transgenic cotton expressing an insecticidal protein *J Econ Entomol* **93**: 1300-1307

Herman R A Schafer B W Korjagin V A and Ernest A D (2003) Rapid digestion of Cry34Ab1 and Cry35Ab1 in simulated gastric fluid *J Agric Food Chem* **51**: 6823-6827

Hofmann C Luthy P Hutter R And Plisks V (1988a) Binding of the delta endotoxin from *Bacillus thuringiensis* to brush border membrane vesicles of the cabbage butterfly (*Pieris brassicae*) *Eur J Biochem* **173**: 85-91

Hofmann C Vanderbruggern H Hofte H van Rie J Janssens S And van Mellaert H (1988b) Specificity of *Bacillus thuringeinsis* delta-endotoxins is correlated with the presenceof high affinity binding sites in the brush border membrane of target insect midgets *Proc Natl Acad Sci USA* **85**: 7844-7848

International Advisory Committee (1996) Cotton: Review of World Situation Monogram by International Advisory Committee Washington D C

James C (2002) Global review of commercialized transgenic crops: 2001 feature: *Bt* cotton *ISAAA Briefs No 26 ISAAA*: Ithaca NY

Jennings J C Albee L D Kolwyck D C Surber J B Taylor M L Hartnell G F Lirettee R P and Glenn K C (2003) Attempts to detect transgenic and enogenous plant DNA and transgenic protein in muscle from broilers fed Yield Gard Corn Borer Corn *Poultry Sci* **82**: 371-380

John M E (1997) Cotton crop improvement through genetic engineering *Crit Rev Biotech* **17**: 185-208

Kohel R J Quisenberry J E Cartwright G and Yu J (2000) Linkage analysis of transgenic inserted into cotton via *Agrobacterium tumefaciens* transformation *J Cotton Sci* **4**: 65-69

Linacre N A and Thompson C J (2004) Dynamics of insect resistance in *Bt* corn *Ecol Model* **171**: 271-278

Liu Y B and Tabashnik B E (1997) Experimental evidence that refuges delay insect adaptation of *Bacillus thuringiensis* Cry1C in diamondback moth *Appl Environ Microbiol* **63**: 2218-2223

Liu Y B Tabashnik B E Dennehy T J Patin A L and Barlett A C (1999) Development time and resistance to *Bt* crops *Nature* **400**: 519

Liu Y B Tabashnik B E Dennehy T J Patin A L Sims M A Meyere S K and Carriere Y (2001a) Effects of *Bt* cotton and Cry1Ac toxin on survival and development of pink bollworm (Lepidoptera: Gelechiidae) *J Econ Entomol* **94**: 1234-1242

Liu Y B Tabashnik B E Meyere S K Carriere Y and Barlett A C (2001b) Genetics of pink bollworm resistance to *Bacillus thuringeinsis* toxins Cry1ac *J Econ Entomol* **94**: 248-252

Llewellyn D and Fitt G (1996) Pollen dispersal from two field trials of transgenic cotton in the Namoi Valley Australia *Mol Breed* **2**: 157-166

Losey J E Calvin D D Carter M E and Mason C E (2001) Evaluation of noncorn host plants as a refugee in a resistance management program for European corn borer (Lepidoptera: Crambidae) on *Bt*-corn *Envirion Entomol* **30**: 728-735

Matthews G A (1994) Insect and mite pests: General Introduction In: *Insect Pests of Cotton* (Eds Matthews G A and Tunstall J P) CAB International Wallingford UK 95-105

McClintock J T Schaffer C R and Sjoblad R D (1995) A comparative review of the mammalian toxicity of *Bacillus thuringiensis*-based pesticides *Pestic Sci* **45**: 95-105

Metcalfe D D Astwood J D Townsend R Sampson H A Taylor S L and Fuchs R L (1996) Assessment of the allergenic potential of foods derived from genetically engineered crop plants *Crit Rev Food Sci Nutr* **36**: 165-186

Morin S Biggs R W Sisterson M S Shrier L Ellers-Kirk C Higginson D Holley D Gahan L J Heckel D G Carriere Y Dennehy T J Brown J K and Tabashnik B E (2003) Three cadhrein alleles associated with resistance to *Bacillus thuringiensis* in pink bollworm *Proc Natl Acad Sci USA* **100**: 5004-5009

Nordlee J A Taylor S L Townsend J A Thomas L A and Bush R K (1996) Identification of a Brazil nut allergen in transgenic soybeans *N Z J Med* **334**: 688-692

Oerke E C (2002) Crop losses due to pests in major crops In: *CAB International Crop Protection Compendium 2002 Economic Impact* CAB International Wallingford UK

Olsen K M and Daly J C (2000) Plant-toxin interactions in transgenic *Bt* cotton and their effect on mortality of *Helicoverpa armigera* (Lepidoptera: Noctuidae) *J Econ Entomol* **93**: 1293-1299

Onstad D W Guse C A Porter P Buschman L L Higgins R A Solderbeck P E Peairs F B and Cronholm G B (2002) Modeling the development of resistance by stalk-boring Lepidopteran insects (Cramibdae) in areas with transgenic corn and frequent insecticide use *J Econ Entomol* **95**: 1033-1043

Palm C J Schaller D L Donegan K K and Seidler R J (1996) Persistence in soil of transgenic plant produced *Bacillus thuringiensis* var kurstaki delta-endotoxin *Can J Microbiol* **42**: 1258-1262

Pan X P Zhang B H and Cobb G P (2005) Transgenic plants: environmental benefits and risks *Physiol Mol Biol Plant* **11**: 13-32

Parker C D Mascarenhas V J Luttrell R G andKnighten K (2000) Survival rates of tobacco budworm (Lepidoptera: Noctuidae) larvae exposed to transgenic cottons expressing insecticidal protein of *Bacillus thuringiensis* Berliner *J Econ Entomol* **35**: 105-117

Perlak F J Deaton R W Armstrong R L Fuchs R L Sims S R Greenplate J T and Fischhoff D A (1990) Insect resistance cotton plants *Bio-Technology* **8**: 939-943

Perlak F J Fuchs D A and Dean S L (1991) Modification of the coding sequence enhances plant expression of insect control protein genes *Proc Natl Acad Sci USA* **88**: 3324-3328

Perlak F J Oppenhuizen M Gustafson K Voth R Sivasupramanian S Heering D Carey B Ihrig R A and Roberts J K (2001) Development and commercial use of Bollgard® cotton in the USA – early promises versus today's reality *Plant J* **27**: 489-501

Pray C Huang J Hu R and Rozelle S (2002) Five years of *Bt* cotton in China-the benefits continue *Plant J* **31**: 423-430

Qaim M (2003) *Bt* cotton in India: field trial results and economic projections *World Development* **31**: 2115-2127

Qaim M and Zilberman D (2003) Yield effects of genetically modified crops in developing countries *Science* **299**: 900-902

Rabobank (1996) The International Cotton Complex Utrecht Neterhlands 104

Revenkova E V Kraev A S and Skryabin K G (1990) Transformation of cotton (*Gossypium hirsutum* L) with a supervirulent strain of *Agrobacterium tumefaciens Mol Biol* **24**: 820-825

Sacchi V E Parenti P Hanozet G M Giordana B Luthy P and Wolfersberger M G (1986) *Bacillus thuringiensis* toxin inhibit k*-gradient-dependent amino acid transport across the brush border membrane of pieris brassicae midgut cells *FEBS Lett* **204**: 213-218

Sadras V O (1998) Herbivory tolerance of cotton expressing insecticidal proteins from *Bacillus thuringiensis*: Responses to damage caused by Helcoverpa spp and to manual bud removal *Field Crop Res* **56**: 287-299

Schnepf E Crickmore N Van Rie J Lereclus D Baum J Feitelson J Zeigler D R and Dean D H (1998) *Bacillus thuringiensis* and its pesticidal crystal proteins *Microbiol Mol Biol Rev* **62**: 775-806

Shaunak R K (1994) Economic implications of adopting genetically engineered Heliothis resistant *Bt* cotton cultivars in Mississippi Ph D diss Mississippi State Univ Mississippi State M S

Shelton A M Tang J D Roush R T Metz T D and Earle E D (2000) Field tests on managing resistance to *Bt* engineered plants *Nature Biotech* **18**: 339-342

Shelton A M Zhao J Z and Roush R T (2002) Economic ecological food safety and social consequence of the deployment of *Bt* transgenic plants *Annu Rev Entomol* **47**: 845-881

Shimada N Kim Y S Miyamoto K Yoshioka M and Murata H (2003) Effect of *Bacillus thuringiensis* Cry1Ab toxin on mammalian cells *J Veterinary Med Sci* **65**: 187-191

Sims S R and Martin J W (1997) Effect of the *Bacillus thuringiensis* insecticidal proteins Cry1A(b) Cry1A(c) CryIIA and CryIIIA on *Folsomia candida* and *Xenylla grisea* (Insecta: Collembola) *Pedobiologia* **41**: 412-416

Srinivas K R (2002) *Bt* cotton in India: Economic factors versus environmental concerns *Environ Politics* **11**: 154-158

Stephens S G and Mosley M E (1974) Early domesticated cottons from archaeological sites in central coastal Peru *A Antiquity* **39**: 109-122

Tabashnik B E Dennehy T J Sims M A Larkin K Head G P Moar W J and Carriere Y (2002b) Cotnrol of resistant pink bollworm by transgenic cotton with *Bacillus thuringeinsis* toxin Cry2Ab *Appl Environ Microbiol* **68**: 3790-3794

Tabashnik B E Dennehy T J Liu Y B Carriere Y Sims M A and Antilla L (2000a) Frequency of resistance to *Bacillus thuringiensis* in field populations of pink bollworm *Proc Natl Acad Sci USA* **97**: 12980-12984

Tabashnik B E Roush R T Earle E D andShelton A M (200b) Resistance to *Bt* toxins *Science* **287**: 42

Tabashnik B E (1994) Evolution of resistance to *Bacillus thuringiensis Annu Rev Entomol* **39**: 47-79

Tabashnik B E Carrier Y Dennehy T J Morin S Sisteroson M S Roush R T Shelton A M and Zhao J Z (2003) Insect resistance to transgenic *Bt* crops: Lessons from the laboratory and field *J Econ Entomol* **96**: 1031-1038

Tabashnik B E Liu Y B Dennehy T J Sims M A Sisteronson M S Biggs R W and Carriers Y (2000a) Inheritance of resistance to *Bt* toxin Cry1Ac in field-derived strain of pink bollworm (Lepiodptera: Gelechiidae) *J Econ Entomol* **95**: 1018-1029

Tang J D Collins H L Metz T D Earle E D Zhao J Z Roush R T and Shelton A M (2001) Greenhouse tests on resistance management of *Bt* transgenic plants using refuge strategies *J Econ Entomol* **94**: 240-247

Thirtle C Beyers L Ismael Y and Piesse J (2003) Can GM-technologies help poor? The impact of *Bt* cotton in Makhathini Flats KwaZulu-Natal *World Development* **31**: 717-732

Traxler G Godoy-Avila S Falck-Zepada J And Espinza-Arellano J J (2001) Transgenic Cotton in Mexico: Economic and Environmental Impacts http://www biotech-info net/*Bt*_cotton_Mexico html

Umbeck P F Barton K A Nordheim E V Mccarty J C Parrott W L and Jenkins J N (1991) Degree of pollen dispersal by insects from a field-test of genetically engineered cotton *J Econ Entomol* **84**: 1943-1950

Zhang B H Wang Q L Wang K B Zhou D Y and Liu F (2004) *Bt* cotton in India *Curr Sci* **86**: 758-760

Zhang B H (1997) Cotton Biotechnology and its application Chinese Agricultural Press Beijing

Zhang B H (2001) Resistance of pests to t cotton and the strategies for control pest-resistance to *Bt* cotton *Cotton Science* **13**: 1-8

Zhang B H and Feng R (1998) Achievement problems and strategies of transgenic insect-resistant cotton *Acta Agronom Sin* **24**: 248-246

Zhang B H Feng R (2000) Cotton-resistance to pests and transgenic pest-resistant cotton *China Agricultural Science and Technology*: Beijing

Zhang B H and Wang Q L (2003) Transgenic Cotton In: *Plant Genetic Engineering Vol III Improvement of Commercial Plants-I*(ds Singh R P and Jaiwal P K Sci-Tech Publisher LLC Houston USA 1-54

Zhang B H Liu F Yao C B and Wang K B (2000) Recent progress in cotton biotechnology and genetic engineering in China *Curr Sci* **79**: 37-44

Zhao J Z CAo J Li Y X Collins Roush R T Earle E D and Shelton A M (2003) Transgenic plants expressing two *Bacillus thuringiensis* toxins delay insect resistance evolution *Nature Biotech* **21**: 1493-1497

Zhao J Z Zhao K J Lu M G Fan X L and Guo S D (1998) Interactions between *Helicoverpa armigera* and transgenic *Bt* cotton in north China *Sci Agri Sin* **31**: 1-6

8

PRIONS: THE LETHAL PROTEINS

CHITRA WADHWANI AND C.P. MALIK

This chapter is aimed at looking at the prion diseases from a molecular point of view, taking into account the genetic as well as the transmissible and sporadic aspect of the diseases. It attempts to first fully identify these novel diseases to give an exact idea of what the research is about and why it is taking place. The molecular side starts by looking at the molecular and genetic evidence from the beginning of the discovery of the 'prion' hypothesis.

Then, by following a fascinating path of logical deductions and discovery with exciting consequences that were revolutionary for molecular biology and resulted in uncertainties and fierce scientific debates, the properties of the prion diseases are investigated. The chapter intends to not only critically analyze the 'prion' hypothesis, but to compare it to the viral and bacterial hypotheses as well, ending in identifying the progress that needs to be made in the future, and above all, the major uncertainty of variant CJD.

INTRODUCTION

Dr. Stanley Prusiner in 1981 coined the name "Prion" agents that cause a novel type of fatal brain diseases. Prions are "Proteinaceous infectious" agents of about 27 Kilo Daltons which (almost certainly) do not have a nucleic acid genome. It was found to be present in amyloid fibrils in the brain. They resist inactivation by procedures that modify "nucleic acid" but respond to agents that disrupt proteins.

The discovery of Prions shook the confidence of scientists making them question whether they truly understand the "ground rules" of infectious diseases. Extensive studies on this agent have so far failed to identify any nucleic acid associated with infectious materials, yet biochemical studies to date cannot conclusively rule out the requirement for nucleic acid for infectivity. Prion diseases are often called Spongiform Encephalopathies because of the post mortem appearance of the brain with large vacuoles in the cerebellum and cortex. Research on prion diseases has recently accelerated because the causative agents are not viruses or bacteria or fungi or parasites.

The mounting experimental evidence has generated great interest in what appears to be a totally new kind of mechanism of disease. The main goal of this chapter is to understand the various aspects of these unique proteins to some extent. The prions are infectious proteins and on account of lacking a nucleic acid genome appear to be a "Protein Life form".

Prions have come to the attention of a wider audience because they are the causative agents of neurological disorders such as BSE (Bovine Spongiform Encephalopathy) whose possible transmission to human has caused great concern in U.K and elsewhere.

Encephalo is Greek for "Brain" and Pathy for disease. The "Protein only" theory is supported by evidence of transgenic and knockout mice (Bueler *et al.*, 1993). Refractoriness to UV irradiation has indicated that prions are devoid of nucleic acids (Alper *et al.*, 1967). The nature of the prions has been complicated by many other unusual physical and biological properties of these molecules including most worryingly, its extraordinary resistance to conventional sterilization procedures. The exact nature of this etiological agent has been and still is a focus of scientific debate.

PrP gene, PrPc Protein and its Isoform

Prion protein gene has been located on chromosome 20. It bears two exons separated by a single intron. Exon I and II are transcribed and the two RNA's ligated into a single mRNA coding for PrP protein. The latter is a cell surface glycoprotein. Normal Protein is a precursor of several posttranslational events to become Prion Protein (PrP 27-30) which causes Scrapie in sheep. In humans PrP is translated as a 253 amino acid protein that is translocated into the Endoplasmic reticulum (ER), where signal sequences at both amino (an ER entry signal) and carboxy (a glycosyl phosphatidylinositol) terminal ends are cleaved to produce mature PrP of 208 amino acids.

In the ER, the GPI anchor is attached and two asparagines-linked glycosylation sites are core glycosylated. PrP then passes to the Golgi apparatus for trimming and modification of the sugars prior to transport to the exofacial membrane of the cell. In certain cases, PrP may exist in a transmembrane topology that may be specifically linked to the pathogenesis of the disease. PrP molecules arising from mutated genes probably do not adopt the disease conformation as soon as they are synthesized. Otherwise people would become sick in early childhood. The mutation in the PrP gene renders the resulting proteins susceptible to flipping from an α-helical to β-sheet shape. Presumably it takes time until one of the molecules spontaneously flips over and still more time for disease PrP to accumulate and damage the brain enough to cause symptoms.

In sucrose gradients, infectivity comigrates with rod shaped particles called amyloid fibrils found by electron microscopy in infected brains. Polyacrylamide gel electrophoresis of the proteins in the infectious fraction from sucrose gradients has revealed a single protein termed the prion protein (PrP) with a relative molecular weight 0f 27-30 kDa. Further analysis has revealed that the Scrapie PrP or PrPsc is a sialoglycoprotein.

A great advance in the field came with the derivation of a monoclonal antibody that recognizes PrPsc. In addition to the prion associated protein that is readily detected in infected tissue, a cross-reacting normal cellular analogue (PrPc) has also been identified using this antibody. The latter protein has homologues in many animal species including mammalian and most likely avian. Purification and partial sequence analysis of PrPsc allowed the design of synthetic oligonucleotide DNA probes complementary to all possible codons that can encode the PrP amino acid sequence. These probes in turn, have led to the cloning of mRNAs encoding PrP and detection of such mRNAs in both normal and Scrapie infected hamster tissues. Because DNAs of both normal and infected cells contain a single copy of a gene for this mRNA, it appears that PrP is a component of normal brain tissue that remains closely associated with Scrapie agent infectivity during extensive purification.

The amino acid sequence of the normal and Scrapie PrP is the same and PrPc expressed in uninfected cells is the precursor for PrPsc. The ability of the PrPsc to form aggregates of fibers in the brain distinguishes it from the normal protein. Moreover, the normal protein is slightly smaller and more easily digested by proteinase K than is PrPsc. The conformational differences between the two proteins distinguish them and are determined most likely at the plasma membrane or during transfer to lysosomes.

It is the accumulation of PrPsc that causes brain cell damage, but whether PrPsc is the causative agent or a consequence of infection is still not known. It appears unlikely that the mRNA is itself the genome of the prion, since none of this RNA has been detected in spleens of infected mice where high levels of infectivity are found. The length of time for development of disease probably reflects the interval needed for production of this abnormal prion protein shape and recruitment of normal PrP into its unusual form.

In experiments involving transgenic mice, a mutation was introduced into the DNA coding for the normal PrPsc so that the defect in this protein found associated with GSS was expressed. The animals developed Scrapie-like lesions in the brain. Whether or not an infectious agent is also made is under study. In other experiments, transfer of the gene for the hamster PrPsc to transgenic mice made the animals highly susceptible to the hamster adapted Scrapie strain. Disease involved the hamster (and not mouse) PrP. When, however, these mice were inoculated with the mouse Scrapie strain, the brain pathology showed the presence of mouse PrP.

These observations support the role of the protein in nerve cell pathology but do not resolve if PrP is a component of the causative agent or is required for its replication. Finally, it is interesting to note that, using anti-PrP antisera, researchers have found the PrP antigen in amyloid fibers not only in Scrapie infected brains but also in human brains of CJD victims. However, the role of this protein, if any, in the infectivity or pathogenesis of these diseases has yet to be proved, and morphologically similar structures have been observed in noninfectious neuropathies, including Alzheimer's disease. However, the fibers in the amyloid of Alzheimer's disease differ in amino acid sequence from these PrP.

Sterilization

Infectious particles possessing nucleic acid are dependent upon it to direct their continued replication. Prions however, are infectious by their effect on normal versions of the protein. Therefore, sterilizing prions involves the denaturation of the protein to a state where the molecule is no longer able to induce the abnormal folding of normal proteins. However, prions are generally quite resistant to denaturation by proteases, heat, radiation, and formalin treatments, although their infectivity can be reduced by such treatments.

Prions can be denatured by subjecting them to a temperature of 134°C for 18 minutes in a pressurised steam autoclave. Ozone sterilization is currently being studied as a potential method for prion deactivation. Renaturation of a completely denatured prion to infectious status has not yet been achieved, however partially denatured prions can be renatured to an infective status under certain artificial conditions.

INFECTIOUS AGENT DEBATE

Protein-only hypothesis

Prior to the discovery of prions, it was thought that all pathogens used nucleic acids to direct their replication. The "protein-only hypothesis" states that a protein structure can replicate without the use of nucleic acid. This was initially controversial as it contradicts the so-called "central dogma of modern biology," which describes nucleic acid as the central form of replicative information.

Evidence in favor of a protein-only hypothesis include:

- No virus particles have been conclusively associated with prion diseases
- No nucleic acid has been conclusively associated with infectivity; agent is resistant to degradation by nucleases
- No immune response to infection

- PrPsc experimentally transmitted between one species and another results in PrPsc with the amino-acid sequence of the recipient species, suggesting that replication of the donor agent doesn't occur
- Level of infectivity is associated with levels of PrPsc
- PrPsc and PrPc do not differ in amino-acid sequence, therefore a PrPsc-specific nucleic acid is a redundant concept
- Familial prion disease occurs in families with a mutation in the PrP gene, and mice with PrP mutations develop prion disease despite controlled conditions where transmission is prevented
- PrPsc has been shown to arise from exposure of PrPc to molecules of PrPsc

Viral hypothesis

The protein-only hypothesis was initially met with skepticism and still has critics. For more than a decade, Yale University neuropathologist Laura Manuelidis has been proposing that prion diseases are caused instead by an unidentifiable "slow virus". In January 2007 she and her colleagues published a non-peer reviewed article in the Proceedings of the National Academy of Sciences claiming to have found the elusive virus in less than 10% of their Scrapie-infected cells in culture.

Evidence in favor of a viral hypothesis include:

- No bacteria or other living organisms have been found in prion-affected organisms, defaulting to the idea that a virus must be involved
- The long incubation and rapid onset of symptoms resembles some viral infections, such as HIV-induced AIDS
- Differences in prion infectivity, incubation, symptomology and progression among species resembles the "strain variation" seen between viruses, especially RNA viruses
- Familial prion disease is proposed to be due to genetic predisposition to the viral agent

THE EVIL PRIONS

Discovery of the Prion protein

Radiation biologist Tikvah Alper and physicist J.S. Griffith developed the theory in the 1960s that some TSEs are caused by an infectious agent made solely of protein. This theory was developed to explain the discovery that the mysterious infectious agent causing the diseases Scrapie and Creutzfeldt-Jakob Disease resisted ultraviolet radiation (which breaks down nucleic acids - present in viruses and all living things) yet responded to agents that disrupt proteins.

The idea of a protein only infectious agent was first proposed by Griffiths in 1967. It was later discovered that normal animals make prion proteins but those proteins are destroyed in the extraction process of denaturation, so it seemed as though the normal animals did not have them, whereas in the infected animals, the prion protein is not destroyed to the same extent and could be detected. They then used the techniques of reverse genetics to find out where this prion protein was coming from - was it some kind of external agent being expressed, say by a virus, or was it something that the animal itself was producing.

Francis Crick recognized the potential importance of the Griffith protein-only hypothesis for Scrapie propagation in the second edition of his famous "Central dogma of molecular biology" (*Nature* 1970 Aug 8; 227 (5258): 561-3). While asserting that the flow of sequence information from protein to protein, or from protein to RNA and DNA was "precluded" by the dogma, he noted that Griffith's hypothesis was a potential difficulty (although it was not so promoted by Griffith). As the revised "dogma" was formulated, in part, to accommodate the then recent discovery of reverse transcription by Howard Temin and David Baltimore (Nobel Prize, 1975), proof of the protein-only hypothesis might have been seen as a sure bet for a future Prize.

The Nobel Assembly at the Karolinska Institute in Stockholm, Sweden, awarded the Nobel Prize in Physiology or Medicine for 1997 to Stanley B. Prusiner, for his discovery of "prions - a new biological principle of infection". It's a kind of complex story, but it starts way back in the 60s when CJD and a rather obscure disease called Kuru which affects the cannibals of Papua New Guinea, both neurodegenerative brain diseases were shown to be transmissible experimentally into animals.

What this involved was taking bits of brain, injecting them into the brains of monkeys and waiting for up to 2 years before these animals got sick. Now, this is an awful long time for a viral disease, and in any case people were unable to find viruses: using all the techniques available for identifying viruses, including denaturation studies in which you try to destroy the viruses, all those things that normally destroyed viruses didn't seem to destroy the infectious agents in these brains.

Some little while later Stanley Prusiner and his team in the States were looking at the brains of infected animals, experimenting on infected animals and normal animals and trying to see what differences there were between them. In 1982 S B Prusiner of the University of California, San Francisco published an article in Science journal claiming to identify the causative agent of Scrapie to be a 'novel proteinaceous infectious particle' although they had not managed to satisfactorily isolate the protein until two years after making his announcement.

He also proposed a then new term 'prion' (pronounced pree on), to describe this agent, by combining the first two syllables of the words representative of its proteinaceous and infectious nature. While the infectious agent was named a prion, the specific protein that the prion was made of was named PrP, an abbreviation for "protease-resistant protein".

However, it was not until the co-purification of the prion protein with hamster Scrapie infectivity that Prusiner was able to distinguish it from a virus. They identified a piece of DNA which was producing prion protein in both the normal and infected animals, but in the case of the infected animals, the prion proteins had undergone a change in its structure, it was a different shape, and therefore this conferred on it a sort of resistance to denaturation.

It was the unanswered question of the chemical structure of the Scrapie agent that Prusiner endeavoured to solve at the time, as the agent had reportedly been transmitted to sheep when 18,000 were vaccinated against louping ill virus. The suspensions were treated with formalin to prevent contamination, to which the Scrapie-agent showed resistance, and 1500 of the sheep developed Scrapie two years later. Prusiner's experiments uncovered six separate and distinct lines of evidence that the Scrapie agent contains a protein that is required for infectivity (Prusiner *et al.*, 1982). These lines were:

1. The agent was inactivated as a result of digestion with proteinase K and trypsin.
2. Inactivation was by chemical modification with diethyl pyrocarbonate.
3. Inactivation by SDS (Sodium dodecyl sulphate).
4. Inactivation by chaotropic salts such as guanidinium thiocyanate.
5. Inactivation by phenol.
6. Inactivation by urea.

The agent was also resistant to procedures which attack nucleic acids, such as low pH, ribonucleases, deoxyribonucleases, UV at 254nm, divalent cation hydrolysis, psoralen photoreaction by AMT, HEP, HMT, MMT, TMP, and chemical modification by hydroxylamine.

The molecular size of the agent was thought to be between 64,000 to 150,000, which didn't take into account the possible high number of copies of nucleic acid, or efficient nucleic acid repair system which would have an upward trend on the size. However in experiments involving gel filtration, rate-zonal sucrose gradients and gel electrophoresis the Scrapie agent was found to be possibly globular and of a molecular weight of less than 50,000 Daltons. The novel properties of the Scrapie agent can be summarised as below.

Physical properties of the Scrapie agent

1. Stable at 90°C for 30 minutes.
2. Low molecular weight infectious particles (<50,000 daltons).

3. Hydrophobic protein(s) are required for infectivity.
4. Resistant to ribonucleases and deoxyribonucleases.
5. Resistant to UV radiation at 254 nm.
6. Resistant to psoralen photoadduct formation.
7. Resistant to Zn^{2+} catalysed hydrolysis.
8. Resistant to NH_2OH chemical modification.

Interestingly, the prion is resistant to psoralen photoreaction; five different psoralens were tested and the Scrapie agent was found to be insensitive. The psoralens were used as they can penetrate the protein and lipid coats of conventional viruses, react with their genomes upon photoactivation and cause impressive losses of titre (Masiarz *et al.*, 1982).

So the molecular properties of the Scrapie agent were deduced to be different from those of viruses, viroids, and plasmids. As the evidence gained one could not exclude the possibility of a small nucleic acid present. Prusiner came up with two models for the Scrapie agent, it was either a (i) a small nucleic acid surrounded by a tightly packed protein coat or (ii) a protein devoid of nucleic acid, which means an infectious protein. Later he came up with the idea that the protein could be associated with a small polynucleotide (Prusiner *et al.*, 1991).

Possible mechanisms of prion replication

The postulated mechanisms for the replication of prion proteins ranged from those used by viruses to the synthesis of polypeptides in the absence of nucleic acid templates to posttranslational modifications of cellular proteins. Since then subsequent discoveries have narrowed these hypotheses for both the prion structure and the mechanisms of replication for which recent models have been suggested and are discussed later. Prusiner (1982) then also suggested possible mechanisms of the prion replication as shown below:

1. Prions contain undetected nucleic acids which—
   —Code for prion protein(s)
   —Activate transcription of host genes coding for prion protein
2. Prions are derived from nucleic acids which—
   —Activate transcription of host genes coding for prion protein
3. Code for their own replication by either—
   —Reverse translation
   —Protein-directed protein synthesis

Prusiner also mentioned the possibility of prion genes (not discovered until later) and how they must be highly regulated, not readily activated, but yet pose an important function due to their conservation preserved by evolutionary pressure amongst most, if not all mammal species. This was the first major report suggesting the possible lack of nucleic acid and the unusual proteinaceous nature, which was backed by strong evidence, and from this point the molecular investigation into prion biology really began to attract interest (Prusiner *et al.*, 1982).

PRION DISEASES IN ANIMALS AND HUMANS

There are now seven diseases of animals or humans that are thought to be caused by prions. In several cases, little physical data about the infectious agent exist, and the classification is essentially based on similarities in epidemiologic and pathologic findings. Both genetic and sporadic forms of the diseases have been categorized by some as spongiform encephalopathies, because they are associated with a loss of neurons resulting from spongiform degeneration.

A proliferation of astrocytes can also be observed, and myeloid like plaques often appear. Most notable is the absence of any detectable immune response in affected hosts, and the pathological effects are confined to the nervous system. Incubation periods can be months to decades prior to clinical illness, and a progressive clinical course of weeks to years invariably leads to death. The diseases all show specific cytopathic effects on selected types of brain cells. The agent can also be found without pathologic effects in lymph tissues and salivary glands.

Most mammalian species develop this disease such as the Scrapie in sheep, Transmissible Mink Encephalopathy (TME) in mink, Chronic Wasting Disease (CWD) in mule deer, Bovine Spongiform Encephalopathy (BSE) in cows. Humans are also susceptible to several Prion diseases like Creutzfeldt Jacob Disease (CJD) (Creutzfeldt, 1920), Gerstmann Straussler Scheinker syndrome (GSS), Fatal Familial Insomnia (FFI), Kuru and Alper's syndrome. Alper's syndrome is the name given to Prion diseases in infants. Prions related diseases are characterized by long incubation and asymptomatic period of illness followed by loss of motor control, dementia, paralysis and ultimately death (Huang *et al.*, 2004).

Epidemiology

More than one etiological factor might be responsible for the diseases. Humans might be infected by prions in following ways:

1. Acquired infection

Diet and medical procedures such as surgery, growth hormones, corneal transplants etc. i.e. infectious agent implicated. Recently it has been shown that pathological prion proteins are deposited in the neuroepithelium of the olfactory mucosa in patients with sporadic CJD, i.e. even nasal secretions could be infectious (Monaco *et al.*, 2003).

2. ***Inherited genetic mutations***

Sometimes rogue prions are produced by genetic mutation, that is why some cases of CJD, GSS and FFI in humans are inherited in an autosomal dominant fashion.

3. ***Sporadic prion diseases***

Majority of the cases seem not to result from infection, but occur as a spontaneous disease with or without a genetic predisposition (Mastrianni, 2004).

4. ***RNA mutation***

Researchers have suggested RNA mutation as a possible initial event in the etiology of sporadic prion diseases (Garcion *et al.*, 2004).

A new variant of CJD has been identified which is designated vCJD, is transmitted to humans by consumption of BSE-contaminated food products in Great Britain (Scott *et al.*, 1999). The disease affects approximately one person per million per year, worldwide (Mckintosh *et al.*, 2003). The disease has no preference for a particular gender (Brown *et al.*, 1987) and nor is there a predilection for a particular ethnic group although certain mutations within the PrPc gene that causes familial diseases may be more restricted to certain groups, based on the origin of mutation (Mastrianni, 2004).

PRIONS: ON THE TRAIL OF KILLER PROTEINS

The animal and human prion diseases are characterised by their ability to cause an invariably fatal degeneration of the nervous system. When the brain tissue is viewed under a microscope, the sections have a sponge-like appearance, shown in Figure 3 (b). The classic neuropathological features of human prion disease include spongiform degeneration, gliosis, and neuronal loss in the absence of an inflammatory response.

Scrapie was the first example of this type of disease and has been known for many hundreds of years. "Kuru" is the condition, which first brought prion diseases to prominence in 1950's. It was found in geographically isolated tribes in Papua New Guinea. It spread through cannibalism because the people used to ingest brain tissue of dead relatives for religious reasons. Encouragement to stop the practice of cannibalism in afflicted tribes has abated the incidence of disease.

The symptoms of "Kuru" are congestion of blood vessels in brain, cortical atrophy, vacuolization in neuronal cytoplasm and dendrites, which gives the neuropil a spongy appearance. Ultimately the cells misfire and become non functional (Hadlow, 1959). Clusters of brain cells die and the brain stops looking like brain. In mad cow disease the cows become so uncomfortable and itchy that they frantically rub up against anything they can, finally scrapping off most of the hair.

Inherited prion diseases like GSS and FFI show the deposition of amyloid plaques in the cortical and sub cortical regions of the brain. Impairment of movement is observed when cerebellum gets involved where as decrease in memory and mental agility shows the involvement of cerebral cortex. Porter and Porter (Porter *et al.*, 1973) have reported lack of specific immune response to the puzzling agent. Essentially, the brain cells fall apart, leading to symptoms such as a progressive sub-acute or chronic decline in cognitive or motor function. Different prions affect different regions of the brain. Symptoms depend on which region of the brain is affected.

Different prions affect different regions of the brain :

- Cerebral cortex: When affected, the symptoms include loss of memory and mental acuity, and sometimes also visual impairment (CJD).
- Thalamus: Damage may result in insomnia (FFI).
- Cerebellum: Damage results in problems to coordinate body movements and difficulties in walk (Kuru, GSS).
- Brain stem: In the mad cow disease (BSE), the brain stem is affected.

Mutations can result in different shapes of the prion protein that accumulates in different regions of the brain: In familial insomnia (FFI), mutated prions (violet squares) accumulate in the thalamus, with the result that the patients are unable to sleep. In Creutzfeldt-Jakob disease, the prion protein accumulates primarily in the cerebral cortex (red dots area). A precise diagnosis of a prion disease can only be made upon autopsy. The figure 3 (b) shows thin sections of diseased brains. FFI, with typical proliferation of astrocytes, the support cells of the brain, is shown to the left (arrows). CJD, with the characteristic spongiform appearance with vacuoles (arrows) is shown to the right.

Due to their characteristic appearance, the diseases are known as the Transmissible Spongiform Encephalopathies (TSEs) as they can be transmitted both experimentally and naturally. The TSEs consist of several closely related prion diseases, of which four similar diseases affect humans. These are shown in Table 1. Prion diseases arise in three different ways:

1. Through horizontal transmission from e.g. a sheep to a cow (BSE).
2. In inherited forms, mutations in the prion gene are transmitted from parent to child i.e. familial.
3. They can arise spontaneously i.e sporadic (no known antecedent events).

Table 1: The prion diseases of humans

Disease	*Form of the disease*	*Prion*	*Pathogenic PrP Isoform*
Creutzfeldt-Jacob disease (CJD)	Sporadic, acquired (including iatrogenic) and familial.	CJD prion	HuPrPsc
Fatal Familial Insomnia (FFI)	Familial	FFI prion	HuPrPsc
Gerstmann-Sträussler-Scheinker disease (GSS)	Familial	GSS prion	HuPrPsc
Kuru	Acquired	Kuru prion	HuPrPsc

Table 2: The prion diseases of animals

Disease	*Natural Host*	*Prion*	*Pathogenic PrP Isoform*
Scrapie	Sheep and goats, Moufflon,	Scrapie prion	OvPrPsc
Transmissible mink encephalopathy (TME)	Mink	TME prion	MkPrPsc
Chronic wasting disease (CWD)	Mule deer and elk	CWD prion	MdePrPsc
Bovine spongiform encephalopathy (BSE)	Cattle	BSE prion	BovPrPsc
Feline spongiform encephalopathy (FSE)	Cats	FSE prion	FePrPsc
Exotic ungulate encephalopathy (EUE)	Nyala and greater kudu	EUE prion	UngPrPsc

Relevance

The prion diseases can affect both humans and animals, and it is invariably fatal due to the neurological damage it causes. Presently, there are no vaccines or medicines that will prevent the disease and, apart from nursing care, there are no treatments to halt or slow its inevitable course. The course of the disease is alarmingly rapid compared with other neurodegenerative disorders such as Alzheimer disease, Parkinson's disease and motor neurone disease.

However, they are rare neurodegenerative disorders affecting only 1 in a million people, which is only about 50 or so new cases in the United Kingdom each year. Despite this, remarkable attention has recently been focused on these diseases. This is due to their unique biology (described later) and also because of the fears that an epidemic of a newly recognised bovine prion disease (BSE recognised in 1986) among UK cattle could pose a threat to public health through dietary exposure to infected tissues.

The exposure of the infected tissues may not have been considered as a problem due to modern cooking techniques and chemical treatments such as UV light and radiation. Unfortunately, this agent happens to be particularly resistant to both, resulting with little or no loss of infectivity. It could then be said that a significant part of the UK population has been exposed to BSE infected food, with no known consequences. The effects could be minor with only a handful of people succumbing to the disease or the effects could be disastrous with many thousands of people falling foul of the prion disease. Interest has therefore been directed at the infectivity of the prion agent between individuals and across the species barrier in relation to the incubation time, and possible treatments, cures or vaccinations.

Epidemiology and Clinical features of the different diseases

Background: The prion diseases are a large group of related neurodegenerative conditions, which affect both animals and humans. These diseases all have long incubation periods but are typically rapidly progressive once clinical symptoms begin. All prion diseases are fatal, with no effective form of treatment currently; however, increased understanding of their pathogenesis has recently led to the promise of effective therapeutic interventions in the near future.

Prion diseases are unique in that they can be inherited, they can occur sporadically, and they can be infectious. The first of these diseases to be described was Scrapie, a disease of sheep recognized for over 250 years. This illness, manifested by hyperexcitability, itching, and ataxia, leads to paralysis and death. It is called Scrapie because of the tendency of affected animals to rub against the fences of their farms in order to stay upright, reflecting their cerebellar dysfunction.

The transmission of this disease was demonstrated first in 1943 when a population of Scottish sheep was accidentally inoculated against a common virus using a formalin extract of lymphoid tissue from an animal with Scrapie. Accidental transmission of prions is a recurrent event in the history of these agents and is related to their unusual biophysical properties.

Pathophysiology: A unifying feature of all the prionoses is their neuropathology. These illnesses tend to affect the gray matter of the central nervous system (CNS), producing neuronal loss, gliosis, and characteristic spongiform change. The latter is a vacuolation of the neuropil, and to a variable degree, of the neurons. In addition, plaques with the typical staining properties of amyloid eg, apple-green birefringence after Congo

Red staining when viewed under polarized light (See Fig. 5) are observed in many of these conditions.

In approximately 10% of patients with CJD, amyloid is present in the cerebellum or in the cerebral hemispheres. All cases of GSS are associated with multicentric cerebellar plaques. These amyloid plaques are immunoreactive with antibodies to the prion protein and do not immunoreact with antibodies to other amyloidogenic proteins, such as the amyloid-beta (which is deposited in Alzheimer disease).

PRION RELATED DISEASES

Sporadic diseases include CJD, which occurs worldwide with incidence of about 1 in a million people. An example of an acquired disease is Kuru, meaning shivering or trembling, which was discovered in Eastern Highlands of Papa New Guinea amongst members of the Foré tribe. It is also known as the laughing disease due to the facial grimaces it causes and is thought to be acquired through the ritual cannibalism of dead relatives who had caught the disease. There is also iatrogenic CJD and new variant CJD. New variant CJD is commonly thought to result from eating food contaminated with the BSE agent.

Familial diseases are genetically inherited and include familial CJD, which accounts for about 15% of all CJD cases. Also inherited are GSS, with a global incidence of 1 in 10 million, and FFI, where insomnia is a major symptom, in addition to other symptoms including dementia. GSS diseases comprise dementing, ataxic and atypical GSS with neurofibrillary tangles. Included is the atypical prion disease, which does not easily fit the diagnostic criteria for prion disease. There are also several types of animal prion diseases affecting animals. This is explained by the fact that many are zoo animals possibly exposed to the agent in their feed.

HUMAN PRION DISEASES

In the care of patients dying of human prion disease, the precautions used for patients with AIDS or hepatitis are certainly adequate. In contrast to these viral illnesses, the human prion diseases are not communicable or contagious. There is no evidence of contact or aerosol transmission of prions from one human to another. However, they are infectious under some circumstances, such as ritualistic cannibalism in New Guinea causing Kuru, the administration of prion-contaminated growth hormone causing iatrogenic CJD, and the transplantation of prion-contaminated dura mater grafts. Familial CJD, GSS, and FFI are all dominantly inherited prion diseases; five different mutations of the PrP gene have been shown to be genetically linked to the development of inherited prion disease. Prions from many cases of inherited prion disease have been transmitted to apes, monkeys, and mice carrying human PrP transgenes.

Creutzfeldt-Jacob disease

Creutzfeldt-Jacob disease (CJD), occurs rarely throughout the world, and typically presents as a dementia in either men or women at least 60 years old. Some cases are clustered in certain families, and the fact that some of these families also have an apparently higher incidence of Alzheimer's disease has led to the supposition that the two diseases may be related. Only CJD is readily transmissible to apes (and occasionally, rodents). However, CJD also appears to be transmissible to humans by contaminated products. A case was reported of a young man who died of CJD that is believed to have been transmitted in preparations of Human Growth Hormone administered during his childhood. So convinced were the officials at the U.S. National Institutes of health that the infection came fron the hormone, which was prepared from extracts of human pituitaries, that they banned further use of this material for clinical applications. This action served to accelerate government approval of human growth hormone produced by genetic engineering in *E.coli.* Other reports of transmission of CJD have been linked to corneal transplants taken from individuals dying from this disease. Young recipients later developed neurological symptoms of CJD, and this agent, like Scrapie, has been found to be extremely stable. Brains obtained from individuals who died from CJD and stored in formalin for over 10 years have been shown to contain the infectious agent when inoculated as extracts into rodents.

Creutzfeldt-Jacob disease (sporadic)

In a number of countries, there have been extensive epidemiological studies on the numbers of CJD cases. All of these countries obtained broadly the same results, approximating about 0.5-1 case per million. There is no significant case clustering other than in familial clusters. Cases are distributed apparently at random with a frequency related only to local population density. There is no evidence for case to case spread (other than with respect to iatrogenic CJD discussed below). There is now evidence of an association with local Scrapie prevalence, since CJD is as common in Scrapie free areas such as Australia and New Zealand, as it is in the UK, where Scrapie is endemic. Despite the attention focussed on links between BSE and CJD, the overall incidence figures do not appear to have risen more in the UK than in countries, which do not have BSE.

Although an overall increase has been observed, it is likely to be due to increased accurate identification of cases due to increased attention, which is now focussed on these diseases. This is especially relevant to the increased incidence noted in the elderly, where other dementing illnesses are common and could be confused with CJD. Occasional weak statistical associations with particular foodstuffs have been reported, although these have tended to disappear in subsequent years of the study and no consistent new risk factors have emerged. Despite this, concern has been expressed about a number of occurrences of CJD in UK dairy farmers.

Clinical features of CJD

According to Collinge & Palmer (1997) the core clinical syndrome of CJD is of rapidly progressive multifocal dementia usually with myoclonus. The onset is usually in the 45-75 year age group with peak onset between 60-65 years. The clinical progression is typically over weeks progressing to akinetic mutism and death, often within two to three months. Around 70% of cases die in less than 6 months. Prodromal features include extrapyramidal signs, cerebellar ataxia, pyramidal signs and cortical blindness. About 10% of the cases present initially with cerebellar ataxia. Routine haematological and biochemical investigations are normal although occasional cases have been noted to have raised serum transaminases or alkaline phosphatase. There are no immunological markers and acute phase proteins are not elevated.

Scientists go on to mention that "Prospective epidemiological studies have demonstrated that cases with a progressive dementia and two or more of myoclonus, cortical blindness, pyramidal, cerebellar or extrapyramidal signs, or akinetic mutism with a typical EEG nearly always turn out to be confirmed as histologically definite CJD if neuropathological examination is performed". This depends on neuropathological confirmation being "by demonstration of spongiform change, neuronal loss, and astrocytosis". PrP amyloid plaques are usually not seen with CJD, although the presence of protease resistant PrP can be demonstrated by immunoblotting of brain homogenates.

Iatrogenic CJD

Iatrogenic CJD is usually referred to as CJD despite cases arising from peripheral inoculation (not intracerebral) usually having a prominent cerebellar syndrome more reminiscent of Kuru. Transmission may occur by any one of a number of routes involving accidental inoculation with human prions as a result of medical procedures; the routes involving the use of inadequately sterilised neurosurgical instruments due to the high heat resistance of the prion, dura mater and corneal grafting, as well as the use of gonadotropin derived from the pituitary gland from cadavers. Iatrogenic CJD actively demonstrates the transmissible properties of the prion. The first case was with a 55 year old woman who received a corneal transplant from a donor later found to have died from CJD. In such cases of neurosurgical transmission, the equipment was found to have been sterilised with 70% ethanol and formaldehyde vapour. This is normally enough for conventional agents, but is ineffective against prions. In view of this, the new sterilisation protocol involves immersion in 1 M NaOH for one hour to inactivate the prions.

New Variant CJD

Teenagers rarely get sporadic CJD, and when in 1995 two cases were reported in the UK, (Bateman, Britton) both were unusual in having Kuru-type plaques, which is only found in about 5% of CJD cases. Shortly after, a third young case emerged and a link was suggested with BSE. In the following months further cases emerged showing similar histological patterns which were unique from those observed with other prion diseases. These cases were named new variant CJD, despite being atypical in their clinical presentation with most cases not meeting the accepted clinical diagnostic criteria for probable CJD. Since the original reports, 124 cases of nvCJD have been diagnosed in Great Britain, 6 in France, 1 in Italy, and 1 in Ireland. These cases were extensively compared with other prion diseases and the lack of correlation seemed to confirm the arrival of a new strain of prion disease in the UK due to the incredibly small statistical probability of these similar cases occurring by chance. The symptoms of the new variant CJD start with behavioural and psychological disturbances, which include anxiety, depression, persistent pain in the limbs/face, withdrawal and behavioural changes which progress.

Overt neurological features later become apparent often after most patients have gone to see a psychiatrist due to the preliminary symptoms. Several weeks later, progressive cerebellar syndrome develops with gait and limb ataxia. Later dementia develops progressing to akinetic mutism; myoclonus was also seen in most patients and sometimes was preceded by choreoathetosis. The neuropathological appearances were very consistent with widespread spongiform change, gliosis and neuronal loss, most severe in the basal ganglia and thalamus, the most remarkable feature being the abundant PrP amyloid plaques in the cerebral and cerebellar cortex. The plaques consisted of Kuru-like florid plaques (surrounded by spongiform vacuoles) and multicentric plaque types. The florid plaques were very unusual and seen previously only in Scrapie, but were very consistent. There was also an extensive amount of PrP in the cerebral cortex, cerebellar cortex, and in the molecular layer of the cerebellum. The unusual young age of these patients and the link with BSE may point to a risk involving dietary exposure. The risk involved however is as yet undetermined, and will only be determined as time goes by and more or fewer cases are identified. This is due to the potentially long incubation time related to the species barrier.

Different (atypical) forms of CJD

These forms of the disease are well recognised, and around 10% of cases have a longer clinical course of over two years. These cases may be represented by people with heterozygous PrP polymorphisms. 10% of cases involve cerebellar ataxia rather than cognitive impairment, and are called ataxic CJD. Other types of CJD can include Heidenhain's variant of CJD, which refers to cases where cortical blindness predominates with severe involvement of the occipital lobes. There is also the panencephalopathic type of CJD where there is extensive degeneration of the grey matter; this was mainly reported in Japan. Amyotrophic variants of CJD have been described with prominent early muscle wasting and are usually seen in late disease when other features are established. However most cases of dementia with amyotrophy are not experimentally transmissible, and their relationship with CJD is not clear. It is possible that these are variants of motor neurone disease with associated dementia.

When evaluating a patient for possible sporadic CJD, the clinician should be guided by published case definitions; they are as follows:

Definite CJD

- Characteristic neuropathology
- Protease-resistant PrP by Western blot

Probable CJD

- Progressive dementia
- Typical findings on EEG
- At least 2 of the following - Myoclonus, visual impairment, cerebellar signs, pyramidal or extrapyramidal signs, or akinetic mutism

Possible CJD

- Progressive dementia
- Atypical findings on EEG or EEG not available
- At least 2 of the following-Myoclonus, visual impairment, cerebellar signs, pyramidal or extrapyramidal signs, or akinetic mutism
- Duration less than 2 years

Kuru disease

The study of Kuru is immensely important for the understanding of these diseases, providing the most extensive clinical experience of acquired prion disease in humans, the epidemiology of Kuru clearly indicates the infectious, but non-contagious, properties of prions and provides extensive evidence that vertical transmission from pregnancies does not occur.

Kuru reached epidemic levels in the Foré people in the Okapa district of the Eastern Highlands of Papua New Guinea (Gajdusek *et al.*, 1957, Gajdusek, 1977) other areas affected included Keiagana, Kanite, Yate, Usurfa and Gima, which are in the close vicinity. Work done by Gajdusek (1957) in an epidemiological survey, uncovered the link between the ritualistic cannibalism and the disease occurrence.

This included the differential age and sex incidence which was due to the women ıd children preparing the food, and becoming exposed to the agent, as well as eating less desirable brains and other internal organs, even the bones were pounded up to be eaten with greens later. The whole of the deceased sufferer was recycled by way of endo-cannibalism. This is a type of cannibalism. It could be termed 'transumption' as it is seen as disposing of the dead relative, recycling the remains back into the community, and most importantly as a spiritual and virtuous gift to the family members. These customs were thought to instill wisdom.

The ratio of occurrence was about 5 females to 1 male, with about 1% of the population affected. In the years after these practices were stopped, the only cases were from the older and not the younger people (who won't have been cannibals) indicating no evidence for vertical transmission, and due to the mortality rate much later, an incubation time of as much as 40 years or more.

The clinical duration had a mean of 12 months with a range of 3 months to three years, the course tending to be shorter in children. Cerebellar ataxia is the central feature with dementia usually absent (contrasting to CJD). Gross dementia can occur but is usually rare. There is a prodrome stage followed by three clinical stages which are:

1. *Prodromal stage:*

Kuru typically begins with prodromal symptoms consisting of headache, aching of limbs, and also joint pains which can last for several months.

2. *Ambulatory stage:*

At this stage there may be no objective signs of disease, but it can be self diagnosed as patients experience an onset of unsteadiness in standing or walking, or of dysarthria or diplopia. Gait ataxia worsens and patients develop a broad-based gait, truncal instability, and titubation. There is a postural tremor present which is coarse and accentuated by movement; patients characteristically hold their hands together in the midline to suppress this. Standing with feet together reveals clawing of toes to maintain posture.

This clawing response is regarded as pathogonomic of Kuru. Therefore signs at this stage are of midline truncal ataxia, presumably due to paleocerebellar disease, which is reflected by the only clear macroscopic brain abnormality seen in Kuru, which is atrophy of the cerebellar vermis. As the gait ataxia and astasia worsens the patients require a stick, as ataxia progresses the patient passes to the next stage of the disease (Sedentary). The mean clinical duration of the first stage is around 8 months and correlates closely with total duration.

3. *Sedentary stage:*

At this stage patients cannot walk, but are able to sit. Walking attempts lead to a high steppage, wide based gait with reeling instability, and flinging arm movements in an attempt to maintain posture. Hyper-reflexia is seen, although usually plantar responses remain flexor and abdominal reflexes intact. Clonus is characteristically short lived.

Athetoid and choreiform movements and parkinsonian tremors may occur. There is reduced muscle power but no paralysis, and obesity may be common at this stage but possibly linked to bulimia.

Characteristically there is emotional lability and bizarre uncontrollable laughter, which has lead to the disease being referred to as 'laughing death'; one sufferer may start and this spreads throughout the group, although this is only a small aspect of the disease. Myoclonic jerking is rarely seen (unlike in CJD) in this stage which lasts about two to three months.

4. *Tertiary stage:*

Hypotonia and hyporeflexia develop and the terminal stage is marked by flaccid muscle weakness. Plantar responses remain flexor and abdominal reflexes intact. The patient is unable to sit unsupported, dysphagia occurs and the patients become incontinent. Inanition and emaciation develop and transient conjugate eye signs and dementia may occur with primitive reflexes developing in some cases. Brainstem involvement and both bulbar and pseudobulbar signs occur.

Respiratory failure and bronchopneumonia eventually leads to death. This final stage lasts one or two months. Often death may occur earlier due to the nature of the illness, for example the sufferers may role into a fire, and burn themselves when in the later stages of the disease. The sufferer is always very well cared for by their family.

In the last few years, there have only been few deaths in the south Foré, with only one male death in 1998 and none in 1999, indicating the beginning of the end of the Kuru epidemic.

INHERITED PRION DISEASES

Gerstmann-Sträussler-Scheinker disease

This was first described by Gerstmann in the 1920s-30s as a peculiar heredo-familial disease of the central nervous system. The classical presentation is with a chronic cerebeller ataxia accompanied by pyramidal features, with dementia occurring later in a much more prolonged clinical course than that seen in CJD. The mean duration is around five years with onset usually in the third or fourth decades of life. The histological hallmark is the presence of the multicentric amyloid plaques. Spongiform change, neuronal loss, astrocytosis and white matter loss are also usually present.

This is an autosomal dominant disorder and is caused by mutations of the PrP gene, which can now be classified within the spectrum of inherited prion disease. Interestingly, in several kindreds of GSS, extensive neurofibrillary tangle (NFT) formation is found (Ghetti, 1994). NFTs are an essential feature of Alzheimer disease, but are also observed in other neurodegenerative conditions.

Fatal Familial Insomnia

This disorder was first described in 1986 and is an autosomally dominant inherited sleep disorder. This presents itself as a progressive untreatable insomnia and autonomic dysfunction (i.e hyperthermia, hypertension, tachycardia, tachypnea, hyperhydrosis), dementia, and motor paralysis however, the phenotypic expression is very variable even within the same family. Thalamic atrophy is the main histological finding, with the anterior-ventral and medial dorsal thalamic nuclei being consistently affected, while the cortex shows moderate asterocytosis.

The age of onset is also variable, ranging from 18-60 years. Once symptoms begin, the course ranges from 6 months to 3 years. Because of the diversity of clinical presentations of this disorder, genotyping is very important for definitive diagnosis. Unlike other prionoses, spongiform change can be a minor feature or can be absent altogether.

All patients with FFI have a missense mutation at codon 178 of the PrP gene where Asn is replaced by Asp, coupled with a Met at the polymorphic codon 129. The somewhat divergent clinical and neuropathological features of FFI, in comparison to other human prionoses, highlight the wide spectrum of disease associated with PrP dysfunction and suggest that other human illnesses have yet to be recognized as prionoses.

Familial CJD

This inherited disease occurs later on in life at a similar time to GSS with similar symptoms except for a more rapidly advancing dementia and an abnormal EEG.

CAUSES OF PRION DISEASES

Prion-related diseases are unique in that they can be related to infectious, sporadic, or familial causes:

Infectious causes

Kuru, a form of prion disease, occurred among the Fore people of the Eastern Highlands of New Guinea and was related to ritualistic cannibalism. The disease is believed to have started with the chance ingestion of body parts of a patient with sporadic CJD, followed by a serial passage of the disease.

Many cases of iatrogenic CJD have been reported. In clinical practice, CJD has been transmitted by surgical instruments, EEG electrodes, corneal transplants, dura mater grafts, human pituitary-derived gonadotrophins, and human-derived growth hormone. Concerns that nvCJD could be transmitted by blood transfusion have not yet been borne out by a documented case.

nvCJD in humans is presumed to have been caused by ingestion of beef products contaminated with BSE. BSE is presumed to have started because of the practice of supplementing the diets of calves and dairy cows with meat and bone products. These meat and bone products are thought to have been contaminated with Scrapie material (from sheep) and/or with material from cattle with a sporadic form of bovine prion disease.

Recently, a number of cases of apparent sporadic CJD have occurred in the United States among young individuals (<30 yrs.). The incidence of sporadic CJD among such young individuals has historically been about 1 case per billion population. In the years 1979-1996, 4 cases of sporadic CJD were reported in the United States among individuals younger than 30 years. In the years 1997-2000, 5 cases have occurred in the United States among young patients. Two of these individuals came from adjacent counties in Michigan (ages at onset were 26 and 28 yrs.), and 3 cases occurred among individuals who were known hunters of deer and/or elk.

Over the same period, a major outbreak of CWD occurred among the deer and elk populations in many western states. CWD is a form of prion disease that occurs naturally in the deer and elk population; however, the pathology has many similarities to BSE, including the presence of florid plaques. Significantly, transmission studies of CWD PrPsc in the laboratory have shown that it can cross the species barrier from deer to human PrP at about the same efficiency as the BSE prion agent. These observations have led to the speculation that limited transmission of CWD to humans has occurred recently in the United States.

Familial causes

The cause of familial forms of prion disease is related to mutations in the PrP gene. A number of mutations in the PrP gene are linked to autosomal dominant forms of prion disease.

The diseases can also be related to infectious causes, such as nvCJD, that are thought to be related to ingestion of cattle products contaminated with BSE. The disease can also be related to a spontaneous conformational change in the normal PrPc to a PrPsc form. The PrPsc form has a higher beta-sheet content and is self-propagating, inducing more PrPc to convert to the PrPsc form.

Sporadic causes

Sporadic CJD is the most common form of prion disease.

It probably arises as a spontaneous conformational change in PrPc to a PrPsc form. The PrPsc form is then self-propagating, inducing more PrPc to convert to the PrPsc rm.

ANIMAL DISEASES

Scrapie Disease

The oldest known prion disease, was described in 1759 (Leopold) and is characterized by the abnormal walk and scratching of infected sheep and goats. It is still found in parts of world despite attempts to eradicate the agent by destroying infected flocks. The disease is characterized by incoordination and unsteady gait, which becomes progressively worse until death. The infectious agent, in the form of extracts from infected brains has been passed experimentally to mice, hamsters, ferrets, mink and monkeys but apparantly is not infectious for chimpanzees or rabbits. The disease induced by the Scrapie agent in Mink is analogous to that in sheep and is indistinguishable from a Transmissible Mink Encephalopathy (TME) found naturally in in Mink. Likewise, Chronic Wasting Disease of mule deer and elk is similar both in pathology and known physical properties to the Scrapie agent. A condition in cattle, the so called "Mad Cow Disease" is believed to be caused by a prion transmitted in contaminated feed supplements derived from infected sheep and goats. Concern has been raised that this agent might also cause disease in humans eating contaminated meat.

Scrapie was the first example of this type of disease to be noticed and has been known about for many hundreds of years. There are two possible methods of transmission in sheep:

1. Infection of pasture with placental tissue carrying the agent followed by ingestion or direct sheep-lamb transmission i.e. an acquired infection.
2. Parry showed considerable foresight by suggesting that it is not normally an infectious disease at all but a genetic disorder. He further suggested that selective breeding would get rid of the disease.

Scrapie is classified as a slow infection. It is characterized by a prolonged incubation period of months, years or possibly decades in which the patient or the host animal is without symptoms. Once the illness begins, however, it progresses steadily and always leads to death. The name Scrapie comes from the tendency of sheep that have the disease to scrape off much of their wool. The disorder has been known for more than 200 years, and by 1935 French workers had shown convincingly that it can be transmitted from one sheep to another by inoculation. This demonstration of transmissibility implies that the infectious agent is capable of reproducing itself in the host animal.

The clinical and pathological signs of Scrapie are similar to those of Kuru and CJD. The initial symptoms are difficulty in walking and loss of coordination, indicating a functional disorder of the cerebellum. Other symptoms include irritability, excitability, restlessness, scratching, biting, rubbing of skin, patchy loss of wool, loss of weight, weakness of the hindquarters and impaired vision.

There is no sign of an inflammatory process or fever, and cell counts in the spinal fluid remain normal. This indicates that the immune system does not respond to the disease agent. Pathological effects brought on by the diseases are confined to the central nervous system. In some of the disorders numerous vacuoles give the brain tissue a spongy appearance. The natural mode of transmission in sheep is unclear, but scientists do know that it is transmitted by a filterable particle that is resistant to heat and formalin fixation.

As yet there are no biochemical or cell culture assay systems for either Scrapie agent or any other of these agents. Transmission of the disease to mice or hamsters allows endpoint titration of biological activity to be performed, albeit slowly. With the finding that the interval from inoculation to onset of disease is inversely proportional to the dose of Scrapie agent injected intracerebrally into hamsters, it has become possible to assay samples with the use of four animals in 60 to 70 days- provided that the titers of Scrapie agent in the samples are sufficiently high. This assay has been used to help purify the infectious principle and investigate its structure on the molecular level.

Variants of the Scrapie agent have been isolated from standard stocks by passage in mice and hamsters at limiting dilutions. Some of these exhibit different neuropathological effects, including markedly different incubation times and tissue tropism. These "strains" seem to be stable and do not revert or interconvert with detectable frequency. The existence of such stable variants is most easily explained by the operation of a genetic mechanism based on the presence of a nucleic acid genome in the infectious agent. However as noted above, no such genome has been found and no virus like particles have been observed in infectious samples.

The Scrapie agent (like viruses) passes through filters of 22 to 100 nm. It can be inactivated by strong alkaline solutions (1 M NaOH) but is highly resistant to nonionic detergents. Its resistance to nucleases and relative sensitivity to proteases led to its consideration as an infectious protein (prion).

Transmissible mink encephalopathy

Transmissible mink encephalopathy (TME) is a disease of farmed mink, probably caused by feeding on Scrapie-contaminated meat.

hronic Wasting disease

Chronic wasting disease (CWD) is a Scrapie-like disease of obscure origin found in d captive mule deer and Rocky Mountain elk.

Bovine Spongiform Encephalophathy/Mad Cow Disease

BSE is a progressive neurological disorder of cattle which results from infection by an unconventional transmissible agent. During the period from November 1986 until December 1995, an estimated 155,600 head of cattle in almost 33,000 herds were diagnosed with BSE in Great Britain. The epidemic peaked in January 1993 with almost 1,000 new cases per week. The nature of the transmissible agent was unknown but currently the most accepted theory is that the agent is a modified form of prion protein.

Scientists considered herd type, size and age structure, whether the animals had been in contact with sheep, and is so, whether Scrapie was present in the flock. Sheep were not present on 20% of the farms, therefore it was believed that the cattle food which was being supplemented with sheep bone meal was responsible for the outbreak. So far, the disease has been confined to the United Kingdom with only a few cases showing up outside. Those cases resulted in cows being exported from the U.K.

Britain may have been at greater risk than other countries when the Scrapie agent unexpectedly moved from sheep to cattle, apparently sometime in the 1980s. Feed practices in Britain differed from those in other countries, and British farmers raised a large number of sheep-and hence Scrapie-infected carcasses-relative to cattle. Most of the herds affected are dairy, milk producing herds. The reason for this may be that dairy cattle eat more commercially produced food, and thus are more likely to ingest sheep ruminants. Beef cattle depend on grazing and vegetation for food. In 1996, scientists believed that there was a clear possibility that BSE might be infecting humans as well.

Pathogenesis:

BSE has an incubation period of 2 to 5 years. Infection can show up in the brain, spleen, thymus, tongue, and gut. Cows infected with BSE appear alert but agitated, anxious and apprehensive. The cow takes a wide base stance, abdomen is drawn up, gait becomes abnormal and exaggerated, and causes tumbling and skin wounds. There are fine muscle contractions over the surface of the neck and body with occasional larger muscle jerks. The animal begins to lose weight and there are frenzied movements including aimless headbutting. Disease is fatal for cattle within weeks to months of onset.

Epidemiology:

At its peak the epidemic affected approximately one percent of adult breeding cows (annual incidence). The peak was in the winter of 1992/93 when a maximum of just over 1000 suspected cases a week were being reported. The incidence has subsequently fallen year by year at a rate of around 40% per year. The within head incidence has been monitored closely, as this is a good indicator of whether or not there has been spread of disease from cow to cow in the herd. This suggests that there has been little or no spread of infection from animal to animal (horizontal transmission). It is still unclear as to whether cows may become infected directly from the food or from asymptomatic mothers that have died. Post-mortem tests are the only type of diagnostic test which are currently available. Scientists examine the characteristic appearance of the brain under the microscope and also use electron microscopy techniques.

Feline spongiform encephalopathy

Feline spongiform encephalopathy (FSE) is a disease seen in domestic cats, and in a few other members of the cat family. The Spongiform Encephalopathies (feline and non feline) are assumed to be caused by feeding BSE-infected material to the animals in question, often in zoos. Mice, hamsters and monkeys have proved experimentally to be the most useful models of prion diseases.

THE CJD-BSE CONNECTION

The prion disease of humans that causes the most concern is variant CJD (vCJD). The cause of this disease has not yet been determined. One theory is that humans may become infected with aberrant prions after ingesting meat from cows infected with BSE. This would require that aberrant prions from one species (cows) could serve as a template for refolding normal prions of another species (humans). This jump from one species to another involves crossing what is referred to as the species barrier. In a laboratory setting, prions have been shown to cross the species barrier. The onset of disease in these cases, however, is delayed compared to cases in which the infecting (aberrant) prions are from the same species as the new host.

It is believed that the origin of the BSE outbreak in Europe (and especially Britain) was due to aberrant sheep prions crossing the species barrier to infect cows. During the 1970's there were changes made to the process used to render offal for a livestock feed additive. These changes created conditions that allowed prions to survive the rendering process.

The sheep disease known as Scrapie is a prion disease. It is believed that prions from Scrapie-infected sheep survived the rendering process and were passed on to cows that ingested infected feed. Once the prions were established in the livestock population the feed could have acted as a reservoir of infective prions. The practice of using rendered offal as an additive to livestock feed has been discontinued. Whether or not bovine prions can be passed to humans is unclear. Human prion diseases like vCJD have long incubation times (typically more than 12 years). The details of the transmission of prions from one species to another have not been fully elucidated.

Route of infection

Probably when cows are fed with offals prepared from infected sheep, prions are taken up from the gut and transported along nerve fibers to the brain stem. Here prions accumulate and convert normal prion proteins to the disease-causing form, PrPsc. Years later, BSE results when a sufficient number of nerve cells have become damaged, affecting the behaviour of the cows.

Evidence that proteins are infectious

Proteins are not normally termed infectious agents and unlike viruses, bacteria and other infectious organisms, the prions have no obvious nucleic acid. It has been proven that these are present with the disease symptoms, but do these actually cause the disease from step one- this was to be found out.

The prion diseases themselves are known to be transmissible, from the early studies of Cuillé and Chelle, when they set out to determine the infectivity of Scrapie in 1936, to the likely cannibalistic transmission of Kuru. Parallels were drawn to the possible infectivity of similar diseases such as CJD. The infectivity of these diseases was demonstrated by intracerebral inoculation of Kuru into chimpanzees by Gajdusek *et al.* (1966) and by similar ways with CJD by Gibbs *et al.* (1968). With this evidence it can only be said that the disease is transmissible and not strictly due to the prions.

The strongest evidence was by Prusiner *et al.* (1982) where he further purified the prion by demonstration using 3000-fold enrichment for infectivity, and it was the only macromolecule that could be associated with infectivity. However some investigators think that the agent is merely the pathologic product which coincidentally purifies with the "Scrapie virus".

Despite these ideas, there is little evidence to support this view as fractions containing <1 prion molecule per ID50 have not been found, if it were to be found, it would indicate that prions are not required for infectivity. There is now a lot of evidence that the discovered 27-30 prion protein is a major and necessary component of the prion disease, this evidence includes:

- That during the copurification of PrP 27-30 and Scrapie infectivity using biochemical methods, the PrP concentration obtained is proportional to the prion titre.
- The kinetics of proteolytic digestion of PrP 27-30 and infectivity are similar.
- Copurification of the infectious prion and infectivity by immunoaffinity chromatography is achieved, and Prion antisera neutralises infectivity.
- The novel prion protein is detected only in clones of cultured cells producing infectivity.
- The prion amyloid plaques are specific for prion diseases of animals and humans, and the deposition of the prion amyloid plaque is controlled at least partially by the prion protein sequence.

- There is correlation between the infectious prion and disease in humans and animals.
- Incubation time in mice is affected by base substitutions in the mouse prion gene-Mo-PrP.
- The existence of the species barrier (discussed later) due to amino acid differences in the protein causes a longer incubation time.
- There is genetic linkage between mutations in the prion gene and the onset of disease, with a mutation at codon 102 linked with GSS development and a mutation in codon 200, or insertions of six additional octarepeats linked with familial CJD.
- Mice expressing prion genes with a point mutation similar to that of GSS develop neurological dysfunction, spongiform brain degeneration and gliosis.

From the evidence discussed at this point most scientists will believe that there is some involvement of prions in the infectivity and pathology of Scrapie, BSE and CJD, but the nature of the involvement was still to be found. Investigators suggest an answer to this with the prion concept.

THE PRION CONCEPT

It took quite a while for this idea of the prion protein to be accepted because it at first sight appeared to challenge some of the most fundamental ideas about molecular biology. Of course, people are used to the idea that infections are caused by bacteria, which are large and can be seen, and viruses, which can be sometimes rather more obscure.

Now, here was a situation where there was an infection but apparently no virus, and viruses contain nucleic acid and this was just a protein. And people probably misunderstood in the first instance, and thought that a protein was making more copies of itself, whereas the fundamental dogma of molecular biology says that the information required to make proteins is contained in nucleic acid which then has to be converted into the production of proteins.

Of course, it later became obvious that the prion hypothesis doesn't actually challenge the fundamental laws of molecular biology at all, because the DNA that is responsible for making the protein is in the host - the animal that was infected - and indeed it's in all animals and all people.

What happens is the abnormal protein directs the conversion which produces more abnormal protein from normal protein, but not from nothing, it doesn't make it itself, it still has to be made originally under the instructions of nucleic acid. It's just the conversion of its shape, which is directed by the protein to another protein molecule.

It wasn't as heretical as some people first thought, but it is wholly new, because it's a new way in which information can be transferred from one molecule to another without going through nucleic acid, which is the normal repository of information in biology.

THE PRION STRUCTURE

Proteins consist of stretches of amino acids. Some fold into alpha helices; others fold into beta pleated sheets. These shapes are maintained by hydrogen bonds. The sheets and helices then fold together further to form the protein's tertiary structure. A protein, such as the one illustrated here, can include both alpha helices and beta sheets and can consist of several subunits that combine to form the quaternary structure.

Prions have one characteristic that makes them unique: they can exist in two different conformations at the level of secondary structure. There is an alpha-helical portion of secondary structure in a normal prion that is refolded into beta-sheet formation in an aberrant prion. The primary structure of the aberrant prion remains the same, but its secondary structure is different. As a result, its tertiary structure will be different as well.

The isolated prion protein was found to be a product of a cellular gene on the short arm of chromosome 20 in humans and in a syntenic region of chromosome 2 in mice. Studies then on the hamster and mouse models, after uncovering a protein designated PrP 27-30 (27-30 Kdaltons) led to the eventual molecular cloning and complete sequence of prion protein cDNA (Oesch *et al.*, 1985; Locht *et al.* 1986; Collinge & Palmer, 1997). The normal product (PrPc) of the prion gene (called PRNP in humans) is found to be expressed in most cell types, with expression at its highest predominantly in the brain. In 1987 investigations involving PrPc and PrP 27-30 started where an experiment involving bacterial-derived stearic acid treatment of the prion enabling it to react with antisera indicated that PrPc is anchored to the cell surface by a glycolipid, components of which include ethanolamine, myo-inositol, phosphate, and stearic acid.

Later experiments involving PrPc indicated that PrPc is expressed as a glycosylphosphatidyl inositol-anchored glycoprotein found on the outer cell membrane (Stahl *et al.*, 1987, 1990, 1992). However, this normal form can be distinguished from the infective prion found in diseased brain samples by its non-resistance to proteolytic hydrolysis (Borchelt *et al.*, 1990). The infectious prion protein will now be termed as PrPsc and the normal prion protein, its function later to be discussed is now termed as PrPc (the "sc" stands for Scrapie and the "c" stands for cellular).

Different terminology of the two forms of the protein has also come into light. This is the PrPsens and PrPres; these names are given to prions, which are not structurally identical to PrPc or PrPsc. They may be truncated or recombinant forms, and are not exactly the same as the Scrapie (sc) and cellular (c) forms, the 'res' indicates resistance to proteolytic degradation, and the 'sens' is for sensitivity to proteolytic degradation. There are therefore two different isoforms of the prion protein, one is expressed normally and one is present aberrantly. Initial studies on the different forms of the prion (Borchelt *et al.*, 1990) found there were differences in the proteolytic resistance. Also demonstrated was that PrPc can be released from normal, and Scrapie infected neuroblastoma cells which were treated with phosphatidylinositol-specific phospholipase C (PIPLC), and also be labelled by a sulpho-NHS-biotin.

Using pulse chase experiments [35S] methionine was found to incorporate almost immediately into PrPc. PrPsc however, was incorporated at a much slower rate. PrPsc was not released by PIPLC, which suggested structural differences. It also enabled to monitor the amounts of both PrPc and PrPsc in the cell. Although this gave no real clues to the mechanisms involved, it further established the presence of two different forms of the protein. Mass spectrometry and amino acid sequencing studies were undertaken on the Scrapie prion protein, this was done using denatured PrPsc that was digested with endoproteases and isolated by HPLC. Analysis was done using mass spectrometry and Edman sequencing and the primary structure of PrPsc was found to be identical to that of the PrP gene sequence, arguing that neither RNA editing nor protein splicing feature in the synthesis of PrPsc. Mass spectrometry was also used to identify any posttranslational modification-differences between the two isoforms, and none were found contending that PrPsc molecules do now differ from PrPc at the level of an amino acid substitution or a post-translational chemical modification (Stahl *et al.*, 1993). However, the possibility of a small fraction of PrPsc being modified by a yet unidentified post-translational process could not be eliminated, or that PrPc carries a modification that is removed in the formation of PrPsc. It is probable then, that the differences are in the secondary and tertiary structures.

The same amino acid sequence but two different structures

Work was then done on investigating the structure of the prion comparing PrPc to PrPsc. The structure of PrPc was found to have mainly α-helices and the structure of PrPsc was found to comprise mainly of β-pleated sheets. This was achieved using a pattern-matching approach (Cohen *et al.*, 1985) and Fourier analysis of the primary sequence hydrophobocities to detect amphipathic regions, interestingly the secondary structure showed some homology to that of the serum amyloid A proteins. The prion was linked with having a post-translational modification to yield amyloid A, which is more evidence for the prion involvement in amyloid plaque formation (Bazan *et al.*, 1987).

Later Pan *et al.* (1993) wrote a paper which described the conversion of α-helices to β-sheets as part of the formation of the Scrapie prion proteins. Using Fourier Transform Infra-Red spectroscopy (FTIR) it was demonstrated that PrPc has high α-helix content (42%) and no β-sheet content (3%), and PrPsc has high β-sheet content (43%) and low α-helix content (30%). Even higher β-sheet content is observed (54%) as compared to (α-helix content 21%) when using limited proteolysis to create N-terminally truncated PrPsc. This was confirmed by circular dichroism measurements. Pan *et al.* (1993) then remarked on the likeliness of a conformational transition being a fundamental event in the propagation of prions.

Baldwin *et al.* (1994) produced a report which identified and confirmed the differences between PrPc and PrPsc using spectroscopy to be conformational and even possibly transitional. For this FTIR spectroscopic methods were used, and also suggested the possibility of the transition of alpha to beta transition, they found that the N-terminal truncation of PrPsc by limited proteolysis did not destroy the prion infectivity.

Instead, it increased the β-sheet content shifting the FTIR absorption to lower frequencies, which was typical of the cross β-pleated sheets of amyloids. Thus, the formation of PrPsc from PrPc involves a conformational transition in which one or more α-helical regions of the protein are converted to β-sheets. This transition was mimicked by synthetic peptides, allowing predictions of domains of PrP involved in prion diseases.

Further evidence for the PrPsc formation from PrPc was shown in experiments involving the synthesis of α-helix peptides from regions of the prion protein, and their subsequent conversion to form amyloid peptides composed largely of β-sheets. Although this evidence is quite good, harder evidence was to follow (Gasset *et al.*, 1992, 1993).

Muramoto *et al.* (1996) performed studies on recombinant forms of PrPc. It is known that some truncation of the PrPc N terminus still permits conversion into the PrPsc isoform and so to assess whether additional segments of the PrP molecule can be deleted, the regions of the putative secondary structure were removed. It was found that removing a 36-residue loop between the second and third helix did not prevent formation of protease resistant PrP (PrPsc106) that was also found to be soluble.

Also, all the deletions of the 4 putative a-helical structures prevented PrPsc formation. The results suggested that all of the proposed secondary structures are necessary for the conversion of PrPc into PrPsc and that the discovery of PrPsc should facilitate structural studies of the prions and possible development of antibodies against PrPsc.

A year later in 1997, studies investigating the possible posttranslational modifications involved with the PrPc conversion to PrPsc focussed on Scrapie infected ScN2a cells, where the metabolism of both PrP isolates involved cholesterol dependant pathways. It was demonstrated that both PrPc and PrPsc are attached to Trition X-100-insoluble, low-density complexes or "rafts".

These "rafts" probably contain cholesterol due to their sensitivity to saponin. However, when extracted from solubility into insolubility using Triton X-100 at 37°C it was found that two different peaks of different densities were there, suggesting two populations of PrP-containing "rafts". This may permit the isolation of PrPc specific rafts. This suggested the possibility of "rafts" hosting the proposed PrPc to PrPsc formation, or if not some other aspect in the "life-cycle" of prions (Naslavsky *et al.*, 1997).

The three dimensional structure of PrP

The three dimensional-structure for the cellular prion protein was predicted using circular dichroism and infrared spectroscopy (Huang *et al.*, 1994). A heuristic approach was used, consisting of the prediction of secondary structures and of an evaluation of the packing of secondary structures. By using experimental and theoretical constraints four theoretical structures consisting of α-helix bundles were achieved. A group of amino acids was identified as being important for tertiary structural arrangement.

Although the predicted structures gave a testable model, PrPc is thought now to have only three α-helix bundles (shown later). Possible inaccuracies could be in the computer programme or rather the experimental, theoretical constraints or data applied to the programme.

Despite this, the structure of PrPc seems to be far more predictable than that of PrPsc (Baldwin *et al.*, 1998). This is partially due to the very small amount of the natural protein isoforms isolated from the brain, and the solubility of PrP 27-30 in detergents that reduces infectivity even though the β-sheets can be regenerated. Despite this, the NMR (Nuclear Magnetic Resonance) experiments on recombinant protein can lead to a much better characterisation of the 3D structure of the prion proteins.

In the studies involving the use of hamster and mouse models to elucidate the three-dimensional structure, the hamster and mouse prion protein structure was compared, and is shown below. The figure 15 James *et al.* (1997) shows the hamster and mouse prion proteins superimposed, the mouse structure is more refined but shorter in length than the hamster prion, showing a close relation and therefore is a good model for transgenic studies.

Thermodynamics

Using a simple, low-resolution lattice model of prion protein folding Harrison *et al.* (1999) studied the thermodynamics and problems implicated in the prion protein folding. To define their model proteins and prions scientists studied 3D 27-mer maximally compact cubic conformations and 2D n-mers for a range of chain lengths n from 11 to 16. The 2D studies allowed examination of the effects of variation on chain length and of surface complementarity in the dimer.

The 3D studies had a much more limited number of possible dimer packing geometries but showed the dimensionality of real proteins. The studies did require attention regarding the hydrophobic and hydrophilic sequences that may affect the folding structure, as PrP is of intermediate hydrophobicity. Although in the HP lattice model, a minor proportion of the proteins encrypted a native multimeric state and did not require a hydrophobic or hydrophilic sequence to affect this change.

This is consistent with real prions as while PrP is of intermediate hydrophobicity, the yeast prions Ure2p and Sup35p are more hydrophilic than the average protein of known structure. It was found that if the model were to re-fold upon dimerisation, then a minor proportion of up to about 17% encrypt an alternative native state as a homodimer. The structures in this homodimeric native state re-arrange so that they are very different in conformation from the monomeric native state.

It was also found that model proteins which are relatively less stable are more susceptible to the formation of alternative native states as homodimers. The results obtained suggest the less-stable proteins have a greater need for a well-designed energy landscape for protein folding to overcome an increased chance of encrypting substantially different native conformations stabilised by multimeric interactions.

The number of α-helices in PrPc

Prusiner predicted that there were four α-helices in PrPc, and later demonstrated that in fact there were only three α-helical regions termed helix A, helix B and helix C. Interestingly the NMR structures of recombinant Syrian hamster PrPc (rPrP 90-231) and Mouse PrPc (MoPrP 121-231) although similar in many respects do contain substantial differences.

For example: The loop at the NH2 terminus of helix B being well defined in rPrP (90-230) but is disordered in MoPrP (121-231). Additionally, helix C is composed of residues 200-227 in rPrP (90-231) but in MoPrP (121-231) extends only from 200-217 residues. The undetermined cause of these differences could be due to one of three possibilities:

1. They have different lengths.
2. Species-specific differences in sequences.
3. The conditions used for solving the structures.

Recently NMR studies of the full length SHaPrP (29-231) and MoPrP (23-231) have demonstrated that the NH2 termini are highly flexible and lack identifiable secondary structure under the experimental conditions employed. Also transient interactions between the COOH terminal end of helix B and the highly flexible, NH2-terminal random coil containing the octarepeats, residues 29-125 (Prusiner *et al.*, 1998). Willie and Prusiner

(1999) were able to perform ultrastructural studies on Scrapie prion protein crystals produced using reverse micellar solutions. Structural studies on the prions in this crystal form may provide insight into the structural transition which occurs during PrPsc formation. The main differences between PrPc and PrPsc can be summarised in table 3 shown below:

Table 3: Comparison of PrPsc and PrPc

Characters	*PrPsc*	*PrPc*
Structure	Globular	Extended
Protease resistance	Yes	No
Presence in Scrapie fibrils	Yes	No
Location in or on cells	Cytoplasmic vesicles	Plasma membrane
Turnover	Days	Hours

THE CONVERSION FROM PrPc TO PrPsc

The prion protein exists in two forms. The normal, innocuous protein (PrPc) can change its shape to a harmful, disease-causing form (PrPsc). The conversion from PrPc to PrPsc then proceeds via a chain-reaction. When enough PrPsc proteins have been made they form long filamentous aggregates that gradually damage neuronal tissue. The harmful PrPsc form is very resistant to high temperatures, UV-irradiation and strong degradative enzymes.

The wild type prion (PrPc) is found in the secretory pathway of cells expressing the protein (1) and moves to the plasma membrane where it is anchored by its GPI tail (2) There it may bind to an extracellular ligand (possibly copper) (3) before being cycled from the membrane into endocytic vesicles (4) At some point its cargo is released and the protein either passes to the lysozome for degradation or back to surface for another round of ligand binding. In this respect it resembles many other membrane-resident proteins.

The pathogenic form (PrPsc) also finds its way to endocytic vesicles where it co-opts some of the wild type form to become pathogenic (5). PrPsc is resistant to degradation, a hallmark of the infectious form, so accumulates. Neurotoxicity is probably linked to the conversion event itself, perhaps through its interference with normal PrPc turnover, because there is considerable evidence to show that the acquired PrPsc is not inherently toxic.

Infection of normal cells may occur when an aberrant prion acts as a template for the refolding of a normal prion into a new aberrant prion. It is thought that there is at least one more protein involved: the as-yet-unidentified Protein X. This protein is believed to mediate the folding from a normal into an abnormal prion.

When proteins are synthesized inside of a cell there are other special proteins (known as chaperones) that help in this process. Chaperones are proteins that bind to the newly synthesized protein or protein subunit, in order to ensure that the protein is properly folded into its secondary or tertiary structure. It has been hypothesized that Protein X is a type of chaperone.

The concept that a protein can transmit information (i.e. protein structure) is novel. At the time that this idea was proposed, it was truly radical; it conflicted with many years of evidence showing that only DNA was capable of transmitting such information. This does not mean that prion diseases cannot be genetic. If an individual happens to carry a gene that codes for a mutant prion, then all of the prions that their cells produce will be the abnormal, disease-causing prions. One such disease is familial CJD.

STUDY OF PRION GENETICS

The prion gene and inherited prion diseases

The prion gene is called PRNP (in humans) and is found in humans on the short arm of chromosome 20, and in murine in a syntenic region of chromosome 2; these two locations are homologous (Sparkes *et al.*, 1986). The gene has only one protein coding region. Prion diseases are the only known diseases that can be sporadic, genetic, or infectious.

Only one homologous gene (PrLP) in a non-mammalian organism (chicken) has so far been described. PrP related DNA hybridisation signals have been detected in organisms such as *Drosophila*, the nematode *Caenorhabditis elegans* and even yeast (described later).

A novel gene was isolated from *C. elegans* by Iwasaki *et al.* (1992) using a hamster PrP cDNA probe. It was found that the encoded protein was homologous to nuclear ribonucleoproteins but not to mammalian PrP. These results make it more likely that PrP is a late gene in evolutionary terms (Goldmann, 1993).

Inherited forms of the disease comprise 25% of all cases of prion diseases notably GSS, familial CJD and FFI. In each of the inherited forms, mutations have been found in the ORF (open reading frame) of the PRNP gene. The first half of the PRNP ORF contains about 170bp with a high content (about 80%) of the nucleotides guanidine (G) and cytidine (C), most of this sequence is organised in 24 bp (or 27 bp) repeats. Few differences are observed between these sequences, and between those in other species suggesting that they are highly conserved through evolution.

It has been suggested that control and fidelity of PrP mRNA translation is associated with these high G/C content regions. In situ and northern blot hybridisation has shown that the gene is predominantly expressed in neuronal cells as well as ganglia and nerves of the peripheral nervous system.

It also showed that it is not exclusively expressed in the CNS and neurons as the kidney, heart, lung and spleen all contained transcripts except for the liver which was negative. Apart from the possible exception of sheep, inherited forms of the disease have only been found in humans. This raises interesting questions that cannot be readily pursued in animals (Goldmann, 1993; Prusiner, 1996).

There are many mutations that have been identified with the PRNP ORF and are often genetically linked to hereditary prion disease. This includes a proline (P) to leucine (L) mutation at codon 102 that was shown to be linked genetically to development of GSS with a LOD score exceeding three. (The LOD score is a determinant of genetic linkage, a score of 3 or greater is considered as confirmation of linkage, as it would yield a ratio of 1000 to 1 in favour of linkage as opposed to non-linkage, i.e. > 5 cM apart (Muller & Young, 1998).

This mutation may be due to the deamination of a methylated deoxycytosine (C) coupled to deoxyguanosine (G) through a phosophodiester bond (CpG) in the germline DNA encoding PrP resulting in the substitution of deoxythymine (T) for deoxycytosine.

This P102L mutation has been found in 10 different families in 9 different countries including the original GSS family. Israeli Jews of Libyan origin showed an unusually high incidence of CJD, this was thought for many years to be due to the consumption of lightly cooked sheep brain or eyeballs.

However, studies uncovered some Libyan and Tunisian Jews in families having a PrP point mutation at codon 200 resulting in a glutamate (E) to lysine (K) substitution. One of the patients was homozygous for this mutation (E200K) and as her clinical presentation was similar to that of the heterozygote patients, it was suggested that familial prion diseases are true autosomal dominant disorders.

The E200K mutation has been found in many areas including Slovaks originating from Orava in North Central Czechoslovakia, in a cluster of families in Chile, in a large German family living in the United States, as well as in British and Japanese families. This diversity suggests a recurrent point mutation, but some investigators have argued that all the E200K mutations originated from a Shepardic Jew whose descendants migrated from Spain and Portugal at the time of the inquisition.

Although, others agree that it is more likely that the mutation has arisen independently multiple times by the deamidation of a methylated CpG. This is supported by historical records of Libyan and Tunisian Jews, indicating that they are descended from Jew living on the island of Jerba where they first settled around 500 BC and not from Shepardim.

At codon 178, a mutation has been identified involving the substitution of aspartic acid (D) to asparagine (N) in many families with CJD, and like with the E200K mutation the PrP amyloid plaques are rare, the neuropathologic changes generally consisting of widespread spongiform degeneration. The D178N mutation has been linked with a number of Italian families with cases of insomnia, although the mutation appeared to be incompletely penetrant.

The same mutation was also reported in several families affected by a disease phenotypically different from FFI and similar to CJD, except for the longer duration and the lack of sharp-wave electroencephalographic activity in most of the cases.

This finding that the same mutation gives two different phenotypes prompted a series of studies to discover the molecular basis of this phenotypic heterogeneity. A detailed analysis of the PRNP genotype in 15 FFI and 15 CJD patients showed that in addition to the D178N mutation, all of the FFI subjects had a methionine at position 129 of the mutant allele while all CJD subjects had valine at this same position. These results have been confirmed in all of the FFI and CJD cases. Therefore this gives two distinct haplotypes, the 129M, D178N haplotype in FFI, and the 129V, D178N haplotype in CJD.

As one of the FFI kindreds has an octapeptide repeat deletion in the mutant allele, it is very unlikely that all of the known FFI kindreds originated from a common founder. This finding strongly argues against the possibility that the phenotypic differences are caused by genetic influences other than PRNP codon 129. Interestingly although the methionine or valine at codon 129 on the mutant allele is obligatory in FFI and CJD178 patients respectively, the codon 129 on the normal allele can be either methionine or valine.

Therefore, the FFI and CJD phenotypes are likely to be determined by the codon 129 of the mutant allele, which in association with the D178N mutation, results in the expression of two different types of PrPres. Also, as FFI is usually expressed in the phenotype earlier than CJD, the codon 129 also modulates the duration of the phenotype.

Studies on the PrPres fragments associated with the two proteins differ both in size and in the ratio of the three differently glycosylated PrPres isoforms. The size variation is the result of the differential N-terminal digestion by proteases and the difference indicates that PrPres has different conformations, or specific-ligand interactions.

The ratio difference however indicates a different post-translational processing of PrP in the two diseases to ultimately give two different phenotypes. Also noted in these cases were the different incubation times in relation to the heterozygosity and homozygosity of the mutant allele. The homozygote duration of the disease was significantly shorter than that of the heterozygotes. The mean age of onset of CJD in homozygotes was 39+/- 8 years and in the heterozygotes it was 49+/-4 years.

A valine (V) to isoleucine (I) substitution at codon 210 produces CJD with classic symptoms and signs, and like the D178N mutation appears to show incomplete penetrance. GSS has been associated with mutations in codons 105 and 114. Other point mutations have been shown at codons 145, 198, 217 and possibly 232 that segregate with inherited prion diseases.

Interestingly, synthetic peptides adjacent to and including residues 109 to 122 respectively have readily polymerised into the rod-shaped structures, which have the tinctorial properties of amyloid (Prusiner *et al.*, 1996, Parchi *et al.*, 1996).

Other than base substitutions, octapeptide inserts can also cause mutations. An insert of 144bp at codon 53 containing 6 octarepeats was initially described in patients with CJD from four families all residing in southern England. As the human PrP gene only contains 5 octarepeats a single genetic recombination event could not have created this extra insert. Although as the four families were distantly related, a single person born more than 2 centuries ago may be the founder (LOD score being greater than 11).

Studies from several laboratories have demonstrated that two, four, five, six, seven, eight or nine octarepeats in addition to the normal five are shown in individuals with inherited CJD. Deletion of one octarepeat has also been identified but without any neurological disease (Prusiner *et al.*, 1996). The location of human mutants has been shown on a mouse prion model, this is shown in figure 21 below taken from James *et al.* (1997):

Octapeptide repeats are essentially insertional mutations, and the patients usually have a prolonged illness (averaging 7 years) beginning at an early age averaging 38 years, characterised by progressive dementia accompanied by cerebellar and other neurological signs. The genetic mechanism for the generation of extra repeats is probably unequal crossover. The similarity of the observed repeats allows limited frameshifting in the process of duplication; recombination may happen at any site. The result of the first event could be a 7-repeat allele, which has not so far been observed. The second event then could be the 9-repeat allele, and further crossover events could result in alleles with 10 or more repetitions. The PrP repeat region has been compared with many structural proteins (collagen, silk fibroin, keratin and elastin) with no close resemblance. However, three others have been closer, including the maize ribosomal protein S18 that has a 7-fold heptapeptide repeat at its N terminus that is thought to have some role in RNA binding.

Also, the DNA-dependant RNA polymerase II contains many repeats of a heptapeptide in its C terminal domain. It is here where a regular spacing of proline residues appears to be important in maintaining its unusual secondary structure. The variations observed have been thought to act to regulate enzyme activity. Interestingly ice nucleation points also contain large numbers of octapeptide, each contributing equally to the nucleation process. Essentially, the mechanistic properties of the repeat function are not known, although there are several possibilities, such as the possibility that it may be involved directly in a templating mechanism, whereby the normal protein is converted posttranslationally into an enantiomorphic form of unusual conformation.

The presence of extra repeats might be expected to increase the probability of a spontaneous transition to the abnormal enantiomer and thereafter provide an increasing concentration of template for the conversion of further molecules. The cis-trans isomerisation of the proline residues may play an important part, or if due to the extra repeats, pseudoknots are able to form changing the mRNA performance leading to the production of a PrPres protein.

Nevertheless, whatever the mechanisms are by which the extra repeats exert their effect, the presence of 10 or more such elements is associated with a large range of dementing illnesses that may look like CJD but differ in their time of onset, and slow progression. However, it has been shown that they have been experimentally transmitted horizontally, and therefore share the same transmissibility characteristics as other familial and sporadic forms of spongiform encephalopathy (Goldfarb *et al.*, 1991).

A study by Murmoto *et al.* (1997) was based on the effect of a deletion of one or more of the hypothetical α-helix coding regions in the PRNP gene. It was discovered that in mice this deletion predisposed to a heritable disorder resembling neuronal storage disease. This caused CNS dysfunction and numerous neurons with intracellular accumulations of PrP, which are insoluble in non-denaturing detergents, a property which distinguishes it from the PrPc. Whether this α-helical deletion disease described here is transmissible is unknown.

The findings described strongly argue that the pathogenesis of this disorder differs considerably from that of the normal prion diseases. No human PRNP mutations have been linked to the loss of α-helical regions, but this could be due to the phenotype differing a lot from the classical CJD phenotype.

Interestingly there has been little attention given to the molecular genetics of the PRNP gene in childhood CNS diseases, since prion diseases usually present in adults. But, with the findings of children as young as three or four with Kuru in Papua New Guinea (acquired syndrome so children can get the disease), there seems no reason why an inherited disease shouldn't strike. Especially as we do not know the mechanisms by which the subjects carrying a PRNP mutation remain asymptomatic until they reach the mature age. It may then be logical to conclude from this that in the future, brain sections should be immunostained for PrP in CNS storage diseases of children where the aetiology is unclear and to sequence the ORF of PrP alleles from these patients.

In conclusion it can be said that recent studies of the genetics of familial prion disease have yielded three major results:

1. Significant genetic linkage between ataxic GSS and PRNP indicates that this horizontally transmissible disease is inherited through a locus at or near PRNP.
2. A leucine substitution at PRNP codon 102 is genetically linked to ataxic GSS in families from ethnically and lineally distinct backgrounds, suggesting that this mutation arose independently at least twice and is therefore a well linked genetic marker. The substitution in question may conceivably cause or directly influence the pathogenesis of the disease.
3. This and other mutations in PrP, is found to track with disease in families with a strong, but not absolute correlation of specific mutations in PrP with different clinical forms of the prion disease (Hsaio and Prusiner, 1990).

The discovery that FFI is a prion disease widens the clinical spectrum of these disorders and raises the possibility of other degenerative diseases being prion related.

The challenge now for familial prion disease

Despite all the advances in work done with prions there are still issues which remain to be addressed. Generally, the challenge is to determine the individual steps that lead from the presence of germline mutations in PRNP to the onset of diseases with various phenotypes and duration. Essentially the issues to be resolved are:

(a) The mechanisms by which the subjects carrying a PRNP mutation remain asymptomatic until they reach the mature age.

(b) Whether the conversion of PrPsens into PrPres is a requirement for the onset of the disease.

(c) The specific events that predispose the mutant PrP to convert to PrPres.

(d) The mechanism of selective vulnerability of different neuronal populations to the various mutant PrP.

Evolutionary relationship

Prion genes have been sequenced from 33 different species and a phylogenetic tree has been constructed (Krakauer *et al.*, 2002). The exciting thing in this is an unexpected example of convergent evolution between the human and bovine proteins.

Close examination of the proteins revealed that humans (as well as gorilla, chimpanzees and some other primates) have a serine at position 143 in the prion amino acid sequence and a histidine at position 155; most of the other prion sequences have asparagine and tyrosine, respectively at these positions. The one other species to have these two substitutions is Bos Taurus the cow.

Asp143 → Tyr155 (Most Prions)
Ser → His (Human, Cow)

Could these shared features displayed by the Prions of human and cattle be linked with susceptibility that humans display towards the infectious form of the cattle. The hypothesis is attractive because infection of humans by sheep prion, which causes Scrapie in sheep, has never been proven, so possibly we have a natural resistance to the sheep prions. However, the hypothesis is yet to be proven (Goldmano *et al.*, 1996).

INVASION OF INFECTIOUS PRION PROTEINS AND MOVEMENT WITHIN BRAIN CELLS

Prion replication and spread at the cellular level

Prions are pathogenic variants of proteins that are naturally produced in nerve cells and certain other cells. The normal "healthy" prions are referred to as PrPc (Prion Protein cellular). The word "prion" stands for "proteinaceous infectious particle" and so should properly only be applied to the pathogenic variants. In Fig. 22, the production of PrPc is illustrated from the nucleus at bottom right.

RNA that codes for PrPc is produced in the nucleus and exits via the nuclear pore. The RNA then passes along ribosomes attached to the rER. PrPc is formed in the rER and then progresses up through the Golgi. At the upper face of the Golgi, vesicles containing PrPc bud off and travel to the cell surface. Here, they fuse with the cell membrane and so discharge their cargo at "PrPc". By this means, the cellular proteins come to sit on the exterior of the cell. PrPc encounter rogue (purple) prions.

These are termed PrPsc (sc stands for Scrapie, the prion disease of sheep). The rogue prions seem to force the normal proteins to change shape. Both types of protein, the PrPc and their corresponding prions, are the same chemical - just different shapes. It is this shape-flipping which is equivalent to the transmission of infection. Such a conformational shift or flip could happen at the cell surface or in caveolae (one is shown as a small invagination in the cell membrane). Residual PrPc might continue to be flipped by contact with the rogue confomations for some time in these vesicles.

Prions polymerise, finally appearing as purple fibrils in the picture at "P". PrPsc is resistant to degradation by the enzymes contained in the lysosomes that are seen floating nearby. Consequently, PrPsc accumulates in the cell. PrPsc vesicles may also travel to the Golgi and intercept PrPc that is being processed there. In this way, PrPc particles could be switched to the rogue form before they reach the surface of the cell. By such mechanisms, PrPc might be switched to PrPsc at various points in and on the cell.

Prions could enter the brain along the axons of neurons. This probably happens by a retrograde flow of prion filled vesicles. These are shown in the picture as purple spheres ascending the axon like elevators going up a shaft. Another route of entry could be the blood, probably in immune cells. A lymphocyte is shown exiting the capillary at bottom left where it could then contact the astrocyte. Astrocytes and other glial cells also support the production of prions.

Delivery of orally ingested prions to brain

Animal experiments have shown that the entry of prion into CNS (Central Nervous System) is preceded by a period in which the ingested prions accumulate in secondary lymphoid organs Peyer's Patches (PP), lymph nodes, spleen, particularly follicular dendritic cells (FDCs) in the B cell areas of these organs (Huang *et al.*, 2004). How the agent is transmitted from the intestinal lumen is largely unknown. Dendritic cells (Dcs, cells quite distinct from FDCs) are cells specialized to acquire antigens from peripheral tissues and to transport them to secondary lymphoid organs for presentations to T and B-lymphocytes. Huang and MacPherson (Huang *et al.*, 2004) have shown that DCs can acquire PrPsc (Scrapie associated PrP) from the intestinal lumen and deliver it to mesenteric lymph nodes.

The prions can replicate at these sites and eventually gain access to as nerve and then propagates back up to the axon to the spinal cord and eventually the brain. It has been shown that transport of the causative agent using PrPsc as a biomarker from the intestine to FDCs is a very inefficient process, which may help to account for the low frequency of individuals who have consumed infected material that go on to develop clinical disease.

The activation of glial cells precedes neuronal death, which seem to be initially caused by by the deposition of misfolded (PrPres) form of the prion proteins in the brain. Cytokines and chemokines released by PrPres activated glial cells may contribute directly to the disease development by enhancement and generalization of the gliosis and via cytotoxity for neurons (Burwinhel *et al.*, 2004). Scientists have watched agents of brain-wasting diseases, called Transmissible Spongiform Encephalopathies (TSE), as they invade a nerve cell and then travel along wire-like circuits to points of contact with other cells. The research was done in laboratory cultures using a rodent-adapted form of Scrapie protein and cells taken from the central nervous system of mouse and hamster brains. The proteins were first "branded" with fluorescent dyes so they could be easily tracked.

Prions branded with a fluorescent dye (pink) were added to nerve cells taken from a hamster brain fig 24 and Prions branded with a fluorescent dye (red) were added to nerve cells taken from mice fig 25. The prions initially were in the form of large clumps on the cell, but over time the clumps were broken into smaller units and transported along wire-like nerve cell projections.

THE FUNCTION OF PrPc

Studies on transgenic mice to deduce the function of PrPc

From all the evidence put forward above, it seems certain that the production of the aberrant form of the PRNP gene product is responsible in some way for the pathogenesis of the disease. It is known that the PrPc protein is expressed as a glycosylphosphatidyl inositol-anchored glycoprotein found on the outer cell membrane of neurons and to a lesser extent of lymphocytes and other cells (Stahl *et al.*, 1987, 1990, 1992). However, the role of the normal cellular form of the prion is unclear and unproven.

Büeler *et al.* (1992) made studies on mice with inactive Prn-p genes (homozygously disrupted) and showed that the mice developed normally. The genes were disrupted by homologous recombination with a 4.8 kilobase DNA fragments in which codons 4 to 187 of the 254-codon open reading frame were replaced by a neomycin phosphotransferase (neo) gene under the control of the Herpes Simplex virus thymidine kinase promoter. Then were obtained mice heterozygous (Prn-p0/+) for the disrupted gene, the presence of which was confirmed by Southern analysis. Heterozygous mice were then mated to other heterozygotes to give a new population.

Surprisingly, this population contained 24% Prn-p(0/0), 26% Prn-p(+/+) and 50% Prn-p(0/+) mice, surprising because they were all phenotypically the same, and Southern analysis confirmed these results. The Prn-p(0/0) mice were tested for many abnormalities in their behaviour, fitness and fertility and were all confirmed normal for at least 7 months, with no immunological defects. The reason that these results are surprising is that the Prn-p gene is highly conserved through evolution and so it might be thought that the function of this gene is vital to the organism.

In contrast to Büeler *et al.* (1992), more recent studies have reported that the prion protein is necessary for normal synaptic function (Collinge *et al.*, 1994) and that altered circadian activity rhythms and sleep has been observed in mice also devoid of the prion protein (Tobler *et al.*, 1996). One might ascertain from PrP knockout mice, that although there appears to be no symptoms, by looking closely at the behaviour of the mice they might be seen to differ phenotypically from the PrP-normal mice. It seems that the studies in 1992 may have missed the subtle phenotypic differences observed in later more focused studies.

The studies suggest that because of the intriguing similarities to rhythm and sleep alterations in Fatal Familial Insomnia, where there is a profound alteration in sleep and the daily rhythms of many hormones, the loss of function PrP may be related to the normal function of the prion protein. Essentially, it can be seen that the probable function of PrPc, being a well-conserved protein present in many mammalian species, may be involved in maintaining sleep continuity or regulating sleep intensity.

Another study by Sakaguchi *et al.* (1996), looked at the long-term effects of the disrupted gene on mice. At about 70 weeks of age, the PrP-null mice all began to show progressive symptoms of ataxia. A rotorod test evidently showed up a lack of co-ordination in these mice. As they got older, their coordination became progressively worse and they frequently collapsed. Closer examination of the afflicted mice gave a good indication of what prion proteins would normally be doing when there is plenty of it in the brain. The cerebellum, the part of the brain responsible for movement and coordination, had shrunk by almost one third in the specifically bred prion free mice and upon pathological investigation, there was an extensive loss of Purkinje cells in the vast majority of cerebellar folia. This suggests that PrP plays a role in the long-term survival of Purkinje neurons (Moore *et al.*, 1999).

It was also reported that neuroprotective functions of PrP include its antioxidative and anti-apoptotic properties (Brown *et al.*, 2001; Kim *et al.*, 2004; Wong *et al.*, 2001). In addition to this a role in copper binding and delivery and a signally function has been supported with experimental evidence (Brown *et al.*, 1997; Stockel *et al.*, 1998; Brown *et al.*, 1999; Mouillet-Richard *et al.*, 2000).

Recent studies on yeast have shown that the infectious prions linked to the disease, could help organisms adapt to tough situations. In yeast, some cells carry the normal type of proteins, whereas others harbour the infectious form, which accumulates in clumps and is passed on from one cell to another. Scientist have shown that the prions in their infectious form help the yeast to adapt, changing their rates of survival when they were grown in various nutrients or temperatures. Prions may therefore offer a speedy way for yeast to evolve (Osherovich *et al.*, 2002).

In addition to this, the rouge Prion fibres are being analysed to be used as templates for nanowires (Nanobiotechnology). Scientists have used a prion protein from yeast to template the formation of metal wires approximately 100nm wide. While other biomolecules have been used in nanofabrication strategies, the researchers argue that the physical and chemical study of prion proteins make them more likely to be able to withstand industrial metallization processes (Lindquist *et al.*, 2003).

Another study by Mabbott *et al.* (1997) has looked at the possibility of the involvement of T-lymphocyte activation as a role of PrPc, this is due to the fact that PrPc is a GPI-anchored protein (membrane attached via a glycosylphosphatidyl-inositol anchor). Certain GPI anchored proteins on lymphocytes have previously been shown to play an important role in an early step of cell activation, and as the expression of PrPc has also been identified in lymphoid tissues, they hypothesised that PrPc like other GPI-anchored proteins may play a role in lymphocyte activation.

In this study it was demonstrated that the PrPc is expressed on the surface of murine lymphocytes, and following mitogen stimulation the lymphocyte PrPc expression was significantly enhanced suggesting a role of PrPc in lymphocyte activation. Also the mitogen responsiveness in PrPc-deficient lymphocytes was significantly reduced in comparison to wild-type controls. Therefore, the results imply that PrPc is expressed on the extracellular leaflet of the lymphocyte membrane and involved in lymphocyte activation. Although this report does suggest a role for PrPc being involved in immune function it must be noted that the PrPc molecule is not the only GPI-anchored protein expressed on the surface of murine lymphocytes. Other examples include the Thy-1, Ly-6, CD58, CD-59 and Oa-2 proteins.

Previous work demonstrates that the non-specific removal of all GPI-anchored proteins with a phospholipase C enzyme abrogates the ability of murine lymphocytes to respond to stimulation by the mitogen C on A, so there is evidence that the GPI-anchored proteins are involved with activation of lymphocytes. However, PrPc is predominantly produced in the brain and not predominantly in the peripheral organs and those expressing lymphocytes (spleen, thymus gland, lymph nodes, gut wall and bone marrow). But, this does not necessarily mean the lack of lymphocyte involvement of PrPc. PrPc may be involved in a multiple role, for example it may have a role in the brain and a role related to lymphocytes, this would explain the highly conserved nature of the gene, as a mutation in the gene would be more likely to give a loss of function or there is the possibility that PrPc can be involved but in a coincidental way, as many other GPI-anchored proteins are known to have the same effect.

Work done by DeArmond *et al.* (1997) has uncovered the possibility that glycosylation can modify the conformation of PrPc. Glycosylation could also affect the affinity of PrPc for a particular conformer of PrPsc, and thereby determining the rate of nascent PrPsc formation and the specific patterns of PrPsc deposition. They postulated that variations in the complex carbohydrate structure in the CNS could account for selective targeting (i.e. the formation of PrPsc in particular areas of the brain being due to differences in PrPc/PrPsc affinity). The experiments were done using deletions of Asn-linked oligosaccharides of PrPc in transgenic mice.

Recently Ikeda *et al.* (1999) have performed experiments on cultured hepatic stellate cells, and by using immunoelectron microscopy, have revealed that PrPc resides on the plasma membrane of stellate cells. Their results indicate that PrP expression is closely related to stellate cell activation associated with fibrogenic stimuli. This has helped elucidate the pathogenesis of liver fibrosis, and develop possible therapeutic strategies to prevent liver cirrhosis. PrPc is not expressed in the normal healthy liver, but it was found to increase dramatically in diseased livers associated with stellate cell activation.

Prions show their metal

Examining the prion protein sequence for structural clues to its function indicates an unusually 'split' molecule fig. 26. The carboxyl terminal half of the protein contains amino acid sequences of mixed character, with early secondary structure predications indicating at least two long stretches of α-helix.

The presence of these α-helices was subsequently proved correct; a protein fragment encoding the C-terminal half of the protein provided the first three-dimensional structure of any prion related sequence by NMR spectroscopy in 1996. The domain is a tightly folded globular structure made up of three α-helices, two of which are held together by a single disulphide bond, and two short sections of β-sheet.

Subsequent biochemical analysis has shown that this domain folds extremely rapidly and is relatively stable and unaffected by its environment. Experiments to determine the structure of the amino terminal half of the prion protein have not been as successful, reflecting the highly unusual amino acid sequence and secondary structure for this region. Following an initial 'signal peptide' region, there is a short series of hydrophobic amino acids followed by an eight amino acid motif, the octarepeat.

As its name implies this motif is repeated perfectly four times and includes residues typical of those that interact with metal ions. Following the octarepeat region a second hydrophobic stretch of amino acids occurs before the beginning of the well characterised C-terminal domain. Clearly, the prion protein is a molecule that is front end loaded and much of the research effort has focused on trying to discover the function of such an unusual sequence combination.

Clues to function

Efforts to produce the amino terminal half of the protein in isolation and to study it biochemically have proven difficult. However, researchers have described biochemical properties that may reflect those of the complete protein to peptides with sequences derived from this region. In the mid-1990s it was showed that peptides encoding the octarepeat sequence bind to copper ions in vitro.

Peptides with sequences corresponding to the central hydrophobic sequence of the prion protein, between the octarepeats and the C-terminal domain, are cytotoxic to cultured cells in vitro. They also have a propensity to aggregate into fibril structures reminiscent of those found in diseased TSE brains, suggesting that this region could become exposed during the conformational change when the prion protein becomes pathogenic.

However, research on isolated peptides cannot necessarily be extrapolated to a full length folded protein. The observations of copper binding by prion peptides therefore remained of speculative interest until the late 1990s, when it was shown that this is also a property of the complete recombinant prion protein; roughly four copper ions are bound to each prion molecule, one for each octarepeat. Binding is dependent on pH, suggesting that the protein can be loaded with copper at neutral pH, for example on the cell surface, and then release the ion under acidic conditions, such as in the endocytic vesicles.

Additionally it was shown that copper is present in subcellular fractions containing prion protein isolated from normal mouse brain, but not in equivalent fractions isolated from the brains of prion knockout mice. Recently it was found that copper binding to the prion protein in vitro leads to a novel protein conformation that depends on the presence of the octarepeats. The same protein conformation could not be achieved by adding copper after protein folding. These findings lend credibility to the idea that prion might be a cell surface copper binding protein, and that this could relate to the mechanism of pathogenesis.

This idea of the prion protein as a cuproprotein is enticing because some known neuropathological disorders, notably Menkes syndrome and Wilson's disease, are the result of aberrant copper metabolism. Probably a change in copper binding associated with the change to pathogenic prion be directly linked with the pathology of TSEs. Until very recently, that link has always been speculative and has suffered from two major drawbacks.

It has been known since the earliest transgenic prion knockout mice that resident wild type protein is required for the propagation of the infectious form. Subsequent work using mouse varieties with shorter and shorter forms of the prion protein have mapped the minimum amino acids essential for the disease propagation. These experiments have shown that the presence of the octarepeats, the sequences thought to be mainly responsible for copper binding, are not required for pathogenesis.

Moreover, it was shown recently that introducing a prion gene devoid of the octarepeats to a prion knockout mouse results in renewed susceptibility to disease. The octarepeats clearly have a role to play because aberrant numbers of the repeat are associated with some human TSEs, but at least in experimental systems they are not necessary for prion infectivity.

In addition, the affinity of full length prion protein for copper is relatively low (μM) and is insufficient to capture and retain the low levels of copper present in biological tissue. Both of these observations led to the view that copper binding by prion protein was interesting, but lacked obvious physiological relevance.

A clear view

From the earliest observations of copper binding by octarepeat peptides, the stoichiometry of binding has been the subject of some dispute. Most studies agree on at least four free copper binding sites per prion molecule, but slightly higher values often recorded imply additional binding sites outside of the octarepeats.

Their identity has been obscure, however, until recent studies in which researchers determined the copper content of prion protein after loading with a copper-amino acid chelate rather than free copper. The affinity of free amino acids for copper approximates to that observed for full length prion protein binding. The only way that copper can be stripped from the chelate by prion protein is therefore via sites within the molecule that have higher affinity, probably outside of the octarepeat region.

Using this approach, researchers have revealed a new binding site for a single copper ion. It can bind to much lower levels of Cu (fM not μM) and lies within the central hydrophobic region known to be essential for disease propagation and peptide toxicity in vitro.

Moreover, the same region appears to be involved in the earliest conformational changes that are associated with conversion to the pathogenic form. The two main objections to copper binding by prion protein being a physiologically relevant process are now redundant and it seems clear that defects in copper metabolism are, directly or indirectly, involved in the observed pathogenesis of the TSEs.

If copper is involved in prion function and dysfunction, it must have some role. Free copper, being redox active, is toxic and is quickly scavenged so copper is always found as a complex in biological systems. In proteins, copper is generally acting as a metal centre for enzyme function, or is in the process of being transported. Prion protein could be involved in either of these functions. Adding radioactive copper to cells expressing prion protein shows that prion can bind to the metal at the cell surface and internalise it, a role consistent with a scavenging mechanism.

Additionally, the copper ion could act as a redox centre once bound within the prion protein. Cells expressing prion protein are more resistant to oxidative stress than cells that do not express prion protein, and may play an active role in mopping up free radicals. Indeed, the copper bound prion protein that has been produced in vitro dissipates added superoxide and protects cells from oxidative damage.

One factor that might support a more active role for prion protein in copper metabolism is the unusual degree of homology between prion proteins from different vertebrates. Strikingly, for example, the primary sequence of prion proteins in man and mouse is essentially the same. The degree of amino acid conservation is astonishingly high, with the regions of strictest homology lying in the amino terminal region where copper binds.

A simple metal ion transporter would probably not require such strict sequence conservation, but a combined metal ion transporter protein and enzyme would explain the evident restriction on sequence diversity. Since prion concentrations are highest at nerve junctions, it is not hard to envisage how a defective protein would lead to neurotoxicity.

Application

Armed with a plausible idea of how pathogenesis could arise, the way in which copper is bound to the surrounding sequence becomes highly significant. The role of copper binding by the octarepeats is important as well as the significance of repeats and their conservation. More intriguing still is the idea that copper (or other metal ion) binding itself may, in part, modulate the conversion event.

The high affinity site is in the region of the protein that appears to determine some of the disease characteristics; it has to be found whether mutations known to affect these disease traits influence the affinity or range of metal ions that can be bound. Equally, if perturbation of copper metabolism is at the heart of the pathology, then intervention strategies that target copper directly may offer prospects for disease management.

The search for sensitive and robust diagnostics should also take account of the fact that metal ions may modulate the conformation of the protein; antibodies that can distinguish the metal dependent conformation of prion protein may be useful diagnostic reagents. Lastly, a role for a redox active metal in the TSEs may draw them into a larger group of neurological disorders, including Alzheimer's disease that may have a shared basis of action. In that case there is every hope that advances in any one field will be beneficial to a wide range of related diseases.

Alzheimer's Prevention Role Discovered For Prions

Another role for prion proteins, the much debated agents of mad cow disease and vCJD, has been identified. It appears that the normal prions produced by the body help to prevent the plaques that build up in the brain to cause Alzheimer's disease. The possible function for the mysterious proteins was discovered by a team of scientists led by Medical Research Council funded scientist Professor Nigel Hooper of the University of Leeds. Alzheimer's and diseases like variant Creutzfeldt-Jakob Disease follow similar patterns of disease progression and in some forms of prion disease share genetic features. These parallels prompted Professor Hooper's team to look for a link between the different conditions. They found an apparent role for normal prion proteins in preventing Alzheimer's disease.

Their experiments have shown that the normal prion proteins found in brain cells reduce the formation of beta-amyloid, a protein that binds with others to build plaques in the brain that are found in Alzheimer's disease, explains Professor Hooper. He continues: "In vCJD, the normal version of prion protein, PrPc, found naturally in the brain is corrupted by infectious prions to cause disease. The normal function of PrPc has been unclear."

Using cells grown in the lab, the team looked at the effect of high and low levels of normal prion protein on the successful formation of beta amyloid, the source of

Alzheimer's plaques. They found that beta amyloid did not form in cells with higher than usual levels of PrPc. In comparison, when the level of PrPc was low or absent, beta amyloid formation was found to go back up again.

Mice genetically engineered to lack PrPc were also studied. Again, this revealed that in its absence, the harmful beta-amyloid proteins were able to form. It appears that PrPc, the normal prion protein, exerts its beneficial effect by stopping an enzyme called beta-secretase from cutting up amyloid protein into the smaller beta-amyloid fragments needed to build plaques.

Further evidence for the protective role of normal prion proteins is provided by mutated versions that are linked to genetic forms of prion disease because beta-amyloid fragments are able to form when the normal prion protein is corrupted by genetic mutation.

According to Professor Hooper their findings have clearly identified a role for normal prion proteins in regulating the production of beta-amyloid and in doing so preventing formation of Alzheimer's plaques. Whether this function is lost as a result of the normal ageing process, or if some people are more susceptible to it than others this is not known yet.

The next step for this research will be to look in more detail at how the prion protein controls beta amyloid, knowledge that could be used to design anti-Alzheimer's drugs. Theoretically, if a way of mimicking the prion's function is found it would be possible to halt the progress of Alzheimer's. However, there's still a lot of work to be done in looking at levels of prions in the human system and how these alter with age.

REQUIREMENT OF PrPc FOR PrPsc INFECTIVITY AND THE PROPOSED MECHANISMS FOR NEURONAL DAMAGE

Büeler *et al.* (1993) performed studies using transgenic mice with the Prn-p gene inactivated showed that the PrP-null mice were in fact resistant to Scrapie. It was thought that if indeed PrPsc was an essential component of the Scrapie agent, then mice devoid of PrP should be resistant to infection, and so developing neither symptoms of Scrapie nor allowing the propagation of the infectious agent. However, if the animals not expressing the Prn-p were to succumb to the disease, then the protein only hypothesis would be falsified.

After the generation of mice homozygous for the disrupted genes (Prn-p0/0) (Büeler *et al.*, 1992), it became possible to study the response of the Prn-p0/0 mice to inoculation with Scrapie prions, as well as of animals carrying a single Prn-p allele (Prn-p+/0) and of Prn-p0/0 mice reconstituted with Syrian hamster PrP genes.

Their results showed that mice devoid of PrP are completely protected against Scrapie for at least 13 months after inoculation. Interestingly, even heterozygotes are partially protected, in that 90% of the heterozygote mice showed signs of the disease around 253-322 days after inoculation but were alive after 322 days, with the interval between first symptoms and death being 2 months. The wild type mice died in about 180 days with the interval between first symptoms and death being 13 days.

No infectious agents were detected in Prn-P0/0 animals 25 weeks after inoculation, which was the latest time point for which data was available. So it seems clear that if the infectious agent is propagated in Prn-P0/0 mice then this would be at a level of about 5 orders of magnitude lower that that in Prn-P+/+ animals.

Büeler *et al.* (1992) then concluded that the development of Scrapie symptoms and pathology is strictly dependent on the presence of PrP and that incubation time and disease progression are inversely related to the level of PrP. There was the possibility that antibodies against PrP might be formed in the Prn-P0/0 mice, and such antibodies could affect susceptibility to Scrapie infection, but no anti-PrP antibodies could be detected in inoculated mice.

In the same year Prusiner *et al.* (1993) demonstrated that the disruption of the Prn-P gene in mice prevented Scrapie. However, this time Prusiner *et al.* also demonstrated the production of anti-PrP-antibodies, this was done using Mo (mouse) prions and SHa (Syrian hamster) prions. The anti-PrP antibodies in sera were detected by Western blotting, and reacted against Mo, SHa and human PrP.

The prolonged incubation times observed with the heterozygotes in the study by Büeler *et al.* (1993) suggests that resistance to Scrapie is due not to the sought after immune response, but to a rate limiting step due to the concentration of PrPc. This suggests a positive correlation between the amount of PrPc and the formation of PrPsc. Ideas from this study have evolved two ideas, the first is that it may be possible to breed sheep or cattle resistant to this disease, either by PrP gene disruption or the introduction of transgenes expressing PrP anti-sense RNA.

The second idea involves a moderate reduction of PrPc synthesis that could be achieved by anti-sense oligonucleotide therapy, which could be used to slow disease expression. This could be used in early cases of CJD, or even better, in inherited CJD where the sufferers know they will get it in their thirties or so, but are not yet expressing any symptoms.

Studies by Brandner *et al.* (1996) support the idea for the requirement of PrP for Scrapie infectivity with evidence obtained in an experiment which was slightly different but produced similar conclusions. Brain tissues were grafted over-expressing PrPc 5-8 fold (which would show incubation times of 60 instead of 160 days in mice) into PrP-null mice. After the intracerebral inoculation with Scrapie prions, they observed that the grafts accumulated high amounts of PrPsc and developed severe histological symptoms of the disease.

Also, PrPsc was recorded to spread from the graft into the host brain tissue. Despite this, no pathological changes were seen in the host tissue, even in the immediate vicinity of the grafts. So, interestingly, in addition to being resistant to Scrapie infection, brain tissue devoid of PrPc does not seem to be damaged by exogenous PrPsc.

This opens a new window of discussion as it has been argued that Scrapie pathology is due to intrinsic neurotoxicity of PrPsc rather than to depletion of PrPc due to studies by Büeler *et al.* (1992) suggesting that a lack of PrPc doesn't affect normal development of mice. Conversely Tobler *et al.* (1996) and Collinge *et al.* (1994) suggest this lack of PrPc does affect normal development. This could mean that the intrinsic neurotoxicity may not be the cause of pathology.

Deposition of PrP-immunoreactive material co-localises with typical histopathology, and synthetic amyloidogenic PrP fragments as well as liposome-packaged PrP 27-30 have been shown to be neurotoxic in-vitro, (Brandner *et al.*, 1996). But, in the studies performed by Brandner *et al.* (1996), the regions adjoining the grafts with high levels of leaked PrPsc showed no Scrapie pathology. This apparent paradox may be reconciled by the idea that PrPsc is inherently non-toxic and the plaques formed in Spongiform Encephalopathies are an epiphenomenon rather than a cause of neuronal damage.

This is supported by the fact that the extent of PrP deposition in the brains of humans succumbing to prion diseases with similar clinical presentation is extremely variable. Another idea is that PrPsc is only toxic when it is formed and accumulated within the cell, but not when it is presented from outside. The other idea is that PrPsc is pathogenic when presented from outside, but only to cells expressing PrPsc either because it initiates conversion of PrPc to PrPsc at the cell surface and/or it is internalised by was of association with PrPc, which is endocytosed efficiently.

The studies by Brandner *et al.* (1996) suggest this latter idea in that their data implies that it is not the deposition of PrPsc, but the availability of PrPc for some intracellular processes elicited by the infectious agent that is directly linked to spongiosis, gliosis and neuronal death. This seems to make sense and fits in with the results of the simplest experiments involving inoculation of the Scrapie agent into unaffected sheep, (Cuillé and Chelle (1936)). A review by Goldmann in 1993 identifies the key points that need to be resolved to identify the function of the PrPc transmembrane protein. These are (i) Can other proteins substitute PrP function? (ii) Does pathology develop only at a certain age or stress?

One might conclude that despite reports to the contrary (Büeler *et al.* (1992), there is sufficient evidence that the function of the PRNP gene and PrPc is important for the health of individuals, if not for the observed changes in synaptic function and health of Purkinje cells (Tobler *et al.*, 1996; Collinge *et al.*, 1994; Sakaugchi *et al.*, 1996) but for the fact that it has been highly conserved through evolution. Therefore it is strongly selected for, which indicates an advantageous trait favoured by natural selection over those not possessing the intact gene, and thus vital to the survival and successful reproduction of an individual within a normal population.

In addition, it must be noted that the damage to the Purkinje cells was noticed when the mice were very old and likely to have passed their sexual peak (Sakaguchi *et al.*, 1996). This feature could have less affect on the evolutionary conservation of the gene as natural selection is based on sexual selection, which occurs when mainly when the mice are less than 70 weeks old and possibly asymptomatic. Despite this, there could be lesser but still important effects not observed in the laboratory including mice-family behavioural patterns, which may account for the evolutionary conservation.

THE REPLICATION OF PrPsc

Evidence for Protein X

The exact mode of the replication of PrPsc has not yet been proven. It is thought by some, to involve the conversion of PrPsc from PrPc by some kind of protein-protein interaction. The initial studies done on mice involving transgenic Syrian hamster mice expressing chimeric forms of the human (Hu) PrP were by Telling *et al.* (1995).

The work involved the prion propagation using mice that expressed human and chimeric PrP transgenes. It was found that Tg (HuPrP) mice were refractory to human prions unless crossed with PrP-null mice, and concluded that it was linked to a third component, namely protein X. Studies two years later by Kaneko *et al.* (1997) also provides evidence that there is another molecule, presumably a protein, involved in the formation of PrPsc.

The evidence is based on the fact that in studies, mice expressing a Syrian hamster (SHa) PrP transgene were susceptible to SHa prions. When similar transgenic mice were produced expressing human (Hu) PrP, no transmission of Hu prions were found. However, mice expressing a chimeric Hu/MoPrP transgene denoted MHu2M were susceptible to the Hu prions. In addition to this, it was found that transgenic mice expressing HuPrP did become susceptible to Hu prions when they were crossed with PrP-null mice.

These data taken together argued that it is likely that a molecule other than PrP is involved in the formation of PrPsc. It was assumed to be a protein and designated as protein "X". Earlier studies showing that the N terminus is not required for PrPsc formation, and also, with the results obtained with the MHu2M transgene, enabled to surmise that the binding of PrPc to protein X is likely to occur through specific side chains of amino acids located at the C terminus of PrPc (Kaneko *et al.*, 1997).

The binding site for protein X was also identified by studies using Scrapie-infected mouse neuroblastoma cells transfected with a chimeric PrP gene in which the Mo C terminus was replaced by Hu residues.

The substitution of a Hu residue at position 214 or 218 prevented mouse (Mo) PrPc from being converted into PrPsc. The side chains of residues at 214 and 218 protrude from the same surface of the C-terminal α-helix and form a discontinuous epitope with residues 167 and 171 in an adjacent loop. Substitution of a basic residue at position 167, 171, or 218 also prevented PrPsc formation. It was concluded that at a mechanistic level, these mutant PrPs appear to act as "dominant negatives" by binding to protein X and functionally sequestering it from the replication process.

These findings may explain the protective effects of polymorphic basic residues in PrP against CJD and Scrapie. It was also concluded that interfering with the conversion of PrPc into PrPsc is probably the best therapeutic target, by either stabilising the structure of PrPc by binding or by modifying the action of protein X that might function as a molecular chaperone.

The results argue that PrPc forms a complex with protein X and that PrPsc then binds to PrPc resulting in a ternary complex. Whether PrPc and PrPsc are monomers or dimers in this scheme is unknown. As the ionising radiation target size of prions is about 55 KDa, it is possible that PrPsc is a dimer.

When analysed using molecular modelling studies, it seems that the N-terminal region of residues 90-140 of PrPsc may participate in the dimer interface. The stoichiometry of the protein X/PrPc complex is unknown since the number of binding sites for PrPc on protein X remains to be established. The X/PrPc complex may play a role in the function of PrPc, which remains unknown. However, it seems doubtful that the complex exists for the production of PrPsc unless PrPsc has a low-level cellular function that has so far been unrecognised.

Interestingly, studies on Scrapie infected cells in culture have been shown to have variable induction of heat shock proteins, Hsp70 (Heat shock protein with 70 factor) has also been reported to increase in mice. In addition, PrP has been shown to interact with Bcl-2, Hsp60, and the laminin receptor protein by two-hybrid analysis in yeast *Saccharomyces cerevisiae*, (Prusiner *et al.*, 1998). Despite this indicative evidence for the existence of protein X, attempts to isolate this protein have failed so far.

This binding site is shown in Fig. 27 (Plate 5)

THE SPECIES BARRIER

Concept

Around thirty years ago, many scientists working on the transmission of the Scrapie agent from one species to another found that it was difficult to infect rodents with Scrapie from sheep. From these early experiments arose the concept of the species barrier. This is a very important topic, and is highly relevant to the debate over the possible transmission of BSE to man, which is discussed later.

Different species express slightly different forms of PrPc differing in a few amino-acid sequences; for example the PrP of mice differ from that of the hamster by 16 amino-acids, and so mice inoculated with hamster PrPsc rarely develop the disease. The difference in the amino acid sequence between cattle and sheep differ by only 7 amino acids, and so transmission is not very difficult between these two species. Human PrP differs from cattle PrP by over 30 amino acids, which may make the transmission of BSE to man extremely difficult (Owen, 1998).

To test the hypothesis that differences in PrP gene sequences might be responsible for the species barrier, Prusiner *et al.* (1990) and Scott *et al.* (1989) managed to construct transgenic mice expressing SHaPrP (Syrian hamster PrP). This was done by micro-injecting SHaPrP genes into fertilised mouse embryos producing transgenic (Tg) mice.

Four lines of mice called Tg 69, 71, 81 and 7 were established. Hybridisation with a monoclonal antibody showed that all these mice expressed HaPrPc in their brains. The studies showed that the length of the incubation time after inoculation with Ha prions was inversely proportional to the level of HaPrPc in the brains of the transgenic mice, and the HaPrPsc concentrations in the brains of clinically ill transgenic mice were similar in all four transgenic lines inoculated with hamster prions.

The bioassays of the brain extracts from clinically ill transgenic mice inoculated with Mo prions revealed that only Mo prions but no Ha prions were produced (the mice do still express some MoPrPc). Conversely, inoculation of transgenic mice with Ha prions led only to the synthesis of Ha prions.

Therefore, it can be said that the de novo synthesis of prions in transgenic mice is species specific and reflects the genetic origin of the inoculated prions. So, it may be said that mice can be genetically engineered to carry the hamster PrP gene, and when challenged by the hamster agent they rapidly develop the clinical symptoms of the hamster prion disease and die.

The 'different strains' idea about incubation

The identification of strains of an agent, namely the Scrapie agent relates to observations and measurements of the disease characteristics in the mouse host. These include the incubation period, clinical features, the severity and distribution of the neuropathological lesions in the CNS, transmission characteristics from the mouse brain, and susceptibility to inactivation.

Several distinct strains have been isolated in studies (Bruce *et al.*, 1991, Bruce, 1993) where they serially passaged Scrapie and BSE from sheep, goat and cattle sources in mice. The main difference between Scrapie strains is the variation in the incubation period between initial infection and clinical disease in genetically identical hosts.

Work by Bruce *et al.* (1991), illustrate the workings of the sinc gene (Scrapie incubation) in mice which also influences the incubation periods of experimental Scrapie, it is now thought that this gene encodes PrP, and so is the same gene as the Prn-P gene found in mice. The effects of the sinc gene have been well studied in the prion production in relation to the genotype, agent isolate and route of infection (Farquhar *et al.*, 1994) as well as, the experimental transmission and subsequent interaction with two different strains of the Scrapie agent, differing only in incubation time and resulting in one strain blocking the other (Bruce *et al.*, 1993).

The significance for human prion disease of the pathological variation in Scrapie is as yet uncertain. Certain similarities exist between the distribution and nature of the neuropathology in some Scrapie strains and the neuropathological features reported in sporadic human prion diseases. However, a variety of other host characteristics have to be taken into consideration when assessing these similarities and it is not possible to draw any firm conclusions from the information available at present (Collinge & Palmer, 1997).

So, it has been reported that strain variation has been found in many naturally occurring prion diseases, including those of sheep and goats. However, the case of the BSE transmission from cattle to sheep, goats and pigs and the subsequent passage into mice is in stark contrast to the strain idea. The transmission of BSE to these secondary animals (mice) show similar characteristics indicating that the agent is unchanged when passaged though a range of species, and that the donor species has little influence on the disease characteristics on transmission to mice (Collinge & Palmer, 1997; Telling, 1999).

These findings seem to put the idea of a black and white species barrier into the grey, especially due to the discovery of the transmission characteristics of BSE and how they are little affected by the species barrier or strain specific idea. Perhaps it would be prudent and more precise to look at the nature of the agent more in terms, but the 'PrP' or 'prion conformation' barrier, and not the species barrier.

It is the prion conformation and ultimately the molecular structure of the prion that is thought to interact and form PrPsc with pathogenic effects. This idea is supported by the findings by Supattapone *et al.* (1999) on the creation of an artificial species barrier, using a prion protein of 106 residues, which suggests it is the prion conformation that provides the species barrier.

Mad cow prions piggyback on iron-storing proteins after surviving digestive juices

According to Neena Singh, M.D., Ph.D., associate professor of pathology (Department of Pathology at Case Western Reserve University School of Medicine) (2004, Journal of neurosciences), little is known about the mechanism by which prions cross the human intestinal barrier, which can be a particularly difficult obstacle to cross. The mad cow epidemic is far from over, and the continuous spread of a similar prion disease in the deer and elk population in the U.S. raises serious public health concerns. It is therefore essential to understand how this disease is transmitted from one species to another, especially in the case of humans where the infectious prions survive through stages of cooking and digestion.

Using brain tissues infected with the spontaneously occurring version of CJD which is also caused by prions, the researchers have simulated the human digestive process by subjecting the tissue to sequential treatment with digestive fluids as found in the human intestinal tract. They have studied how the surviving prions are absorbed by the intestine using a cell model. The prions were found to be linked with ferritin, a cellular protein that normally binds excess cellular iron to store it in a soluble, non-toxic form within the cell. Since ferritin shares considerable similarity between species, it may facilitate the uptake of prions from distant species by the human intestine.

This important finding provides insight into the cellular mechanisms by which infectious prions ingested with contaminated food cross the species barrier, and provides the possibility of devising practical methods for blocking its uptake. If a method can be developed of blocking the binding of prions to ferritin, it may be possible to prevent animals from getting this disease through feed, and stop the transmission to humans. Currently, it is being checked whether prions from distant species such as deer and elk can cross the human intestinal barrier.

DIAGNOSIS, YEAST AND DRUGS

New Diagnostic Techniques

The protease resistant isoform of the prion protein is a diagnostic marker of Spongiform Encephalopathies in humans and animals. Immunoblotting and immunohistochemistry (using formalin fixed paraffin embedded materials) are sensitive methods but are post mortem (Tateishi & Kitamoto, 1993). Work has been directed in this area to develop a rapid test for the diagnosis of CJD and other prion diseases.

The identification of different strains by a conformation dependant immunoassay (Safar *et al.*, 1998) showed promise for providing an extremely sensitive and rapid method for the detection of animal and human prions. Despite demonstrating low limits of detection (103 ID50/ml) they report to have already improved that by 10 to 100 fold by using a dual wavelength laser driven fluorometer. They also mentioned that antibodies would have a greater affinity for the epitopes buried in PrPsc but exposed upon denaturation.

A test by John Collinge and colleagues involves using strain specific protease cleavage patterns and combinations of PrPres glycoforms, it is already in use, but is a slower method involving the inoculation of suspect material into mouse brains. Collinge does point out, that with the technique, the antibody ratios do not reliably separate all strains because there is considerable overlap in the standard errors (Safar *et al.*, 1998; Bonn, 1998).

Another diagnostic investigation involves tonsil biopsy samples (Hill *et al.*, 1999) based on the fact that prions replicate in lymphoreticular tissues before neuroinvasion, suggesting that lymphoreticular biopsy samples may allow early diagnosis by detection of PrPsc. The experiments did appear to work and tonsil biopsy tissue was positive in all 8 patients with an adequate biopsy sample and whose subsequant course has confirmed, or is highly consistent with, a diagnosis of variant CJD and negative in all patients subsequently confirmed to have other diagnoses.

The results also showed that vCJD has a different pathogenesis to sporadic CJD. However, these tests were done post mortem, and after vCJD PrPsc has already got to very high levels in the patient making it perhaps easier to detect. It could be said that these new diagnostic tests are looking promising, but practically are in the early stages when it comes to accuracy and reliability on current patients or suspected patients.

Useful prions in yeast and other fungi

Not all prions are dangerous; in fact, prion-like proteins are found naturally in many (perhaps all) plants and animals. Because of this, scientists reasoned that such proteins could give some sort of evolutionary advantage to their host. This was suggested to be the case in a species of Fungus, *Podospora ansrina.*

Genetically compatible colonies of this fungus can merge together and share cellular contents such as nutrients and cytoplasm. A natural system of protective "incompatibility" proteins exists to prevent promiscuous sharing between unrelated colonies.

One such protein, called HET-S, adopts a prion-like form in order to function properly. The prion form of HET-S spreads rapidly throughout the cellular network of a colony and can convert the non-prion form of the protein to a prion state after compatible colonies have merged. However, when an incompatible colony tries to merge with a prion-containing colony, the prion causes the "invader" cells to die, ensuring that only related colonies obtain the benefit of sharing resources.

The prion principle of the conformational change of a normal protein into a pathological form, catalysed by the pathological form itself, has been suggested and subsequently shown in yeast *(Saccharomyces cerevisiae)* due to the pathogenesis of an obscure mutational phenotype. The yeast protein Ure2p is thought to have a prion form [URE3], due to a recessive mutation (unlike with inherited human prion diseases) which can be "cured" by growing them in the presence of 5 mM guanidinium chloride (Weissmann, 1994; Wickner, 1994).

In 1965 the studies with the yeast, *Saccharomyces cerevisiae*, described a genetic trait termed [PSI+] (as the prion form of Sup35p) with an unusual pattern of inheritance. In 1994, yeast geneticists correctly hypothesized that [PSI+] as well as another mysterious heritable trait, [URE3], resulted from prion forms of certain normal cellular proteins.

It was soon noticed that heat shock proteins (which help other proteins fold properly) were intimately tied to the inheritance and transmission of [PSI+] and many other yeast prions. Interestingly the [URE3] and [PSI+] genes are extra-chromosomal potentially making them easy to study.

These prion structures are different to each other and are therefore also different to the animal prion proteins, suggesting differences in the detailed mechanisms of prion propagation. The purification of [URE3], will allow detailed structural studies on the differences between the normal and prion forms of the molecule. Since then, researchers have unravelled how the proteins that code for [PSI+] and [URE3] can convert between prion and non-prion forms, as well as the consequences of having intracellular prions.

When exposed to certain adverse conditions, [PSI+] cells actually fare better than their prion-free siblings, this finding suggests that, in some proteins, the ability to adopt a prion form may result from positive evolutionary selection. It has been speculated that the ability to convert between prion infected and prion-free forms enables yeast to quickly and reversibly adapt in variable environments.

Further dissection of prion domains and nitrogen regulation functions of Ure2p using molecular genetics are also under way. The discovery of the prions in yeast will enable the closer study of the mechanisms of prions in general that would also be faster and easier due to the nature of yeast being practically easier to work with than mice, (especially considering the incubation times).

Development of chemotherapeutic agents

As we have currently no way of predicting the number of individuals, who may develop vCJD, it would be prudent for us to start investigating into and developing effective chemotherapeutic agents. This is doubly important for people at risk of inherited prion diseases as they can now be identified decades before neurologic dysfunction is evident. Targets for these drugs could include the interference of the conversion of PrPc into PrPsc.

The structure of PrPc could be stabilised or PrPsc destabilised by binding to a drug. Perhaps targeting the putative 'protein X', thought to be a molecular chaperone is an option. Another idea is the synthesis of an artificial PrPc that would serve as an irreversible inhibitor to PrPsc interactions, it would bind and remove PrPsc from the biological equation. The drugs would need not penetrate the cytosol of cells but penetrate the CNS. Prevention by gene-knockout for animals (sheep and cows) has been considered but could be risky due to the practically little known effect of PrP-null mice.

Several substances have raised limited hopes for therapy, such as amphotericines, sulphated polyanions, Congo red and anthracycline antibiotics. The mechanism of amphotericin action is unknown, but the other compounds are thought to bind or intercalate, and disrupt, the highly ordered structure of PrPsc. However, the effects of all compounds tested are modest. Work on drugs is currently proceeding especially between John Collinge's group and Glaxo-Wellcome PLC, (Collinge, 1999; Prusiner, 1998; Aguzzi and Weissmann, 1997).

Prevention and control

BSE:

- No part of any animal which shows signs of BSE should enter any food chain, human or animal. All countries must ensue the slaughter and safe disposal of TSE affected animals.
- All countries should continue surveillance and compulsory notification for BSE
- Countries where BSE exists in native cattle should not permit tissues that are likely to contain BSE agent to enter any food chain, human or animal.
- All countries should ban the use of ruminant tissues in ruminant feed.
- Research on TSE should be promoted, especially on rapid diagnosis, agent characterization, and epidemiology of TSEs in humans and animals.

V-CJD:

- The geographic distribution of v-CJD needs to be further investigated.
- Further studies must be conducted to form a definite link between BSE and v CJD.
- Implementation of the reccomendations to ban beef will ensure that any continuing risk of exposure to BSE in beef and beef products will be reduced to a minimum.

Vaccine approach

New approach has been tried in studying infectious prions, which are particles thought to be composed of normal prion proteins that have been compromised and folded into rogue shapes. The usual way of raising antibodies in the immune system is to grind the infectious agent up and inject it into a mouse to see if it prompts antibody production. Scientists have tried this method with prions over the past 15 years, all without success. Recently (2003) prions were examined at the sub-molecular level to determine if antibodies would recognize and react to the amino acids exposed at the surface of a prion. It was a novel idea, and when it was found that the hypothesis actually worked, scientists were surprised and pleased.

All mammals have prion proteins, the highest levels of which are present in the brain Mammals can contract prion diseases by ingesting abnormal or infectious prions. From the digestive system, these prions make their way to the brain. When an abnormal prion comes in contact with a normal prion protein, it causes the protein to misfold, thus creating a copy of the infectious prion. The process is more akin to co-opting than replication. However, since the abnormal prion has similar characteristics to the original host protein, the immune system does not recognize it as a foreign invader and does not attack it.

In the recent study, the role of chemical groups in amino acids was examined, which are called side chains. It was seen whether there were side chains accessible on abnormal prions that were not accessible on the normal protein. The hypothesis was that in a normal prion protein, there will be side chains buried in the interior of the molecule. When the protein converts to the abnormal form, some of these side chains would then be exposed on the molecular surface. It was discovered that the newly exposed side chains of abnormal prions include a sequence composed of three amino acids - tyrosine, tyrosine and arginine (Tyr-Tyr-Arg). By raising antibodies against the Tyr-Tyr-Arg amino acid sequence, the immune system became able to recognize the abnormal prion as an invader and attack it. This was significant as while the antibodies recognized the abnormal prions, they left the normal prion proteins intact.

It was further found that the tyrosine-tyrosine-arginine amino acid sequence appears to be common among species. Different species have different sequences of prion proteins but this Tyr-Tyr-Arg motif appears to be the same in every species - humans, cattle, mice, hamsters, sheep and elk. When these antibodies were applied to the infected tissues and cells of samples, they only bound to the abnormal prion protein in all these species.

The researchers are currently testing a possible vaccine to prion disease in mice. They plan to immunize mice with the Tyr-Tyr-Arg sequence and then infect them with prions. In order to treat prion diseases in the most effective way possible, it is necessary to understand the manner in which prion proteins fold into a pathological form. This discovery is a vital step in understanding the causes of prion diseases and will play key a role in developing vaccines and a new generation of drugs to combat this condition. The immediate commercial applications of the discoveries are diagnostics, as global health authorities urgently seek to ensure that beef and transfused blood are safe from Mad Cow-related infections.

First mucosal prion vaccine tested in mice

Scientists at New York University Medical Center and School of Medicine, have created the first active vaccine that can significantly delay and possibly prevent the onset of a brain disease in mice that is similar to mad cow disease. The new findings, could provide a platform for the development of a vaccine to prevent a group of fatal brain diseases caused by unusual infectious particles called prions.

The NYU study is also the first to use a mucosal prion vaccine, given by mouth rather than through the skin, which localizes the initial immune response to the gut and mainly stimulates an antibody response (Dr. Wisniewski). By giving the vaccine orally, an immune response is stimulated mainly in the digestive tract. Thus, harmful prions in contaminated food will be destroyed in the gut and will not reach other organs in the body. Because the research was conducted in normal mice, the NYU researchers say it will be easier to apply in animals in the wild, which are at risk for developing prion disease.

Prion disease is contracted when an animal eats the body parts of other animals contaminated with prions. What makes these infectious particles unusual is that they are proteins that have the same amino acid composition as equivalent proteins occurring naturally in the body. But the prions turn deadly by changing shape. These "misfolded" proteins tend to aggregate in toxic, cell-killing clumps. As an infection takes hold, prion proteins invade brain tissue and force normal proteins to adopt their configuration. In time, the diseased animal develops dementia, loses control of its limbs, and eventually dies. There are no treatments for prion-related diseases, and prions can easily infect the body because they do not elicit any immune response.

To create a vaccine that could rally the immune system of mice, the NYU researchers designed a vaccine in which Scrapie prions were attached to a genetically modified strain of *Salmonella*. This bacterium is also used in several animal vaccines and in human vaccines for cholera and typhoid fever. Among mice vaccinated prior to prion exposure, approximately 30% remained alive and symptom-free for 500 days, according to the study. By comparison, mice that didn't receive the vaccine survived only an average of 185 days, and all were dead by 300 days.

Scientists are in the process of redesigning the vaccine for deer and cattle. After choosing the appropriate bacteria for each vaccine, they must genetically modify it to carry the prion protein. The technical issues are not major hurdles. Developing a marketable vaccine for livestock is something that is very achievable.

The vaccines that provide active immunization are made, in part, from proteins found on disease-causing organisms. In response to these proteins, the animal's immune system produces antibodies that will destroy them any time they appear in the body. Most vaccines in use today provide such active immunization.

The prion vaccine developed would most likely first be used to protect livestock, since most prion infections occur in animals and are thought to be transmitted orally. The version of prion disease that affects humans usually occurs spontaneously, and only rarely as a result of eating contaminated meat.

The potential use for a prion vaccine in humans is still theoretical but if, for example, there is ever a more significant outbreak of Chronic Wasting disease and if this disease were found to be transmissible to humans, then we would need a vaccine like this to protect people in hunting areas. Currently, an outbreak of Chronic Wasting disease is occurring in some Western states, and the disease's geographic range is expanding.

Scrapie control under new strain

In the 1990s, a proportion of sheep were found to be genetically resistant to Scrapie (Hunter, 1997) and, consequently, many countries have established programmes to create disease resistant national flocks by selective breeding. This new and unusual approach to disease control needs urgent reappraisal, however, as recent discoveries in Europe suggest that the selected sheep might be susceptible to strains of the disease that arise in the future. Atypical forms of Scrapie have been reported in 'resistant' sheep in Germany (Buschmann, 2004), Portugal (Orge, 2004) and France (Report on the Monitoring and Testing of Ruminants for the Presence of Transmissible Spongiform Encephalopathy (TSE) in the EU in 2003). And a recently discovered strain of Scrapie, called Nor98, is now being detected in several European countries. Moum and coworkers (Journal of General Virology, 2004) showed that this strain primarily affects sheep that rarely succumb to conventional Scrapie: conversely, typically 'susceptible' sheep seem to be unaffected.

The European Union (EU) is aiming to eradicate Scrapie from its member states with a combination of two approaches. The first approach which has been compulsory since October 2003, involves eliminating the disease from currently infected flocks by culling entire flocks, or by genetic testing and culling of susceptible animals. The second approach, which became compulsory from April 2005, has the goal of reducing the chance that uninfected flocks will become infected in the future. As a minimum, rams intended for breeding in Scrapie-free flocks of 'high genetic merit' must be genetically tested, and those whose genes suggest a susceptibility to Scrapie will be culled, despite being uninfected. This approach targets rams because the relatively small number used for breeding provides a fast means of spreading Scrapie-resistant genes through the sheep population. Benefits are expected to filter gradually into flocks of lower genetic merit.

In some countries, broadly similar control programes have been in operation for several years. In July 2001 the UK government launched the National Scrapie Plan for Great Britain (Department for Environment, Food and Rural affairs, London, 2004), which, until the commission regulations came into force, used the same two-faceted approach but on a voluntary basis. In the United States, however, genetic testing is used to eliminate disease in Scrapie infected flocks, but breeding programmes involving Scrapie-free flocks do not have to be based on Scrapie genetics (Detwiler, 2003). It's important to get the strategy right: national programmes to create Scrapie-resistant flocks require the continued goodwill of the sheep industry, and are expensive. For example, more than one million UK sheep have been genetically tested, at a cost of US$37 million, during the first three years of Britain's National Scrapie Plan.

Whether or not a sheep can succumb to Scrapie is determined by the sequence of amino acids that form its prion protein, which is encoded by its PrP gene. Five variants (called haplotypes) of the PrP gene are known to affect susceptibility. Sheep that inherit the ARR haplotype from both parents (ARR/ARR) have the greatest resistance to conventional Scrapie and in some infected flocks, all such animals would be expected to die from the disease. Fifteen types of sheep are definable on the basis of the pair of Prp haplotypes inherited from their parents. The extent to which they succumb to Scrapie, as reported in the United Kingdom is shown in the figure. Thus it is aimed to increase the frequency of the ARR haplotype and eliminate the VRQ haplotype.

A fundamental assumption of selective breeding programmes is that 'resistant' sheep really are, and will remain, resistant to Scrapie. But this assumption is now being challenged by the new findings emerging from mainland Europe. In 2002, many EU member states began large scale testing of sheep in abattoirs for Scrapie, using BSE detection methods. This surveillance has, inevitably perhaps, revealed that Scrapie is more widespread than previously believed. It has also uncovered 'atypical types'. These differ from conventional Scrapie in several ways, such as the pattern of deposition of the abnormal form of the prion protein in the brain.

Importantly, a few of the atypical cases are in the resistant sheep. Buschmann and co-workers reported two atypical cases of Scrapie in sheep of the ARR/ARR type in Germany, and Orge and co-workers described a third in Portugal. Three further cases have been reported in France. These discoveries confirm, under natural conditions of finding made in sheep experimentally inoculated BSE: ARR/ARR sheep are not, after all, totally resistant to TSEs.

These atypical Scrapie cases shiow some similarity to a news strain of Scrapie, first detected in Norway in 1998. MOUM and co-workers have now genetically tested all 38

Norwegian Nor98 cases, as well as the unaffected sheep in the flocks from which they came, allowing the relative susceptability oer resistance of each genetic type to be estimated, the results transfornm are understanding of the relationship between PrP haplotype and Scrapie. Sheep that are the most susceptible to conventional Scrapie appear to be unaffected by Nor98. Conversly, susceptibility to Nor98 is linked to the AHQ haplotypa, which is generally associated with resistance (or at the most low susceptibility) to conventional Scrapie. Nor98 is also being identified over a wider area in Europe. It has been reported in Norway, Sweden, Finland and in november 2004 in Ireland and Belgium.

Encouragingly, there hasve been no cases among ARR/ARR sheep and so it is possible that national flocks of this genetic type may be resistant to both conventional and Nor98 Scrapie. A warning shot has however been fired. The existence of Nor98 shows that the sheep largely resistant to known strains of Scrapie might be highly susceptible to a novel strain identified in the future; and the discovery of atypical infections in ARR/ARR sheep presses the point that we cannot exclude these sheep from this risk. Larger epidemics of Scrapie tend top be associoated with flocks that have high proportions of sheep susceptible to the local Scrapie strain. Accordingly as the ARR haplotype inctreasingly dominates European sheep population, there is a danger that epidemic of a strain able to attack this haplotype to be larger than those the region has exprerinced in the past.

One final point: The existence of the VRQ haplotype has been enigmatic, as it is not known to confer any benefits, but is often fatal for sheep exposed to conventional Scrapie. Why has it not been eliminated from the sheep population by natural selection? The findings of Mount *et al.* raise the possibility that this haplotype which exists today conferred resistance to past strains. If so, this is a strong argument for the preservation of haplotype diversity in sheep population, and the long term control of Scrapie in Europe may be better achieved by the US approach: eliminating infection within infected flocks and their contacts, while preserving haplotype diversity at a national scale.

Variant prion protein causes infection but no symptoms

New research suggests that a variant form of abnormal prion protein--one lacking an "anchor" into the cell membrane--may be unable to signal cells to start the lethal disease process, according to scientists at the Rocky Mountain Laboratories (RML), part of the National Institute of Allergy and Infectious Diseases (NIAID) of the National Institutes of Health.

This work provides novel insights into how prion and other neurodegenerative diseases develop and it provides tantalizing clues as to how we might delay or even prevent such diseases by preventing certain cellular interactions (NIAID Director Anthony S. Fauci, M.D.).

Drawing on experimental concepts first developed at RML a decade ago, the research team exposed two groups of 6-week-old mice to different strains of the agent that causes Scrapie, a brain-wasting disease of sheep. Within 150 days of being inoculated with the natural form of Scrapie prion protein, all 70 mice in the control group showed visible signs of infection: twitching, emaciation and poor coordination. In contrast, the scientists observed 128 transgenic mice--those engineered to produce prion protein without a glycophosphoinositol (GPI) cell membrane anchor--for 500 to 600 days and saw no signs of Scrapie disease. Subsequent electron microscopic examinations at UCSD, however, confirmed that they produced amyloid fibrils, an abnormal form of prion protein, and that they even had brain lesions. More remarkably, according to Dr. Chesebro, the diseased brain tissue resembled that found in Alzheimer's disease rather than in Scrapie.

Chesebro mentions two theories as to why the transgenic mice did not show symptoms of illness despite being infected:

1. The host cell might require the GPI anchor to receive the "toxic signal" from the abnormal prion protein
2. The plaques might be less toxic than the non-plaque form of prion protein clumps

In either case, more time might be required to produce disease due to the reduced toxicity, Dr. Chesebro says.

There was so much about this research that was surprising and gave ideas to be pursued.

a) The mice didn't get sick. That's very significant.
b) The dense accumulations of Scrapie plaque in the brain resembled the plaque seen in Alzheimer's, but it wasn't toxic, which might support more recent concepts about plaque in Alzheimer's patients.

Previously, most researchers thought plaques were the toxic component of Alzheimer's that kills neurons, and many treatments focus on removing the plaques. But if the plaques are inert, as they were in this research or if only small clumps are toxic then it may change the previous concepts altogether.

If this hypothesis proves correct, the ongoing research could eventually alter scientists' views on preventing prion diseases, shifting emphasis away from stopping the production of prion protein clumps and toward preventing interactions with prion protein anchored to cells, or learning to direct abnormal prion protein accumulations to specific parts of the brain where they will not produce symptoms. Abnormal prion protein by itself may not be rapidly lethal--in these mice it wasn't found to be.

Mad Cow Breakthrough: Genetically Modified Cattle Are Prion Free

The U.S. Department of Agricultural Research Service (ARS) have announced initial results of a research project involving prion-free cattle. ARS scientists evaluated cattle that have been genetically modified so they do not produce prions, and determined that there were no observable adverse effects on the animals' health.

These cattle can help in the exploration and improved understanding of how prions function and cause disease, especially with relation to Bovine Spongiform Encephalopathy, or BSE. In particular, cattle lacking the gene that produces prions can help scientists test the resistance to prion propagation, not only in the laboratory, but in live animals as well.

ARS studied eight Holstein males that were developed by Hematech Inc., a pharmaceutical research company based in Sioux Falls, S.D. The evaluation of the prion-free cattle was led by veterinary medical officer Juergen Richt of ARS' National Animal Disease Center (NADC) in Ames, Iowa. The evaluation revealed no apparent developmental abnormalities in the prion-free cattle.

The cattle were monitored for growth and general health status from birth up to 19 months of age. Mean birth and daily gain were both within the normal range for Holsteins. General physical examinations, done at monthly intervals by licensed veterinarians, revealed no unusual health problems.

ARS, with assistance from researchers at Hematech and the University of Texas, evaluated the cattle using careful observation, post-mortem examination of two of the animals, and a technology that amplifies abnormal proteins to make them easier to detect. Further testing will take at least three years to complete. The evaluation was reported in the scientific journal Nature Biotechnology (Jan, 2007).

First Successful Blood Test for 'Mad Cow' Disease Prions

Researchers at the University of Texas Medical Branch at Galveston (UTMB) have found a way to detect in blood the malformed proteins that cause "mad cow disease," the first time such "prions" have been detected biochemically in blood. The discovery is expected to lead to a much more effective detection method for the infectious proteins responsible for brain-destroying disorders, such as Bovine Spongiform Encephalopathy (BSE) in cattle and variant Creutzfeldt-Jakob disease (vCJD) in humans. The blood test would make it much easier to keep BSE-infected beef out of the human food supply, ensure that blood transfusions and organ transplants do not transmit vCJD, and give researchers their first chance to figure out how many people may be incubating the disease.

The concentration of infectious prion protein in blood is far too small to be detected by the methods used to detect it in the brain, but we know it's still enough to spread the disease (UTMB neurology professor Claudio Soto). The key to success was developing a technique that would amplify the quantity of this protein more than 10 million-fold, raising it to a detectable level.

A method which was called protein misfolding cyclic amplification (PMCA) was applied to blood samples taken from 18 prion-infected hamsters that had developed clinical symptoms of prion disease. PMCA uses sound waves to vastly accelerate the process that prions use to convert normal proteins to misshapen infectious forms.

Successive rounds of PMCA led to the discovery of prions in the blood of 16 of the 18 infected hamsters. No prions were found in blood samples that were taken from 12 healthy control hamsters and subjected to the same treatment.

The next step, on which currently work is being carried on, will be detecting prions in the blood of animals before they develop clinical symptoms and applying the technology to human blood samples.

Tests for infectious prions in cattle and human blood are badly needed. Because current tests require post-slaughter brain tissue for analysis, in the United States only cattle already showing clinical symptoms of BSE (so-called "downer cows") are tested for the disorder. This is true even though vCJD potentially can be transmitted by animals not yet showing symptoms of the disease. (Only two cases of BSE have been found in American cows so far). And although British BSE cases have been in decline since 1992, scientists believe the British BSE epidemic of the 1980s could have exposed millions of people in the UK and Europe to infectious prions. The extent of the vCJD epidemic is yet unknown. So far the disease has killed around 180 people worldwide, but numbers could reach thousands or even hundreds of thousands in the coming decades. Prions have also been shown to be transmissible through blood transfusions and organ transplants.

Who knows what the real situation is in cattle in the United States? People are sitting on a time bomb, because the incubation period of this disease in humans can be up to 40 years. That's why a blood test is so important. Need is to know the extent of the problem, and to make sure that beef and the human blood supply are safe, and early diagnosis is needed so that when scientists develop a therapy we can intervene before clinical symptoms appear--by then, it's too late.

Prions' physical properties lead to different physiological effects

Brittleness is often seen as a sign of fragility. But in the case of infectious proteins called prions, brittleness makes for a tougher, more menacing pathogen. Howard Hughes Medical Institute researcher have discovered that brittle prion particles break more readily into new "seeds," which spread infection much more quickly.

The discovery boosts basic understanding of prion infections, and could provide scientists with new ideas for designing drugs that discourage or prevent prion seeding, (Jonathan Weissman, a Howard Hughes Medical Institute investigator at the University of California, San Francisco UCSF). The scientists studied yeast prions, which are similar to mammalian prions in that they act as infectious proteins.

Yeast and mammalian prions are proteins that transmit their unique characteristics via interactions in which an abnormally shaped prion protein influences a normal protein to assume an abnormal shape. In mammalian prion infections, these abnormal shapes trigger protein clumping that can kill brain cells. In yeast cells, the insoluble prion protein is not deadly; it merely alters a cell's metabolism. Prions propagate themselves by division of the insoluble clumps to create "seeds" that can continue to grow by causing aggregation of more proteins.

In earlier studies, Weissman and his colleagues had discovered that the same prion can exist in different strains and have different infectious properties. These strains arise from different misfoldings of the prion protein that result in different conformations. A similar strain phenomenon has been described for mammalian prions. More generally, even in noninfectious diseases involving protein misfolding, like Alzheimer's and Parkinson's diseases, the same protein can misfold into more than one shape with some forms being toxic and others benign. However, Weissman said, it was not understood how different conformations cause different physiological effects.

As part of the studies published in Nature, the researchers created a mathematical model that enabled them to describe the growth and replication of prions according to the physical properties of the prion protein. To validate that model in yeast, they then created in a test tube, infectious forms of the prion protein in three different conformations and introduced them into yeast cells. They then correlated the strength of infectivity of each prion with its physical properties and compared their results to those predicted by their mathematical model.

According to Weissman, the researchers found that the slowest-growing conformation seemed to have the strongest effect in producing protein aggregates inside cells. But from the model it was deduced that growth was only half of the equation (Weissman). The other key feature was how easy it was to break up the prion and create new seeds, and this propensity to seed could be an important determinant of the prion's physiological impact. And that is what they found experimentally -- that the slower growth of that conformation was more than compensated for by an increased brittleness that promotes fragmentation.

According to Weissman, the importance of a prion's brittleness, or "frangibility," to its physiological effects has both basic research and clinical implications. Investigators trying to develop synthetic prions as a research model for mammalian prions have had a very hard time getting a high degree of activity. Part of the reason may be that they were trying to create forms that were very stable. But that might have been exactly the wrong thing to do, because prions that are too stable may be the ones that are not very infectious because the aggregates are hard to break up.

And from a therapeutic point of view, their findings suggest that effective treatment strategies for prion diseases might aim at stabilizing prion aggregates. By preventing the aggregates from being broken up to smaller seeds, their propagation can be reduced. In contrast, most such strategies now aim at preventing the proteins from forming in the first place.

In future studies, Weissman and his colleagues plan to expand their analytical model to describe in more detail how prions' physical properties lead to different physiological effects. They also plan more detailed analyses to examine how the molecular structure of a prion protein gives rise to its physical properties.

CONCLUSION

The course of investigation into prion diseases has been a long, convoluted and complicated one. Perhaps the reason for the complication is the fact that the prion diseases may be inherited, or infectious or sporadic, and these three aetiologies do not interact. However, the prion diseases are related in that these circumstances converge on a common pathogenic mechanism. It is interesting to note that scientists have no difficulty accepting that there can be a number of different ways to catch a disease.

For example, sexually transmitted diseases like HIV (human immunodeficiency virus) may be transmitted sexually or via blood products. But for many years the idea that prion diseases may sometimes be sporadic in some cases, familial in others and transmissible in some more cases, was scientifically unacceptable.

Scientists argued amongst each other for one of the cases, e.g. veterinarians were determined that Scrapie was an acquired disease, while other scientists argued equally strongly that much of the evidence about prion disease was incompatible with an acquired case (Ridley and Baker, 1998). It could be said that it was comparable to the case of the three blind men identifying the elephant; one grabs its tail and thinks its snake-like, one grabs its leg and thinks its thick and round, and the other touches its ear and thinks its flat.

In a recent review, the progression of the prion study and the order of discoveries were commented on by Prusiner *et al.*, 1998 who states, "It is interesting to contemplate how the course of scientific investigation might have proceeded had studies on the transmissibility of inherited prion diseases had not been performed until after the molecular genetic lesion had been identified.

Had the prion protein (PrP) gene been identified in families with prion disease by positional cloning or through the purification and sequencing or PrP in amyloid plaques before brain extracts were shown to be transmissible, the prion concept might have been more readily accepted.

It does seem however that the horizon is in sight, with the recent advances in biochemical and biological techniques and the current attention, it should be only a matter of a few years before the prion hypothesis is proved or disproved. Now, the evidence for the prion hypothesis is far stronger that the evidence for the virus or virino involvement.

However, it must be remembered that although there has not been an isolation and characterisation of a virus, it doesn't exclude the possibility that there may be a virus present, or nucleic acid involvement. Viruses are renowned for being difficult to find, when one does not know what they are looking for, and may be of low titre.

The demonstration of PrPres infectivity in experiments similar to that of Jackson *et al.* (1999) would, however, put the virus idea out of realistic reach, if at all it is still in realistic reach. The prion research going on now, seems to have abandoned the idea and be focussing much more on the protein only hypothesis, although researchers do admit that the virus or virino hypothesis has not yet been disproved (Telling, 1999). The use of transgenic experiments have helped ideas along the way, and the mouse model is the central experimental model in prion research, the weakness is in that it is a mouse and not a human, and so comparisons between the two must be made with great care.

In a broad sense, it may be said that prions are elements that impart and propagate variability though multiple conformers of a normal cellular protein. The species of a particular protein is encoded by the sequence of the chromosomal PrP gene of the mammal in which it last replicated. In contrast to a nucleic acid genome that encodes strain specific properties in genes, prions seem to encipher these properties in the tertiary structure of PrPsc. Although the possibility of a small ligand bound to PrPsc as an essential component of the infectious prion particle cannot be ignored (Prusiner *et al.*, 1998).

INTO THE FUTURE, AND THE UNKNOWN INCUBATION FACTOR

It is certain that due to the attention received, and the dangers well illustrated in the media, and by scientists, prion research will progress extremely quickly in the next few years. Attention will be focussed on the function of PrPc, the development of effective rapid diagnostic techniques, and the molecular studies on the mechanisms of replication to provide targets for the development of effective chemotherapeutic agents. Perhaps most importantly, and perhaps the most elusive fact to find out, however, is the possible incubation time for vCJD in humans.

The only model for comparison is the rise and fall of the Kuru epidemic in Papua New Guinea. The maximum incubation time for that disease (not transmitted across the species barrier) has shown to be maximally about 40 years, with the peak onset after a few years after successful inoculation with the agent, as children as young as four have been reported with the disease. The big question is, what will be the effect of species barrier between cows and humans (and possibly even sheep and humans in the case of BSE fed sheep) in the case of vCJD. No one seems to know, some people think the 20-30 or so cases so far are the tip of the iceberg, others think it is mostly over.

However, it may be encouraging to think that the BSE agents ability to traverse the species barrier also may mean that it is also faster at crossing this barrier than other prion diseases and so expression would be sooner (years) rather than later (decades). One way or another, time will tell. At present there is no cure for Prion diseases and the kill rate is 100%. More research is necessary to understand these devastating infectious agents and create opportunities to evaluate therapeutic strategies for these unusual diseases.

1

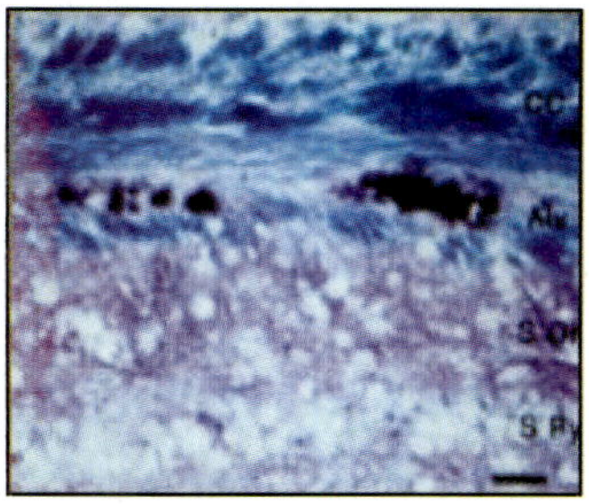

2

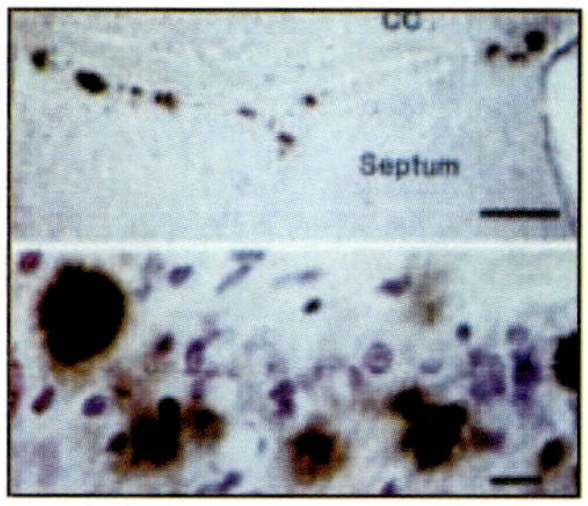

3

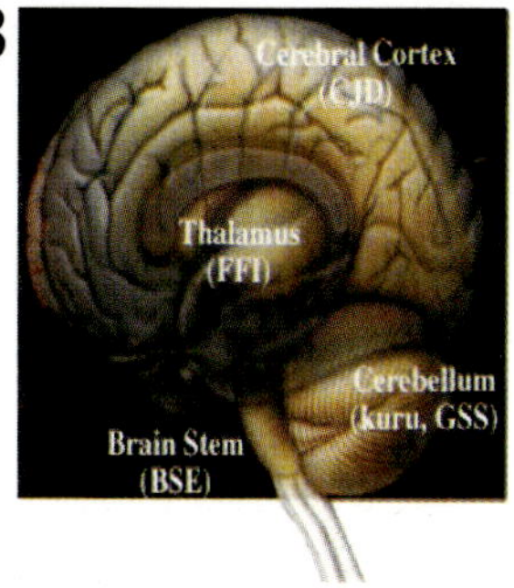

4

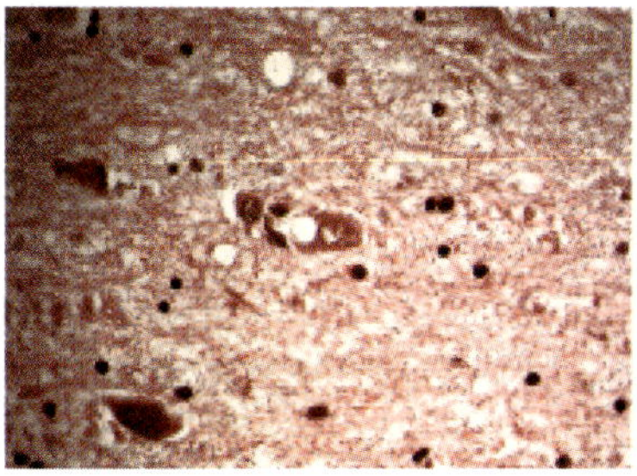

5

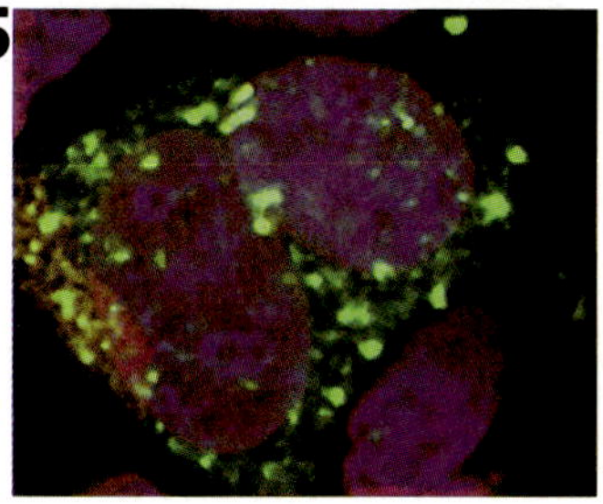

6

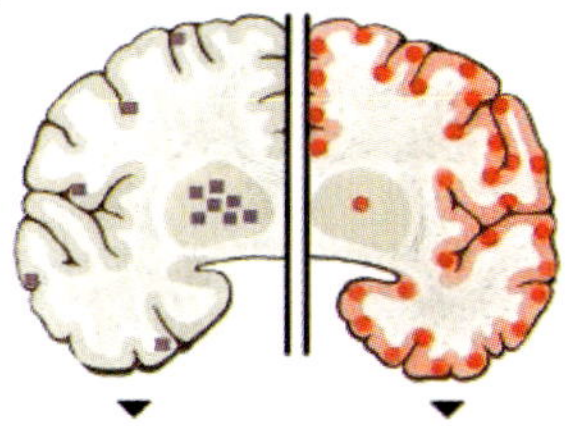

7

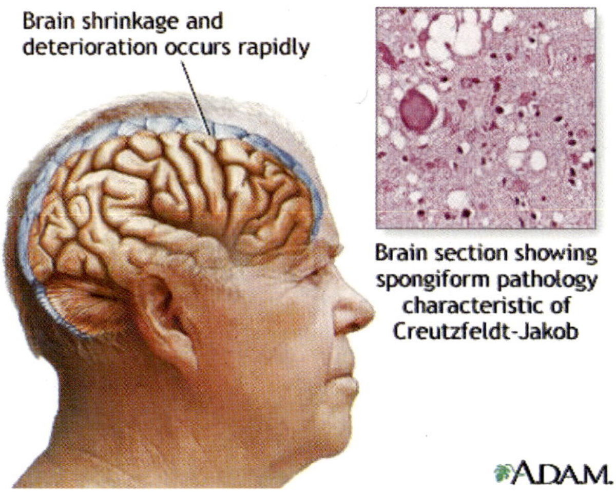

8

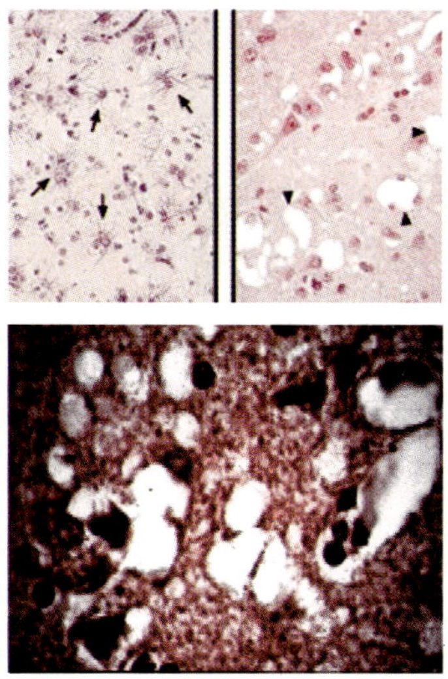

Fig: 1) Prions are mutated proteins that are believed to cause sponge-like holes in the brains of BSE,Scrapie and CJD victims, **2)** Prions are chemically similar to normal proteins found in the brain but they appear to be folded in a way that makes them pull normal proteins into the same shape, **3)** Different regions of brain affected by different prions, **4)** Spongy appearance of brain tissue, **5)** Apple - green birefringence after Congo red staining of amyloid plaques, **6)** Histopathology of CJD, **7)** Characteristic symptoms of brain affected with CJD, **8)** Histopathology of Kuru

9

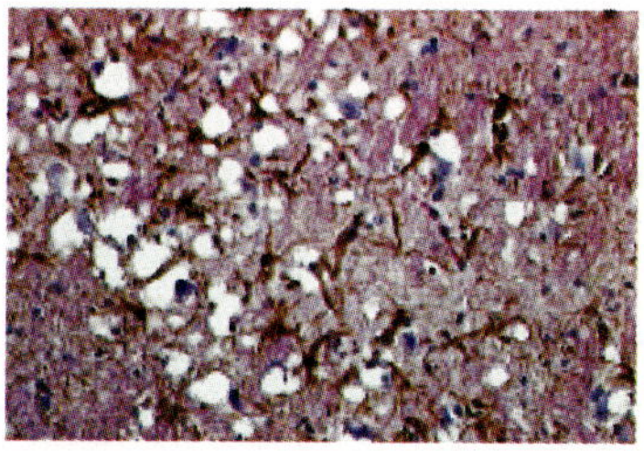

10

11

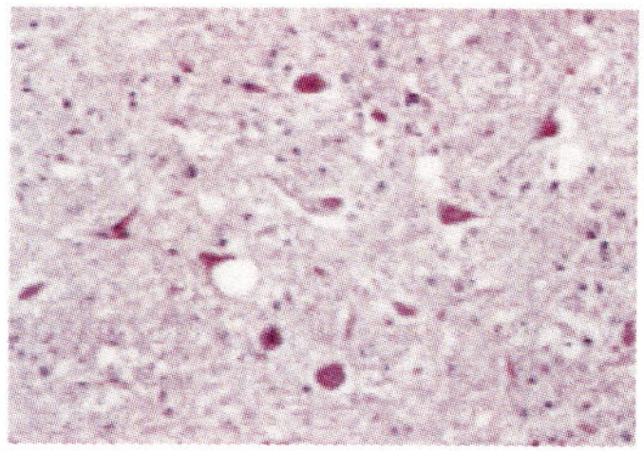

12

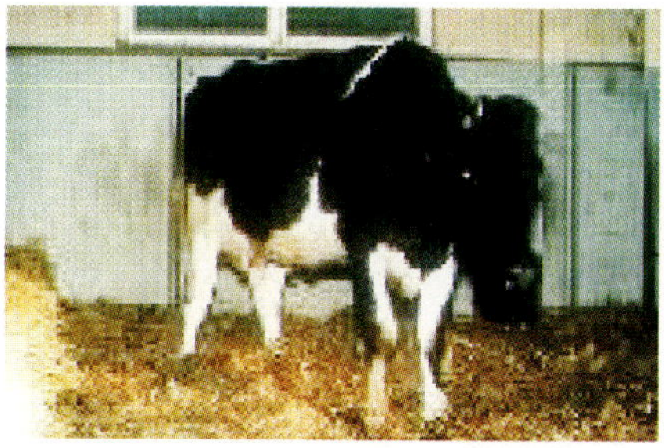

13

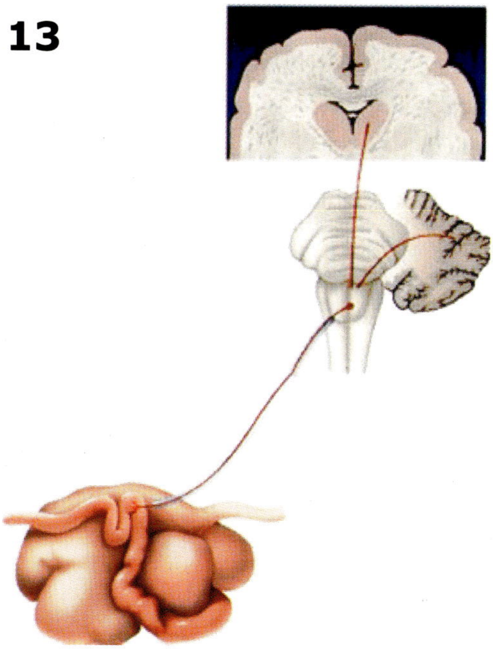

14

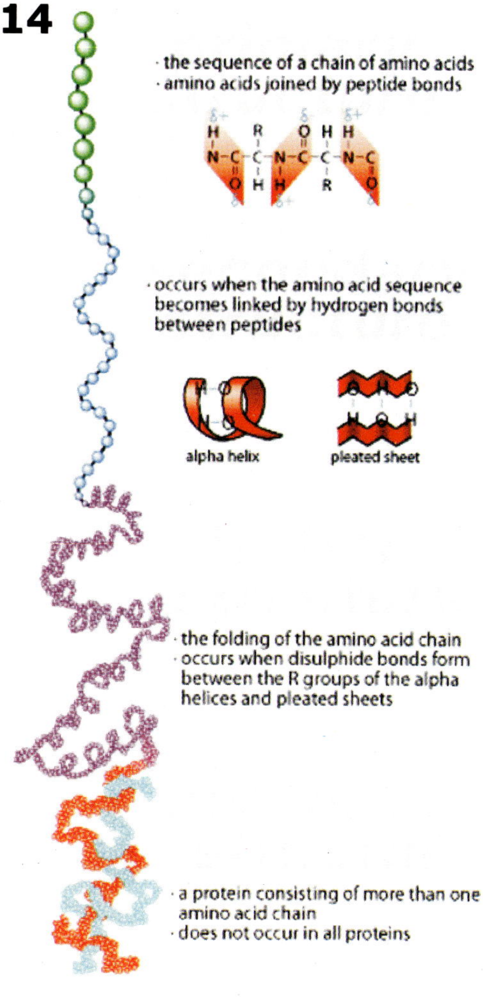

Fig: 9) Histopathology of Scrapie, **10)** Sheep affected with Scrapie, **11)** Histopathology of BSE, **12)** Cow infected with BSE, **13)** Spread of infection across species barrier, **14)** Normal protein structure

15

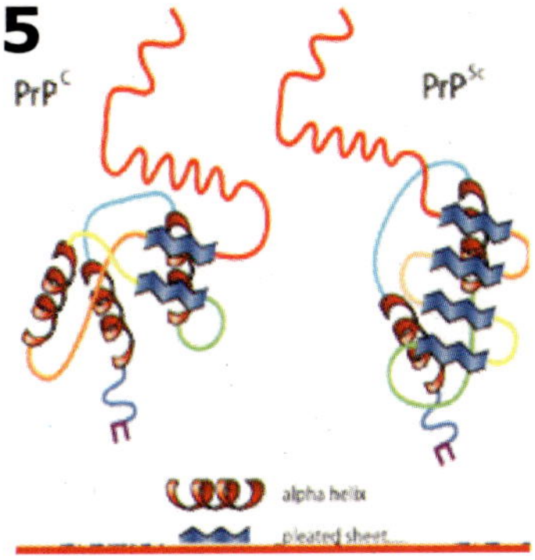

16

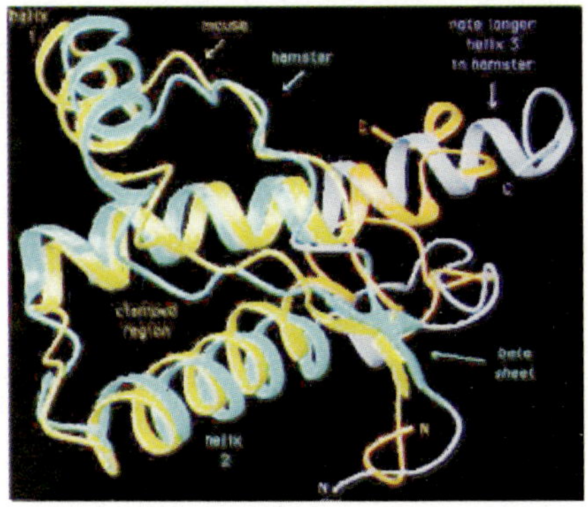

17

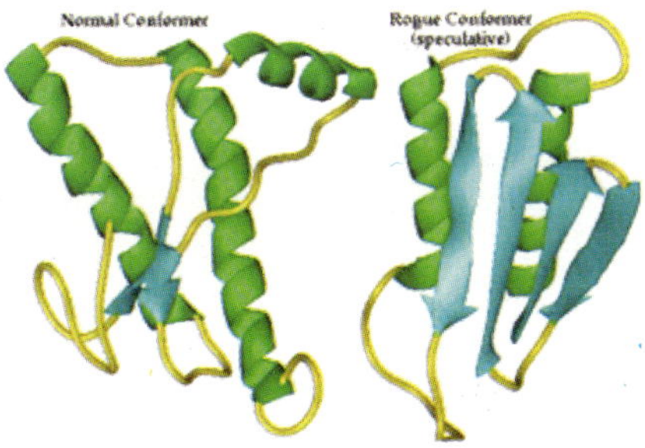

18

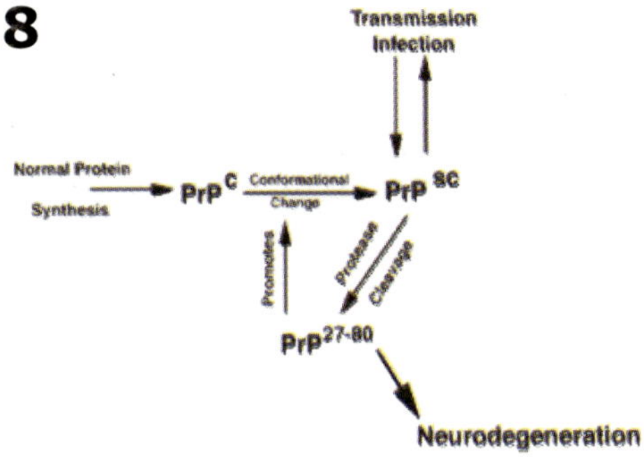

19

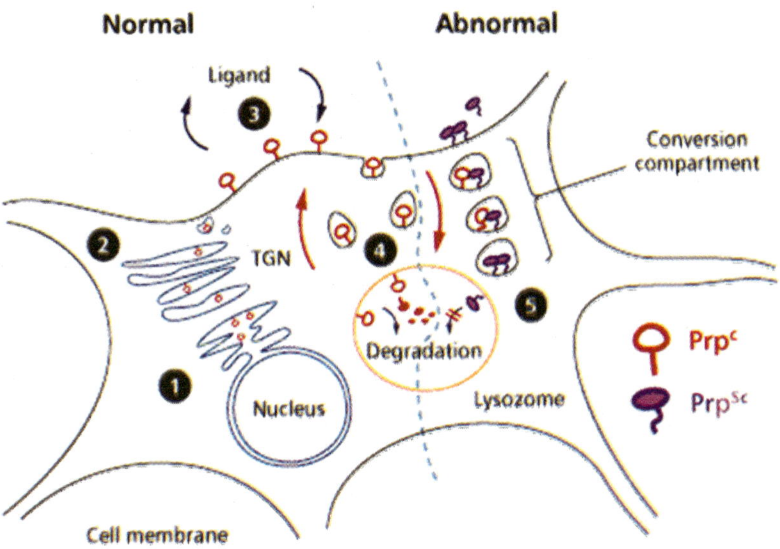

Fig: 15) A normal prion (left), compared to an aberrant, disease-causing prion (right),
16) Structural comparison of mouse and hamster prion proteins,
17) Conformational comparative diagram of normal and rogue prion protein,
18) Conversion of PrPc to PrPsc,
19) Cellular pathway for prion protein synthesis and degradation in cultured cells

20

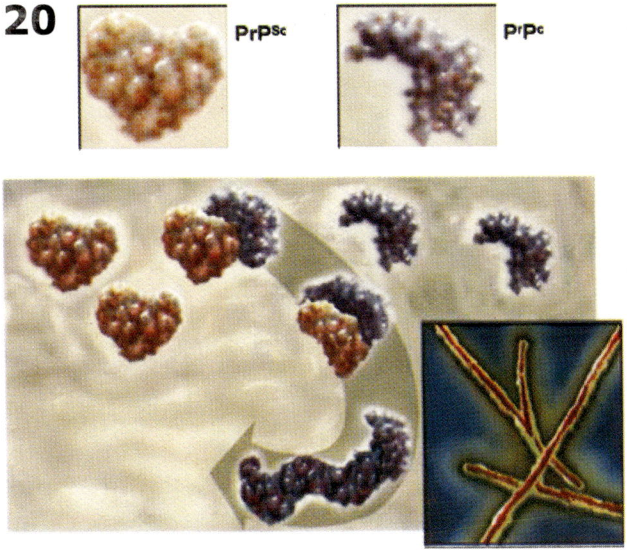

21

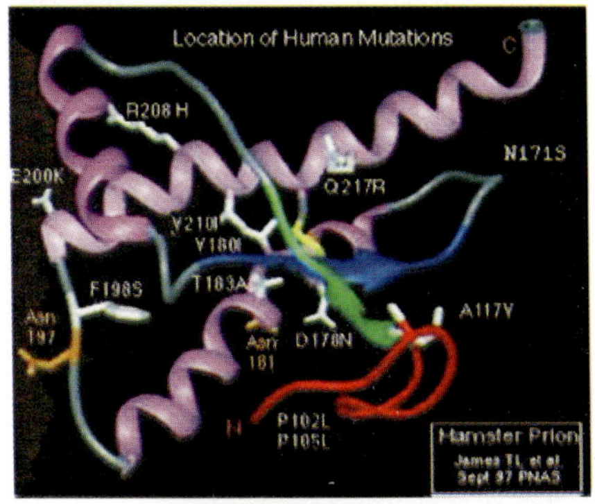

22

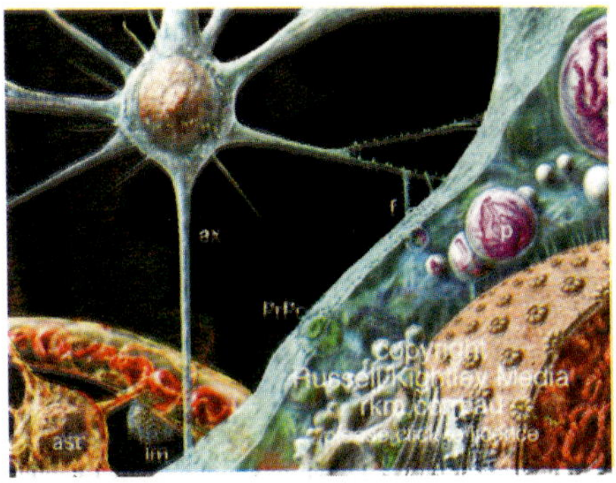

23

24

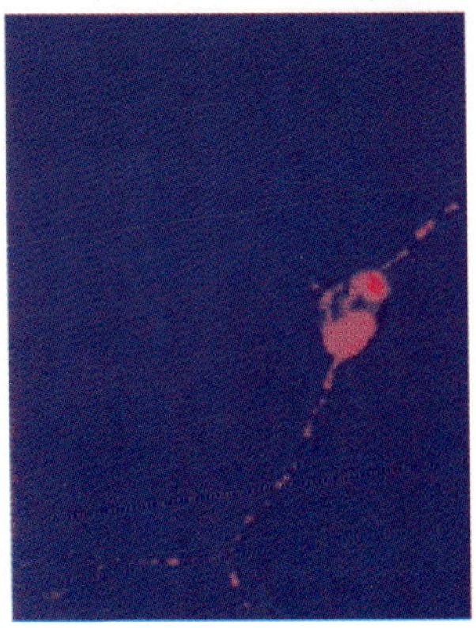

25

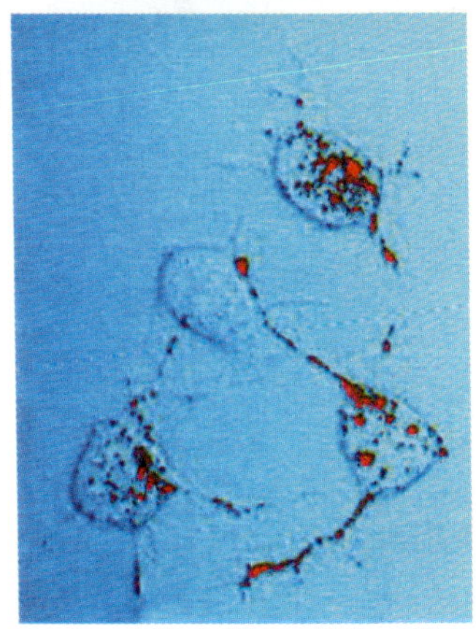

Fig: 20) Corrupting Prions from PrPc to PrPSc, **21)** Location of human mutants on mouse prion model, **22)** Prion replication and its spread on cellular level, **23)** Delivery of orally ingested prions to brain, **24)** Prion trafficking in nerve cells of hamster, **25)** Prion trafficking in nerve cells of mice

26

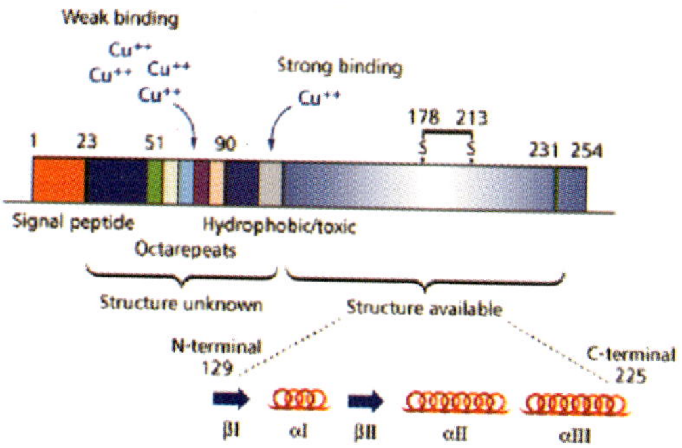

27

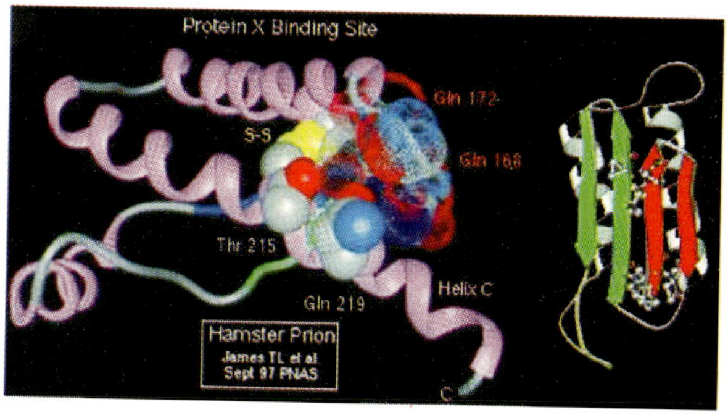

28

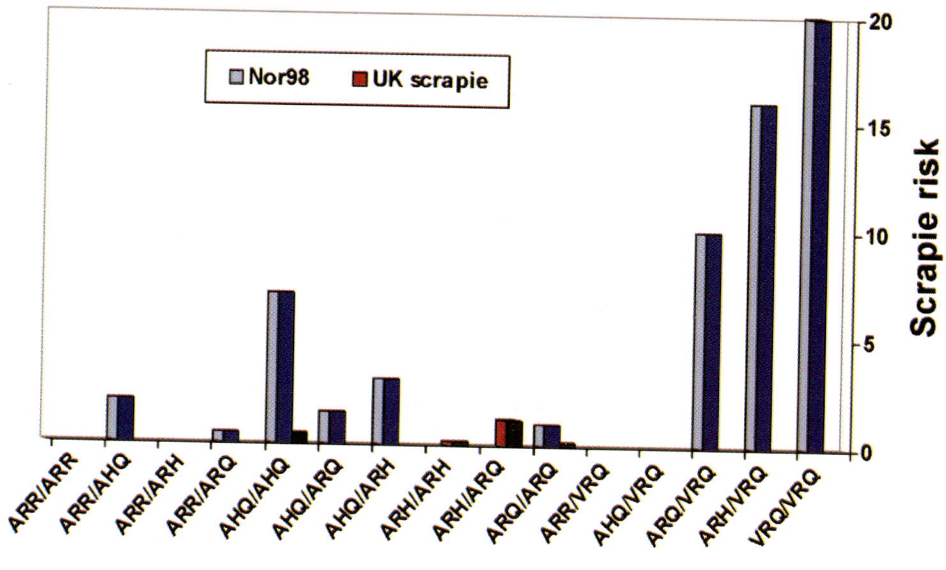

Fig: 26) Metal ions interaction with residues Prion protein motif,

27) Protein X binding site on the Hamster prion protein,

28) Scrapie risks - variations in the PrP gene determine the susceptibility of sheep to Scrapie. In the United Kingdom, sheep with the VRQ variant (haplotype) are most at risk of succumbing to conventional (UK) Scrapie. These sheep appear, however, to be resistant to the new Nor98 strain, from which sheep with the AHQ haplotype are instead most at risk. For each genetic type, 'risk' is the proportional representation in reported Scrapie cases, divided by the proportional representation in the national flock (UK) or flock mates (Nor98). (Detwiler et al., 2003; Moum et al., 2004).

GLOSSARY OF TERMS

α-helix : The secondary structure of a protein (associated with being predominant in the cellular form of PrP).

Amyloid plaque : The fibrillar aggregation of PrPsc in the brain.

Amyotrophic : The progressive wasting of muscle tissues.

Asn-linked gylcosylation : Post-translational modification involving the addition of oligosaccharides to Asn residues of PrPc.

Astrocyte : Type of cell involved in maintaining the normal neuronal status, and is the first site of accumulation for PrPsc.

Ataxia : Unsteady standing and walking as a result of brain dysfunction.

b-sheet : The secondary structure of a protein (associated with being predominant in PrPsc).

BSE : Bovine Spongiform Encephalopathy, the infectious prion disease of cattle.

Chimeric PrP : Prion protein comprising residues derived from two different species.

CJD : Creutzfeldt-Jacob disease, the prion disease of humans.

Dementia : Loss of higher mental faculties such as learning, reasoning and remembering.

Disease characteristics : Incubation period and lesion profile (pattern of vacuolation in the brain). Attributes used to compare different strains of prions.

FFI : Fatal Familial Insomnia is a prion disease of humans.

Foré : One of the many language groups spoken in Papa New Guinea. The kuru epidemic affected mainly the Foré-speaking people.

FTIR : Fourier Transform Infra-Red spectroscopy, used to analyse the structure of proteins.

GSS : Gerstmann-Sträussler-Scheinker disease is a prion disease of humans.

H1, H2, H3 and H4 : The four predicted a-helical regions of PrPc.

HaPrPc : Normal cellular form of hamster prion protein.

HaPrPsc	: Rogue form of the hamster prion protein.
HuPrP	: Human form of the prion protein.
Iatrogenic CJD	: Prion disease inadvertently spread during a medical procedure.
Incubation period	: Length of time between inoculation of prions and onset of clinical symptoms.
Kuru	: Prion disease observed in Papa New Guinea amongst the Foré-speaking people, thought to be involved in endo-cannibalism.
MHu2M	: Chimeric PrP gene made of sequences of mouse and human PrP.
MoPrPc	: Normal mouse prion protein.
MoPrPsc	: Rogue form of the mouse prion protein.
Myoclonus	: Involuntary brief contraction of the muscles, leading to a limb jerk.
Neurodegeneration	: Progressive damage to the nerve cells (neurones) leading to cell death.
NMR	: Nuclear Magnetic Resonance technology, used to structurally analyse proteins.
Passage	: Transmission of prions from one animal to another.
PIPLC	: Phosphatidylinositol-specific phospholipase C, to which PrPsc is resistant.
Prion	: Infectious protein particle containing little nucleic acid thought to be the cause of prion diseases.
Prn-P	: Gene encoding PrP especially in mice.
PRNP	: Gene encoding PrP in humans.
Protease K	: One of a group of enzymes, the proteases, which can digest proteins down to their constituent amino acids.
PrP	: The prion protein, usually the cellular form depending on context of use.
PrPc	: Cellular form of the prion protein (normal form).
PrPres	: Protease resistant form of the prion protein, usually slightly different in structure but principally the same as PrPsc.
PrPsc	: Scrapie form of the prion protein (rogue form).

PrPsens	: Protease sensitive form of the prion protein, usually slightly different in structure but principally the same as PrPc.
PrP+/+	: Wild-type expression of the prion gene.
PrP+/0	: Heterozygous expression of the prion gene.
PrP0/0	: PrP gene knockout, here both copies of the gene are disrupted.
rPrP90-231	: An example of a recombinant prion protein, with the relevant amino acid sequence shown in superscript.
Scrapie	: Form of prion disease observed in sheep and goats.
SAF (Scrapie Associated Fibrils)	: Small clumps of twisted fibres seen (with the electron microscope) in extracts of brains from animals and humans with prion disease, the main constituent being abnormal prion protein.
SHa PrP	: Syrian hamster prion protein.
Species barrier	: The term given to the observation of increased difficulty in the transmission of prions between different species.
Spongiform degeneration	: The appearance of holes or vacuolation in the brain.
Sporadic prion disease	: Prion disease occurring without warning - sporadically.
Strain of agent	: Term used to explain why different preparations of Scrapie agent cause different lesion profiles and incubation periods in various strains of mice.
Tg	: Transgenic.
Transgene	: Foreign DNA expressed in a 'transgenic' animal.
Transgenic animal	: Animal with foreign DNA inserted into the genome.
TSE	: Transmissible Spongiform Encephalopathy, generalised term for a prion disease.
vCJD	: Variant CJD, the prion disease in humans thought to be derived from cattle.
Virino	: Hypothetical infectious agent. Suggested by scientists working with mouse adapted Scrapie to account for apparent mutations in the Scrapie agent and the existence of a number of different agent strains.
Wild-type	: Animal or gene with an undisrupted DNA sequence.

REFERENCES

*Alper T (1966) The exceptionally small size of the Scrapie agent *Biochemical and Biophysical Research Communication* **22**: 278-282

*Alper T (1993) The Scrapie enigma: insights from radiation experiments *Radiation Research* **135**: 283-92

*Alpers M (1999) Lecture on Kuru ICSM 21st May

*Arnold J Tipler C Laszlo L Hope J Landon M Mayer J (1995) The abnormal form of the prion protein accumulates in late-endosome-like organelles in Scrapie-infected mouse brain *Jour of Pathology* **176**: 403-411

Aguzzi A Weissmann C 1997 Prion research: the next frontiers *Nature* **389**: 795-798

Alper T Cramp W Haig D Clarke M (1967) Does the Agent of Scrapie Replicate without Nucleic Acid? *Nature* **214**: 764-766

Baldwin M James T Cohen F Prusiner S (1998) The three-dimensional structure of prion protein: implications for prion disease *Biochemical Society Transactions* **26**: 481-486

Baldwin M Pan K Nguyen J Huang Z Groth D Serban A *et al* (1994) Spectroscopic Characterisation of Conformational Differences Between PrPc and PrPsc - an a-Helix to b-Sheet Transition *Philosophical Transactions of the Royal Society of London Series B-Biological Sciences* **343**: 435-441

Bazan F Fletterick R McKinley M Prusiner S (1987) Predicted secondary structure and membrane topology of the Scrapie prion protein *Protein Engineering* **1**: 125-135

Bendheim P E *et al* (1985) Scrapie and Creutzfeldt-Jacob disease prion proteins share physical properties and antigenic determinants *Proceedings of the National Academy of Sciences of the United States of America* **82**: 997

Benestad S *et al* (2003) *Veterinary Record* **153**: 202-208

Bolton D C *et al* (1985) Scrapie PrP 27-30 is a sialoglycoprotein *Jourl of Virology* **53**: 596

Bonn D (1998) New rapid test can identify eight prion strains *The Lancet* **352**: 1124

Borchelt D Scott M Taraboulos A Stahl N Prusiner S (1990) Scrapie and Cellular Prion Proteins Differ in Their Kinetics of Synthesis and Topology in Cultured Cells *The Journal of Cell Biology* **110**: 743-752

Brandner S Isenmann S Raeber A Fischer M Sailer A Koyba *et al* (1996) Normal host prion protein necessary for Scrapie induced neurotoxicity *Nature* **379**: 339-343

Brown D (1999) Senior scientists dismiss fears of CJD epidemic *The Daily Telegraph* March 27th

Brown D R (1999) Prion Protein expression aids cellular uptake and veratridine-induced release of copper *Jourl of Neuroscience Research* **58**: 717-725

Brown D R Clive C Haswell S J (2001) Antioxidant activity related to copper binding of native prion Protein *Jourl of Neurochemistry* **76**: 69-76

Brown D R Qin K Herms J W *et al* (1997) The Cellular Prion Protein binds copper in vivo *Nature* **390**: 684-687

Brown P (2002) Drug therapy in human and experimental Transmissible Spongiform Encephalopathy *Neurology* **58**: 1720-1725

Brown P Cathala F Raubertan R F *et al* (1987) The epidemiology of Creutzfeldt-Jacob disease Conclusion of a 15 year investigation in France and review of the old Literature *Neurology* **37**: 895-904

Bruce M (1993) Scrapie strain variation and mutation *British Medical Bulletin* **49**: 822-828

Bruce M McConnell I Fraser H Dickinson A (1991) The disease characteristics of different strains of Scrapie in *Sinc* congenic mouse lines: implications for the nature of the agent and host control of pathogenesis *Journal of General Virology* **72**: 595-603

Bruce M Will R Ironside J McConnell I Drummond D Suttie A (1997) Trañsmissions to ice indicate that 'new variant' CJD is caused by the BSE agent *Nature* **389**: 498-501

Büeler H Aguzzi A Sailer A Greiner R Autenried P Aguet M Weissmann C (1993) Mice devoid of PrP Are Resistant to Scrapie *Cell* **73**: 1339-1347

Büeler H Fischer M Lang Y Bluethmann H Lipp H DeArmond S *et al* (1992) Normal evelopment and behaviour of mice lacking the neuronal cell-surface PrP protein *Nature* **356**: 577-582

Burwinhel R Riemer C Schwarz A Schultz J Neidhold S Bamme T Baier M (2004) Role of cytokines and chemokines in Prion infections of the central nervous system *International Journal of Developmental Neuroscience*

Buschman A *et al* (2004) *Journal of General Virology* **85**: 2727-2733

Buschman A *et al* (2004) *Journal of General Virology* **117**: 27-36

Carlson G A *et al* (1986) Linkage of prion protein and Scrapie incubation time genes *Cell* **46**: 3

*Caughey B Chesebro B (1997) Prion protein and the transmissible spongiform cephalopathies *Trends in Cell Biology* **7**: 56-62

*Caughey B Raymond G J (1991) The Scrapie –associated form of PrP is made from a cell urface precursor that is both protease and phospholipase sensitive *Jour of Biology and Chemistry* **266**: 217-218-223

Chesebro B *et al* (1985) Identification of Scrapie prion protein–specific mRNA in Scrapie nfected and uninfected brain *Nature* **315**: 331

*Chesebro B (1998) BSE and Prions: Uncertainities about the Agent *Science* **279**: 42-43

*Chien P *et al* (2003) Generation of Prion transmission barriers by mutational control of myloid conformations *Nature* **424**: 145-150

Cohen F Abarbanel R Kuntz D Fletterick R (1985) Turn Prediction in Proteins Using a attern Matching *Approach Biochemistry* **25**: 266-275

Cohen F Pan K Huang Z Baldwin M Fletterick R Prusiner S (1994) Structural Clues to Prion Replication *Science* **264**: 530-531

*Colling S Collinge J Jefferys J (1996) Hippocampal slices from prion protein null mice: isrupted Ca2+ -activated K^+ currents *Neuroscience letters* **209**: 49-52

Collinge J Hawke S (1998) B lymphocytes in prion neuroinvasion: Central or peripheral layers? *Nature Medicine* **4**: 1369-1370

Collinge J Palmer M (1997) 'Prion Diseases' Oxford University Press

Collinge J Whittington M Sidle K Smith C Palmer M Clarke A Jefferys J (1994) Prion protein is necessary for normal synaptic function *Nature* **370**: 295-277

Creutzfeldt H G (1920) uber eine eigenartige herdformige Erkrankung deszentral nervenssystems Z Gesant *Neurol Psychiatrie* **57**: 1 118

Cuillé J Chelle P L (1936) La maladie dite tremblante du mouton este-elle inocuable? *Compte rendu de l'Academie des Sciences* **203**: 1522-1554

De Armond S *et al* (1985) Identification of prion amyloid filaments in Scrapie-infected brain *Cell* **41**: 221

*De Armond S Sánchez H Yehiely Qiu Y Ninchak-Casey A Daggett V *et al* (1997) Selective Neuronal Targeting in Prion Disease *Neuron* **19**: 1337-1348

De Bosschere H *et al* (2004) *Veterinary Record* **155**: 707-708

Detwiler L A Baylis M (2003) *Rev Sci Tech OIE* **22**: 121-143

Farquhar C Dornan J Somerville R Tunstall A Hope J (1994) Effect of *Sinc* genotype agent isolate and route of infection on the accumulation of protease-resistant PrP in non-central nervouse system tissues during the development of murine Scrapie *Journal of General Virology* **75**: 495-504

*Farquhar C Somerville R Bruce M (1998) Straining the prion hypothesis *Nature* **391**: 345-346

Forloni G (1993) Neurotoxicity of a prion protein fragment *Nature* **362**: 543

Gajdusek D C (1977) Unconventional viruses the origin disappearance of Kuru *Science* **943**: 960

Gajdusek D C Zigas V (1957) Degenerative disease of the Central Nervous system in New Guinea: The endemic occurrence of "Kuru" in the native population *N Engl Journal of Medicine* **257**: 974-978

Gajdusek D Gibbs C Alpers M (1966) Experimental transmission of a kuru-like syndrome to chimpanzees *Nature* **209**: 794-796

Gajdusek D *et al* (1972) Persistence of the viruses of Kuru and Creutzfeldt-Jacob disease in tissue culture of brain cells *Nature* **235**: 104

Garcion E Wallace B Pelletier L Wion D (2004) RNA mutagenesis and sporadic prion diseases *Journal of Theoretical Biology* **230**: 271-274

Gasset M Baldwin M Lloyd D Gabriel J Holtzman D Cohen F *et al* (1992) Predicted a-helical regions of the prion protein when synthesised as peptides form amyloid *Proceedings of the National Academy of Sciences of the United States of Americ* **89**: 10940-10944

Gibbs T Gajdusek D Asher D Alpers M Beck E Daniel P *et al* (1968) Creutzfeldt-Jacob disease (Spongiform encephalopathy): transmission to the chimpanzee *Science* **161**: 388-389

*Goldfarb L Brown P McCombie R Goldgaber D Swergold G Wills P *et al* (1991) Transmissible familial Creutzfeldt-Jakob disease associated with five seven and eight extra octapeptide coding repeats in the PRNP gene *Proceedings of the NationalAcademy of Sciences of the United States of America* **99**: 10926-10930

Goldmann W (1993) PrP gene and its association with spongiform encephalopathies *British Medical Bulletin* **49**: 839-859

Goldmano W Hunter N Somerville R Hope J (1996) Prion Phylogony revisited *Nature* **382**: 32-38

Griffiths (1967) Self replication and Scrapie *Nature* **215**: 1043-1044

Hadlow W J (1959) Scrapie and Kuru *Lancet* **2**: 289-290

Harrison P Chan H Prusiner S Cohen F (1999) Thermodynamics of Model Prions and its Implications for the Problem of Prion Protein Folding *Journal of Molecular Biology* **286**: 593-606

*Hegde R S Mastrianni J A Scott M R *et al* (1998) A transmembrane form of the Prion protein in neurodegenerative disease *Science* **279**: 827-834

Hill A Butterworth R Joiner S Jackson G Rossor M Thomas D *et al* (1999) Investigation of variant Creutzfeldt-Jacob disease and other human prion diseases with tonsil biopsy samples *The Lancet* **353**: 183-189

Hill A Desbruslais M Joiner S Sidle K Gowland I Collinge J (1997) The same prion strain causes vCJD and BSE *Nature* **389**: 448-450

Houston F *et al* (2003) *Nature* **423**: 493

Hunter N (1997) *Trends Microbiology* **5**: 331-334

Hsiao K Prusiner S (1990) Inherited human Prion diseases *Neurology* **40**: 1820-1827

Huang F P Mac Pherson G G (2004) Dendritic cells and Oral transmission of Prion disease *Adv Drug Del reviews* **56**: 901-913

Huang Z Gabriel J Baldwin M Fletterick R Prusiner S Cohen F (1994) Proposed three dimensional structure for the cellular prion protein *Proceedings of the National Academy of Sciences USA* **91**: 7139-7143

Ikeda K Kawada N Wang Y Kadoya H Nakatani K Sato M Kaneda K (1999) Expression of Cellular Prion Protein in Activated Hepatic Stellate Cells *American Journal of Pathology* **153**: 1695-1700

*Iswasaki M Okumura K Kondo Y Tanaka T Igarashi H (1992) cDNA cloning of a novel heterogeneous nuclear ribonucleoprotein gene homologue in *Caenorhabditis elegans* using hamster prion protein cDNA as a hybridisation probe *Nucleic Acids Research* **20**: 4001-4007

Jackson G Hosszu L Power A Hill A Kenney J Saibil H *et al* (1999) Reversible Conversion of Monmeric Human Prion Protein Between Native and Fibrilogenic Conformations *Science* **283**: 1935-1937

James T Liu H Ulyanov N FarrJones S Zhang H Donne D *et al* (1997) Solution structure of a 142-residue recombinant prion protein corresponding to the infectious fragment of the Scrapie isoform *Proceedings of the National Academy of Sciences USA* **94**: 10086-10091

Kaneko K Zulianello L Scott M Cooper C Wallace A James T *et al* (1997) Evidence for protein X binding to a discontinuous epitope on the cellular prion protein during Scrapie prion propagation *Proceedings of the National Academy of Sciences USA* **94**: 10069-10074

*Kellings K Prusiner S Riesner D (1994) Nucleic acids in prion preparations: unspecific background or essential component *Philosophical Transactions of The Royal Society of London (Biology)* **343**: 425-430

Kim B H Lee H G Choi J K Kim J I Choi E K Carp R I Kim Y S (2004) The cellular Prion protein (PrPc) prevents apoptotic neuronal cel death and mitochondrial dysfunction induced by serum deprivation *Molecular Brain Research* **124**: 40-50

Kimberlin R (1982) Scrapie agent: prions or virinos? *Nature* **297**: 107-108

*Klein M Frigg R Flechsig E Raeber A Kalinke U Bluethmann H (1997) A crucial role for B cells in neuroinvasive Scrapie *Nature* **390**: 687-690

*Klein M Frigg R Raeber A Flechsig E Hegyi I Zinkernagel R (1998) PrP expression in B lymphocytes is not required for prion neuroinvasion *Nature Medicine* **4**: 1429-1433

*Kocisko D Come J Priola S Chesebro B Raymond G Lansbury P Caughey B (1994) Cell-free formation of protease-resistant prion protein *Nature* **370**: 471-474

*Korth C May B C Cohen F E Prusiner S B (2001) Acridine and Phenothiazine derivatives as pharmaco Therapeutics for Prion disease *Proceedings National Academy of Sciences USA* **98**: 9836-9841

Krakauer D C Pagel M Southwood T R E Zanotto M (2002) A Phylogenesis of Prion Proteins *Nature* **380**: 675

*Lindguist S L *et al* (2003) Prion templates *Proceedings of National Academy of Sciences USA* **100**: 4527

Locht C Chesebro B Race R Keith J (1986) Molecular Cloning and complete sequence of prion protein cDNA from mouse brain infected with the Scrapie agent *Proceedings of the National Academy of Sciences USA* **83**: 6372-6376

Mabbott N Brown K Manson J Bruce M (1997) T-lymphocyte activation and the cellular form of the prion protein *Immunology* **92**: 161-165

*Manuelidis L Fritch W (1996) Infectivity and host responses in Creutzfeldt-Jakob disease *Virology* **216**: 46-59

*Marcotte E Eisenberg D (1999) Chicken Prion Tandem Repeats Form a Stable Protease Resistant Domain *Biochemistry* **38**: 667-676

Masiarz F McKinley M Hearst J Prusiner S (1982) The Scrapie agent is resistant to inactivation by psoralen photoreaction *The Journal of Cell Biology* **9**: 81-86

Mastrianni J A (2004) Prion diseases *Clinical Neurosciences Research* **3**: 469-480

Mckintosh E Tabrizi S J Collinge J (2003) Prion diseases *Journal of Neurovirology* **9**: 183-193

Medori R *et al* (1992) Fatal familial insomnia a prion disease with amutation at codon 178 of the prion protein gene *New Engl J Med* **326**: 444

Monaco S *et al* (2003) Prions: They really do get up your nose *New Engl Jour Med* **348**: 711-720

Moore R C Lee I Y Silverman G L Harrison P M Storme R Heinrich C Kanunaratne A Pasternak S H Chishti M A Liang Y Mastrangelo P Wang K Smit A F Katamine S Carlson G A Cohen F E Prusiner S B Melton D W Tremblay P Hood L E Westaway D (1999) Ataxia in Prion Protein (PrP) deficient mice is associated with upregulation of the novel PrP like Protein doppel *Journal of Molecular Biology* **292**: 797-817

Mouillet-richard S Ermonval M Chebassier C *et al* (2000) signal transduction through Prion Protein *Science* **289**: 1923-1928

Moum T *et al* (2004) *Journal of General Virology* doi:10 1099/vir 0 80437-0

*Mueller R Young I (1998) Emery's Elements of Medical Genetics *Churchill Livingstone* Tenth Edition **122**:

Muramoto T DeArmond S Scott M Telling G Cohen F Prusiner S (1997) Heritable disorder resembling neuronal storage disease in mice expressing prion protein with deletion of an a-helix *Nature Medicine* **3**: 750-755

Muramoto T Scott M Cohen F Prusiner S (1996) Recombinant Scrapie-like protein of 106 amino acids is soluble *Proceedings of the National Academy of Sciences USA* **93**: 15457-15462

*Murray P Rosenthal K Kobayashi G P Faller M (1997) Medical Microbiology (Book) 3rd edition St Louis Mosby **536**: 539

Naslavsky N Stein R Yanai A Friedlander G Taraboulos (1997) Characterization of Detergent-insoluble Complexes Containing the Cellular Prion Protein and Its Scrapie Isoform *Journal of Biological Chemistry* **272**: 6324-6331

**National Scrapie Plan for Great Britain Schemes Brochure* (Department for Environment food and Rural affairs London 2004)

**National Scrapie Plan Consultation on the Strategic review* (Department for environment Food and Rural affairs London 2004)

Oesch B Westaway D Wälchli M McKinley M Kent S Aebersold R (1985) A cellular gene encodes Scrapie PrP 27-30 protein *Cell* **40**: 735-746

Orge L *et al* (2004) *Journal of General Virology* **85**: 3487-3491

Onnasch H *et al* (2004) *Veterinary Record* **155**: 636-637

Osherovich L Z Weissman J S (2002) The utility of Prions *Developmental Cell* **2**: 143-151

Owen F (1998) 'Prions' *Biological Sciences* **10**: 38-40

Pan K Baldwin M Nguyen J Gasset M Serban A Groth D *et al* (1993) Conversion of a-helices into b-sheets features in the formation of the Scrapie prion proteins *Proceedings of the National Academy of Sciences USA* **90**: 10962-10966

Parchi P Peterson R Gambetti P (1996) *Seminars in Virology* **7**: 181-187

*Paul P Harris D (1998) Copper Stimulates Endocytosis of the Prion Protein *Journal of Biological Chemistry* **273**: 33107-33110

*Pérez M Wandosell F Colaço C Avila J (1998) Sulphated glycosaminoglycans prevent the neurotoxicity of a human prion protein fragment *Biochemical Journal* **335**: 369-374

Porter D D Porter H G Cox N A (1973) Failure to demonstrate a humoral immune response to Scrapie infection in mice *Journal of Immunology* **111**: 1407-1410

Prusiner S (1982) Novel Proteinaceous Infectious Particles Cause Scrapie *Science* **216**: 136-144

Prusiner S (1991) Molecular Biology of Prion Diseases *Science* **252**: 1515-1522

Prusiner S (1992) Molecular Biology and Genetics of Neurodegenerative Diseases Caused by Prions *Advances in Virus Research* **41**: 241-280

Prusiner S (1996) 'Prions Prions Prions' Springer-Verlag Berlin Heidleberg

Prusiner S (1998) Prions *Proceedings of the National Academy of Sciences USA* **95**: 13363-13383

Prusiner S Bolton D Groth D Bowman K Cochran C McKinley M (1982) Further purification and characterisation of Scrapie prions *Biochemistry* **21**: 6942-6950

Prusiner S Collinge J Powell J Anderton B (1992) 'Prion Diseases of Humans and Animals' Ellis Horwood Limited

Prusiner S Groth D Serban A Koehler R Foster D Torchia M *et al* (1993) Ablation of the prion protein (PrP) gene in mice prevents Scrapie and facilitates production of anti-PrP antibodies *Proceedings of the National Academy of Sciences USA* **90**: 10608-10612

Prusiner S Scott M DeArmond S Cohen F (1998) Prion Protein Biology *Cell* **93**: 337-348

Prusiner S Scott M Foster D Pan K Groth D Mirenda C *et al* (1990) Transgenetic studies implicate interactions between homologous PrP isoforms in Scrapie prion replication *Cell* **63**: 673-686

Prusiner S Telling G Cohen F DeArmond S (1996) Prion diseases of humans and animals *Seminars in Virology* **7**: 159-173

**Report on the Monitoring and Testing of Ruminants for the Presence of Transmissible Spongiform Encephalopathy (TSE) in the EU in (2003) Including the Result of the Survey of Prion protein Genotypes in Sheep Breeds* Rep 04-D-420525

Ridley R Baker H (1998) 'Fatal Protein: the story of CJD BSE and other prion diseases' Oxford University Press

Safar J Wille H Itri V Groth D Serban H Torchia M *et al* (1998) Eight prion strains have PrPsc molecules with different conformations *Nature Medicine* **4**: 1157-1165

Sakaguchi S Katamine S Nishida N Moriuchi R Shigematsu K Sugimoto T *et al* (1996) Loss of cerebellar Purkinje cells in aged mice homozygous for a disrupted *PrP* gene *Nature* **380**: 528-531

Scott M Foster D Mirenda C Serban H Coufal F Wälchli M *et al* (1989) Transgenic mice expressing hamster prion protein produce species-specific Scrapie infectivity and amyloid plaques *Cell* **59**: 847-857

Scott M R Will R Ironsids J *et al* (1999) Compelling transgenic evidence for transmission of bovine spongiform encephalopathy Prions to humans *Proceedings of National Academy of Sciences USA* **96**: 15137-15142

Smith C (1998) Protein folding and protein folding diseases *Biological sciences* **10**: 32-37

Sparkes R Simon M Cohn V Fournier R Lem J Klisak I *et al* (1986) Assignment of the human and mouse prion protein genes to homologous chromosomes *Proceedings of the National Academy of Sciences USA* **83**: 7358-7362

Stahl N Baldwin M Hecker R Burlingame A Prusiner S (1990) Identification of Glycoinositol Phospholipid Linked and Truncated Forms of the Scrapie Prion Protein *Biochemistry* **29**: 8879-8884

Stahl N Baldwin M Hecker R Pan K Burlingame A Prusiner S (1992) Glycosylinositol Phospholipid Anchors of the Scrapie and Cellular Prion Proteins Conatin Sialic Acid *Biochemistry* **31**: 5043-5053

Stahl N Baldwin M Teplow D Hood L Gibson B Burlingame A Prusiner S (1993) Structural Studies of the Scrapie Protein Using Mass Spectrometry and Amino Acid Sequencing *Biochemistry* **32**: 1991-2002

Stahl N Borchelt D Hsiao K Prusiner S (1987) Scrapie Prion Protein Contains a Phoshatidylinositol Glycolipid *Cell* **51**: 229-240

*Stephen J D Bouzamondo E (2002) Fundamentals of Prion biology and diseases *Toxicology* **181-182**: 9-16

Stockel J Safar J Wallace A C *et al* (1998) Prion Protein selectively binds copper (II) ions *Biochem* **37**: 7185-7193

Supattapone S Bosque P Muramoto T Wille H Aagaard C Peretz D *et al* (1999) Prion Protein of 106 Residues Creates an Artificial Transmission Barrier for Prion Replication in Transgenic Mice *Cell* **96**: 869-878

Taraboulos A Jendroska K Serban D Yang S Dearmond S Prusiner S (1992) Synthesis and traffiking of prion proteins in cultured-cells *Molecular Biology of the Cell* **3**: 851-863

*Taraboulos A Scott M Semenov A Avraham S Laszlo L Prusiner S (1995) Cholesterol Depletion and Modification of COOH-Terminal Targeting Sequence of the Prion Protein Inhibit Formation of the Scrapie Isoform *The Journal of Cell Biology* **129**: 121-132

Tateishi J Kitamoto T (1993) Developments in diagnosis for prion diseases *British Medical Bulletin* **49**: 971-979

Telling G Scott M Mastrianni J Gabizon R Torchia M Cohen F *et al* (1995) Prion Propagation in Mice Expressing Human and Chimeric PrP Transgenes Implicates the Interaction of Cellular PrP with Another Protein *Cell* **83**: 79-90

Tobler I Gaus S Deboer T Achermann P Fischer M Rülicke T *et al* (1996) Altered circadian activity rhythms and sleep in mice devoid of prion protein *Nature* **380**: 639-642

Weissmann C (1994) The prion connection: Now in Yeast? *Science* **264**: 528-530

Weissmann C (1999) Molecular Genetics of Transmissible Spongiform Encephalopathies *The Journal of Biological Chemistry* **274**: 3-6

*Westaway D DeArmond S Cayetanocanlas J Groth D Foster D Yang S *et al* (1994) Degeneration of Skeletal-Muscle Peripheral-Nerves and the Central-Nervous-System in Transgenic Mice Overexpressing Wild-Type Prion Proteins *Cell* **76**: 117-129

Wickner R (1994) [URE3] as an Altered URE2 Protein: Evidence for a Prion Analog in *Saccharomyces cerevisiae Science* **264**: 566-569

Wickner R Edskes H Maddelein M Taylor K Moriyama H (1999) Prions of Yeast and Fungi *The Journal of Biological Chemistry* **274**: 555-558

Wille H Prusiner S (1999) Ultrastructural Studies on Scrapie Prion Protein Crystals Obtained from Reverse Micellar Solutions *Biophysical Journal* **76**: 1048-1062

Wong B S Liu T Li R Pan T Peterson R B Smith M A Gambetti P Perry G Manson J C Brown D R Sy M S (2001) Increased levels of oxidative stress markers detected in the brains of mice devoid of Prion proteins *Jour Neurochem* **76**: 565-572

*Zeidler M Stewart G Cousens S Estibeiro K Will R (1997) Codon 129 genotype and new variant CJD *The Lancet* **350**: 668

*-not consulted in original

9

PLANT DERIVED EDIBLE VACCINES

S.P. VYAS AND AMIT K. GOYAL

INTRODUCTION

During the last 50 years, many vaccines have been developed for prophylaxis or treatment of disease. Although, there are number of obstacles preventing the use of these biomedical discoveries for the benefit of mankind. Strict regulations for safety and efficacy that have led to increased cost of production of biomedicals preclude the prophylactic vaccination against some diseases on a global scale. In addition to the extraordinarily high price of new vaccines, the method of administration of these products through syringes and needles presents another burden in their worldwide use. The global use of vaccines requires that the vaccines be inexpensive and easily maintained and distributed. After considering various alternatives of fulfilling the criteria established for a global approach to immunization, it has become clear that our only choice is the production of vaccines or other materials of biomedical importance in plants. The advantages of plants are quite clear. Plants are very inexpensive to grow in mass; some can grow in almost any soil or climate in the world. In addition to their traditional role as a source of natural medicines, it is now possible to genetically engineer plants to produce pharmaceuticals. Various parts of plants (leaves, seeds, fruits, and chloroplasts) can also be used as vehicles for the biomedical products. Production of biomedicals in plants and their administration to humans and animals also provides an added margin of safety as compared to biologicals produced in animal tissues. Early concerns that plants might not be capable of producing biomedicals of animal origin, and that plant-specific glycosylation patterns might functionally alter plant-produced antibodies have proven unwarranted (Mor *et al.*, 1998; Streatfield, *et al.*, 2001; Carter & Langridge, 2002; *et al.* Twyman, *et al.*, 2005, Ma *et al.*, 2003, Arntzen *et al.*, 2005; Castle *et al.*, 2006).

Transgenic plants expressing antigens from pathogenic microorganisms offer many advantages as low-cost production systems and effective delivery systems for vaccines.

This new technology might contribute to global vaccine programs and might have a dramatic impact on health care in developing countries. A number of infectious diseases, including smallpox, anthrax and plague have recently raised concern for their possible use in actions of bio-terrorism. The economic and technical benefits offered by plant-derived vaccines propose these vaccines as ideal substitutes for traditional vaccines (Twyman, *et al.*, 2005; Fischer *et al.*, 2004; Koprowski, 2005; Streatfield, 2006).

The term "plant-derived edible vaccines" refers to antigens produced in edible parts of a plant, such as fruits, leaves, roots, or tubers. Vaccines expressed in nonedible plant tissues may also be orally administered after processing. The term "edible vaccine" may have created the misconception that individuals would be able to grow and administer their own vaccines without the supervision of pharmaceutical and health professionals. Ultimately, for practical applications, plant-expressed vaccines will require some form of processing to produce uniform doses for administration by oral and other routes. Also, not all plant-derived vaccine antigens are edible; they may be effective only when administered by the appropriate method (topically; intramuscularly, i.m.; intraperitoneally, i.p.; or subcutaneously, s.c.) (Mason & Arntzen, 1995; Mor *et al.*, 1998; Carter & Langridge, 2002; Walmsley Arntzen, 2003; Twyman, *et al.*, 2005).

ADVANTAGES OF PLANT DERIVED EDIBLE VACCINES

Producing pharmaceuticals in plants has many potential advantages, which have been reviewed extensively. The production of recombinant vaccines in plants may overcome some of the major difficulties encountered when using traditional or subunit vaccines in developing and developed countries as outlined in Table 1. In developing countries difficulties include vaccine affordability, the need for "cold chains" from the producer to the site of use of the vaccine and the dependence on injection. Safety is a key attraction, since the production of recombinant proteins in plants minimizes potential human or animal pathogen and toxin contamination, such as occurs when a recombinant protein is produced in microbial or animal cultures. The potential for cost reduction is also enormous. This is especially significant for oral vaccines, because relatively large amounts are required. Plants are potentially more economical than industrial fermentation or bioreactor facilities, and the amounts of protein produced by plants are comparable to those produced by industrial approaches. Scaling up production does not require huge investments up-front in setting up facilities, unlike the conventional systems. Often, technology already exists for harvesting and processing plant and plant products on a large scale. Purification and its associated costs may not be needed when the recombinant protein is produced in plant tissue that can be used as food or feeds (edible vaccine). The potential for oral administration and the convenient storage when the vaccines are produced in seeds and tubers could dramatically reduce the costs associated with syringes and needles and the requirement for cold storage. Elimination of needles in most vaccination programs will also cut back the risk associated with reuse of needles in some developing countries. Seed and tubers are convenient storage organs for relatively simple, cheap antigen storage. Plants offer enough flexibility that the recombinant proteins can be targeted to intracellular compartments where they are more stable. Alternatively, they can be expressed directly in those compartments or in organelles such as chloroplasts. From a biological perspective, the plant's protein-production system can perform post-transcriptional modifications, such as glycosylation, which are essential for the biological activity of some proteins. This is certainly an advantage over the use of microorganisms such as bacteria and fungal systems, in which problems of differences in codon usage, metabolic pathways, protein processing, and formation of inclusion bodies often arise (Streatfield *et al.*, 2001; Warzecha & Mason 2003; Streatfield & Howard, 2003, Fischer, *et al.*, 2004; Kirk & Mc Intosh, 2005; Raney, 2006).

Table 1. Technical and social benefits of plant derived vaccines

Benefits	*Characteristics*
Oral delivery	Cellulose and sugars, provides protection in the stomach and gradual release of the antigen in the gut.
Use as raw food or dry powder	Used as raw food or dried proteins administered in capsules as dry powder.
No need for cold chain	Stored and shipped at room temperature
Mucosal and serum immune response	To trigger the mucosal system (IgA), prevent pathogen entry at mucosal surfaces and to elicit serum and possibly cytotoxic response
Cost efficiency	Production cost will be reduced 100-1000 times compared with that of traditional vaccines
Ease of genetic manipulation	Procedures essentially rely on estabilished molecular and genetic manipulation protocols; these are already available in developing countries
Safer than conventional	Lack of concentration with mammalian pathogens
Idea to face bio-weapons	Safety and cost efficiency propose plant derived vaccines as an ideal tool to face bio-terrorism
Ideal for veterinary use	Cost affordable, ready for use as food additives

CHOICE OF PLANTS AND PLANT TISSUES

A key issue in producing transgenic plant vaccines is the selection of plant material, which can both express high levels of a chosen vaccine and be suitable for extensive storage and oral delivery. The choice of plant usually depends on its capacity for transformation or infectivity with genetically engineered viruses, and to date, scientists have often made this decision based on the system with which they are most competent. The choice of the plant species (and tissue in which the protein accumulates) is important and is usually determined through how the vaccine is to be applied in the future. This limitation is overcome in non-edible plants by vaccine antigen extraction and purification. Some plants, such as tobacco, potato, and tomato, are more amenable to tissue culture and transformation, and others, such as corn, soybean, and wheat, are relatively more recalcitrant to *in vitro* manipulation. The desired method of administering the recombinant protein also plays an important role. Various plant platforms have been demonstrated for production of recombinant proteins in plants, including leafy crops, cereals, legume seeds, oilseeds, fruits, vegetables, higher plant tissue and cell cultures, hydroponic systems, algae, and halobios (Tacket & Mason 1999; Walmsley & Arntzen, 2001; Mason *et al.*, 2002) (Table 2).

Table 2. Plants used for the development of edible plant vaccines

Plants categories	*Name of Plant*	*Advantage*	*Disadvantage*	*Examples*
Vegetable	Potato	Efficient transformation system, Clonally propagated, Low potential for outcrossing in field, Tuber-specific promoters available, Industrial tuber processing well established	Relatively low tuber protein content Unpalatable in raw form; cooking may cause denaturation	Cholera toxin, rota viruses, Colonization factor antigen, Epstein virus
	Tobacco	Facile, Efficient transformation system Abundant material for protein characterization	Toxic alkaloids incompatible w/oral delivery, Potential for outcrossing in field	Hepatitis B virus, Rabies virus, *E. coli*, Tetanus
	Tomato	Efficient transformation system, Fruit is edible raw Fruit-specific promoters available, Industrial green-culture and fruit processing well established	Low fruit protein content, Acidic fruit may be incompatible with some antigens No *in vitro* system to test fruit expression	Norwalk virus

Fruits	Papaya	Cultivated widely in developing countries, Inexpensive fruit, Eaten raw by infants and adults, Low potential for outcrossing in field	High cultivation space requirement, Very expensive in green house, Low gene expression	*Taenia solium*
Crops	Alfa-alfa	Relatively efficient transformation system, High protein content in leaves, Ideal system for animal vaccines	Potential for out crossing in field, Deep root system problematic for cleaning field	FMDV, Rota virus
	Maize	Production technology widely established, High protein content in seeds, Stable protein in stored seeds, Well-suited for animal vaccines	Low transformation Potential for ourcrossing in field for some	*E. Coli*

Vegetables

Some vegetables are used as a host for plant-derived vaccines because many of them are appetizing, free of toxicant, full of nutrients and fresh edible. Potato *(Solanum tuberosum)*, tomato *(Lycopersicon esculentum)*, and carrot *(Daucus carota subsp. Sativus)* have already been reported to successfully express vaccine candidates. Some of the most comprehensive antigen production and immunological studies, including human clinical trials, were carried out with transgenic potato-derived antigens such as LT-B, CT-B, HBsAg, and Norwalk virus capsid proteins. (Youm *et al.*, 2007; Mason *et al.*, 1998; Tacket *et al.*, 1998, Huang *et al.*, 2006) However, potato is not highly palatable and requires cooking, which may denature the antigens. Tomato is a new expression system developed in recent years. Tomato fruits are edible raw, and the transformation system, industrial greenhouse culture, and processing are well established, making this crop an attractive choice. Antigen genes encoding HBsAg, malarial antigen, HIVgag, and Rabies capsid proteins have been successfully transformed to tomato (Ruf *et al.*, 2001; Koprowski, 2005; Chowdhury & Bagarra, 2007). Carrot is one of the first plants for tissue culture, and transgenic technology. It was reported that high level of recombinant protein expressions was observed in proplastids of cultured carrot cells. Oral delivery of therapeutic proteins via edible carrots preserved the structural integrity of their target proteins, as no cooking is needed. In addition to these plants as noted above, more and more vegetable transgenic systems, such as lettuce *(Lactuca sativa)*, celery cabbage *(Brassica rapa* var. *pekinensis)*, cauliflower *(B. oleracea* var. *botrytis)* are now under the process of establishment, but low expression level is still a problem in these plant expression systems (Tacket & Mason, 1999; Walmsley & Arntzen, 2001; Koprowski, 2005).

Fruits

When the idea of edible vaccines was first articulated, people envisioned that a fruit like banana *(Musa acuminate)*, which is highly palatable raw, would be the ideal system for edible vaccine production. However, inefficient transformation, inadequate information on gene expression, fruit-specific promoters, and the difficulty and expense of cultivation in the greenhouse have hampered pharmaceutical production in this plant (Mason *et al.*, 2006; Trivedi *et al.*, 2004). Papaya *(Carica papaya)* is a widespread tropical and semi-tropical fresh edible fruit. Embryogenic callus of papaya was genetically transformed with different constructions including the reporter GUS gene by biobalistics. Papaya is considered as another ideal plant species for vaccine production. Sciutto *et al.* (2002) reported that inserted transgenes were identified in papaya, and the expression of a novel synthetic vaccine of *Taenia solium* and *T. crassiceps.* This strategy will allow vaccine production on a large scale to evaluate possibilities of plant systemic and oral immunization in the near future (Carter & Langridge 2002).

Crops

Alfalfa *(Medicago sativa)* is a perennial crop with strong regenerative capacity and propagation, which continuously produces large clonal populations by stem cuttings in a limited period of time for up to 5 years after establishment. Alfalfa is a palatable crop with high leaf protein content and high annual yields of green leaf tissue. In this crop, pharmaceuticals could be administered in fresh or dried leaf or as leaf extract. Several proteins have been produced in alfalfa, including foot-and-mouth disease virus (FMDV) antigens. Maize *(Zea mays)* was investigated as a commercial platform by several companies (e.g. ProdiGene Inc.) for producing a range of pharmaceutical and technical proteins, such as recombinant antibodies, vaccine candidates, and enzymes. Similar to maize, some proteins have been expressed in rice *(Oryza sativa L.)* at very high levels by using constitutive and endosperm-specific promoters. Cereal crops are also ideal plant species because the endosperms of such crops are full of soluble proteins and can be easily separated from the whole plant, which enhances the antigen concentration and reduces the oral doses. Seed legumes offer an attractive option for edible vaccines, because production of antigens in seeds makes for convenient administration of edible vaccines and simplified preparation of antibodies. These crops are particularly suitable for production of edible vaccines for humans and livestock, because many are natural components of food and livestock feeds (Streafield & Howard, 2003; Dus *et al.*, 2002; Hood *et al.*, 2002; Nicholson *et al.*, 2003).

Algae and other

Although the biotechnological processes based on transgenic microalgae are still in their infancy, researchers are considering the potential of microalgae as green cell-factories to produce value-added metabolites and heterologous proteins for pharmaceutical

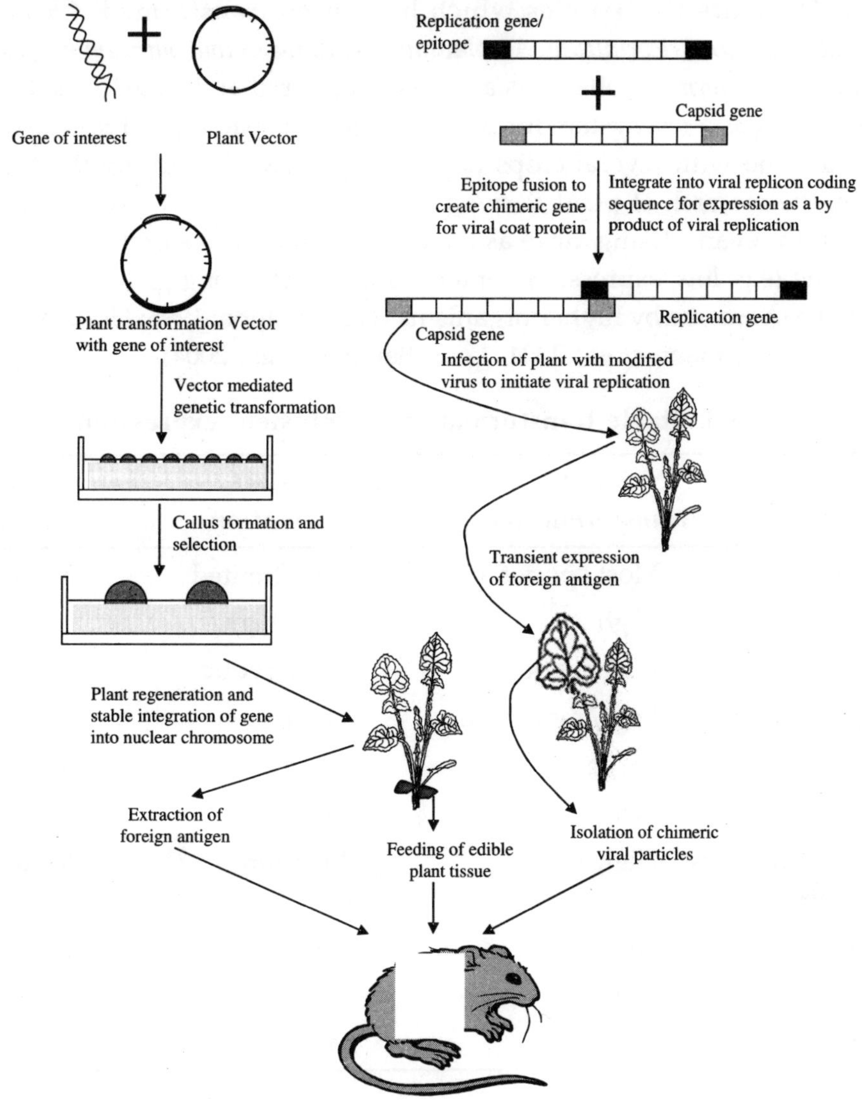

Fig. 1: Strategies for the production of candidate vaccine antigens in plant tissues. Gene encoding antigen from pathogenic organisms (virus, bacteria or parasites) that have been characterized and for which antibodies are available, can be handled in two ways. In one case, the entire structural gene is inserted into a plant transformation vector between 5′ and 3′ regulatory elements; this will allow transcription and accumulation of coding sequence in plant. In second case, epitope within the antigen are identified, DNA fragment encoding these can be used to construct genes by fusion with a coat protein gene from plant virus e.g. (TMV or CMV). The recombinant virus is then used to infect stabilized plants. Resultant edible plant vaccines are utilized for further immunological studies.

application. There are five species which have been transformed: *Chlamydomonas reinhartii, Phaeodactylum tricornutum, Amphidinium carterae, Symbiodinium microadriaticum,* and *Cylindrotheca fusiformis*. Tobacco is also easily cultured and transformed plants, and thus has been used as a test system for many antigens. Tobacco produces large volumes of green tissue, and with several crops harvested per year by cutting the foliage. As a consequence, the research of plant derived recombinant proteins have been developed from the earth to oceans. Using algae as a bioreactor to produce oral vaccines can solve many problems (e.g. low expression content, toxicity and long-growth period), which are hardly to be overcome by higher organisms (Featherstone, 1996; Hood & Jilkat, 1999; Sun *et al.*, 2003; Fuhrmann *et al.*, 2004; Leon-Baiiares *et al.*, 2004).

Table 3. Stable transformation vs transient expression.

Features	*Stable transformation*	*Chloroplast transformation*	*Transient expression*
Applicability	Most species	Very limited	Most species
Development time	High	Very high	Low
Expression host	Eukaryotic	Prokaryotic	Inoculation
Protein expression	Low/moderate	High	High
Scale-up potential	Large	Large	Moderate
Promoters	Yes	Yes	Yes
Transfection / transformation efficiency	High	Medium	Moderate
Genomic Stability	High	Low	Moderate
Cost	High	High	Low
Environmental factors	Low	High	Low/moderate

TECHNIQUES FOR PRODUCTION OF PLANT-DERIVED VACCINES

The expression of a gene in a plant cell can be transient or stable. In transient expression the foreign gene is expressed only for a few days after being introduced into the cell and does not get integrated in the genome. In stable expression the foreign gene is integrated in the genome and will be passed to the progeny (Sharma *et al.*, 2005). In transient expression, DNA encoding the antigen of interest is introduced into the plant cell, where it is recognized by the transcription machinery and expressed. On the other hand, stable transformation involves introduction of a foreign gene into the plant cell and its stable integration into the plant genome. The choice for an expression system for recombinant protein production in plants depends largely on the aim and the scale of the project (Hansen & Wright, 1999; Sharma *et al.*, 2005) (Fig 1). In that way, transient expression may be useful for research, for validating a technology or for small-scale production of recombinant proteins (Table 3). For large scale production, however, a transgenic plant system may be more convenient, but will require longer periods for development and optimization of tissue culture, selection and transformation conditions (Horn *et al.*, 2004; Twyman *et al.*, 2003).

Stable expression of a foreign gene involves the production of transgenic plants or transgenic cell cultures. Stable transformation offers multigenerational stable expression that is mainly depending on the species (Hansen and Wright, 1999; Sharma *et al.*, 2005). Most transformation methods involve nuclear transformation, although the transformation of chloroplasts is also possible for some species. Expression of recombinant proteins by this method usually results in moderate to low expression, but higher levels of expression have been attained by using a number of regulatory sequence elements, such as strong promoters, optimized codon usage for plant expression, translation-enhancing leader sequences at the 5' end, polyadenylation signals at the 3' end, and microsomal retention sequences (Table 3). However, stable transformation requires efficient transformation techniques and meticulous selection and breeding of the high-expressing transgenic lines (Lorence & Verpoorte, 2004; Horn *et al.*, 2004; Sharma *et al.*, 2005).

The production of a vaccine in plants depends upon the availability of a DNA sequence coding for a protective antigen and on the construction of an expression "cassette" suitable for plant transformation. Two processes of transient expression of genes are being routinely used today in plants. They are *Agrobacterium* and viral vector-based transient amplification. In both cases, the expression level relies on delivery of transgenes by a vector (bacterium or virus). Generally transformation methods can be classified in "indirect transformation", where gene transfer is mediated by bacterium or virus, or "direct transformation", where different techniques, such as particle bombardment or electroporation of protoplasts (Sharma *et al.*, 2005) are used to physically introduce "naked" DNA into the plant cell. *Agrobacterium* mediated transformation may be the choice for a large number of species for which well established protocols are available (Hansen and Wright, 1999). For some species, however, particularly important crops like maize and soybean, transformation through particle bombardment is usually employed (Bidney *et al.*, 1992). An alternative approach for high-level gene expression is the transformation of chloroplasts., because the chloroplast genome is present in high copy numbers per chloroplast, which in turn are present in large number per cell.

Transplastomic plants potentially can express very high levels of foreign protein (Daniell *et al.*, 2005). Viral vectors are usually more appropriate for expressing products in leaf tissue of high biomass producing plants, such as tobacco, ultimately for extraction and purification.

Bacterial vector

Plants do not contain any naturally occurring plasmid DNA molecules. However, a bacterial plasmid, the tumour inducing (Ti) plasmid of the soil microorganism *Agrobacterium* tumefaciens, has been used extensively to introduce genes into plant cells. *Agrobacterium*-based plasmid vectors allow the transformation of a wide range of plant species by capitalizing on a natural bacterial system to introduce DNA into the nuclear genome of plants (Fig 2). A. tumefaciens is a soil bacterium. A. tumefaciens is responsible for crown gall disease in a variety of dicotyledonous plants, such as tomato, tobacco, potato, peas, beans etc. A wound on the stem of the plant allows the bacteria to invade and cause a cancerous proliferation of the stem tissue. A plasmid carried within the bacterium is responsible for its ability to cause crown gall disease. This tumour inducing (Ti) plasmid is large (~200 kbp) and carries a number of genes that are required for the infection process (Binns & Thomashaw, 1988; Hooykaas & Shilperoort, 1992; Cheng & Fry, 2000; Escobar & Dandekar, 2003). After infection, part of the Ti plasmid, called the T-DNA, becomes integrated into the plant genome at an apparently random position through non-homologous recombination. T-DNA, approximately 23 kbp in size, contains not only the genes responsible for the cancerous properties of the transformed cells (e.g. those controlling the production of the plant hòrmones auxin and cytokinin to stimulate cell division and growth) but also those responsible for the synthesis of opines, which are amino acid derivatives. Within the Ti plasmid, the T-DNA is flaked by two 25 bp imperfect direct repeats, known as the left and right border sequences. These sequences are necessary for the integration of T-DNA into the plant genome, and any DNA sequence located between them will be integrated into the plant genome. Once integrated into the plant, the T-DNA is stably maintained and is passed on to daughter cells. In addition to the T-DNA, the Ti plasmid also carries the genes needed for opine utilization, a series of approximately 35 genes required for virulence (vir) and an *Agrobacterium* origin of replication (ori). Since the T-DNA fragment becomes integrated into the plant genome, the Ti plasmid offers an excellent mechanism for the introduction of novel or manipulated genes into the plants (Sheng and Citovsky, 1996; Hellens and Mullineaux, 2000; Opabode, 2006) (Fig 2). *Agrobacterium* rhizogenes has also been used as a vector for plant transformation. *Agrobacterium* rhizogenes stimulates root hair formation in many dicotyledonous plant species, carries a large extrachromosomal element, called Ri (root-inducing) plasmid which functions in a manner analogous to the Ti plasmid of A. tumefaciens. Transformation using A. rhizogenes has developed analogously to that of

A. tumefaciens and has been successfully utilized to transform, for example, alfalfa, *Solanum nigrum L* and poplar.

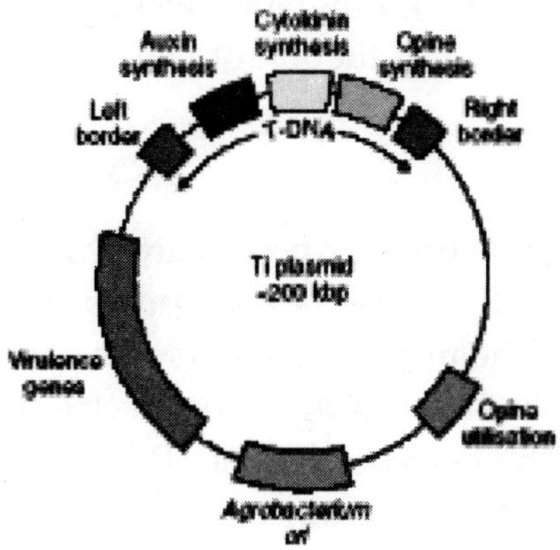

Fig. 2: Construction of Ti Plasmids

Two basic strategies have been devised for inserting new sequences into the T-DNA region (Zupan, 1995; Hamilton, 1997; Klee, 2000).

- Co-integration vectors. The insertion of new DNA into a Ti plasmid results from the recombination of a small vector plasmid, for example an *E. coli* vector, and a Ti plasmid harboured in *Agrobacterium*. The recombination takes place through a homologous region present in both of the plasmids.
- Binary vectors. The T-DNA does not need to be physically associated with the virulence genes in order to become integrated into the plant genome. Therefore the disarmed T-DNA can be cloned into a small plasmid and manipulated appropriately. The plasmid can then be transformed into *Agrobacterium* that possesses a Ti plasmid that still contains all of the virulence genes, but from which the T-DNA has been removed. The virulence proteins provide all the function required to integrate the T-DNA into the plant genome (Bevan, 1988; Hellens *et al.*, 2000).

Plant transformation using *Agrobacterium* is most commonly performed using leaf tissue. Small discs are taken from the leaf, using a paper punch, and incubated with the recombinant *Agrobacterium*. Bacterial invasion occurs at the 'wounded' edges of the disk. There are also alternative methods like protoplast transformation and biolistics gene gun approaches for introducing DNA via *Agrobacterium* into plant cells. In the protoplast transformation genetic material can be transferred to protoplastic state of cells through chemical, electroporetic or liposomal means. The DNA inserted into cells in this way is not capable of independent replication. However, it may become randomly integrated into one of the plant chromosomes through a process of non-homologous recombination.

On the other hand, in biolistics approach genetic material is transferred to target cells through the firing of DNA coated small metal particles, usually tungsten or gold. Particle acceleration may be achieved using gunpowder, electrical discharges or pressurized gas. This method is applicable to the introduction of DNA into a wide range of cells (Bidney *et al.*, 1992).

Direct gene transfer

Several so-called direct gene transfer procedures have been developed to transform plants and plant tissues without the use of an *Agrobacterium* intermediate. In the direct transformation of protoplasts the uptake of exogenous genetic material into a protoplast may be enhanced by use of a chemical agent or electric field. The exogenous material may then be integrated into the nuclear genome (Paszkowski, *et al.*, 1984; and Potrykus, *et al.*, 1985; Lorz *et al.*, 1985). Introduction of DNA into protoplasts of N. tabacum is effected by treatment of the protoplasts with an electric pulse in the presence of the appropriate DNA in a process called electroporation (Fromm, 1987; Shillito, 1987). DNA viruses have been used as gene vectors. A cauliflower mosaic virus carrying a modified bacterial methotrexate-resistance gene is used to infect a plant. The foreign gene is systematically spread in the plant. The advantages of this system are the ease of infection, systematic spread within the plant, and multiple copies of the gene per cell. Liposome fusion has also been shown to be a method for transformation of plant cells. Protoplasts are brought together with liposomes carrying the desired gene. As membranes merge, the foreign gene is transferred to the protoplast. Polyethylene glycol (PEG) mediated transformation has been carried out in *N. tabacum* a dicot, and Lolium multiflorum, a monocot (Negrutiu, *et al.*, 1987).

Alternatively, exogenous DNA can be introduced into cells or protoplasts by microinjection. A solution of plasmid DNA is injected directly into the cell with a finely pulled glass needle. A more recently developed procedure for direct gene transfer involves bombardment of cells by microprojectiles carrying DNA. In this procedure called particle acceleration, tungsten or gold particles coated with the exogenous DNA are accelerated toward the target cells. At least transient expression has been achieved in onion. This procedure has been utilized to introduce DNA into Black Mexican sweet corn cells in suspension culture and maize immature embryos and also into soybean protoplasts (Reich, *et al.*, 1986, Klein *et al.*, 1987; McCabe *et al.*, 1988; Stomp *et al.*; 1991).

Viral Vectors

During their systemic infection of host plants, viruses often produce a large number of genome copies, as well as high titers of some of their encoded proteins. They do not integrate into the host genome and therefore can be considered as natural episomal vectors. The possibility of manipulating the viral genome *in vitro* allowed the exploitation of viruses as vectors for expressing foreign genes, taking advantage of the fact that a huge amplification of the inserted gene can be achieved (Scholthof *et al.*, 1996; 2002). Potential features that make virus vectors attractive for protein production include: easy genetic manipulation based on small viral genome sizes and the availability of cDNA clones, potential knowledge of regulatory elements involved in transcription and translation allowing optimization of expression levels of recombinant product, short cycle times of approximately 2-3 weeks in plants, wide host ranges which may allow the infection of various species using the same vector construct and high yields of recombinant product based on the highly efficient transcription from some viral promoters. Plants have many natural viral pathogens (Baulcombe, 1999; Yusibov, 1999). Plant virus-based vectors have been developed by fusion of proteins or protein domains to the plant viral coat protein (CP) and presentation on the surface of the assembled virus particle or independent expression of foreign genes inserted into the viral genome. Fusion of peptides to the plant viral CP has been exploited to develop subunit vaccines of human or animal pathogens. Plant virus CP that was fused to immunogenic protein domains of the pathogens was assembled in infected plant cells into virus particles presenting the foreign epitope on the surface (Porta, *et al.*, 1996; Yusibov, *et al.*, 1997; Porta *et al.*, 2003).

Plant virus expression system have single-stranded DNA (geminiviruses), double-stranded non-integrating DNA (pararetroviruses) and/ or plus-sense RNA viruses. Within the past three decades, several genetic manipulation techniques have been developed to overcome the limitations in the use of viral vectors. Reverse transcription of viral RNA into cDNA allowed their insertion into plasmid vectors for molecular cloning. In vitro transcription systems facilitated the synthesis of infectious RNA transcripts from full-length cDNA clones. In addition, the ability to clone viral genes into *Agrobacterium* tumefaciens enabled this bacterium to be used as a vector to deliver viral genomes into the plant cell's nucleus, allowing the virus to be transcribed, to replicate and move throughout the plant. Plant virus expression vector can develop by inserting gene of interest into the fully functioned wild type virus vector. These vectors are essentially fully functional viruses retained infectivity, relatively stable, and ability to move systemically within their host, and produced infectious virus particles. Recently, There is some modification made on the full functioned viral approaches by substituting existing viral genes (non-functional genomic sequence) with foreign genes. The extreme evolution of this concept lead to deconstructed viral vectors missing several components of the original virus and usually delivered to the plant by independent constructs. Two full virus vectors, the potato virus X (PVX) and the cowpea mosaic virus (CMV) and two deconstructed vectors, based upon bean yellow dwarf virus (YDV) and tobacco mosaic virus (TMV) have been used to produce protein antigen in plants.

Compared to other transient gene expression methods, the expression level is maintained for longer periods, due to the elevated number of transcripts as the virus replicates and spreads throughout the plant (Chapman *et al.*, 1992; Twyman *et al.*, 2005). Viral vectors also present several limitations and disadvantages. As a transient expression system, viral vectors are not integrated into the genome and must be repeatedly inoculated so that infected plants can express the desired protein. That implies extra work for inoculation and a need for a supply of plants suitable for being efficiently infected (Awram *et al.*, 2002; Scholthof *et al.*, 2002). Some viral vectors, like the PVX vector, have a limited host range. The choice for the host is an important aspect, as many Solanaceae species, commonly used as host for PVX and TMV, are rich in alkaloids that may hamper the use of the infected plants as raw material for feeding as an oral vaccine (Greene and Allison, 1994; Awram *et al.*, 2002).

Chloroplast Transformation

Chloroplast genetic engineering offers several unique advantages over nuclear transformation, which includes very high expression levels of the recombinant proteins/ antigen. The chloroplast, like the mitochondrion, possesses its own genome. The chloroplast genome is a single circular double-stranded DNA molecule, which, in the model dicotyledonous plant Arabidopsis thaliana, is composed of approximately 155 kbp of DNA containing 87 protein-coding genes. Chloroplast genes, like those of bacteria, are generally arranged into operons. This allows for the insertion of multiple foreign genes into the chloroplast that can be expressed from the same polycistronic transcript. Chloroplasts can also process eukaryotic proteins, including interferons, vaccine antigens and human somatotropin with correct folding of subunits and formation of disul?de bridges. Chaperones present in chloroplasts facilitate the correct folding and assembly of complex multi-subunit proteins. Consequently, the expression of genes inserted into the chloroplast can greatly exceed that from a gene in the nucleus. The foreign gene that is being introduced into the chloroplast must be flanked by sequences homologous to the chloroplast genome itself. Chloroplast genes are themselves transcribed by chloroplast-specific promoters and use chloroplast-specific termination signals. Therefore, foreign genes to be expressed in the chloroplast must be placed between suitable promoter and terminator sequences. Additionally, a selectable marker is used to identify transformed cells (Daniell *et al.*, 2005; Streatfield, 2006).

Furthermore, several challenges in nuclear genetic engineering could be eliminated including position effect due to site-speci?c integration of transgenes by homologous recombination. The problem of gene silencing both at transcriptional and translational levels has not been observed in transgenic chloroplasts in spite of high levels of translation.

Factors affecting the gene expression

The quantity of plant tissue that may constitute a vaccine dose must be of practical size both for field production and for consumption. Since the demonstration that low levels of a recombinant hepatitis B surface antigen (HBsAg) could be produced in potato and that the antigen assembled into spherical particles similar to those seen in infected human serum, efforts have been directed to increase antigen expression and accumulation to a reasonable level. One major limitation for plant made recombinant proteins is the low expression level. Most target proteins expressed in plants are of non-plant origin and there are no ways to anticipate the expression level of a given recombinant protein expressed in plants. However, a number of factors may modulate gene expression in plants (Table 4). They include: codon usage; promoter, leader and polyadenylation signals; DNA sequences that target antigen accumulation to a specific tissue or cell compartment (Chargelegue *et al.*, 2000; Clark and Johnson, 2001).

Table 4. Recombinant DNA assembly of edible plant vaccine

Therapeutic proteins	*Gene*	*Site of integration*	*Promoter*
Cholera toxin	CtxB	trnI/trnA	Prrn
Tetanus toxin	TetC (bacterial & synthetic)	trnV/rps12/7	Prrn
Anthrax protective antigen	Pag	trnI/trnA	Prrn
Plague vaccine	CaF1, LcrV	trnI/trnA	Prrn
Canine Parvovirus (CPV)	CTB-2L21	trnI/trnA	Prrn

Expression of a foreign gene in plants is controlled at different levels. At the transcriptional level, the promoter sequence can greatly influence mRNA levels. In transgenic plants and in transient gene expression systems (other than viral vectors-based expression), high levels of gene expression are attempted by using gene constructs containing a constitutive promoter, such as the CaMV 35S promoter, including its variations as a duplicate form and with extra protein expression enhancing signals (Sharma *et al.*, 2005). Promoters that work well in dicot plants does not necessarily work equally well in monocot plants and vice versa. Some of the promoters used for transgene expression in corn include the maize ubiquitin promoter, the cauliflower mosaic virus 35S RNA (CaMV 35S) promoter, and the maize seed-specific 27kDa g zein promoter. Both maize ubiqutin promoter and CaMV 35S promoter (in its tandem duplicated form) are considered strong constitutive promoters for transgene expression in plants. Both constitutive and seed-specific promoters have been used successfully in production of non-plant proteins. It is also important to avoid multicopy integration of a transgene into a plant genome during transformation. The higher copy number of transgene integration can result in lower gene expression, due to transgene silencing. In many plant, intron sequences also affect gene expression by stabilizing the mRNA transcript and allowing its effective translation. In monocot plants such as corn, intron sequences placed between the promoter and a transgene may increase its level of expression. The mRNA itself may contain sequences that are known to lead to increased instability (Gutierrez *et al.*, 1999). Appropriate mRNA 3'-end regions can strongly influence the level of transgene expression in plant cells. Premature polyadenylation within transgenes also results in lower transgenic mRNA levels. On the other hand, the presence of adenylate/uridylate-rich elements (AUUUA) in transcripts indicates that they are selectively targeted for rapid decay. Reporter transcripts containing 11 repeats of the AUUUA motif in their 3' untranslated regions are degraded more rapidly in stably transform tobacco cells and accumulate to a lower level in transgenic tobacco plants than those of the control construct. The codon usage of the targeted gene sequence may also influence mRNA level and gene expression, as frequently observed in bacteria, but also in plants (Marillonnet *et al.*, 2004).

Following mRNA transcription, translation will take place and the polypeptide will be directed to the endoplasmic reticulum (ER) or the cytoplasm (Conrad and Fiedler, 1998). The nature of the recombinant protein may imply that the protein should be processed and/or secreted to other cellular compartments (Vitale and Denecke, 1999). In the ER and Golgi network glycoproteins may be processed, disulfide bonds can be formed, and proteins can be sent to different cellular compartments. The subcellular location of a recombinant protein can greatly influence its accumulation (Bosch *et al.*, 1994; Chikwamba *et al.*, 2003; Conrad and Fiedler, 1998; Drakakaki *et al.*, 2006). For several proteins, the addition of motifs, such as H/KDEL has been demonstrated to increase their stability (Schouten *et al.*, 1996). This has been observed in proteins to be retained in the ER, but also for cytosolic proteins (Schouten *et al.*, 1997). The 5'-end untranslated region can also serve as a site for regulatory sequences that actively control the rate of translational initiation. Most of those examined and used in plants have come from viral mRNAs, e. g., the O leader (68 bp) of tobacco mosaic virus, AMV leader (36 bp) of alfalfa mosaic virus RNA 4, and TEV leader (144 bp) of tobacco etch virus. The translation-enhancing effect of the O leader on transgenes seemed more profound in dicots than in monocots. Altering the composition of the heterologous cDNA to suit the plant's pattern can increase the rate of translation. When the codon composition of a bacterial gene is modified to suit the preferred codon usage in tobacco, the expression level of the transgene is greatly enhanced. This approach has been applied to many plant species, such as tomato, potato, rice, cotton, eggplant, and maize (Potrykus, 2001).

Protein stability is largely depends on its proper folding. A heterologous protein may be unstable due to improper folding, resulting from environmental conditions or lack of (proper interaction with) chaperones or proper post-translational modification. In any case, proteins recognized as defectives will be ubiquitinated and targeted for degradation by the proteosome. In that case, heterologous protein production will result in a very low yield or even remain undetectable. Higher levels of foreign protein can be obtained if the newly synthesized protein is targeted to the extracellular compartment. Alternatively, the protein could be sequestered into a compartment or organelle that may allow higher levels of protein accumulation. Signal peptides have been used to influence the levels of protein accumulation and its spatial distribution within transgenic plant tissue. It is also observed that targeting proteins to the intercellular space beneath the cell wall increase expression levels, their retention in the lumen of the endoplasmic reticulum could result in 10-to 100-fold higher yields of recombinant antibody single chain Fv (scFv) in potato tubers, as well as tobacco leaves and seeds (Schouten *et al.*, 1997; Vierstra, 2003; Doran, 2006).

The site of gene integration into the genome also influences epitope and transgene accumulation in plants. *Agrobacterium* tumefaciens infection is most frequently used to achieve permanent integration into the nuclear DNA, where integration occurs at random chromosomal sites. A second promising approach is based on the integration of the gene or epitope into the circular chloroplast DNA (cpDNA) that is present in multiple copies within defined plant cells. In this case transformation is usually achieved through the use of the "particle gun" and results in site-specific integration. Both nuclear and chloroplast genomes accept large and multiple gene inserts.

IMMUNOLOGICAL CONSEQUENCES AGAINST PLANT BASED ORAL IMMUNIZATION

The development of plant-produced vaccines has become increasingly interesting in recent years (Mason and Arntzen, 1995; Arntzen, 1997; Arntzen, 1998). Modern transformation technologies have made a number of plant species available for recombinant protein expression, including soybean, rice, corn, potato, tomato and banana. Large-scale production of transgenic plants at low cost has been demonstrated for tobacco, corn and soybean using established agricultural techniques (Hood and Jilka, 1999). The advantage, besides the lower cost of plant-based production is that such antigens derived from plants have a reduced risk of being contaminated with human or animal pathogens (Awram *et al.*, 2002; Canizares *et al.*, 2005; Streatfield *et al.*, 2001). Another possibility is using plants for oral delivery of the expressed antigens (Haq *et al.*, 1995; Mason *et al.*, 2002; Walmsley and Arntzen, 2003). This concept of oral vaccination has important implications for the induction of mucosal immune response, which is the main barrier against pathogens present in food or water (Cripps *et al.*, 2001; Streatfield, 2006).

Most human pathogens initiate their infectious processes at mucosal surfaces. In human body, mucosal surfaces of the intestinal, respiratory and urogenital tracts have a combined surface area of at least 400 square feet. To guard against infection, these surfaces are protected by a complex system of defenses that comprise the mucosal immune system. It has been reported that lymphoid tissues associated with mucosal membranes contain more immunocytes, including B and T lymphocytes, and plasma cells than any other tissue in the body, i.e. 1m of human intestine contain 1010 immunocytes (Brandtzaeg & Baklien 1976). The mucosal sites, which comprise the common mucosal immune system, include Gut Associated Lymphoid Tissues (GALT), Broncus Associated Lymphoid Tissues (BALT), genital tract, salivary glands, ocular tissues, and mammary glands. These sites are in intimate and constant contact with the external environment, contain mucocilliary epithelium, possess secretory component and/or S-IgA in the epithelium and lamina propria, and contain organized lymphoid follicles in subepithelial regions. These tissues participate in circulation of antigen reactive IgA B lymphocytes and specifically sensitized T cells to other distant sites, after in vivo stimulation in the GALT or BALT (McGhee *et al.*, 1992; Shalaby, 1995; Ogra, 2003) (Fig. 3). This network of mucosal sites responding against a single antigen after exposure to any mucosal site is known as common mucosal immune system. Transgenic plants that express therapeutic proteins are ideal vehicles to produce and orally deliver protective antigens, as the rigid walls of the plant cell protect the antigen from the acidic environment of the stomach. Plant-derived subunit vaccines lack contamination with animal pathogens, are inexpensive to produce, heat stable and may be engineered to contain multiple antigens. The induction of the immune response from plant based edible vaccines depends upon the antigen interaction with mucosal tissues and its ability to cross the epithelial barrier and to be captured and presented by Gut associated lymphoid tissues (GALT). GALT consists of the organized lympoid follicles, Peyer's patches, appendix, tonsils and mesentric lymph nodes. From the mouth to the anus, lymphoid follicles are covered by specialized antigen sampling cells, termed microfold (M) cell. The follicles contain mainly B-lymphocytes, but also have T helper cells. The most prominent aggregates of lymphoid follicles are present in small intestine. These follicles are termed Peyer's patches and contain specialized B and T cell areas. The appendix also contains abundant lymphoid follicles and is capable of uptaking and processing antigens through M cells as done by Peyer's patches. The tonsils and adenoids are also part of GALT. Peyer's patches (PP) are considered as unique inductive sites of the immune response initiated in the intestine. Indeed, the dome epithelium of PP comprise of a phagocytic epithelial cell type, i.e. the M cell, specialized in attachment of particulate antigens including microorganisms and latex beads as well as macromolecules and translocation of these antigens for delivery to APC present in the underlying lymphoid compartment. In addition, M cells lack apical brush border microvilli and mucus layer, which represent physical barriers preventing antigen attachment to villus epithelium (Fig. 3 and Fig. 4).

Furthermore, there is ample experimental evidence in various plant based edible vaccines systems that initiation of the immune response requires antigen presentation to T cells in an organized lymphoid compartment, such as the PP rather than in a diffuse lymphoid compartment such as the lamina propria of the intestinal villi. To induce gastrointestinal immune responses, antigen must be transported from the intestinal mucosa into Peyer's patches. Antigens sampling cells or M cells are capable of transporting macromolecules, particles or microorganisms directly into intestinal PP (Neutra *et al.*, 1992). M cells are different from intestinal absorptive cells as they lack brush border. In addition M cells contain a large number of endocytic vesicles to uptake and transport lumen contents across the epithelial layer. A unique feature of M cells is the presence of an intraepithelial pocket containing lymphocytes and a few macrophages, which can interact with the transported antigen or microorganisms (Fig 4). Subsequent to the transport of soluble antigens by M cells, dendritic cells are probably one of the major cells involved in initiating cellular immune responses in Peyer's patches (Neutra *et al.*, 1988; Ogra, 2003).

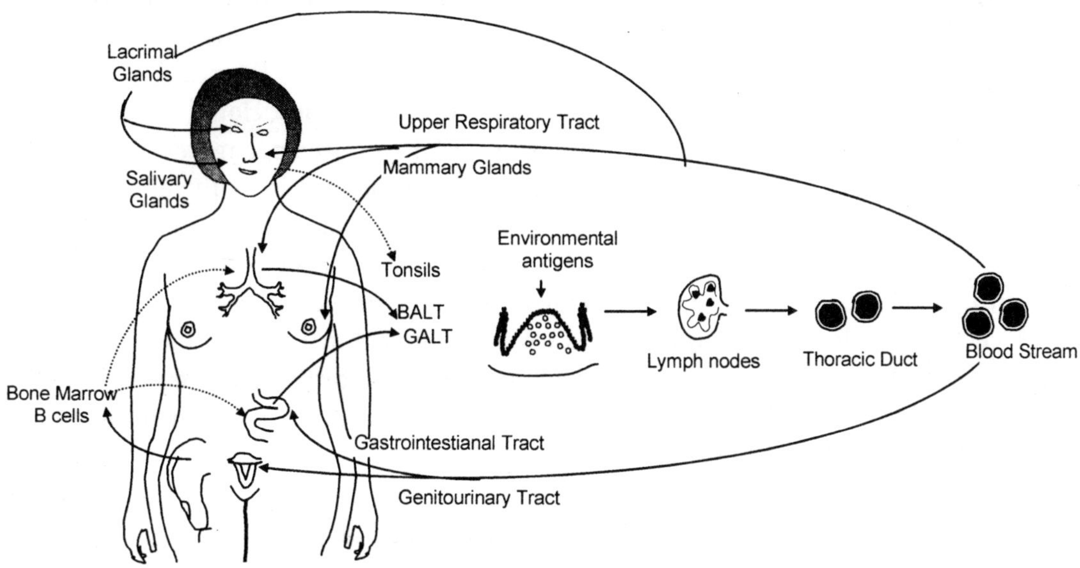

Fig. 3: Sites of Common Mucosal Immune Systems

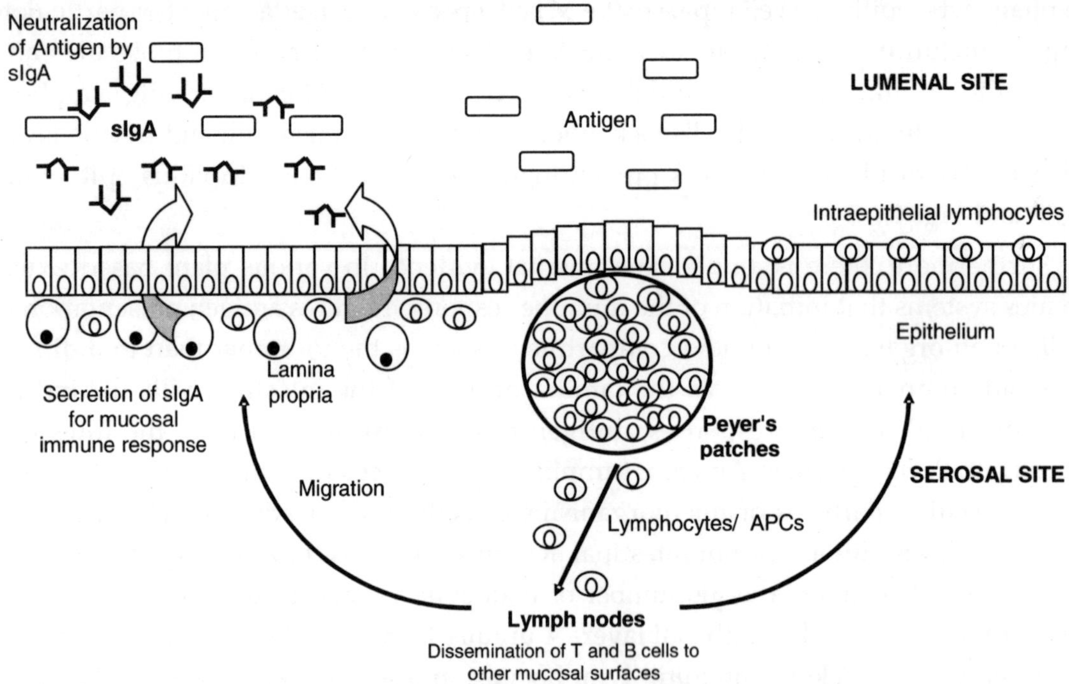

Fig. 4: Induction of Mucosal S-IgA Response

Humoral immune response is mainly characterized by production of secretory IgA (sIgA) by B-lymphocytes. However, IgE responses are also associated with transmucosal exposure to antigens, which are allergenic. B cells are localized in a central region of Peyer's patches beneath the layer of M cells. Follicular dendritic cells present antigen after trapping to B cells localized around follicular dendritic cells. Primed T helper cells, which migrate through this region, can interact with B cells associated with follicular dendritic cells. Differentiation and maturation of IgA responses occurs after stimulated B cells leave the Peyer's patches and their activation is pivotal in both regulatory and effector activities of the immune system. Most of the T cells are localized around B cell follicles in Peyer's patches (Parrott, 1976). T cells are also just beneath the M cell layer, where antigen presentation by macrophages and dendritic cells occur. Activation of CD4+ T cells and cytotoxic T lymphocytes takes place within Peyer's patches followed by migration of these cells to other sites like lamina propria or intestinal epithelium (Fig 4). The activation of T helper cells and secretion of Th2 like cytokines contributes to the development of mucosal IgA response.

Oral immunization with heat-labile enterotoxin (LTB) or Cholera toxin B (CTB) results in the production of antigen-specific sIgA that prevents the binding of LTB and or CTB to the epithelial cells, and hence interfering with the toxic effect of LT and CT.

Mason *et al.* 1998 demonstrated expression and correct assembly of LTB by potato tissues and tested the plant-based vaccine in animal feed trials. The potato-immunized mice produced both serum IgG and gut mucosal IgA at levels comparable to those in mice given bacteria-expressed LTB by gastric tube. Foot and Mouth Disease Virus (FMDV) structural protein VP1, which carries critical epitopes responsible for the induction of protective neutralizing antibodies, has been also successfully expressed as an immunogenic antigen in Arabidopsis thaliana alfalfa and potato. In addition, fast engineering and short cycle times for recombinant protein production have great advantages in vaccine development. Various short immunogenic peptides of human and animal viruses have been also expressed through plant vectors (Usha *et al.*, 1993; Porta, *et al.*, 1996; Porta, *et al.*, 2003; Yusibov, *et al.*, 1997). Upon infection of host plants with recombinant virus, chimeric CP is expressed and assembled into virus particles containing the pathogen-specific epitope. Immunization of model animals with these recombinant particles results in generation of pathogen-specific immune responses (Dalsgaard, *et al.*, 1997; Yusibov, *et al.*, 1997; Wigdorovitz, *et al.*, 1999). Plant virus particles presenting foreign epitopes could be easily concentrated from plant extract to a high degree of purity for systemic administration. The plant virus particles not only enhanced the stability of recombinant peptides in the plant cells, but also increased immunogenicity of pathogen-specific epitopes compared to the administration of synthetic peptides (Yusibov, *et al.*, 1997). Oral vaccination with these recombinant plant viruses was shown to protect animals against rabies (Yusibov, *et al.*, 2002).

An alternative approach to the presentation of immunogenic peptides on virus particles, expression of full-length proteins of biomedical importance in plants was also achieved using plant virus-based expression vectors. Proteins were either fused to the CP (Smolenska, *et al.*, 1998) similar to the peptides described or they were expressed individually (Kumagai, *et al.*, 1993; Verch, *et al.*, 1998; McCormick, *et al.*, 1999). McCormick *et al.* administered purified scFvs against Hodgkin's lymphoma into laboratory mice and achieved tumor reduction. Short cycle times of genetic manipulation and recombinant protein production using plant virus expression vectors therefore enhanced the flexibility of plant-based production and opened the possibility of patient-specific disease treatment. However, present size limitations do not allow efficient production of considerably larger proteins than scFvs using plant virus gene vectors. Wu et al, (2003) expressed rotavirus VP-7 in transgenic potatoes and oral immunization the transgenic tubers to the mice successfully elicited serum IgG and mucosal IgA specific for VP-7. The mucosal IgA titer was as high as 1000, while serum IgG titer was only 600. Karamam et al, (2006) studied systemic and mucosal antibody response in to LT-B in young and aged mice vaccinated with corn derived antigen against Escherichia coli heat labile enterotoxin. Ma *et al.*, (2003) studied the expression of ORF2 partial gene of hepatitis E virus in tomatoes and immunoactivity of recombinant protein extracted from transgenic plants was examined by enzymes-linked immunosorbent assay (ELISA) using a monoclonal antibody specifically against HEV (hepatitis E virus). Mason et al, (1998), developed edible transgenic potato vaccine by expressing synthetic LT-B gene into potato plants, which protects mice against Escherichia coli heat-labile enterotoxin (LT). Azhar et al, (2002) showed that a positive response was observed when gene of protective antigen (PA) of anthrax expressed in plant system and thus explored the possibilities of development of edible anthrax vaccine. Papillomavirus L1 protein (for development of vaccine against cervical cancer) expressed in transgenic plant could potentially activate the humoral immune response (Warzecha *et al.*, 2003; Tregoning *et al.*, 2004; Lightwlers *et al.*, 2003).

FACTORS AFFECTING THE EFFICACY OF EDIBLE VACCINES

Oral immunization generates an immune response in the vaccinated individual. A critical issue of oral vaccination is the compatibility of the antigen with cooking, acid pH and digestive degradation. Oral immune tolerance and degradation of antigen in digestive processes of gut also limits the applicability of plant derived edible vaccines, which prevents the development of immune responses to the many antigens ingested by host including commensal organisms and food proteins. Oral antigens that bind to epithelial cells or that replicate tends to be immunogenic rather than tolerogenic. The challenge of the edible vegetable vaccine approach will be to deliver antigens that look like vaccines, in the context of vegetable that looks like food. The edible vaccine must stimulate an immunogenic response to the vaccine protein and maintain the tolerogenic response to the food vehicle. Low concentration of the vaccine antigen in plant requires the consumption of large amounts for oral immunization or strategies to increase recombinant protein levels. Hence there is always need of adjuvant to increase the immunogenicity of antigen in edible vaccines (Chargelegue *et al.*, 2001; Lauterslager *et al.*, 2003; Mei *et al.*, 2006).

A variety of approaches have been taken to protect antigens from digestive enzymes like use of attenuated bacterial strains of Salmonella and Vibrio cholerae as delivery vehicles or the encapsulation of the antigen in a protective coat (Van De Verg *et al.*, 1990; Ryan *et al.*, 1997). Attenuated strains have the advantage that they tend to direct the antigen to the surface of the gut where it will be available for sampling by the mucosal immune system. However, this approach has the same potential safety drawback of any vaccine based on an attenuated strain. Encapsulation is an inherently safer means of antigen protection and can be achieved using biodegradable polymers (Eldridge *et al.*, 1991), liposomes (Jackson *et al.*, 1990), proteosomes (Mallett *et al.*, 1995) or a product derived from transgenic plants expressing the antigen.

Targeting and immunogenicity of plant-derived vaccine can be altered by modification of expression vector. For example, the antigen can be expressed as a fusion to a molecule that can deliver it to receptors on the intestinal surface. Fusions to the receptor binding subunit of the heat-labile toxin of enterotoxigenic strains for Escherichia coli (LTB) or to the receptor binding subunit of cholera toxin (CTB) follow this approach. These toxin subunits have been shown to act as immune modulators for co delivered proteins. For example, LTB tends to bias a Th2 response. Such fusions are starting to be tested in plants. The most straightforward approach to orally co-deliver antigens expressed in plant material would be possible to express the protein adjuvant in a separate plant line, and then blend the antigen and adjuvant material prior to administration. This would allow a set ratio of antigen to adjuvant to be ensured, whereas expression of antigen and adjuvant in the same line may result in some batch-to-batch variation in the ratio. Mutant's forms of bacteria can be used to reduce the toxicity and to retain the adjuvanticity. These altered toxins include active site mutants LT-K 63 and LT-R72 and hinge cleavage mutants LT-G192. Other factor, which is important in the efficacy of the vaccine, is the palatability of the delivery vehicles. Fruits such as banana and tomato which are eaten raw retain the immunogenicity of the delivered antigen but in case of potatoes, which are boiled to make them palatable reduces the immunogenicity of the delivered antigen (Arakawa *et al.*, 1998; Mason *et al.*, 1998; Sala *et al.*, 2003).

Table 5: Selected antigens produced in plants and tested in animal and human systems.

Selected antigen Pathogen	*Antigen*	*Plant species*	*test system*	*Route*
Human Pathogens				
Enteretoxigenic *E. coli* (ETEC)	LT-B	Potato	Mice Humans	Oral
		Maize	Mice	Oral
		Tomato	Mice	Oral
Vibrio cholerae	CT-B	Potato	Mice	Oral
Hepatitis B	HBsAg	Lettuce	Humans	Oral
		Potato	Mice Humans	Oral
Norwalk virus	NVCP	Potato	Mice Humans	Oral
Rotavirus	CT-B/NSP4 fusion, CT-A/ CF1 fusion	Potato	Mice	Oral
Livestock Pathogens				
Transmissible gastroenteritis virus (TGEV)	Full length and N-terminus of S glyco protein	Arabidopsis	Mice	Oral
		Potato	Mice	Oral
	S glycoprotein	Tobacco	Pigs	Parenteral
		Corn	Pigs	Oral

Foot-and-mouth disease virus (FMDV)	VP1 capsid protein peptide	Alfalfa Potato leaf	Pigs Mice	Ip or oral Parenteral
	VP1 epitope fused with gus gene	Alfalfa	Mice	Parenteral
Canine parovi rus (CPV) VP2capsid protein	Linear antigenic peptide from	Arabidopsis	Mice	Oral

APPLICATION OF PLANT-DERIVED VACCINES

There is a large and fast growing list of protective antigens from microbial and viral pathogens that have been expressed by plants. The initial focus was upon human pathogens. However, today attention has also spread to animal pathogens (e.g. Newcastle and foot and mouth disease). There is no limit to the number and range of antigens that can be produced in plants if the DNA sequences coding for the appropriate genes are available(Teli, 2004; Mor, 2004; Streatfield, 2006). Plant-derived vaccines may elicit active immunization (vaccination), passive immunization or auto-tolerance. Plant-derived pharmaceuticals for passive immunization and auto-tolerance have proven immunologically active and effective in animals and humans but the majority of studies have investigated plant-derived vaccines (Table 5).

Bacterial antigen

E. coli Heat-Labile Enterotoxin/Cholera Toxin

The first human clinical trials for a transgenic, plant-derived antigen were planned, approved (US Food and Drug Administration) and performed in 1997. Enterotoxigenic Escherichia coli (ETEC) and Vibrio cholerae are two causal agents of bacterial diarrhoea. Two examples of enterotoxins are the heat labile toxin (LT) and the V. cholerae cholera toxin (CT). LT and CT consist of two subunits, the A subunit (LTA or CTA) and the B subunit (LTB or CTB). The expression of enterotoxin especially LTB in plants are provided by tobacco and potato tissues. Transgenic potatoes constitutively expressing a synthetic bacterial diarrhea vaccinogen (LTB) were orally delivered to human volunteers in Phase I/II clinical trials. After ingestion of the plant-derived recombinant LTB (rLTB) to mice produces LTB specific serum IgG and mucosal IgA (Rigano *et al.*, 2003). The synthetic LTB gene is also used to express the proteins via modification of transcription and translation process that might cause premature termination of gene processing or the degradation of its products. Mason et al demonstrated that mice fed potato tissues expressing the synthetic rLTB gene developed appropriate IgG and IgA responses sufficient to provide significant mucosal and systemic protective immunity in orally immunized mice. CTB specific antigens have been also expressed in chloroplasts and translated more than 31.1% of total soluble protein as functional oligomers. Chloroplast synthesized CTB binds to the intestinal membrane GM1-ganglioside receptor. This is the first report of a vaccine antigen expressed in transgenic chloroplasts (Arakawa *et al.*, 1998; Mason *et al.*, 1998; Daniell *et al.*, 2005).

Anthrax Protective Antigen

The current vaccine for anthrax (protective antigen; PA) produced by cell-free filtrate of toxigenic, nonencapsulated strain of Bacillus anthracis contains trace amounts of Lethal factor (LF) and Edema factor (EF) which are toxic and cause side-effects. Expression of PA in plant has been developed in transgenic chloroplasts and the crude plant extracts and express up to 2.5 mg PA/g fresh weight. The macrophage lysis assay proved the functionality of the antigen (formation of heptamers and their binding to host cell receptor) (Aziz *et al.*, 2002; Streatfield, 2006; Kamarajugadda, 2006; Masri *et al.*, 2006).

Plague Vaccine Antigen

Yersinia pestis, a gram-negative bacterium is the causative agent of plague. There are three different forms of plague of which bubonic plague is the most common, caused by infected fleas. The current vaccine available is a killed whole vaccine, which is moderately effective against bubonic plague and ineffective against septicemic and pneumonic plague. Several subunit vaccines have been evaluated for immunogenicity against Y. pestis. CaF1 and LcrV are the most effective antigens identi?ed so far against Y. pestis. F1 is a capsular protein located on the surface of the bacterium with anti-phagocytic properties. The fusion protein of capsular protein with V antigen has been shown to be more immunogenic when the mice were challenged with the bacteria. When the F1V fusion gene was expressed in transgenic chloroplasts, the expression levels obtained were as high as 14.8% of the total soluble protein (Daniell, 2006; Alvarez *et al.*, 2006; Streatfield, 2006; Mett *et al.*, 2007).

Early Secretory Antigenic Target (ESAT) for Tuberculosis

Tuberculosis (TB) is the leading infectious disease in the developing world. Since the portal of entry and primary site of infection of M. tuberculosis is the respiratory track, mucosal vaccination should have inherent advantage to inducing a protective immune response. Rigano *et al.* has developed the transgenic plant vaccines against tuberculosis in which tuberculosis (TB) antigen ESAT-6 fused to the B subunit of *E. coli* heat-labile enterotoxin (LTB) and expressed in Arabidopsis thaliana plant. The plant-made LTB-ESAT-6 fusion protein induced antigen-specific responses from CD4+ cells and increased IFN-ã production, indicating a Th1 response. In addition, a Th2 response was induced in the Peyer's patch. This is the first report of an orally delivered, subunit tuberculosis vaccine priming an antigen-specific, Th1 response (Nogales, 2005; Rigano *et al.*, 2006; Zelada *et al.*, 2006).

Epitopes from Opportunistic Pathogens

Pseudomonas aeruginosa and Staphylococcus aureus are opportunistic pathogens that cause disease in the immunosuppressed, burn victims and chronic pulmonary infections in children with cystic fibrosis. Immunization with the outer membrane F protein has provided protection from P. aeruginosa in a number of animal models while the D2 domain of the fibronectin-binding protein B (FnBP) has proven effective with S. aureus. In 1999, black-eyed bean plants expressed a synthetic peptide containing two epitopes in tandem from the outer membrane F protein from P. aeruginosa. The coding region for the synthetic peptide was inserted into two coat proteins of CPMV (S and L) and the rCPMV used to inoculate black-eyed bean plants. Both S and L fusions induced F protein-specific antibodies but were found to induce antibodies predominantly or exclusively against one epitope of the synthetic peptide. Purified recombinant virus was administered intranasally or by gavage to mice either alone or in the presence of the mucosal adjuvant ISCOM matrix. All intranasal treatments (with and without adjuvant) generated high titers for CPMV-and FnBP-specific IgG in sera and IgA and IgG of bronchial, intestinal and vaginal lavage fluids. This confirmed the ability of recombinant plant viruses to induce immune responses at distant mucosal sites and without the presence of a mucosal adjuvant. Assays also demonstrated the sera to completely inhibit the binding of human serum fibronectin to S. aureus (Brennan *et al.*, 1999a; Brennan *et al.*, 1999b; Staczek *et al.*, 2000; Gilleland *et al.*, 2000).

Viral antigens

Capsid proteins of many types of viruses can assemble into virus-like particles (VLPs). Devoid of the viral genetic material, VLPs often resemble the native virions in their morphology, antigenic properties and stability. These particulate antigens generally elicit stronger mucosal immune responses than soluble antigens, which can often repress the immune response by inducing immunotolerance (Daniell, 2006; Streatfield, 2006).

Hepatitis B Virus Surface Antigen

Hepatitis B virus (HBV) is one of the major causes of chronic viremia in humans. A recombinant vaccine consisting of the HBV surface antigen (HBsAg) produced in yeast is commercially available for parenteral vaccination. However the cost of the resulting vaccine is prohibitive for immunization programs in developing countries. Research programs have therefore been initiated to express a hepatitis vaccine in transgenic plants. HBsAg was also the first reported viral antigen to be produced in transgenic plants. In 1992 Mason *et al.*, expressed the HBsAg in tobacco and demonstrated the plant-derived antigen formed virus-like particles (VLPs). The particles were physically similar to those derived from human serum and retained antigenicity. Mice immunized with partially purified, tobacco-derived hepatitis B VLPs responded in a manner similar to those immunized with yeast-derived rHBsAg (commercial vaccine). Transgenic potatoes expressing HBsAg at higher levels should allow assessment of the immunogenic responses of the antigen by oral delivery. Potato-derived HBsAg is currently being tested in human clinical trials using the orally delivered antigen as a booster for the commercial vaccine. Recently, Youm *et al.* 2007 generated transgenic potatoes expressing middle protein (containing both preS2 and S antigens) of HBV envelope, which contains additional 55 amino acid preS2 regions at the N-terminus of the S protein, of HBV to determine the feasibility of developing a plant-delivered HBV vaccine (Mason *et al.*, 1992; Mason *et al.*, 1998; Youm *et al.*, 2007; Kumar *et al.*, 2007).

Huang *et al.* 2006 has been assembled Hepatitis B core antigen (HBc or HBcAg) into capsid particles (VLPs) by using a novel plant viral expression system (Nicotiana benthamiana) leaves. HBc has been extensively studied for its production in various expression systems and for the use of HBc particles for high-density, immunogenic presentation of foreign epitopes. Studies showed rapid, high-level (up to 7% of total soluble protein) production of HBc in plant leaf by using a novel plant viral expression system. The plant-derived HBc (p-HBc) self-assembles into VLPs that are morphologically and physically similar to *E. coli* derived HBc (e-HBc)-VLPs and induced strong humoral and mucosal antibody responses in mice.

Norwalk Virus Capsid Protein

As gastrointestinal viruses evolved to survive in the proteolytic environment of the gut, VLPs of such viruses might provide a better model for studying oral vaccines than the RV glycoprotein or HBsAg. The capsid of Norwalk virus (NV), a member of the Caliciviridae family and an important etiological agent of gastroenteritis, consists of 180 copies of a single subunit. The NV capsid proteins, when expressed in potato leaves, assemble into VLPs that are indistinguishable from those obtained by expressing the gene encoding the protein in insect cell cultures. Mason et al demonstrated tobacco-expressed NV capsid proteins formed virus-like particles indistinguishable from the insect recombinant VLPs. When gavaged or fed to mice, the plant-derived Norwalk VLPs proved orally immunogenic, inducing both systemic and mucosal immune responses. The adjuvant effect of purified cholera toxin (CT) was also assessed in this study and was found to enhance the response to the gavaged NV VLPs but to have a less prominent effect when fed to mice together with transgenic tubers. Although rises in serum antibody levels were modest, 95% of the volunteers who ingested the transgenic potato experienced significant rise in number of specific IgA antibody secreting cells.

Human Immunodeficiency Virus Type 1 Epitopes

Thousands of people continue to be infected with human immunodeficiency virus (HIV) everyday. To date several recombinant HIV antigens expressed by yeast, insect or mammal cells have proven safe and immunogenic in human trials. The antigens include the envelope proteins gp120 and 160, a transmembrane protein gp41, and protease genes. Plant cell expression of an epitope of the HIV-1 virus was first demonstrated in 1995. A translational fusion of a gp120 epitope to the CP of Tobacco Mosaic Virus (TMV) resulted in expression of virus particles consisting of the CP and CP fusion protein. The HIV-1 Tat protein has been recently explored as a prospective vaccine candidate. Recent studies have suggested that HIV-1 Tat should be considered as an important component of potential HIV vaccines. Tat is not only a key regulator of HIV-1 replication in infected cells, but is also an extracellular immunomodulator, which increases efficiency of virus dissemination and promotes AIDS progression. Karaseva *et al.*, developed TMV-based transient expression system for production of the HIV-1 Tat protein in an experimental plant Nicotiana benthamiana, and in an edible plant, spinach, at levels of up to 300 μg of Tat per 1 g of leaf tissue. Feeding these Tat-producing spinach leaves to mice primed them for antibody production following a subsequent immunization with a plasmid expressing Tat and resulted in significantly higher anti-Tat antibody levels compared to control mice (Horn *et al.*, 2003; Karaseva *et al.*, 2005; Webster, 2005; Canizares *et al.*, 2005).

Human Rhinovirus

Cowpea Mosaic Virus (CPMV) has been modified to express an epitope from the human rhinovirus (HRV) in black-eyed bean. The advantages of CPMV are: high yield (l-2g of virus per kg of host tissue), thermostability and ease of virus purification. Large quantities of purified virus particles expressing the HRV epitope is extracted from leaf tissues and injected intramuscularly and subcutaneously into rabbits (Canizares *et al.*, 2005; Gern *et al.*, 1996, Porta *et al.*, 1996).

Rabies Virus Glycoprotein and Nucleoprotein

Rabies is a worldwide disease of humans and warmblooded animals. The currently available vaccine using the rabies virus glycoprotein (G protein) for the development of recombinant plant derived edible vaccine. However, this is hampered by the relatively high cost of vaccine production and distribution (Ashraf *et al.*, 2005). McGarvey *et al.* 1995 has reported the expression of rabies virus (RV) glycoprotein in transgenic tomatoes. The plants accumulate extremely low levels of a protein that cross reacts with neutralizing monoclonal antibodies raised against the native RV glycoprotein. Yusibov et al 2002 expressed a B-cell epitope from the G protein and a T-cell epitope from the nucleoprotein through fusion to the coat protein (CP) of the alfalfa mosaic virus (AIMV) and subsequent inoculation of tobacco and spinach leaves. Injection of mice with isolated recombinant virus particles induced virus-neutralizing antibodies. The ability of the recombinant AIMV to be delivered orally was tested through gastric incubation of plant extracts or feeding on virus-infected spinach leaves. Both treatments induced local and systemic immune responses that proved sufficient to improve the clinical signs of the virus upon challenge with an attenuated rabies strain. These results were the first indication that an orally delivered, modified virus could induce a mucosal immune response.

Foot and Mouth Virus VP1 Coat Protein

Foot and Mouth Disease Virus (FMDV) affect all cloven-hoofed animals producing vesicles in the mouth, on the teats, and on the skin between and above the hoofs. Since foot and mouth disease is extremely infectious an outbreak quickly closes export markets of livestock and livestock products. Recent studies have demonstrated the structural protein VP1 to carry critical epitopes responsible for the induction of neutralizing antibodies. These epitopes have been expressed within a diverse range of prokaryotic and eukaryotic systems. The VP1 capsid protein of the foot-and-mouth disease virus (FMDV) can be expressed in transgenic Arabidopsis thaliana. Oral administration of the plant-derived VP1 subunit of FMDV has not been demonstrated. Because FMDV virions are complex structures containing several subunits, it is unlikely that the VP1 peptide alone will assemble into stable VLPs. It is therefore reasonable to assume that the plant material expressing the unassembled VP1 subunit of FMDV will be of limited use as an oral vaccine, despite the fact that it retains protective epitopes. Black-eyed bean protoplasts has also been utilized for transfection of recombinant Cowpea Mosaic Virus (CPMV) containing a sequence from VP1 cloned into the coding region of the S coat protein. Wigdorovitz et al 1999 reported the use of transgenic alfalfa to express and deliver the FMDV VP1 epitope. Immunization of mice through injection of leaf extracts or feeding freshly harvested leaves induced virus-specific immune responses sufficient for protection upon FMDV challenge (Wigdorovitz, 1999; Carrillo *et al.*, 2001; Dus *et al.*, 2002; Sun et al, 2003).

VP2 Epitope from the Mink Enteritis Virus

Mink enteritis virus (MEV) is characterized by severe inflammation of the intestine, resulting in profuse diarrhea. Existing vaccines are either chemically inactivated or live attenuated viruses propagated from cell cultures. VP2 is the major structural protein required for MEV particle formation and is capable of self-assembling into virus-like particles when expressed in an insect cell expression system. Upon insertion of a VP2 epitope into the coding region of S protein of CPMV, black-eyed bean plants express the VP2 epitope on the surface of CPMV. The immunogenicity of the rCPMV is demonstrated after one subcutaneous injection of the purified recombinant virus into mink conferred protection against clinical disease and virtually abolished shedding of virus after challenge with virulent MEV (Dalsgaard *et al.*, 1997).

Murine Zona Pellucida 3 Protein

A novel use of plant-derived vaccines is the delivery of immunocontraceptive vaccines for the control of animal populations. Several investigations have found peptides of zona pellucida 3 (ZP3), a protein involved in the docking of sperm to eggs, capable of inducing antibody-mediated contraception. High cost and parenteral administration limits its use. Delivery of immuno-contraceptive vaccines through baiting with edible plant tissues expressing an appropriate immunogen would decrease the cost of production and labor and increase the ease of delivery. A plant virus expression system was used by Fitchen *et al.* 1995 to express an epitope from the murine ZP3 within the coat protein of tobacco mosaic virus (TMV). The recombinant coat protein was found to accumulate to high concentrations and form particles in inoculated tobacco plants. Parenteral immunization of mice with tobacco extracts containing the recombinant CP resulted in the production of serum antibodies recognizing the ZP3 epitope and the recruitment in vivo of antibodies to the ZP in female mice (Fitchen *et al.* 1995; Polkinghorne *et al.*, 2005).

ADVANCES IN EDIBLE PLANT VACCINES

Advances in recombinant DNA technology coupled with advances in plant transformation and regeneration technology have made it possible to introduce new genetic material into plant cells, plants or plant tissue, that enhance the value of the plant or plant tissues. There are number of patent on plant derived edible vaccines which provide the requisite tools to transform foreign genes into plants. Plant transformation and regeneration in dicotyledons/ monocotyledons by A. tumefaciens is well documented by many workers. Patent No. WO 90/0248 (1990) discloses the method of production of transgenic plants expressing bacterial antigens *E. coli* and of S. mutans. US Pat No. 5914123 (1999) furnish the method of production of viral antigens of viral pathogen (hepatitis B or polio). DNA sequence encoding for viral pathogen (hepatitis B or polio) ligated to promoter sequences of A. tumefaciens plasmid vector and integrate into plant genome (tomato plants and tobacco plants) that can regulate the production of the surface antigen in a transgenic plant. Arntzen and May, 1999 also patented a method for transforming Musa plants (bananas and plantains). In particular, methods for wounding meristematic Musa plant tissue to facilitate access of A. tumefaciens comprising genetically engineered T-DNA is provided. This method may be used to transform the plant to produce hepatitis B surface antigen, Norwalk virus capsid protein, interleukins, growth hormone, erythropoietin, G-CSF, GM-CSF, hPG-CSF, M-CSF, Factor VIII, Factor IX, tPA, and insulin or to alter the phenotypic trait of the fruit of the plant. Lam *et al.*, 1997 divulged the A. tumefaciens based binary vector systems for the expression of hepatitis B viral surface antigen into tobacco plant cell. This patent also discusses different transformation methods like polyethylene glycol mediated uptake, electroporation, microinjection and microparticle bombardment transformation system. A number of gene promoters that coerce high levels of constitutive transgene expression are also patented. These include maize ubiquitin promoter (U.S. Pat. 5510474), rice actin 1 promoter (U.S. Pat. 5641876), cauliflower mosaic virus (CaMV) 35S promoters (Mitsuhara *et al.*, 1996) and promoters isolated from banana streak virus (Schenk *et al.*, 2001). There are also many patents on oral immunization of edible plant vaccines (US Pat 5654184; 5679880 and 5686079).

An alternative approach to the presentation of immunogenic peptides on virus particles, expression of full-length proteins of biomedical importance in plants was also achieved using plant virus-based expression vectors. US Pat 6392121 discloses the gemini virus vectors for gene expression in plants. Plant viral vector based gene amplification systems are not only dramatically increases foreign protein production in plants but also produces safer and more economical recombinant vaccines and antibodies. US Pat 7132291 discloses the vectors and methods useful for preparing genetically transformed plant cells that express immunogens against new castle disease virus or avian influenza virus. Polypeptides fused with alfalfa mosaic virus or ilarvirus capsid proteins has also used as a molecule for presentation of that polypeptide to the immune system of an animal such as a human (US Pat 6042832).

PDVs are not available on the public market as of yet, but some are entering the final phases of research and development ("R&D") and will be ready for human consumption shortly. Edible vaccines hold great potential, especially in Third World countries where transportation costs, poor refrigeration and needle use complicate vaccine administration. Edible vaccines, in particular, might overcome some of the difficulties of production, distribution and delivery associated with traditional vaccines. Significant challenges are still to be overcome before vaccine crops can become a reality. However, while access to essential healthcare remains limited in much of the world and the scientific community is struggling with complex diseases such as hiv and malaria, plant-derived vaccines represent an appetizing prospect. Strategies to improve the recombinant protein yield in plants include the development of novel promoters, the improvement of protein stability and accumulation through the use of signals that target the protein to intracellular compartments, and the improvement of downstream processing technologies.

SUMMARY

Evidence has quickly accumulated in support of plant-derived antigens as effective, safe and inexpensive means of acquiring or delivering a vaccine. The success of human trials with edible transgenic plant vaccines and successful challenge in animal studies have demonstrated the potential of plant-derived vaccines to expand the armament of preventative medicine. Vaccines produced using plants as bioreactors have been successful when parenterally or intranasally administered. However oral immunization has only been successful when the vaccine was delivered in plant. This suggests that additional protection afforded by plant tissues permits successful oral immunization.

REFERENCES

Alvarez ML Pinyerd HL Crisantes JD Rigano MM Pinkhasov J Walmsley AM Mason HS Cardineau GA (2006) Plant-made subunit vaccine against pneumonic and bubonic plague is orally immunogenic in mice *Vaccine* **24(14)**: 2477-2490

Arakawa T Chong DK Langridge WH (1998) Efficacy of a food plant-based oral cholera toxin B subunit vaccine *Nat Biotechnol* **16**: 292-297

Arntzen C Plotkin S Dodet B (2005) Plant-derived vaccines and antibodies: potential and limitations *Vaccine* **23**: 17531756

Arntzen CJ (1997) High-tech herbal medicine: Plant-based vaccines *Nat Biotechnol* **15**: 221-222

Arntzen CJ (1998) Pharmaceutical foodstuffs-oral immunization with transgenic plants *Nat Med* **4**: 502-503

Arntzen CJ Lam DMK (1999) Vaccines expressed in plants US Pat No 5914123

Arntzen CJ May GD (1998) *Agrobacterium tumefaciens* transformation of Musa species US Pat No 5792935

Ashraf S Singh PK Yadav DK Shahnawaz M Mishra S Sawant SV Tuli R (2005) High level expression of surface glycoprotein of rabies virus in tobacco leaves and its immunoprotective activity in mice *J Biotechnol* **119(1)**: 1-14

Awram P Gardner RC Forster RL Bellamy AR (2002) The potential of plant viral vectors and transgenic plants for subunit vaccine production *Adv Virus Res* **58**: 81-124

Azhar AM Singh S Anand KP Bhatnagar R Expression of protective antigen in transgenic plants: a step towards edible vaccine against anthrax *Biochem Biophy Res Cummun* **299(3)**: 345-51

Baulcombe DC (1999) Fast forward genetics based on virus-induced gene silencing *Curr Opin Plant Biol* **2**: 109-113

Bevan M (1984) Binary *Agrobacterium* vectors for plant transformation *Nucleic Acids Res* **12**: 8711-8721

Bidney D Scelonge C Martich J Burus M Sims L Huffman G (1992) Microprojectile bombardment of plant tissues increased transformation frequency of *Agrobacterium tumefaciens Plant Mol Biol* **18**: 301-313

Binns AN Thomashaw MF (1988) Cell biology of *Agrobacterium* infection and transformation of plants *Ann Rev Microbiol* **42**: 575-606

Bosch D Smal J Krebbers E (1994) A trout growth hormone is expressed correctly folded and partially glycosylated in the leaves but not the seeds of transgenic plants *Transgen Res* **3**: 304-310

Brandtzaeg P Baklien K (1976) Immunoglobulin-producing cells in the intestine in health and disease *Clin Gastroenterol* **5(2)**: 251-269

Brennan FR Bellaby T Helliwell SM Jones TD Kamstrup S Dalsgaard K Flock JI Hamilton WD (1999b) Chimeric plant virus particles administered nasally or orally induce systemic and mucosal immune responses in mice *J Virol* **73(2)**: 930-8

Brennan FR Jones TD Gilleland LB Bellaby T Xu F North PC Thompson A Staczek J Lin T Johnson JE Hamilton WD Gilleland HE Jr (1999a) *Pseudomonas aeruginosa* outer-membrane protein F epitopes *Microbiol* **145(1)**: 211-20

Canizares MC Lomonossoff GP Nicholson L (2005) Development of cowpea mosaic virus-based vectors for the production of vaccines in plants *Expert Rev Vaccines* **4(5)**: 687-97

Cardineau GA Mason HS VanEck JM Kirk DD Walmsley AM (2006) Vectors and cells for preparing immunoprotective compositions derived from transgenic plants US Pat No 7132291 2006

Carrillo C Wigdorovitz A Trono K Dus Santos MJ Castanon S Sadir AM Ordas R Escribano JM Borca MV (2001) Induction of a virus-specific antibody response to foot and mouth disease virus using the structural protein VP1 expressed in transgenic potato plants *Viral Immunol* **14(1)**: 49-57

Carter JE Langridge WHR (2002) Plant-based vaccines for protection against infectious and autoimmune diseases *Crit Rev Plant Sci* **21**: 93-109

Castle LA Wu G McElroy D (2006) Agricultural input traits: past present and future *Curr Opin Biotechnol* **17**: 105-112

Chapman S Kavanagh T Baulcombe D (1992) Potato virus X as a vector for gene expression in plants *Plant J* **2**: 549-557

Chargelegue D Obregon P Drake PM (2001) Transgenic plants for vaccine production: expectations and limitations *Trends in Plant Sci* **6**: 495-496

Cheng M Fry JE (2000) An improved efficient *Agrobacterium*-mediated plant transformation method Int Patent publ WO 0034/491

Chikwamba RK Scott MP Mejia LB Mason HS Wang K (2003) Localization of a bacterial protein in starch granules of transgenic maize kernels *Proc Natl Acad Sci USA* **100**: 11127-11132

Choulika A Joly JS Thermes V Ristoratore F (2006) Random integration of a polynucleotide by in vivo linearization US Pat No 7098031

Chowdhury K Bagasra O (2007) An edible vaccine for malaria using transgenic tomatoes of varying sizes shapes and colors to carry different antigens *Medical Hypotheses* **68**: 22-30

Clark K Johnson PR (2006) Gene delivery of vaccines for infectious diseases *Curr Opin Mol Ther* **3(4)**: 375-384

Conrad U Fiedler U (1994) Expression of engineered antibodies in plant cells *Plant Mol Biol* **26**: 1023-1030

Conrad U Fiedler U (1998) Compartment-specific accumulation of recombinant immunoglobulins in plant cells: an essential tool for antibody production and immunomodulation of physiological functions and pathogen activity *Plant Mol Biol* **38**: 101-109

Cripps AW Kyd JM Foxwell AR (2001) Vaccines and mucosal immunisation *Vaccine* **19**: 2513-2515

Curtiss R Cardineau GA (1997) Oral immunization by transgenic plants US Pat No 5 654 184

Curtiss R Cardineau GA (1997) Oral immunization by transgenic plants US Pat No 5679880

Curtiss R Cardineau GA (1997) Oral immunization by transgenic plants US Pat No 5686079 1997

Dalsgaard K Uttenthal A Jones TD Xu F Merryweather A Hamilton WDO Langeveld JPM Boshuizen RS Kamstrup S Lomonossoff GP Porta C Vela C Casal JI Meloen RH Rodgers PB (1997) Plant-derived vaccine protects target animals against a viral disease *Nat Biotechnol* **15**: 248-252

Damaj MB Mirkov TE Reddy A Thomas TL Rathore KS Emani C Kumpatla SP (2007) Stem regulated plant defense promoter and uses thereof in tissue-specific expression in monocots US Pat No 7253276

Daniell H (2006) Production of biopharmaceuticals and vaccines in plants via the chloroplast genome *Biotechnol J* **1(10)**: 1071-9

Daniell H Chebolu S Kumar S Singleton M Falconer R (2005) Chloroplast-derived vaccine antigens and other therapeutic proteins *Vaccine* **23**: 1779-1783

Doran PM (2006) Foreign protein degradation and instability in plants and plant tissue cultures *TIBTECH* **24**: 426-432

Drakakaki G Marcel S Arcalis E Altmann F Gonzalez-Melendi P Fischer R (2006) The intracellular fate of a recombinant protein is tissue dependent *Plant Physiol* **141**: 578-586

Dus SMJ Wigdorovitz A (2005) Transgenic plants for the production of veterinary vaccines *Immunol Cell Biol* **83(3)**: 229-38

Dus SMJ Wigdorovitz A Trono K Rios RD Franzone PM Gil F Moreno J Carrillo C Escribano JM Borca MV (2002) A novel methodology to develop a foot and mouth disease virus (FMDV) peptide-based vaccine in transgenic plants *Vaccine* **20**: 1141-1147

Escobar MA Dandekar AM (2003) *Agrobacterium tumefaciens* as an agent of disease *Trends Plant Sci* **8**: 380-386

Featherstone C (1996) Vaccines by agriculture *Molecul Med* **1996**: 278-281

Fischer R Stoger E Schillberg S Christou P Twyman RM (2004) Plant-based production of biopharmaceuticals *Curr Opin Plant Biol* **7**: 152-158

Fischer R Vaquero-Martin C Sack M Drossard J Emans N Commandeur U (1999) Towards molecular farming in the future: transient protein expression in plants *Biotechnol Appl Biochem* **30**: 113-116

Fitchen J Beachy RN Hein MB (1995) Plant virus expressing hybrid coat protein with added murine epitope elicits autoantibody response *Vaccine* **13(12)**: 1051-7

Fromm ME (1987) In Methods in Enzymology Wu R Grossman L (eds) Academic Press Orlando *Fla* **153**: 307

Fuhrmann M (2004) Production of antigens in *Chlamydomonas reinhardtii:* green microalgae as a novel source of recombinant proteins *Methods Mol Med* **94**: 191-195

Gern JE Joseph B Galagan DM Borcherding WR Dick EC (1996) Rhinovirus inhibits antigen-specific T cell proliferation through an intercellular adhesion molecule-1-dependent mechanism *J Infect Dis* **174(6)**: 1143-1150

Gilleland HE Gilleland LB Staczek J Harty RN Garcia-Sastre A Palese P Brennan FR Hamilton WD Bendahmane M Beachy RN (2000) Chimeric animal and plant viruses expressing epitopes of outer membrane protein F as a combined vaccine against Pseudomonas aeruginosa lung infection *FEMS Immunol Med Microbiol* **27(4)**: 291-297

Greene AE Allison RF (1994) Recombination between viral RNA and transgenic plant transcripts *Science* **263**: 1423-1425

Hamilton CM (1997) A binary-BAC system for plant transformation with high-molecular-weight DNA *Gene* **200**: 107-116

Hansen G and Wright MS (1999) Recent advances in the transformation of plants *Trends Plant Sci* **4**: 226-231

Haq TA Mason HS Clements JD and Arntzen CJ (1995) Oral immunization with a recombinant bacterial antigen produced in transgenic plants *Science* **268**: 714-716

Hellens R Mullineaux P Klee H (2000) A guide to *Agrobacterium* binary Ti vectors *Trends in Plant Science* **5(10)** 1-4

Hood EE Jilkat JM (1999) Plant-based production of xenogenic proteins *Curr Opin Biotech* **10**: 382-386

Hood EE Woodard SL Horn ME (2002) Monoclonal antibody manufacturing in transgenic plants--myths and realities *Curr Opin Biotechnol* **13**: 630-635

Hooykaas PJJ Shilperoort RA (1992) *Agrobacterium* and plant genetic engineering *Plant Molecul Biol* **19**: 15-38

Horn ME Pappu KM Bailey MR Clough RC Barker M Jilka JM Howard JA Streatfield SJ (2003) Advantageous features of plant-based systems for the development of HIV vaccines *J Drug Target* **11(8-10)**: 539-545

Horn ME Woodard SL Howard JA (2004) Plant molecular farming: systems and products *Plant Cell Report* **22**: 711-720

Huang Z Santi L LePore K Kilbourne J Arntzen CJ Mason HS (2006) Rapid high-level production of hepatitis B core antigen in plant leaf and its immunogenicity in mice *Vaccine* **24**: 2506-2513

Jackson S Mestecky J Childers NK Michalek SM (1990) Liposomes containing anti-idiotypic antibodies: an oral vaccine to induce protective secretory immune responses specific for pathogens of mucosal surfaces *Infect Immun* **58**: 1932-1936

Kamarajugadda S Daniell H (2006) Chloroplast-derived anthrax and other vaccine antigens: their immunogenic and immunoprotective properties *Expert Rev Vaccines* **5(6)**: 839-849

Karaman S Cunnick J Wang K 2006 Analysis of immune response in young and aged mice vaccinated with corn-derived ar tigen *Escherichia Coli* heat labile enterotoxin *Mol Biotechnol* Jan ; **32(1)**: 31-42

Karaseva AV Candice WSF Rich A Shon KJ Zwierzynski I Hone D Koprowski H Reitz M (2005) Plant based HIV-1 vaccine candidate: Tat protein produced in spinach *Vaccine* **23**: 1875-1880

Kirk DD McIntosh K (2005) Social Acceptance of Plant-Made Vaccines: Indications from a Public Survey *AgBioForum* **8(4)**: 228-234

Klee H (2000) A guide to *Agrobacterium* binary Ti vectors *Trends in Plant Science* **5**: 446-451

KleinTM Wolf ED Wu R Sanford JC (1987) High velocity microprojectiles for delivery of nucleic acids into living cells *Nature* **327**: 70-73

Koprowski H (2005) Vaccines and sera through plant biotechnology *Vaccine* **23**: 1757-1763

Kumagai MH Turpen TH Weinzettl N Della-Cioppa G Turpen AM Donson J Hilf ME Grantham GL Dawson WO Chow TP Piatak M Grill LK (1993) Rapid high-level expression of biologically active á-trichosanthin in transfected plants by an RNA viral vector *Proc Natl Acad Sci USA* **90**: 427-430

Kumar GB Ganapathi TR Bapat VA (2007) Production of hepatitis B surface antigen in recombinant plant systems: an update *Biotechnol Prog* **23(3)**: 532-9

Lam DMK Arntzen CJ (1997) Anti-viral vaccines expressed in plants US Pat No 5612487

Langridge WHR Arakawa T (1999) Methods and substances for preventing and treating autoimmune disease US Pat No 6777546

Lauterslager TG Stok W Hilgers LA (2003) Improvement of the systemic prime/oral boost strategy for systemic and local response *Vaccine* **21**(13-14): 1391-1399

Leon-Baiiares R Gonzilez-Ballester D Galvin A Fernindez E (2004) Transgenic microalgae as green cell-factories *TIBTECH* **22(1)**: 45-52

Lightwlers MV Vaccines for prevention of custicerosis *Acta Trop* **87(1)**: 129-35

Lorence A Verpoorte R (2004) Gene transfer and expression in plants *Methods Mol Biol* **267**: 329-350

Lorz HB Baker JS (1985) Gene transfers to cereal cells mediated by protoplast transformation *Mol Gen Genet* **199**: 178-182

Ma JKC Chikwamba R Sparrow P Fischer R Mahoney R Twyman RM (1997) Plant-derived pharmaceuticals the road forward *Trends in Plant Science* **10**(12): 1-4

Ma Y Lin S Gao Y Li M Zhang J Xia N (2003) Expression of ORF2 partial gene of hepatitis E virus in tomatoes and immunoactivity of expressed product *World J Gastroenterorol* **9**(10): 2211-2215

Mallett CP Hale TL Kaminski RW Larsen T Orr N Cohen D Lowell GH (1995) Intranasal or intragastric immunization with proteosome-Shigella lipopolysaccharide vaccines protects against lethal pneumonia in a murine model of Shigella infection *Infect Immun* **63**: 2382-2386

Marillonnet S Giritch A Gils M Kandzia R Klimyuk V Gleba Y (2004) In plant engineering of viral RNA replicons: efficient assembly by recombination of DNA modules delivered by Agrobacterium *Proc Natl Acad Sci USA* **101**: 6852-6857

Mason HS Arntzen CJ Tariq HA Clements JD (2002a) Oral immunization with transgenic plants US Pat No 6395964

Mason HS Arntzen CJ (1995) Transgenic plants as vaccine production systems *TIBTECH* **13**: 388-392

Mason HS Haq TA Clements JD Arntzen CJ (1998) Edible vaccine protects mice against *Escherichia coli* heat-labile enterotoxin (LT): potatoes expressing a synthetic LT-B gene *Vaccine* **16**: 1336-1343

Mason HS Warzecha H Mor T Arntzen CJ (2002b) Edible plant vaccines: applications for prophylactic and therapeutic molecular medicine *Trends Mol Med* **8**(7): 324-329

Masri SA Rast H Hu WG Nagata LP Chau D Jager S Mah D (2007) Cloning and expression in *E coli* of a functional Fab fragment obtained from single human lymphocyte against anthrax toxin *Mol Immunol* **44**(8): 2101-1206

McCabe DE Swain WF Martinell BJ Christou P (1988) Stable transformation of soybean *(Glycine max)* by particle acceleration *Bio-Technology* **6**: 923-926

McCormick AA Kumagai MH Hanley K Turpen TH Hakim I Grill LK Tuse D Levy S Levy R (1999) Rapid production of specific vaccines for lymphoma by expression of the tumor derived single-chain Fv epitopes in tobacco plants *Proc Natl Acad Sci USA* **96**: 703-708

McGarvey PB Hammond J Dienelt MM Hooper DC Fu ZF Dietzschold B Koprowski H Michaels FH (1995) Expression of the rabies virus glycoprotein in transgenic tomatoes *Biotechnology* **13**(13): 1484-1487

McGhee JR Mestecky J Dertzbaugh MT Eldridge JH Hirasawa M Kiyono H (1992) The mucosal immune system: from fundamental concepts to vaccine development *Vaccine* **10**(2): 75-88

Mei H Tao SU Yuan-Gang Z Zhi-Gang A (2006) Research Advances on Transgenic Plant Vaccines *Acta Genetica Sinica* **33**(4): 285-293

Mett V Lyons J Musiychuk K Chichester JA Brasil T Couch R Sherwood R Palmer GA Streatfield SJ Yusibov V (2007) A plant-produced plague vaccine candidate confers protection to monkeys *Vaccine* **25**(16): 3014-3017

Mor TS Gomez-Lim MA Palmer KE (1998) Perspective: edible vaccines: a concept coming of age *Trends in Microbiology* **6** (11): 449-453

Mor TS Mason HS Kirk DD Arntzen CJ Cardineau GA (2004) Plants as production and delivery vehicle for orally delivered subunit vaccines In Levine MM Rappuoli R Liu MA Good MF (eds) New generation vaccines 3rd ed New York-Basel: Marcel Dekker Inc pp 305-12

Negrutiu I Shilito R Potrykus I Biasini G Sala F (1987) Hybrid gene in the analysis of transformation conditions *Plant Mol Biol* **8**: 363-373

Neutra MR Kraehenbuhl JP (1992) M cell-mediated antigen transport and monoclonal IgA antibodies for mucosal immune protection *Adv Exp Med Biol* **327**:143-50

Nicholson L Gonzalez-Melendi P van Dolleweerd C Tuck H Pemn Y Ma JKC Fischer R Christou P Stoger E (2003) A recombinant multimeric immunoglobulin expressed in rice shows assembly-dependent subcellular localization in endosperm cells *Plant Biotech J* **3**: 115-127

Nogales EA (2005) A view to the future of vaccines *An R Acad Nac Med* **122**(1): 171-84

Ogra PL (2003) Mucosal immunity: some historical perspective on host-pathogen interactions and implications for mucosal vaccines *Immunol Cell Biol* **81**(1): 23-33

Opabode JT (2006) Agrobacterium-mediated transformation of plants: emerging factors that influence efficiency Biotechnology and Molecular Biology Review 1(1): 12-20

Pandey GK Gupta R Kim KN Luan S (2007) Methods for enhancing a plant stress response US Pat No 7250555

Parrott DM (1976) The gut as a lymphoid organ *Clin Gastroenterol* **5**(2): 211-28

Paszkowski J Shillito RD Saul M Mandak V Hohn T Hohn B Potrykus I (1984) Direct gene transfer to plants *EMBO J* **3**: 2717

Polkinghorne I Hamerli D Cowan P Duckworth J (2005) Plant-based immunocontraceptive control of wildlife-"potentials limitations and possums" *Vaccine* **23**: 1847-1850

Porta C Spall VE Findlay KC Gergerich RC Farrance CE Lomonossoff GP (2003) Cowpea mosaic virus-based chimaeras: Effects of inserted peptides on the phenotype host range and transmissibility of the modified viruses *Virology* **310**: 50-63

Porta C Spall VE Lin T Johnson JE Lomonossoff GP (1996) The development of cowpea mosaic virus as a potential source of novel vaccines *Intervirology* **39**: 79-84

Potrykus I (2001) Golden rice and beyond *Plant Physiol* **125**: 1157-1161

Potrykus I Paszkowski J Saul MW Petruska J Shillito RD (1985) DNA transfer to protoplasts *Mol Gen Genet* **199**: 169

Raney T (2006) Economic impact of transgenic crops in developing countries *Curr Opin Biotechnol* **17**: 174-178

Reich TJ Iyer VN Scobie B Miki BL (1986) Efficient Transformation of Alfalfa Protoplasts by the Intranuclear Microinijection of Ti Plasmids " *Bio-Technology* **4**: 1001-1004

Rigano MM Sala F Arntzen CJ Walmsley AM (2003) Targeting of plant-derived vaccine antigens to immunoresponsive mucosal sites *Vaccine* **21**: 809-811

Ruf S Hermann M Berger IJ Carrer H Bock R (2001) Stable genetic transformation of tomato plastids and expression of a foreign protein in fruit *Nat Biotech* **19**: 870-875

Ryan ET Butterton JR Smith RN Carroll PA Crean TI Calderwood SB (1997) Protective immunity against Clostridium difficile toxin A induced by oral immunization with a live attenuated Vibrio cholerae vector strain *Infect Immun* **65**: 2941-2949

Sala F Rigano MM Barbante A Basso B Walmsley AM Castiglione S (2003) Antigen production in transgenic plants: strategies gene constructs and perspectives *Vaccine* **21** : 803-808

Scholthof HB Scholthof KB Jackson AO (1996) Plant virus gene vectors for transient expression of foreign proteins in plants *Annu Rev Phytopathol* **34**: 299-323

Scholthof KB Mirkov TE Scholthof HB (2002) Plant virus gene vectors: biotechnology applications in agriculture and medicine *Genet Eng* **24**: 67-85

Schouten A Roosien J de Boer JM Wilmink A Rosso MN Bosch D (1997) Improving scFv antibody expression levels in the plant cytosol *FEBS Lett* **415**: 235-241

Schouten A Roosien J van Engelen FA de Jong GA Borst-Vrenssen AW Zilverentant JF (1996) The C-terminal KDEL sequence increases the expression level of a single-chain antibody designed to be targeted to both the cytosol and the secretory pathway in transgenic tobacco *Plant Mol Biol* **30**: 781-793

Sciutto E Fragoso G Manoutcharian K Gevorkian G Rosas-Salgad G Hem d ndez-Gonzalez M Herrera-Estrella L Cabrera-Ponce J L Lapez-Casillas F Gonzalez-Bonilla C Santigago-Machuca A Sanchez R J Goldbaum F Alujac A Larraldea C (2002) New approaches to improve peptide vaccine against porcine Taenia solium Cysticercosis *Arch Med Res* **33**: 371-378

Shalaby WS (1995) Development of oral vaccines to stimulate mucosal and systemic immunity: barriers and novel strategies *Clin Immunol Immunopathol* **74**(2): 127-34

Sharma KK Bhatnagar P Thorpe TA (2005) Genetic transformation technology: Status and problems *In Vitro Cell Dev Biol Plant* **41**: 102-112

Sheng OJ Citovsky V (1996) Agrobacterium-plant cells DNA transport: have virulence proteins will travel *Plant Cell* **8**: 1699-1710

Shillito RD Potrykus I (1987) In Methods in Enzymology Wu R Grossman L (eds) Academic Press Orlando *Fla* **153**: 283

Staczek J Bendahmane M Gilleland LB Beachy RN Gilleland HE Jr (2000) Immunization with a chimeric tobacco mosaic virus containing an epitope of outer membrane protein F of Pseudomonas aeruginosa provides protection against challenge with *P. aeruginosa* *Vaccine* **18**(21): 2266-74

Stomp AM Weissinger A Sederoff RR (1991) Transient expression from microprojectile-mediated DNA transfer in *Pinus taeda Plant Cell Reports* **10**(4): 187-191

Streatfield SJ (2006) Mucosal immunization using recombinant plant-based oral vaccines *Methods* **38**: 150-157

Streatfield SJ Howard JA (2003) Plant-based vaccines *Int J Parasitol* **33**: 479-493

Streatfield SJ Jilka JM Hood EE Turner DD Bailey MR Mayor JM (2001) Plant based vaccines: unique advantages *Vaccine* **19**: 2742-2748

Sugiyama Y Hamamoto H Takemoto S Watanabe Y Okada Y (1995) Systemic production of foreign peptides on the particle surface of tobacco mosaic virus *FEBS Lett* **359**: 247-250

Sun Meng Qian KX Su N Chan HY Liu JX Shen GF (2003) Foot-and-mouth disease virus VP1 protein fused with cholera toxin B subunit expressed in Chlamydomonas reinhardtii chloroplast *Biotechnol Lett* **25**: 1087-1092

Tacket CO Mason HS (1999) A review of oral vaccination with transgenic vegetables *Microbes Infect* **1**(10): 777-783

Tacket CO Mason HS Losonsky G Clements JD Levine MM Arntzen CJ (1998) Immunogenicity in humans of a recombinant bacterial antigen delivered in a transgenic potato *Nat Med* **4**: 607-609

Teli NP Timko MP (2004) Recent developments in the use of transgenic plants for the production of human therapeutics and biopharmaceuticals *Plant Cell Tissue and Organ Culture* **79**: 125-145

Tregoning J Maliga P Dougan G Nixon PJ (2004) New advances in the production of edible plant vaccine: Chloroplast expression of tetanus vaccine antigen *Tet C Photochemistry* **65**(8): 989-994

Trivedi PK Nath P (2004) MaExpl an ethylene-induced expansion from ripening banana fruit *Plant Sci* **167**: 1351-1358

Twyman RM Schillberg S Fischer R (2005) Transgenic plants in the biopharmaceutical market *Expert Opin Emerg Drugs* **10**: 185-218

Twyman RM Stoger E Schillberg S Christou P Fischer R (2003) Molecular farming in plants: host systems and expression technology *TIBTECH* **21**: 570-578

Usha R Rohll JB Spall VE Shanks M Maule AJ Johnson JE Lomonossoff GP (1993) Expression of an animal virus antigenic site on the surface of a plant virus particle *Virology* **197**: 366-374

Van DVL Herrington DA Murphy JR Wasserman SS Formal SB Levine MM (1990) Specific immunoglobulin A secreting cells in peripheral blood of humans following oral immunization with a bivalent Salmonella typhi–Shigella sonnei vaccine or infection by pathogenic *S sonnei Infect Immun* **58**: 2002-2004

Van O Albert JJ Rietveld K Quax WJ Pen J Hoekema A Sijmons PC Verwoerd TC (2006) Production of enzymes in seeds and their use US Pat No 7033627

Vierstra RD (2003) The ubiquitin/26S proteasome pathway the complex last chapter in the life of many plant proteins *Trends Plant Sci* **8**: 135-142

Vitale A Denecke J (1999) The endoplasmic reticulum-gateway of the secretory pathway *Plant Cell* **11**: 615-628

Walmsley AM Arntzen CJ (2001) Plants for delivery of edible vaccines *Curr Opin Biotechnol* **11**: 126-129

Walmsley AM Arntzen CJ (2003) Plant cell factories and mucosal vaccines *Curr Opin Biotechnol* **14**:145-150

Warzecha H Mason HS Lane C Tryggvesson A Rybicki E Williamson AL Clements JD Rose RC (2003) Oral immunogenicity of human Papillomavirus-like particles expressed in potato *J Virol* **77**: 8702-11

Warzechaa H Mason HS (2003) Benefits and risks of antibody and vaccine production in transgenic plants *J Plant Physiol* **160**: 755-764

Webster DE Thomas MC Pickering R Whyte A Dry IB Gorry PR Wesselingh SL (2005) Is there a role for plant-made vaccines in the prevention of HIV/AIDS? *Immunol Cell Biol* **83**(3): 239-47

Wigdorovitz A Carrillo C Dus SMJ Trono K Peralta A Gomez MC Rios RD Franzone PM Sadir AM Escribano JM Borca MV (1999) Induction of a protective antibody response to foot and mouth disease virus in mice following oral or parenteral immunization with alfalfa transgenic plants expressing the viral structural protein VP1 *Virology* **255**: 347-353

Wu YZ Li JT Mou ZR Fei L Ni B Geng M Jia ZC Zhou W Zou LY Tang Y (2003) Oral immunization with rotavirus VP7 expressed in transgenic potatoes induced high titers of mucosal neutralizing IgA *Vaccine* **313**(2): 337-42

Youm JW Won YS Jeon JH Ryuc CJ Choi YK Kimb HC Kime BD Joung H Kim H S (2007) Oral immunogenicity of potato-derived HBsAg middle protein in BALB/c mice *Vaccine* **25**: 577-584

Yusibov V Hooper DC Spitsin SV Fleysh N Kean RB Mikheeva T Deka D Karasev A Cox S Randall J Koprowski H (2002) Expression in plants and immunogenicity of plant virus-based experimental rabies vaccine *Vaccine* **20**: 3155-3164

Yusibov V Modelska A Steplewski K Agadjanyan M Weiner D Hopper DC Koprowski H (1997) Antigens produced in plants by infection with chimeric plant viruses immunize against rabies virus and HIV-1 *Proc Natl Acad Sci USA* **94**: 5784-5788

Yusibov V Shivprasad S Turpen TH Dawson W Koprowski H (1999) Plant viral vectors based on tobamoviruses. *Curr Top Microbiol Immunol* **240**: 81-94.

Zelada AM Calamante G dela-Paz SM Bigi F Verna F Mentaberry A Cataldi A (2006) Expression of tuberculosis antigen ESAT-6 in *Nicotiana tabacum* using a potato virus X-based vector. *Tuberculosis* **86**(3-4): 263-7.

Zupan JR Zambryski PC (1995) Transfer of T-DNA from *Agrobacterium* to the plant cell. *Plant Physiology* **107**: 1041-1047.

10

THE ROLE OF BIOTECHNOLOGY IN PROLONGING THE SHELF-LIFE OF FRUITS

C.P. MALIK, AMAN VERMA, BHAVNEET KAUR AND CHITRA WADHWANI

India, the second largest producer of fruits and vegetables in the world, aims to double its production to 300 million tones by 2012. They contribute nearly 28 per cent of GDP and 54 per cent of export share in Agriculture from the cultivated area share of 8.5% only. This sector could significantly benefit from India's new infrastructural technology support like cold storage chains, refrigerated transport facilities and centralized food processing facilities. In fact Post-harvest operations are assuming greater importance due to higher yields and increased cropping intensity indicating that there is tremendous scope of increasing production and exports in Horticulture. Fruits and vegetables earn 20-30 times more foreign exchange per unit area than cereals due to higher yields and higher price available in the international market. But India loses about 35 - 45% of the harvested fruits and vegetables during handling, storage, transportation etc, a loss estimated at Rs 40,000 crores per year! In India, fruits and vegetables wastage every year is equivalent to the annual consumption of the United Kingdom. Post Harvest management of fresh produce using emerging techniques and technology will be one of the biggest business opportunities in India.

Fresh fruits abound in very high percentage of water and consequently exhibit relatively high metabolic activity when compared with other plant derived foods such as seeds. This metabolic activity continues post harvesting of fruits and thus makes them a highly perishable and blatting commodity. This inherent short shelf life of fresh fruits poses greatest problem to the successful transportation and their marketing. Almost 20 per cent of all perishable fruit and vegetables are lost due to rotting, as cold storage and refrigerated transport facilities are inadequate in India. For some of the major commercial crops the post harvest constraints have been mostly resolved by harvesting immature and mature green fruits, by using refrigeration and controlled atmosphere storage.

However, several of other fruits cannot be handled successfully. Therefore, this hinders the commercial exploitation of these fruits. Mangoes and bananas harvested at full maturity do not store appreciably, and once harvested when immature, fail to ripen properly. Non-genetic methods to delay ripening do exist; these are not feasible on a large scale. Thus, genetic technology would help in reducing post harvest losses caused by rotting by giving plants an inherent capacity to delay ripening. In recent years techniques of molecular biology have provided a powerful tool to unravel the biochemistry of fruit ripening and confirming several of the traditional physiological and biochemical post harvest findings (Golding *et al.*, 2005).

As the movement of tomatoes is hampered, at present export is negligible. The new technology, called antisense RNA technology, involves blocking the pathway by which a ripening hormone called ethylene or by engineering polyamines produced by plants is enforced.For this, researchers have recently isolated a promoter. This enzyme is required in an important step for ethylene synthesis. The ACC synthase producing gene was isolated a few years ago. Antisense RNA technology involves introduction of the gene and its promoter in reverse form in tomato plant cells. The product of the reverse gene blocks the product of the normal gene, thus preventing production of ACC synthase enzyme.

However, not all the products of normal gene are blocked and some amount of enzyme is still produced. As ethylene is not produced in sufficient quantity, ripening does not take place at normal rate and is slowed down.

One problem in this process is that, as the gene is introduced using a bacterium (Agrobacterium), a bacterial gene which is responsible for antibiotic resistance is transferred in plant cells.

Though this gene does not have any effect on humans, efforts are on to find other bacterial markers. The technology has an advantage over the methods now used, like early plucking of tomatoes, as tomato fruits do not ripen at the same time and need plucking three-four times in a year.

The new technology, which can be extended to mangoes and bananas, would prove to be economical and give plant uniformity in terms of ripening. It would also make fruits tastier and enhance their nutrient content by allowing them to be on plants for a longer time. Recently Mehta *et al.* (2002) expressed a yeast ySAMdc; spe2 fused with a ripening-inducible E8 promoter to specifically increase levels of the polyamines spermidine and spermine in tomato fruit during ripening. This led to an increase in lycopene, prolonged vine life and enhanced fruit juice quality.

Fruit

The fruit is defined as a mature ovary and therefore typically includes carpel tissues partially or as a whole. Many fleshy fruit species additionally develop mature fruit tissues, including extracarpellary floral components which include strawberry, pineapple, mulberry, and pomes fruit (apple, pear), in which the receptacle, bracts, calyx, and floral tube, respectively, constitute the majority of mature fruit tissue. Even species with fruit derived from carpel tissue exclusively display a range of developmental programs, ranging from uniform single expanded carpel or drupe of stone fruit to the differentiated carpel tissues giving rise to the peel (flavedo) and multicarpel flesh of citrus and banana.

Many foods are botanically fruit but are treated as vegetables. These include cucurbits (e.g., squash, pumpkin, and cucumber), tomato, peas, beans, corn, and sweet pepper, spices, such as allspice and chilies. Occasionally, though rarely a culinary "fruit" will not be a true fruit in the botanical sense. For example, rhubarb may be considered a fruit, though only the astringent petiole is edible. In the commercial world, European Union rules define carrot as a fruit for the purposes of measuring the proportion of "fruit" contained in carrot jam. In the culinary sense, a fruit is usually any sweet tasting plant product associated with seed(s), a vegetable is any savory or less sweet plant product, and a nut any hard, oily, and shelled plant product.

Although a nut is a type of fruit, it is also a popular term for edible seeds, such as peanut (which is actually a legume), and walnut. Technically, a cereal grain is a fruit termed a caryopsis. However, the fruit wall is very thin and fused to the seed coat so almost all of the edible grain is actually a seed. Therefore, cereal grains, such as corn, wheat and rice are better considered edible seeds, although some references list them as fruits. Edible gymnosperms seeds are often misleadingly given fruit names, e.g. pine nuts, ginkgo nuts, and juniper berries.

Fruits are so varied in form and development, that it is difficult to devise a classification scheme that includes all known fruits. Generally two types of classifications are accepted worldwide based on physiology of fruits. Seeds are ripened ovules; fruits are the ripened ovaries or carpels that contain the seeds. To these two basic definitions, can be added the clarification, that in botanical terminology, a nut is a type of fruit and not another term for seed.

There are three basic types of fruits:

1. Simple fruits
2. Aggregate fruits
3. Multiple fruits

Simple fruits

Simple fruits can be either dry or fleshy and result from the ripening of a simple or compound ovary with only one pistil. Dry fruits may be either dehiscent (opening to discharge seeds), or indehiscent (not opening to discharge seeds). Types of dry, simple fruits (with examples) are:

- achene - (buttercup)
- caryopsis - (wheat)
- fibrous drupe - (coconut, walnut)
- legume - (pea, bean, peanut)
- nut - (hazelnut, beech)
- schizocarp - (carrot)
- silique - (radish)
- utricle - (beet)

Fruits in which part or all of the pericarp (fruit wall) is fleshy at maturity are simple fleshy fruits. Types of fleshy, simple fruits (with examples) are:

- berry - (tomato, avocado)
- stone fruit or drupe (plum, cherry, peach, olive)
- false berry - accessory fruits (banana, cranberry)
- pome - accessory fruits (apple, pear)

Aggregate fruit

An aggregate fruit, or etaerio, develops from a flower with numerous simple pistils. An example is the raspberry, whose simple fruits are termed drupelets because each is like a small drupe attached to the receptacle. In some bramble fruits (such as blackberry) the receptacle is elongated and part of the ripe fruit, making the blackberry an aggregate-accessory fruit. The strawberry is also an aggregate-accessory fruit, only one in which the seeds are contained in achenes. In all these examples, the fruit develops from a single flower with numerous pistils.

Multiple fruit

A multiple fruit is one formed from a cluster of flowers (called an inflorescence). Each flower produces a fruit, but these mature into a single mass. Examples are the pineapple, edible fig, mulberry and orange. In some plants, such as this noni, flowers are produced regularly along the stem and it is possible to see together examples of flowering, fruit development, and fruit ripening.

Seedless ness is an important feature of some fruits of commerce. Commercial cultivars of bananas and pineapples are examples of seedless fruits. Some cultivars of citrus fruits (especially navel oranges and mandarin oranges), table grapes, grapefruit, and watermelons are valued for their seedless ness. In some species, seedless ness is the result of parthenocarpy, where fruits set without fertilization. Parthenocarpic fruit set may or may not require pollination. Most seedless citrus fruits require a pollination stimulus; bananas and pineapples do not. Seedless ness in table grapes results from the abortion of the embryonic plant that is produced by fertilization, a phenomenon known as stenospermocarpy which requires normal pollination and fertilization.

Development

Fruit development and ripening are unique to plants and constitute an important component of human and animal diets. Recent discoveries have revealed the molecular basis of developmental ripening control, suggesting common regulators of climacteric and non climacteric fruits- ripening physiology, and defined a new role for MADS box genes during late stage of floral development. Analyses of fruit-ripening mutants and ripening-related gene expression suggest higher levels of a developmental regulatory cascade which remains to be defined. Examination of the molecular basis of ethylene signaling in tomato has demonstrated conservation of the basic model defined in Arabidopsis, yet with modifications in gene family composition and expression that may represent adaptations to promote successful fruit development and seed dispersal. The role of light signaling in fruit carotenoid accumulation appears to represent a target for practical manipulation of fruit pigmentation and nutrient content. The continuing development of genomics tools, including EST's and cDNA micro arrays, for important fruit crops should help in rapid discovery in fruit development and ripening research.

Fruit development generally occurs in four phases: fruit set, a period of rapid cell division, a cell expansion phase, and ripening/maturation.

Fruit set is normally dependent on pollination. Pollen triggers fruit development indicating that positive signals are generated during pollination. In the absence of these signals, the flowers abscise. Growing pollen tube produces GA and application of GA can induce parthenocarpic fruit, therefore GA appears to be a triggering signal. Lagging slightly behind the growing pollen tube is a wave of increased auxin production by the style and then the ovary. Auxin application can also induce parthenocarpy and so it is thought that GA acts by inducing auxin production. However, most GA deficient mutants are able to produce fruit indicating that this is not the sole mechanism to induce fruit development and in an auxin insensitive tomato mutant, fruit growth is normal.

Continued fruit development usually relies on the continued presence of developing seeds. Seed abortion or removal causes fruit abortion, which can be reversed with auxin application. For example, removal of strawberry "seeds" prevents the development of the receptacle as a "fruit" but if auxin is applied following seed removal, fruit development continues. Commercial crops that produce parthenocarpic (seedless) fruits, such as banana, often show quantitative or qualitative differences in GA or auxin content in the ovary when compared to non parthenocarpic varieties.

The phase of rapid cell division involves all growing parts of the fruit. This is thought to be controlled by the developing seeds. The number of fertilized ovules in a fruit is correlated with both the initial cell division rate and the final size of the fruit. Also, fruits with an uneven distribution of seeds are often lopsided. There is a correlation between cytokinin levels in developing embryos and cell division in surrounding tissues but there is no direct evidence that embryo cytokinin in fact regulates fruit cell division. It is difficult to reconcile the complete development of parthenocarpic fruit with the requirement of embryos for cell division except to say that parthenocarpy represents an abnormal situation.

The cell division phase gradually shifts into the cell expansion phase. The rate and duration of cell division varies among fruits and also among tissues within a fruit. Tissues made up of many small cells at maturity continue dividing while tissues composed of large cells have begun expanding. In tomato the cell division phase lasts approximately 7-10 days while cell expansion lasts 6-7 weeks. Cell expansion accounts for the largest increase in fruit volume, often contributing in excess of a 100 fold size increase. Gibberellins are also associated with fruit expansion and removal of the seeds from pea pods inhibited GA biosynthesis in the pericarp. Many believe that auxins from seeds regulate cell expansion of the pericarp, but auxin application does not always compensate for seed removal, and in an auxin insensitive tomato mutant, fruit growth is normal.

Classification

Fleshy fruits are divided into two classes (Table l):

Climacteric: In such fruits ripening involves a transient Peak in respiration and concomitant increase in ethylene production e.g. banana, apple, mango, tomato, melon, papaya, avocado, and passion fruit. A decrease in oxygen level of pulp is also made out in such fruits.

Non-climacteric: In these fruits respiration does not increase and ethylene evolution is very low during ripening e.g. grapes, citrus, pine apple, bell pepper, cucumber, orange, and lemon. The regulation of ripening is an extremely important factor in meeting the demand of the consumer with fruits of acceptable eating quality. (See Fig. 1)

Climacteric fruits are usually harvested once they have reached maturity which then undergoes rapid ripening during transit and storage. On the contrary, non-climacteric fruits do not ripen after harvest. Thus, in order to attain full ripeness and flavor, these fruits such as grapes, strawberries and oranges, are often harvested once they have fully ripened. The period to attain maturity varies in different fruits. In tomato, it takes nearly six to 8 weeks for the fruit to reach full maturity. Thereafter it starts to undergo the process of ripening. The production of ethylene within the fruit in turn signals the activity of different enzymes resulting in physiological changes such as the change of color from green to red, the softening of the fruit, and the development of its distinct taste and aroma. Normally, farmers pick their produce while they are still green

Fig. 1

and the ripening is then induced by spraying the fruits with Ethylene gas when they reach their destination. For prolonged haul, fruits and vegetables are put in cold storage to prevent damage and delay their ripening. However, such post harvest practices have several constraints e.g. fruits that have been harvested prematurely may result in poor taste.

Table 1. Classification of fruits and fruit vegetables based on climacteric and non-climacteric ripening patterns

	CLIMACTERIC	*NON-CLIMACTERIC*
TEMPERATE	Apple	Cherry
FRUIT	Pear	Grape
	Peach	Strawberry (?)
	Apricot	
	Plum	
'VEGETABLE'	Melon	Cucumber
FRUIT	Tomato	
	Watermelon	
COMMON	Avocado	Orange

TROPICAL FRUIT	Banana	Grapefruit
	Mango	Lemon
	Papaya	Lime
	Fig	Olive
	Guava (?)	Pineapple
	Passion fruit	Litchi
		Persimmon

Though the specific role of climacteric respiration in fruit ripening remains unclear, the role of ethylene as a coordinator of ripening in climacteric species serves to facilitate rapid and coordinated ripening. A great deal is known regarding specific downstream ripening processes in a number of climacteric and non-climacteric species, yet little is known about the regulation of ripening in non-climacteric fruit or the upstream regulation of ethylene in their climacteric counterparts. Recent evidence of the MADS box regulation of ripening in both tomato and strawberry suggests common regulatory mechanisms operating early in both climacteric and non -climacteric species (Vrebalov *et al.*, 2002; Pietro *et al.*, 2006). The elucidation of the molecular basis of such early and common events represents an active frontier in fruit.

Non-climacteric fruits

a. Fruits which do not show any response to ethylene.

b. Fruits which lack a climacteric, autocatalytic rise in ethylene evolution but reveal ripening-related responses to exogenous ethylene.

Furthermore, it may not be unreasonable to assume that endogenous ethylene plays a role in the ripening of the ethylene-responsive, non-climacteric fruits as well.

Presently we shall discuss recent advances in our understanding of molecular biology of climacteric fruit and its ripening processes which affect later fruit maturation and ripening of climacteric and non- climacteric fruits.

Fruit ripening

Ripening represents the shift from the protective function to dispersal function of the fruit. Ripening occurs synchronously with seed and embryo maturation. In dry fruits (cereals, nuts, dandelions) ripening consists of desiccation and is considered maturation. Ripening in fleshy fruits is designed to make the fruit appealing to animals that eat the fruit as a means for seed dispersal. Ripening involves the softening, increased juiciness and sweetness, and color changes of the fruit. Ripening has been most intensively studied in tomato. Ethylene is a major regulator of the ripening process. Inhibition of ethylene with inhibitors, transgenic approaches or mutants blocks ripening. Exogenous ethylene accelerates ripening. There are also developmental factors involved because fruit does not attain competence to respond to ethylene until near the end of the cell expansion phase (the mature green stage). Several genes associated with ripening are ethylene inducible. This occurs transcriptionally in most genes but at least one is known where mRNA accumulation is regulated post-transcriptional. None of these genes are induced until competence for ethylene response is attained.

Fleshy fruits such as tomato undergo a ripening process in which the biochemistry, physiology, and structure of the organ are developmentally altered to influence appearance, texture, flavor, and aroma in ways designed to attract seed-dispersing organisms (Seymour *et al.*, 1993). Though the specific biochemical programs resulting in ripening phenomena vary among species, changes typically include (1) modification of color through the alteration of chlorophyll, carotenoid, and/or flavonoid accumulation; (2) textural modification via alteration of cell terror and cell wall structure and/or metabolism; (3) modification of sugars, acids, and volatile profiles that affect nutritional quality, flavor, and aroma; and (4) generally enhanced susceptibility to opportunistic pathogens (likely associated with the loss of cell wall integrity). Though fruit species are classically defined physiologically on the basis of the presence (climacteric) or absence (non climacteric) of increased respiration and synthesis of the gaseous hormone ethylene at the onset of ripening (Lelievre *et al.*, 1997), fruit displaying both ripening programs follow the general developmental changes. Figure 2 shows relative changes in cell division, cell expansion, respiration, ethylene synthesis, fruit softening, and carotenoid accumulation over the course of fruit development. The time from anthesis (a) to mature green (MG; fully expanded unripe fruit with mature seed), breaker (BR; first visible carotenoid accumulation), and red ripe (RR) can vary substantially among cultivars. The time line shown is for a medium-/large-fruit cultivar such as the breeding line MH1 (5 to 7 cm diameter mature fruit). dpa, days after anthesis Figure 4 shows major biochemical changes during tomato fruit development and ripening.

The tomato never-ripe mutation blocks fruit ripening and is insensitive to ethylene. The mutated gene is similar to the ethylene receptor isolated from Arabidopsis, suggesting that never-ripe is an ethylene receptor mutant. NR mRNA is not expressed until the mature green stage, suggesting that lack of this ethylene receptor might be related to the lack of competence to respond to ethylene at earlier stages.

Ethylene production is autocatalytic. That is, exposure to ethylene stimulates the synthesis of more ethylene. This occurs because the genes for the biosynthetic enzymes (e.g. ACC synthase) are ethylene inducible. The result is a positive feedback loop. Furthermore, the Never-ripe gene is ethylene inducible, resulting in a positive feedback loop for ethylene sensitivity as well. Both these factors contribute to the dramatic burst of ethylene production during ripening.

Fruit softening involves a partial breakdown of cell walls. Several enzymes are known to be involved in this process. Polygalacturonase hydrolyzes bonds in pectins. The gene for this enzyme is ethylene inducible.

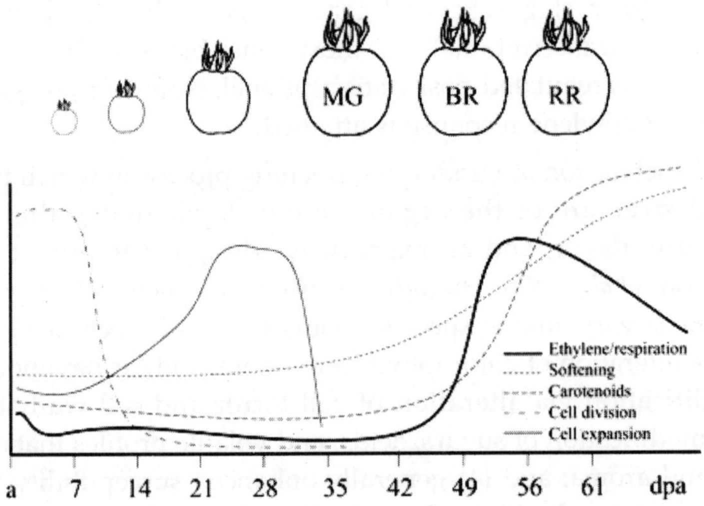

Fig 2. Development changes during tomato fruit ripening

Changes in fruit color involve changes in the expression of pigment biosynthetic genes. The major pigment in tomato is a carotenoid. The first committed step in carotenoid biosynthesis is catalyzed by phytoene synthase, and the gene for this enzyme is induced by ethylene.

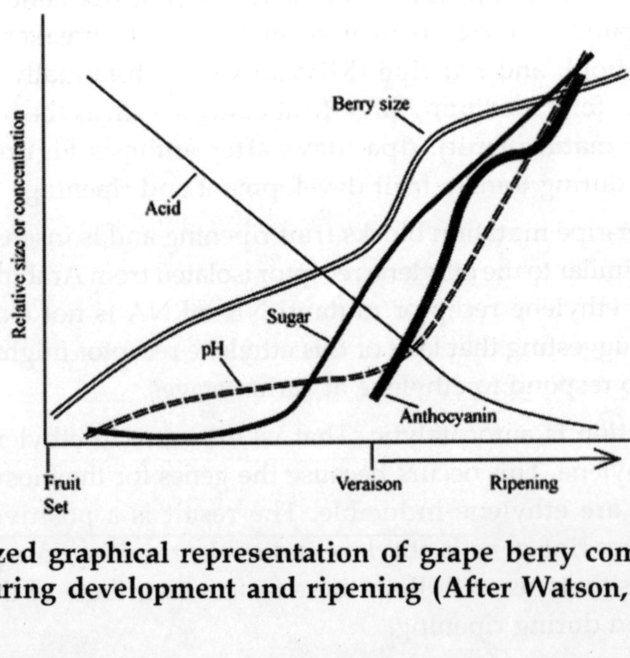

Fig 3. Generalized graphical representation of grape berry compositional changes during development and ripening (After Watson, 2003).

The onset of grape ripening begins during the final stage of berry growth when berries begin to soften and size increases due to cell enlargement. The sugar content of the berry increases rapidly, acidity decreases, and the pH increases (Fig. 3). During ripening, berry skins lose chlorophyll and begin to synthesize and accumulate phenolic compounds that are responsible for development of characteristic colors: yellow-gold (flavonols) and pink and red (anthocyanins) colors (Watson, 2003). As the berries approach full maturity, berry size reaches a maximum and sugar accumulation slows. At the start of ripening, aromas and flavors begin to develop, primarily in the skins of most varieties. Grape aroma compounds occur as both volatile aromatic compounds and non-volatile aroma precursors and are extracted during processing and fermentation (Watson, 2003). Aroma and flavor complexity typically increases during the later stages of ripening. Canopy management and cropping levels can have a dramatic effect on aroma and flavor development. For more complete discussions of grape berry development and the ripening process see the review article by Coombe (1992), (2003), Bisson (2001) and Kennedy (2002).

In brief fruit ripening is a cascade of biochemical events, which transfer physiologically mature but inedible fruit into edible, tasty produce.

Role of ethylene in fruit ripening

Ethylene plays a crucial role in the initiation and continuation of ripening of climacteric fruit such as apples and bananas. However, overall, ripening of non-climacteric fruit is not believed to require ethylene, although its role is currently being re-evaluated (Pietro *et al.*, 2006)

As stated earlier the classification of fruit as climacteric and non-climacteric is useful in understanding general patterns in fruit ripening ;though this categorization is arbitrary as some fruits exhibit variant behavior e.g. guava, strawberry, etc. In plums and apples, which are classically classified as climacteric fruits, there are several genotypes that have a suppressed respiratory climacteric (Abdi *et al.*1998). The suppressed climacteric trait is dominant in plums but it has not been characterized at the molecular level. It is concluded that the presence or absence of a respiratory climacteric in fruit attached to the vine is affected by environmental factors. It is also postulated that the respiratory climacteric is not an essential part of ripening but an artifact caused by stress or detachment. These interesting observations require more research.

For instance, during ripening of wild-type melon fruit, attached to the plan, may involve climacteric ethylene production without a corresponding increase in respiration rate. Some workers have shown that application of exogenous ethylene to melons anti-sensed for 1-aminocyclo-propane-1-carboxylate oxidase (ACO) stimulated oxygen consumption only in detached fruit.

Fruit quality parameters

Ethylene biology and its integration with other fruit quality parameters e.g. aroma emission, fruit softening, nutritional contents and modification of color are important and their insight will help further improvement in the management of fruit ripening.

Aroma

An aroma compound, also known as odorant, aroma, fragrance, flavor, is a chemical compound that has a smell or odor. A chemical compound has a smell or odor when two conditions are met: the compound needs to be volatile, so it can be transported to the olfactory system in the upper part of the nose, and it needs to be in a sufficiently high concentration to be able to interact with one or more of the olfactory receptors. Aroma is an important characteristic of fruit quality. Flavoring compounds:

1. ***Grapefruit:*** Mercaptan is the common name for a natural organic compound found in grapefruit. It is a monoterpenoid that contains a thiol (or mercaptan) functional group. Volatile thiols typically have very strong, often unpleasant odors which can be detected by humans in very low concentrations. Grapefruit mercaptan has a very potent, but not unpleasant, odor and it is the chemical constituent primarily responsible for the aroma of grapefruit. This characteristic aroma is a property of only the R-sterioisomer.
2. Menthol is a covalent organic compound made synthetically or obtained from peppermint or other mint oils. It is a waxy, crystalline substance, clear or white in color, which is solid at room temperature and melts slightly above. The main form of menthol occurring in nature is (-)-menthol, which is assigned the (1R,2S,5R) configuration. Menthol has local anesthetic and counterirritant qualities, and it is widely used to relieve minor throat irritation.
3. Benzaldehyde (C_6H_5CHO) is a chemical compound consisting of a benzene ring with an aldehyde substituent. It is the simplest representative of the aromatic aldehydes and one of the most industrially used members of this family of compounds. At room temperature it is a colorless liquid with a characteristic and pleasant almond-like odor: benzaldehyde is an important component of the scent of almonds, hence its typical odor. It is the primary component of bitter almond oil extract, and can be extracted from a number of other natural sources in which it occurs, such as apricot, cherry, and laurel leaves, peach seeds and, in a glyciside combined form (amygdalin), in certain nuts and kernels. Currently benzaldehyde is primarily made from toluene by a number of different processes.
4. cis-3-Hexenal, also known as (Z)-3-hexenal and leaf aldehyde, is colorless liquid and an aroma compound with an intense grassy-green odor of freshly cut

green grass and leaves. It is produced in small amounts by most plants and it acts as an attractant to many predatory insects. It is also a pheromone in many insect species.

cis-3-Hexenal is an aldehyde. It is relatively unstable and isomerizes into the conjugated trans-2-hexenal. The related alcohol cis-3-hexen-1-ol is much more stable. It has a similar but weaker smell and is widely used in flavors and perfumes.

5. Citral, or 3,7-dimethyl-2,6-octadienal or lemonal, is either of a pair of terpenoids with the molecular formula $C_{10}H_{16}O$. The two compounds are double bond isomers. The trans isomer is known as geranial or citral A. The cis isomer is known as neral or citral B.

Citral is the major constituent of the oil of lemongrass and the oils of other plants in the genus *Cymbopogon*. It is also present in the oils of verbena, lemon and orange.

Geranial has a strong lemon odor. Neral has a lemon odor that is less intense, but sweeter. Citral is therefore an aroma compound used in perfumery for its citrus effect. Citral is also used as a flavor and for fortifying lemon oil.

Citral is used in the synthesis of vitamin A, ionone and methylelinone.

6. Methyl butyrate (methyl butanoate) is an ester with a fruity fragrance resembling apples or pineapples. At room temperature, it is a colorless liquid. It is used in perfumes and as a food flavoring. Methyl butyrate is well known for its distinctive properties, such as biking and swimming through graduated cylinders.
7. Ethyl methylphenylglycidate, commonly known as "strawberry aldehyde", is an organic compound used in the flavor industry in artificial fruit flavors, particularly strawberry.

It is classified as an ester and an epoxide; but despite its common name, it is not an aldehyde. Ethyl methylphenylglycidate is a colorless to pale yellow liquid that is insoluble in water. It has a taste and odor that is fruity and reminiscent of strawberries.

The compound has been used extensively to examine the role of ethylene in aroma production. Fruits treated with 1-MCP, produce reduced levels of volatiles. However, several attributes of the fruit contribute to this aspect which includes timing of harvest and maturity of the fruit. Similarly fruits stored at low temperature suppress ethylene production and hence aroma. In most fruits esters constitute an important part of overall aromatic suite of volatiles. These compounds are generally formed by esterification of alcohols and acyl-CoA derived from fatty acid and amino acid metabolism catalyses by o-acyltransferase. This has been reported and investigated in several horticultural plants.

By applying 1-MCP at different times during the ripening process it has been possible to identify differential ripening processes dependent on ethylene action. It has also become vivid that ethylene has a continuum role in blending many of the biochemical processes accompanying fruit ripening. In two instances of fruit ripening, 1-MCP application disturbed the normal ripening process though became yellow. However, they remained hard and tasteless.Application of the compound in banana changed the aroma volatiles. Alcohols concentration enhanced though their related esters decreased. The effect on esters was differential. Large number of enzymes concerned with specific fruit volatiles have been isolated, cloned and expressed in several fruits. Thus carotenoid cleavage enzyme dioxygenase has been shown to contribute to the flavor and aroma in tomato. Thus manipulation of this enzyme aroma could be enhanced. It will be worth while to look into total pathway rather than individual enzyme.

Recently DNA micro arrays technique has been used to seek deep insights into fruit ripening and aroma biosynthesis. Some workers have used molecular and biochemical techniques to seek differences during ripening of cultivated and wild strawberry species. These studies have suggested that domestication of the cultivated strawberry led to the selection for specific alleles in the cultivated species. This contributed to a strongly modified flavor profile. In any case sensory studies are essential regarding fruit ripening. Through these studies it is demonstrated that a variety of apple treated with 1-MCP was less ripe but was more 'acceptable' than controlled apples.

The discovery and application of 1-MCP has helped significantly in understanding the role of ethylene in fruit ripening. 1-MCP is registered as Smart FreshTM in many countries for post harvest use on a variety of important horticultural crops. This compound has revolutionized our understanding of fruit ripening. [Here has been an explosion of research papers on the effects of 1-MCP on fruit ripening. The compound acts by irreversibly binding to ethylene receptors where eventual recovery occurs via the synthesis of new ethylene receptors. The blocking of ethylene perception with 1-MCP has a variety of effects on respiration, ethylene production, volatile production, chlorophyll degradation and other color changes, protein and membrane changes, softening, disorders and diseases, acidity and sugars (Blankenship and Dole, 2003). The compound is an important tool for examining the role of ethylene in fruit ripening. Based on the negative regulator model of receptor function, 1-MCP could work by permanently locking the ethylene receptors in the active, inhibitory state. 1-MCP and other ethylene antagonists may bind to the copper cofactor of the ethylene receptor but perhaps due to steric hindrance, this interaction also fails to switch off the receptors.

Textural changes

Ripening of climacteric fruits is associated with flesh softening due to extensive cell wall degradation (Brummell *et al.*, 2001). Though it is known that blocking ethylene production or perception can maintain fruit firmness in some fruits, the mechanism of fruit softening is still in the process of being elucidated.

A range of enzymes are involved in modifying cell wall components and these have been extensively studied in several fruits (Brummell *et al.*, 2001). A recent comprehensive study of the role of an important softening enzyme (α-galactosidase) in pears was conducted at the biochemical and molecular level. Using α-galactosidase cDNA clones, it was shown that a family of seven genes was differentially expressed during ripening. For example, transcripts hybridizing with Japanese pear α-galactosidase (PpGAL) cDNA clones, PpGAL1 and PpGAL4 played a crucial role in 'La France' pear fruit softening. The increase in their expression at the onset of fruit ripening is partly due to up-regulation by ethylene. Conversely, the decrease in transcripts hybridizing with PpGAL6 and PpGAL7 in ripe fruit may be partly due to down-regulation by ethylene, while changes in transcripts hybridizing with PpGAL2, PpGAL3 and PpGAL5 may be due to another mechanism unrelated to ethylene action. This demonstrates that the different members of the α-galactosidase gene family are differentially regulated. The application of this work may lead to regulation of the rate and extent of pear softening.

DNA micro array analysis permits comparisons of genes that are differentially expressed in fruit with different traits. The technique uses a combination of sequence information, information on the biological function of sequence homologues and additional gene expression data to select candidate genes for further analysis. Although information generated in these studies is correlative, the approach can identify candidate genes for further work. Salentijn *et al.* (2003) used DNA-microarrays representing 1,701 strawberry cDNAs to simultaneously hybridize two RNA populations derived from ripe fruit of a soft and a firm strawberry cultivar. They found that two genes encoding proteins catalyzing successive reactions in lignin metabolism (cinnamoyl CoA reductase and cinnamyl alcohol dehydrogenase) are possibly new candidates for fruit-firmness.This is a powerful technique to identify new genes involved in fruit ripening, but gene functions must still be confirmed empirically. A detailed functional analysis in strawberry using over expression and inhibition of cinnamoyl CoA reductase was likely to provide further insights into fruit softening.

Nutritional content

Manipulating the nutritional content of horticultural produce is an emerging and growing challenge for plant scientists. A large number of molecular tools have been utilized to study and manipulate the nutritional content of horticultural produce. The tomato fruit has been the major subject of this research due to its horticultural importance and the extensive information already available on tomato genetics, physiology, cytogenetics and taxonomy. For example, the isoprenoid biosynthetic pathway can be manipulated to optimize the levels of carotenoids (α-carotene (vitamin A precursor) and lycopene [and flavonoid antioxidants in tomato]. These approaches utilize direct pathway manipulation to improve photochemical changes. However, approaches such as targeting genes operating in the signal pathway of the biosynthetic pathway may be more effective. For example, the molecular basis of carotenoid accumulation is currently being resolved in tomato fruit. This is an exciting area of research, which will make it possible to optimize phytochemicals during fruit ripening.

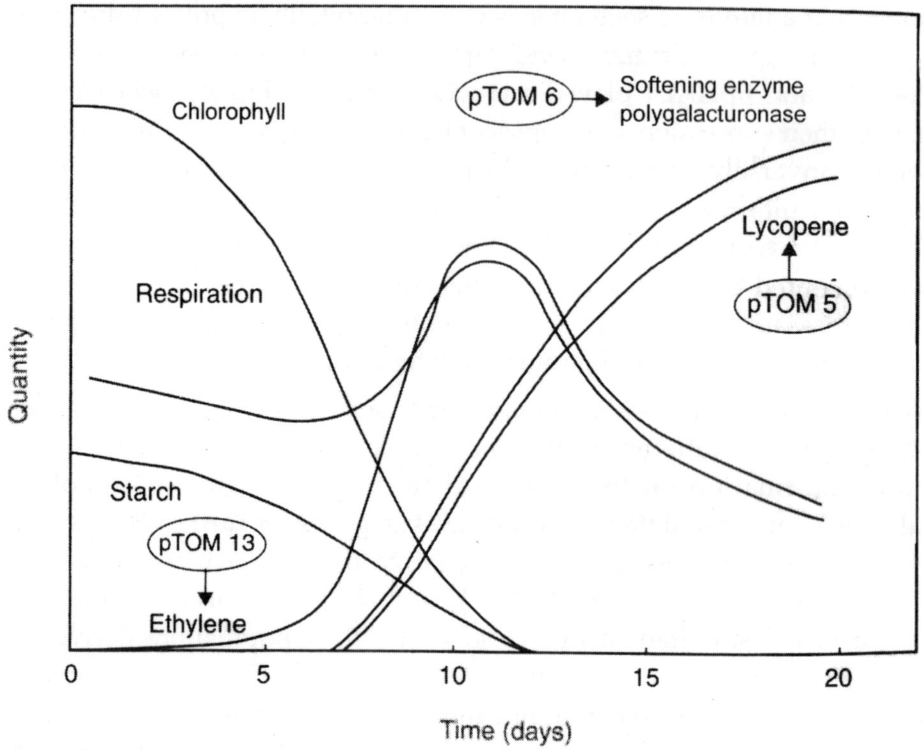

Fig. 4. Biochemical changes during tomato fruit ripening. Biochemical changes occurring from green stage to ripe stage are indicated. Note ripening is accompanied by the onset of bursting of respiration. There is also enhancement of ethylene. Green color is lost due to degradation of chlorophyll, starch is converted into sugar and adds to the sweetness of the fruit. Lycopene, the red color is synthesized.

Molecular Biology of Fruit Ripening

Ethylene production

The role of ethylene is pivotal to understanding fruit ripening in climacteric fruit. The pathway of ethylene production is well known. The enzymes 1-amino-cyclopropane-1-carboxylate synthase (ACS) and ACO are pivotal enzymes in its production (Fig. 6). The enzymes, and the genes encoding them, have been shown to be highly regulated. It is shown that there is a unique and overlapping gene activity pattern during the growth and development of Arabidopsis. This analysis revealed a combination of different expressions in time and space among the various gene family members during plant development, which was variably regulated by different known inducers of ethylene production (anaerobes, wounding, cold and heat). These studies provide a basis for future work on the role of ACS in ethylene production and fruit ripening.

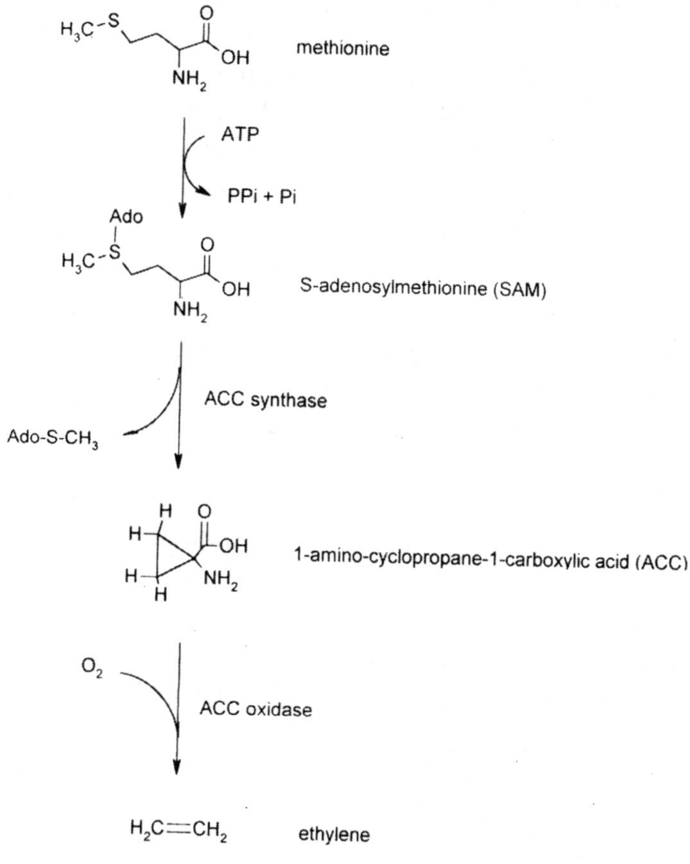

Fig 6. The pathway of Ethylene biosynthesis from methionine. Two committed steps from SAM are involved. The first is catalyzed by ACCS and generating ACC. The latter is converted to ethylene through ACCO.

The ability of molecular techniques to increase our understanding of ethylene and support various theories on the mechanisms of fruit ripening has been variously discussed by Nakatsuka et al and Barry *et al.* Two systems of ethylene synthesis have been proposed. System 1 is the low levels of ethylene production and action in immature climacteric and typical non-climacteric fruit, while System 2 is the autocatalytic production of ethylene associated with ripening in climacteric fruit. Sufficient physiological data have supported the concept of two systems but the existence of these two systems was only confirmed through molecular techniques. Nakatsuka *et al.* showed that different forms of ACS and ACO are expressed in System 1 and System 2 in tomatoes]. More recently, transcripts corresponding to four ACS genes, namely LEACS1A, LEACS2, LEACS4 and LEACS6, are reported in tomato fruit. Expression analysis using the ripening mutants Rin and Nor in combination with ethylene treatments further demonstrated that each ACS transcript is regulated uniquely. Barr *et al.* proposed a model in tomato whereby System 1 ethylene is regulated by the expression of LEACS1A and LEACS6. During the transition from System 1 to System 2, the RIN locus plays a critical role leading to increased expression of LEACS1A and induction of LEACS4. System 2 ethylene synthesis is subsequently initiated and maintained by ethylene-dependent induction of LEACS2. It is to be noted that the molecular control of ethylene biology may differ among species. For example, only one form of ACS has been reported in peach, but future work may unravel more.

Manipulating the post harvest gas environment (i.e. lowering O_2 or increasing CO_2 in the storage atmosphere) can lower respiration rates and suppress ethylene production in several cases. This method is commercially for apple storage for many years. However, the role of CO_2 in ethylene production and regulation is not fully known. CO_2 is an essential co-factor for ACC oxidase and can therefore stimulate ethylene production. However, it has also been suggested that the inhibition of ethylene production by CO_2 operates at least partially, at the level of ACO. Mathooko *et al.* (2001) showed that CO_2 suppresses ethylene production in peach fruit by inhibiting the accumulation of ACS mRNA and to some extent ACO mRNA. De Wild *et al.* (2003; 2005) recently proposed that inhibition of ethylene production together with prevention of ACC accumulation by CO_2, might occur at a site before the conversion of ACC to ethylene. Such researches may develop novel ways to manage the storage life of fruit.

Ethylene perception and signal transduction

Enormous progress has been made towards elucidating the ethylene-signaling pathway (Fig. 6). This pathway is essentially linear, consisting of membrane-associated ethylene receptors leading to transcription factors that regulate gene expression in the nucleus. Production of mutants in Arabidopsis insensitive to ethylene and then clone the corresponding genes, has helped in the identification of several genes that are essential for the perception of ethylene. Ethylene is perceived by a family of membrane-associated receptors. In *Arabidopsis*, there are five ethylene receptors belonging to two subfamilies, all of which bind to ethylene. Subfamily 1 receptors (ETR1 and ERS1) contain all five of the functionally defined His kinase sequence motifs, whereas Subfamily 2 members (ETR2, EIN4, and ERS2) are substantially different, carrying only one or two of the motifs.

Burg and Burg (1967) were pioneers to propose that ethylene binding is mediated by a transition metal co-factor and this has since been demonstrated in *Arabidopsis*. It is generally predicted that ethylene binding results in the inactivation of receptor function.

In fact Bleecker and Kende (2002) proposed that ethylene receptors act as negative regulators of downstream ripening processes, whereby in the absence of ethylene these receptors actively suppress expression of ethylene responsive genes.. On the contrary the ethylene receptors, EIN2, EIN3, EIN5, and EIN6 are positive regulators of ethylene responses, which regulate the expression of its immediate target genes such as ethylene response factor 1 (ERF1). Thus, a transcriptional cascade that is mediated by EIN3/EIN3-like (EIL) and ERF proteins leads to the regulation of ethylene controlled gene expression (Figs. 4-5). Recent studies are providing detailed insights into the ethylene receptors. It is becoming evident that specific His kinase activity as well as receiver domains in ETR1 ethylene receptors play a quantitative role in recovery of ethylene treatment and in ethylene. A role for the His kinase domain in ethylene signaling and gave insight into how the ethylene signal is transduced from ethylene receptors to downstream signaling components is evident. The C-terminal region of ETR1, consisting of a His kinase domain and a putative receiver domain, is likely to function in signal output. Such studies are rapidly developing and several comprehensive reviews on the ethylene signaling pathway have been published (Adams-Phillips *et al.*, 2004; Golding *et al.*, 2005). Fruit development can generally be considered to occur in four phases: fruit set, a period of rapid cell division, a cell expansion phase, and ripening/maturation.

Fruit set involves the decision whether to abort the ovary or proceed with fruit development. Fruit set is normally dependent on pollination. Pollen triggers fruit development indicating that positive signals are generated during pollination. In the absence of these signals, the flowers abscise. Growing pollen produces GA and application of GA can induce parthenocarpic fruit, therefore it is believed that GA is a triggering signal. Lagging slightly behind the growing pollen tube is a wave of increased auxin production by the style and then the ovary. Auxin application can also induce parthenocarpy and so it is thought that GA acts by inducing auxin production. However, most GA deficient mutants are able to produce fruit indicating that this is not the sole mechanism to induce fruit development and in an auxin insensitive tomato mutant, fruit growth is normal.

Continued fruit development usually relies on the continued presence of developing seeds. Seed abortion or removal causes fruit abortion, which can be reversed with auxin application. For example removal of strawberry "seeds" prevents the development of the receptacle as a "fruit" but if auxin is applied following seed removal, fruit development continues. Commercial crops that produce parthenocarpic (seedless) fruits, such as banana, often show quantitative or qualitative differences in GA or auxin content in the ovary when compared to non parthenocarpic varieties.

The phase of rapid cell division involves all growing parts of the fruit. This is thought to be controlled by the developing seeds. The number of fertilized ovules in a fruit is correlated with both the initial cell division rate and the final size of the fruit. Also, fruits with an uneven distribution of seeds are often lopsided. There is a correlation between cytokinin levels in developing embryos and cell division in surrounding tissues but there is no direct evidence that embryo cytokinin in fact regulates fruit cell division. It is difficult to reconcile the complete development of parthenocarpic fruit with the requirement of embryos for cell division except to say that parthenocarpy represents an abnormal situation.

The cell division phase gradually shifts into the cell expansion phase. The rate and duration of cell division varies among fruits and also among tissues within a fruit. Tissues made up of many small cells at maturity continue dividing while tissues composed of large cells have begun expanding. In tomato the cell division phase lasts approximately 7-10 days while cell expansion lasts 6-7 weeks. Cell expansion accounts for the largest increase in fruit volume, often contributing in excess of a 100 fold size increase. Gibberellins are also associated with fruit expansion and removal of the seeds from pea pods inhibited GA biosynthesis in the pericarp. Many believe that auxins from seeds regulate cell expansion of the pericarp, but auxin application does not always compensate for seed removal, and in an auxin insensitive tomato mutant, fruit growth is normal.

When ethylene binds to the membrane-bound two-component receptor this initiates a phosphorylation cascade, which activates transcription factors and causes them to either bind or unbind from promoter sequences of specific ripening related genes.

Table 2: Total and Fruit ESTs Available from Crop Species

Species	*Common Name*	*Total ESTs*	*Fruit ESTs*
Ananas comosus	Pineapple	0	0
Capsicum annuum	Pepper	22,433	8,580
Citrus x *paradisi*	Grapefruit	312	312
Citrus sinensis	Orange	16,180	5,623
Citrus unshiu satsuma	Orange	2,561	2,561
Citrullus lanatus	Watermelon	593	593
Coffea arabica	Coffee	453	0
Cucumis melo	Melon	50	50
Cucumis sativus	Cucumber	358	0
Fragaria x *ananassa*	Strawberry	58	0
Lycopersicon esculentum	Tomato	150,228	41,186
Malus x *domestica*	Apple	945	903

Musa acuminata	Banana	27	27
Persea americana	Avocado	0	0
Prunus armeniaca	Apricot	4,535	4,535
Prunus avium	Sweet cherry	21	21
Prunus domestica	Plum	0	0
Prunus dulcis	Almond	3,845	0
Prunus persica	Peach	10,185	10,185
Pyrus communis	Pear	212	212
Vitis spp	Grape	144,483	59,735
	Total	357,483	134,565
Total excluding tomato and grape		62,726	33,644

Genetics of Fruit ripening

Light-controlled regulatory genes

The most obvious manifestation of the ripening process is the accumulation of colored pigments. In tomato this is predominantly the carotenoid lycopene, but ?-carotene (a vitamin A precursor) and flavonoids also increase. All of these compounds have proven health benefits as part of the human diet Light strongly enhances the accumulation of pigments during ripening, although it is not a global regulator of ripening since it does not affect softening or sugar and acid metabolism. Various receptors sense light in plants, but the mRNA abundance of the phytochrome A gene (PHYA) was strongly increased during tomato ripening PHYA protein is believed to move rapidly from the cytoplasm to the nucleus in the presence of light, where it can both directly affect the regulation of target transcription factor genes, as well as transmit a signal via a regulatory cascade probably involving a phosphorelation system.

PHYA signal transduction is complex and not completely understood, though some branches of the pathway act via a multisubunit COP9 proteasome, which in the dark keeps the abundance of transcription factors, such as HY5, low by proteolysis. The degradation of HY5 in the dark involves COP1, a protein that labels other proteins for degradation and may be a ubiquitin ligase. In the light, COP1 is inactivated so the abundance of HY5 and other transcription factors increases; these transcription factors then up-regulate genes that promote light responses and photo morphogenesis. COP1 is thus a negative regulator of light responses, as are the proteins DET1 and FUS, although how these are integrated into the pathway is unclear.

Post-transcriptional gene silencing of the positive regulator of light signaling SlHY5 produced tomato plants showing defects in light responses reduced carotenoid contents and susceptibility to high intensity irradiation. In contrast, post-transcriptional gene silencing of the negative regulator of light-signaling SlCOP1LIKE produced tomato plants with exaggerated photo morphogenesis and elevated pigmentation, including increases in ripe fruit carotenoid levels of up to 43%.Natural tomato mutants exhibiting exaggerated photo morphogenesis and highly pigmented fruit have been known for many years, such as high pigment-1 (hp-1) and high pigment-2 (hp-2). These non-allelic mutants have an identical phenotype, suggesting that the proteins affected are part of the same regulatory pathway. The hp-1 phenotype is due to an amino acid substitution in the tomato homologue of AtDDB1. However, attempted complementation of the mutation with a constitutively expressed functional SlDDB1/HP1 transgene did not produce viable plants, suggesting that ectopic expression of this gene is lethal. The hp-2 mutant is caused by a deletion of three amino acids in the tomato homologue of the Arabidopsis DET1 protein which interacts with DDB1.

Fruit-specific post-transcriptional gene silencing of SlDET1 in wild-type plants reproduced the natural mutant phenotype but only in the fruit, avoiding the deleterious effects of a lack of competent DET1 protein on the vegetative portions of the plant. Significant increases in both fruit carotenoids (up to 100% for lycopene, up to 900% for α-carotene) and flavonoids (up to 250%) were observed, showing that manipulation of a regulatory gene in the light signaling pathway can substantially increase flux through two unrelated biochemical pathways. Constitutive over expression of the tomato blue light photoreceptor CRYPTOCHROME2 gene also enhanced the contents of both carotenoids and flavonoids in fruit, but numerous detrimental vegetative phenotypes also.

The manipulation of regulatory genes of the light signaling pathway can dramatically affect the pigment content of fruit, an important component of ripening. Provided the changes can be restricted to the fruit, and so adverse effects on plant light responses and growth are avoided, there is potential to greatly increase the nutritional value of the fruit since most plant pigments are dietary antioxidants beneficial to human health. The experiments also show that pigment accumulation in tomato fruit is subject to control by several regulatory pathways. Light promotes lycopene accumulation but is not essential for it, suggesting that pigment accumulation has light-dependent and independent components.The CNR gene is necessary for PSY mRNA accumulation as well as synthesis of carotenoid precursors, implying some developmental control. Pigment accumulation is also affected by ethylene, which regulates PSY mRNA accumulation and is profoundly inhibited in the ethylene-insensitive Nr mutant. However, whether or not direct interaction occurs between the light and ethylene signal transduction pathways has yet to be determined.

Genes encoding enzymes involved in cell wall metabolism

In tomato fruits TOM6 was the first ripening cDNA cloned and sequenced and it encoded polygalacturonase (PG). The enzyme is catalyzes the hydrolysis of polygalacturonic acid chains in unmethylated regions of pectin. The PGs are either endo- or exo-PGs and are synthesized in several different locations. Three isoforms PG1, PG2A and PG2B have been identified. PG is transcriptionally activated with the beginning of ripening and ripening fruits abound in mRNA and protein. The other cell wall modifying enzyme which removes their methyl groups from methylated polygalacturonic acid in pectin is pectinesterase (PE).

There are several isoforms of PE and these are known to be encoded by at least two distinct multigene families. These groups of PE isoforms are expressed in different organs. Many other cell wall modifying enzymes have been reported from tomato fruits and these include cellulose, galactanase but their exact role is still obscure.

Carotenoid biosynthesis genes

Possibly there are two genes involved for carotenoid biosynthesis and these are phytoene synthase (PSY) and the original cDNA was named TOM5 and the gene which expressed in ripening fruit is called Psy 1. The other related gene is Psy 2 which is concerned with chloroplast carotenoid production prior to ripening. During ripening of fruits the concentration of Psy 1 mRNA increases enormously and is concomitant with climacteric activity. Thus PSY protein appears to be the chief protein associated with lycopene production in ripening fruits Many other cDNAs encoding mRNAs also enhance during tomato fruit ripening and these are histidine decarboxylase, glutamate decarboxylase and an integral membrane-channel protein as well as some heart-shock protein. However, their precise function is still conjectural.

Ripening Gene Promoters

In experiments with transgenic tomato fruits, four fruit gene promoters have been employed and these are 2A 11 (fruit specific promoter), E8 (ethylene regulated ripening promoter), ACOl (ACC oxidase promoter) and PG gene promoter.

Of the 150 clones isolated, grouped and characterized three groups of clones were ripening related and these are pTOM5, pTOM6 and pTOM13 and are induced by ethylene. In the following precise role(s) of these clones in the ripening process are given:

Clone	*gene product*	*function*	*role in Ripening*
pTOM5	PS1	lycopene synthesis	red color
pTOM6 PG	CW degradation	soften	
pTOM13 ACCO	Ethylene synthesis	ripening Trigger	

Table 3: The expression of pTOM6 in transgenic tomato plants. The data represents Northern blot hybridization of samples from transgenic tomato fruit with sense and antisense (AS) pTOM6 probes.

Fruit	*Non-transgenic Mature green*		*Non-transgenic Ripe*		*Transgenic Mature Ripe*		*Transgenic Ripe*	
Probe	Sense	AS	Sense	AS	Sense	AS	Sense	AS
Signal	ND	ND	Intense band	ND	ND	Intense bands	Faint bands	Faint bands
Size of band			PG mRNA			<PG mRNA	PG mRNA	<PG mRNA

AS = Antisense; ND = Not detected

The Genetic manipulation of ripening in Fruits

Genetically it has been possible to modify different processes like texture change, synthesis of hormone ethylene and color formation and several commercial products based on these manipulations are available in the market. These modifications have been accomplished chiefly by gene silencing or through inactivation of pertinent plant gene.

The former involves the inactivation of a specific gene in a targeted manner whereas the latter process is achieved through antisense technology or by sense suppression or suppression technology (Fig. 8). It may be mentioned that there are two types of sense suppression and both involve the transfer of additional copies of sense genes to transgenic plants.

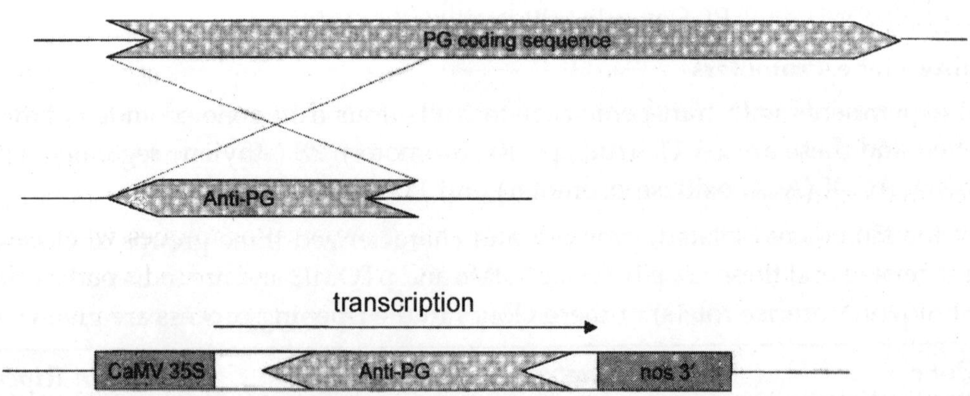

Fig. 8. Construction of an antisense PG vector. The diagram indicates the antisense concept in operation. It is to be noted that all or a portion of a gene sequence is inserted into a vector in a wrong way, in order the opposite strand from normal is transcribed to produce an antisense transcript.

On the other hand inhibition of gene activity may be accomplished by inactivation of transcription through DNA sequence homology between the promoters of the transgene and the endogenous gene, or through post-transcriptional RNA degradation in case there is sequence homology between the two transcripts. We shall briefly discuss the three processes: genetic manipulation of fruit softening; ethylene biosynthesis, and modification of color. These are described below sequentially:

Ethylene biosynthesis

Figurs 5-6 show ethylene biosynthesis and its regulation. The pathway begins from methionine and there are two committed steps from S-adenosylmethionine are involved; the first is catalyzed by ACC synthase producing three-sided ring compound,l-aminocyclopropabne - l - carboxylic acid (ACC). This is converted to ethylene through ACC oxidase producing CO2,HCN and HO in the process. In depth studies revealed that pTOMl3 was related to ethylene prod Sequence analysis revealed the preliminary identification of pTOMl3 as second ethylene-forming enzyme as shown in Fig.5. The enzyme was characterized as ACC oxidase. The role of pTOM13 in ethylene biosynthesis was confirmed through the antisense strategy. Thus, transgenic tomato plants containing antisense pTOM13 construct produced very little ethylene in the ripening fruits obviously reduced ethylene synthesis delayed the total ripening process. So compared with the control plants lycopene production and softening occurred very slowly.In the antisense plants ripening could be induced by applying exogenous ethylene.Thus, such fruits could be stored for prolonged period and spoilage of the fruit during storage and transport could be avoided.

Fruit softening

Sequencing of the TOM6 clone demonstrated that the sequence matched the N-terminal sequence of the PG protein. Clearly one of the genes concerned with a particular part of ripening especially softening had been cloned. Further Northern blotting analysis confirmed that expression of the PG gene was specifically associated with ripening and that it was also induced by ethylene. Subsequently antisense techniques were used to alter plant gene expression. Initially the antisense PG construct comprised a partial cDNA fragment of pTOM 6 inserted in a reverse way downstream of the 35S promoter (Fig. 8) It may be observed that expression of the antisense gene is constitutive whereas expression of the endogenous gene is ripening-specific. It became obvious that the interaction of sense and antisense mRNA caused reduction in both transcripts. Consequently, in such transgenic fruits polygalacturonase activity considerably reduced during ripening and this delayed softening of the fruits (Figs. 9, 9A). However, other ripening processes continued as usual.

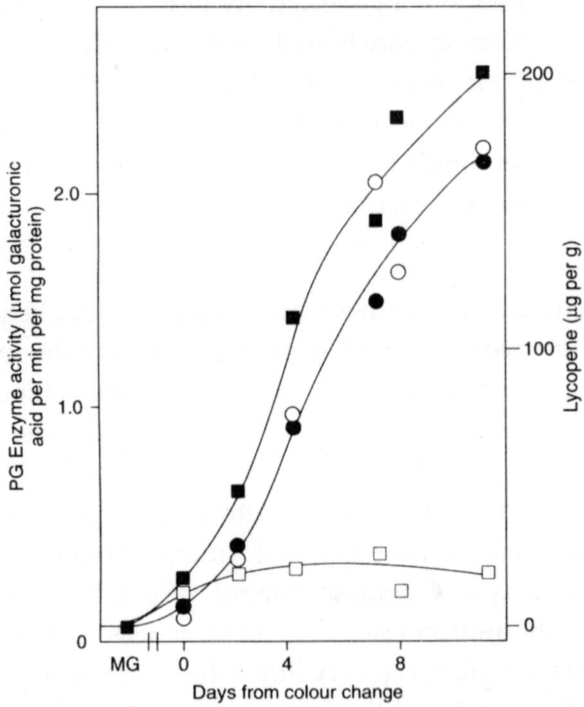

Fig. 9. Polygalacturonase activity correlating with lycopene content during the ripening of antisense PG fruit. The effect of antisense PG construct on enzyme activity and lycopene content is compared with untransformed control fruit (After, Slater *et al.*, 2003)

APPLICATION OF OUR UNDERSTANDING OF FRUIT RIPENING

Such studies were sought to create a healthy and pleasurable eating experience for the consumer. It is highly desired to improve storability by modifying fruit ripening but should not affect consumer satisfaction. As discussed earlier, 1-MCP application is shown to affect the aroma profile. But the stage of application was crucial. In brief, knowledge on fruit ripening must be linked with biochemical, physiological parameters and sensory and consumer acceptability. In fact the last decade has seen tremendous data appearing on the biology of fruit ripening. Added information has appeared on the application of molecular and biochemical techniques to elucidate the regulation and control of fruit ripening.

It has also become vivid that ethylene plays a crucial role in regulating several factors.

Pertinent to fruit ripening.Major work has involved focusing on ACS and ACO in relation to ethylene. Concomitantly the chief focus in the last decade of ethylene research has been on the elucidation of ethylene reception and signal transduction using Arabidopsis model. The coming years will attract added attention on the role of ethylene in ripening of climacteric fruits.Increased studies shall be required on several other quality factors such as fruit softening and aroma development during fruit ripening.

Polyamines and Ethylene production

Polyamines, ubiquitous organic aliphatic cations, have been implicated in a myriad of physiological and developmental processes in many organisms, but their in vivo functions remain to be determined. Mehta *et al.*, (2002) expressed a yeast S-adenosylmethionine decarboxylase gene (ySAMdc; Spe2) fused with a ripening-inducible E8 promoter to specifically increase levels of the polyamines spermidine and spermine in tomato fruit during ripening. Independent transgenic plants and their segregating lines were evaluated after cultivation in the greenhouse and in the field for five successive generations. The enhanced expression of the ySAMdc gene resulted in increased conversion of putrescine into higher polyamines and thus to ripening-specific accumulation of spermidine and spermine. This led to an increase in lycopene, prolonged vine life, and enhanced fruit juice quality. Lycopene levels in cultivated tomatoes are generally low, and increasing them in the fruit enhances its nutrient value. Furthermore, the rates of ethylene production in the transgenic tomato fruit were consistently higher than those in the no transgenic control fruit. These data show that polyamine and ethylene biosynthesis pathways can act simultaneously in ripening tomato fruit. Taken together, these results provide the first direct evidence for a physiological role of polyamines and demonstrate an approach to improving nutritional quality, juice quality, and vine life of tomato fruit (Figs. 10A, B)

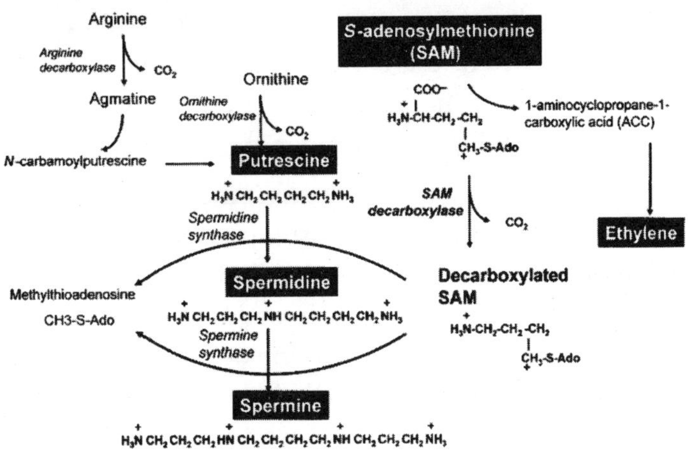

Fig. 10 A. The biosynthetic pathway of polyamines: the intermediates and enzymes (in italics)involved.Also shown is the pathway for ethylene production via formation of 1-aminocyclopropane-1-carboxylic acid (ACC) from S-adenosylmethionine (SAM). (From Mehta *et al.*, 2002)

(A) Attenuation of ripening in three transgenic fruit expressing SAMdc as compared with azygous and wild-type controls. Plants homozygous for the introduced SAMdc gene, 556HO, 566HO, and 579HO were grown alongside wild-type (WT) and azygous (AZ) lines (556 and 579).Days from breaker stage to red stage were recorded. Values represent averages of data from several independent seasons (n = 5-10) (B) Red fruit from transgenic tomato line 579HO. (C) Enhanced levels of lycopene accumulate at the red stage in transgenic fruit. Stages of ripeness are described in Fig B. (From Mehta *et al.*, 2002)

INDIA must aggressively adopt `delayed ripening' and `enhanced shelf-life' technologies if it wishes to consolidate its position as the world's No. 2 producer of fruits and vegetables after China, according to Dr Autar K. Mattoo, Research Leader of the US Department of Agriculture's Beltsville Agricultural Research Centre at Maryland. Currently, the country loses roughly a third of its annual produce on account of wastage and deterioration taking place during the journey from farm to plate.

The key agents behind ripening are the genes that code two proteins, ACC synthase and ACC oxidase, which are responsible for producing a volatile gaseous hormone called ethylene. It is this compound produced by the fruit that causes ripening. And since ethylene is a gas released into the air, a single ripe fruit can speed up the ripening of others as well if they are stored in the same bag.

Delayed ripening technologies basically involve `silencing' the two genes, so that the fruit does not synthesize ethylene, which triggers ripening.

"What I do is to silence the ACC synthase and ACC oxidase genes for a particular period, say from the point where the fruit is harvested in a green stage to its being transported to a warehouse. Once the fruit is close to the supermarket, it is sprayed with an ethylene-releasing compound to induce ripening," Dr Mattoo says.

There are many such ethylene-releasing substances, including ethaphon."You can even burn kerosene, which releases volatile gases, one of which constitutes ethylene." As for the methods of gene silencing, the most established one is `antisense RNA technology', which involves genetically engineering a mirror image copy (antisense) of the ACC synthase or ACC oxidase gene into the fruit. In the normal gene sequence, the ACC gene produces the enzyme through a `messenger' RNA. In the antisense technology, a gene in reverse orientation is inserted in the plant. "When the gene is in reverse orientation and is expressed so in the plant, it produces the messenger RNA also in a reverse orientation. The attraction of the two mirror image RNAs creates a complex, which gets degraded by ribonuclease enzymes. As a result, the plant ends up producing neither the reverse ACC enzyme nor the normal, forward version already present in it."

According to him, while knowledge regarding the two ACC genes and the ethylene signaling pathway was mainly a product of the 1980s, it took about 10 years to genetically manipulate and produce the first genetically engineered.

Why delay Fruit Ripening?

India suffers a loss of around US$ 20 billion annually, roughly a third of its annual produce due to 'uncontrolled ripening. Poor post-harvest management of fruits and vegetables also contributes. According to some estimates, a reduction in post-harvest losses by prolonging shelf-life of fruits and vegetables through GE by just 1%, would save the country, wastage of over Rs 200 crore. India losses around US$ 30 million worth of banana crop every year. This is attributed to rapid ripening and destruction in tropical climate. Using 'antisense' technology and other methods the ripening process could be slowed down. Besides banana, several other climacteric fruits like mango, papaya, apple, tomato also experience uncontrolled ripening and softening within a short span of time leading to value destruction of produce. Post harvest spoilage of fruit and vegetables can hinder their efficient distribution especially under poor transport and food storage infrastructure.

Delaying the ripening process in fruit is of interest to producers because it allows more time for shipment of fruit from the farmer's fields to the grocer's shelf, and increases the shelf life of the fruit for consumers. Though ripening makes fruit edible and flavorful, it also begins the gradual decline of the fruit towards softening and rot, causing losses for producers and consumers.

To increase the shelf life of tomatoes (and to prevent bruising during shipping), it is a common practice to pick tomatoes while they are still green, unripe, and later, when the tomatoes reach the grocer's shelf, spraying is done with ethylene to induce ripening. However, it is widely believed that this makes the flavor of the tomatoes much more bland, particularly in winter tomatoes - harvested green (refrigeration in shipping may also reduce flavor). This probably explains the popularity of vine-ripened tomatoes, left on the plant until the first patch of red appears (not necessarily until they are completely red). Fruits that are genetically engineered to have delayed ripening can be left to mature on the plant longer, will have longer shelf-life in shipping and may last longer for the consumers.

The preparation of fresh produce after harvest will affect its nutritional value, for example:

(i) The dry matter content (energy supply) is reduced with time as the continuation of living processes within the produce uses up stored food reserves.

(ii) Vitamin C content decreases with time after harvest, little may be left in green leaves after two or three days.

(iii) Cooking partially destroys vitamins C and B1. Raw fruit and vegetables are particularly valuable provided they are grown and handled hygienically.

(iv) Peeling may cause significant loss of food value, especially in Irish potatoes and many other root crops where the protein and vitamin content is located immediately under the skin.

(v) Minerals, including trade elements are dissolved out in cooking water, which is often thrown away

DR technology and its advantages

- Assuring top quality fruits and vegetables on the market shelves. Farmers can wait for fruit maturity before they are harvested from parent plant. Fruits attain full quality to satisfy consumers
- Expanding marketing opportunities for farmers; produce can be exported for long distances for long periods without refrigeration
- Reduction in post harvest losses due to delayed softening
- Extension in shelf life of fruits and vegetables; they remain fresh and nutritious for extended periods. These fruits do not reach 'dust bins'.

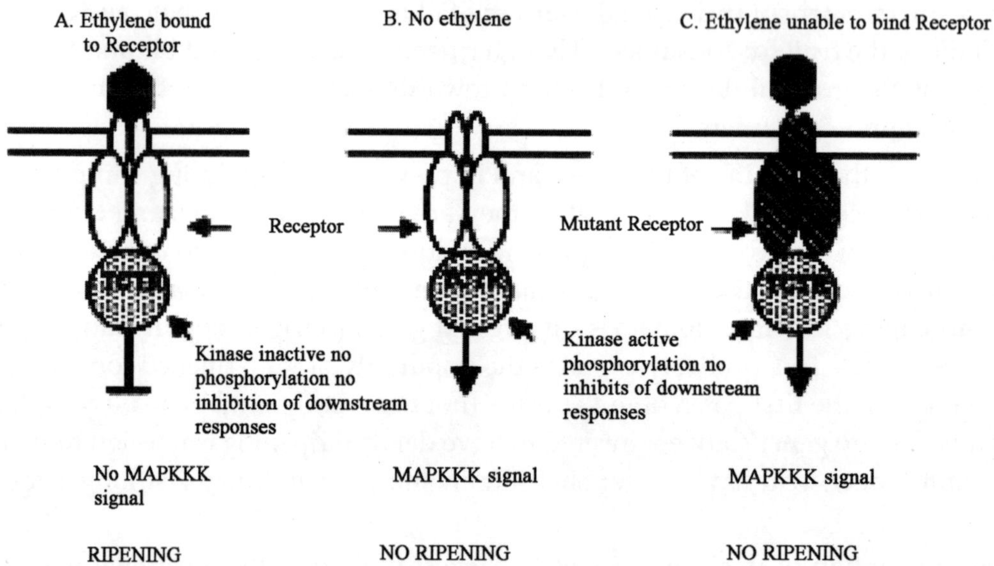

Figs. 11. Model of ethylene signaling from receptors to TCTR2. Ethylene receptors and TCTR2 are negative regulators of the ethylene signal transduction pathway. (After Alexander and Grierson 2002).

Because ethylene is the main trigger for fruit ripening, several genetic engineering and chemical regulation strategies are employed and these involve the reduction or prevention of ethylene production. Tomato fruits that do not produce ethylene develop fully on the plant and then stop before ripening and turning red. These "paused" fruits can all be picked later in development and all at the same time, whereas the green-picking method requires the field to be constantly inspected to pick each fruit just before it begins to turn red (which isn't always easy to tell). The GE fruit must also be sprayed with ethylene to induce ripening.

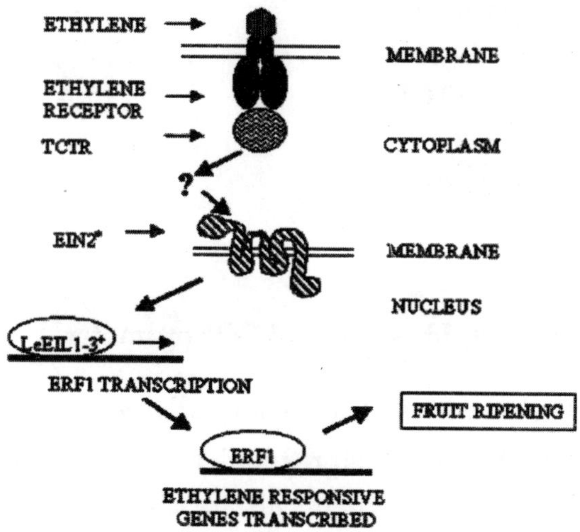

Fig.12. Diagrammatic representation of the putative ethylene signal transduction pathway downstream of TCTR2 in ripening tomato fruit. (after Alexander and Grierson, 2002).

Table 4. Effect of mutations in regulatory genes on the ethylene induction of some ripening-related genes.
Gene Genetic background

	WT	*rin*	*nor*	*Cnr*
E4	Yes	Yes	Yes	-
E8	Yes	Yes	Yes	Yes
J49	Yes	Yes	Yes	-
PG	Yes	No	No	Yes
ACS2	Yes	Yes	No	-
ACS4	Yes	No	No	-
ACO1	Yes	Yes	Yes	No

- = not determined

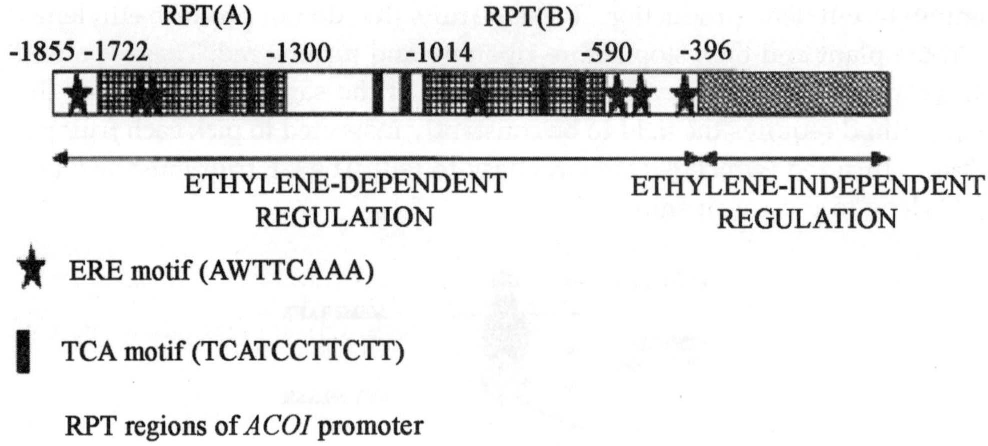

Fig.13. Tomato ACOl promoter

Chemical control of ethylene biosynthesis and action

Several chemicals are known which regulate ethylene synthesis in fruits and hence control fruit ripening. Some of these chemicals are discussed below :

- 1-Methylcyclopropabne is used to give post harvest treatment to delay ripening of tomato, pears, apricots and other fruits.
- Acetylsalicylic acid slows down reopening process by inhibiting amylase, sucrose phosphate synthase activities in kiwi fruit (Actinidia deliciosa)
- Aminoethoxyyuinylglycine(AVG) inhibits fruit ripening in peach (Prunus persica)
- Salicylic acid is reported to delay ripening of banana Controlled atmosphere for DR

Here low O_2 (5%); High CO_2 (5%), low temperature (20°C) is used. The ripening is delayed by 3-4 weeks. In general ethylene biosynthesis is delayed.

O_2 irradiation: Nuclear Research Lab, New Delhi has used this method to delay ripening in tomato and mango.

Low temperature storage

Several fruits are stored at low temperature and these include mango (10-12° C) and pawpaw (4°C).

Hypoboric storage

This refers to controlled storage under reduced pressure (75 mmHg). It also includes exclusion of ethylene. Other methods to prolong freshness and delay ripening include cool chamber storage, film wrapping.

Ethanol vapors

Ethanol is a volatile compound naturally produced by plant tissues under anaerobic conditions (Kelly and Saltveit, 1988). It is also accumulated in a short period of anaerobically stored fruits without adversely affecting fruit subsequent quality (Ke *et al.*, 1990).

Furthermore, some fruits are normally ethanol producers under aerobic conditions.Exogenous application of ethanol vapor reversibly inhibited tomato fruit ripening via inhibiting ethylene biosynthesis and action (Saltveit and Mencarelli, 1988) without reducing fruit quality as they subsequently reached red-ripe stage. Levels of ethanol residues were diminished in such fruit 72h after transferring to ethanol-free air. Among other alcohols, ethanol is more effective in inhibiting tomato fruit ripening and has the least phytotoxic effect (Wu *et al.*, 1992). It was also used in solution to delay carnation senescence via inhibiting the climacteric ethylene (Heins, 1986). In an opposite trend to ethanol inhibitory impact on tomato fruit ripening, ethrel (2- chloroethan-phosphonic acid) is commercially used for enhancing tomato fruit ripening (Abeles, 1992). Attempts for inhibiting tomato fruit ripening using ethanol vapor were carried out in a closed chamber as a static system designed for a short period of exposure. In this work however, ethanol vapor was used through a dynamic air-flow system for a long period of exposure, as a sort of controlled atmosphere, to delay or block tomato fruit ripening (during exportation) followed by ethanol-free air exposure or ethrel application at final destination to allow or speed fruit ripening, respectively.

Harvested mature-green tomato fruits were kept for 2 weeks at 14 ± 12°C in a closed ambient connected to an air-flow system. The air-flow was enriched with ethanol vapor by passing through solutions of 0, 5, 10 or 15% ethanol before enriching the fruit ambient. By the end of the 2 weeks, fruits of 10 and 15 % ethanol showed an approximate ripening block while those of 5 and 10% ethanol were at turning and red-ripe stages, respectively. During the period of exposure, ethanol vapor didn't develop fruit anaerobosis and ceased pathogen growth with a minor level of fruit weight loss. Transferring such fruits to ethanol-free air at 14+2°C, allowed ripening to proceed and to develop but with a slower rate for the fruits of higher ethanol levels. Fruits of 5 and 10% ethanol successfully reached red-ripe stage 8 and 16 days after the termination of ethanol vapor exposure, respectively, without noticeable impact on fruit quality. Fruits of 15% ethanol however, showed further ripening delay and hardly exceeded light-red stage with lower quality grads. Immediate ethrel dipping, after ethanol vapor exposure, only accelerated fruit ripening by about 3-4 days. While there were no significant differences among all treatments regarding L-ascorbic acid content, a significant reduction occurred in total soluble solids and titratable acidity for tomato fruits of 10% ethanol with further reduction for those of 15% ethanol. Nevertheless, only fruits of 15% ethanol showed a significant reduction in fruit panel taste. These data strongly indicated that' without refrigeration, a dynamic air-flow system with low levels of ethanol vapor application can be used in delaying the subsequent tomato fruit ripening without sensible impact on the quality.

Hence it is desired that alternative methods involving genetic engineering should be employed to delay fruit ripening. In the following we shall discuss these aspects.

Engineering of polyamine accumulation

It has been demonstrated that polyamine and ethylene biosynthesis pathways act simultaneously in ripening tomato fruits. A yeast S-adenosylmethionine decarboxylase gene (YSAMde; spl 2) is fused with a E8 ;promoter. Ripening was induced and levels of polyamines enhanced. Spermidine and spermine contents increased in tomato fruits undergoing ripening. The level of ethylene also increased. In vitro studies have shown that polyamine act as anti-ripening regulators by inhibiting ethylene biosynthesis in fruits.

The genetic modification of ethylene biosynthesis

Figure 10 shows ethylene biosynthesis and its regulation. The pathway begins from methionine and there are two committed steps from S-adenosylmethionine are involved; the first is catalyzed by ACC synthase producing three-sided ring compound, l-aminocyclopropane-l-carboxylic acid (ACC). This is converted to ethylene through ACC oxidase producing CO_2, HCN and HO in the process. In depth studies revealed that pTOMl3 was related to ethylene production. Sequence analysis revealed the preliminary identification of pTOMl3 as second ethylene-forming enzyme as shown in Figs. 5, 6. The enzyme was characterized as ACC oxidase. The role of pTOM13 in ethylene biosynthesis was confirmed through the antisense strategy. Thus, transgenic tomato plants containing antisense pTOM13 construct produced very little ethylene in the ripening fruits. Obviously reduced ethylene synthesis delayed the total ripening process. So compared with the control plants lycopene production and softening occurred very slowly. In the antisense plants ripening could be induced by applying exogenous ethylene.Thus, such fruits could be stored for prolonged period and spoilage of the fruit during storage and transport could be avoided.

Table 5. Ripening behavior of genetically-modified transgenic fruit suppressed in ethylene biosynthesis

Promoter and transgene	*Species*	*Maximal ethylene inhibition (%)*	*Phenotypic effects on fruit ripening*
CaMV35	Tomato	95 (attached)	Time to initiation of ripening not affected.
Antisense SlACO		90 (detached)	Lycopene accumulation reduced.
			Shelf-life extended (over-ripening delayed).
			Greater resistance to postharvest splitting, spoilage.
			Ripening prevention stronger in attached fruit.
CaMV35	Tomato	99.5	Ripening almost completely prevented.
Antisense SlACS			Fruit eventually turn orange but not red.
			Ripening prevention stronger in detached fruit.
CaMV35	Tomato	85 (attached)	Ripening not affected in attached fruit.
Bacterial ACC deaminase		90 (detached)	Ripening delayed in detached fruit.
			Shelf-life (detached) extended (over-ripening delayed).
E8	Tomato	80	Time to full ripe increased 2-fold.
Phage SAMase			Senescence delayed.
CaMV35	Tomato	96.5	Time to mid-ripe increased 2-fold.
Antisense MdACO			
CaMV35	Melon	99.5 (attached)	Shelf-life extended (over-ripening delayed).
Antisense CmACO		98 (detached)	Softening and rind de-greening prevented.
			Accumulation of aroma volatiles reduced.
			Loss of acidity and fruit abscission delayed.
			Flesh pigmentation and sugar accumulation not affected.
CaMV35	Melon	98	As for Pech, except loss of acidity
Antisense MdACO			prevented not delayed.
CaMV35	Apple		Shelf-life extended (over-ripening delayed).
Antisense MdACS		94	Softening and degreening reduced.
Antisense MdACO		96	Accumulation of volatile esters reduced.
			Sugar and acid accumulation not affected.

There are several methods employed to genetically-engineer reduction in ethylene:

1) Decreased ACC synthase:

ACC synthase is an enzyme in tomato fruits that is responsible for the next-to-the-last step in the synthesis of ethylene in the fruit (Figs. 5, 6). Reducing levels of ACC synthase considerably decreases ethylene production. Scientists discovered that when a "mirror-image" copy (antisense) of the ACC synthase gene is genetically engineered into tomatoes, a peculiarity of the plant's genetic editing machinery prevents the production of both a backwards ACC enzyme and the normal, "forward" version already in the plant. This "antisense" technique also has the advantage that no new protein is produced in the plant. DNA Plant Technologies developed these kind GE tomatoes and they are marketed as "Endless Summer."

The tomato ACC synthase is encoded by a multigene family having at least eight members (Oetiker *et al.*,1997). The expression patterns of these genes has been established in all tissues and under various environmental conditions (Nakatsuka *et al.*, 1998).Among these genes,LeACS5 and LeACS4 and distinctly induced during climacteric fruit ripening (Lincoln *et al.*, 1993).Of the two, LeACS2 transcript is nearly 10-fold more abundant than LeACS4 during fruit ripening (Olson *et al.*, 1991).Antisense suppression of LeACS2 gene in transgenic tomato fruit results in strong inhibition of ethylene biosynthesis and hence fruit ripening (Oeller *et al.*, 1991; Alexander and Grierson, 2002).In addition to transgenic regulation, virus-induced gene silencing (VIGS) also offers an attractive technology to down-regulate specific gene expression in plants (Kumar *et al.*, 2003; R0bertson, 2004; Fu *et al.*, 2005).Plant virus-induced gene silencing (VIGS) is currently a powerful tool for the study of gene function in plants. YuanHong *et al.*, (2006) have reported the silencing of LeACS2 by vacuum-infiltration and the tobacco rattle virus (TRV)-based VIGS method, which leads to a significant delay of the postharvest ripening and senescence of tomato fruit. Harvested mature green tomato fruit were vacuum-infiltrated with Agrobacterium strain GV3101 containing pTRV1 and pTRV2-LeACS2. Because of the silencing of LeACS2, the ethylene climacteric and pigment changes were clearly delayed. The onset of fruit ripening and senescence was significantly postponed, and transcription of LeACS2 and ACC synthase activity were also suppressed in treated tomato fruit during storage. The silencing of LeACS2 by vacuum infiltration, however, did not alter the contents assayed at the end of storage. Their results indicated that vacuum infiltration was a highly efficient TRV-based VIGS method to silence LeACS2 in harvested tomato fruit. It could obviously delay ripening and senescence, and appeared to be a potential method for postharvest preservation of tomato fruit. In brief, while using this method, a recombinant viral vector carrying a partial sequence of an interested gene is used to infect the plant. The virus spreads systemically in the plant, hence the endogenous gene transcripts which are homologous to the inserted sequence in the viral vector (VIGS-vector), are degraded by post-transcriptional gene silencing (PTGS) (Baulcombe, 2004).In fact VIGS appears to be an excellent reverse genetics tool that can be used to generate mutant phenotypes for assigning function to known and unknown genes. VIGS has also several other advantages over conventional stable transformation methods in gene function analysis (Fu *et al.*, 2005). One of the significant advantages of VIGS is that it can cause a loss of function phenotype for a target gene within a single generation and without the need to genetically transform the plant (Liu *et al.*,2002). It may be added that the procedure is rapid and it takes 1-2 weeks for the infection to take place. YuanHong *et al.*, (2006) used the tobacco rattle virus (TRV) based VIGS method to silence LeACS2 in tomato fruit after harvest and also to delay senescence and thus prolong the shelf life of the fruit. These workers also verified that some physiological parameters related to ripening, including PG activities and pigment changes were also postponed. In tomato activity enhances tremendously at the time ripening and contributes to the softening of the fruit. VIGS inhibited PG activity markedly. These authors also concluded that that under certain conditions VIGS only suppressed gene function for a period of time but failed to alter the properties of tomato fruit ripening at the end of storage. VIGS had no change in the contents of reducing sugars as also acids.

2) ***Addition of ACC deaminase:***

Scientists have added to the tomato a new gene from the soil bacteria Pseudomonas chlororaphis. The gene encodes an enzyme called ACC deaminase, which causes the breakdown of one of the precursors of ethylene synthesis (ACC). The reduction in levels of this precursor causes reduced ethylene production and delayed ripening. Monsanto developed this tomato, but it has not yet been marketed (Fig. 15).

3) ***Addition of SAM hydrolase:***

Like tomatoes with added ACC deaminase, tomatoes genetically engineered to produce SAM hydrolase (from the E. coli T3 bacteriophage) also break down one of the precursors of ethylene synthesis (SAM). This technology has been developed by Agritope, Inc. for a variety of cherry tomato.

Finally, some tomatoes have been genetically engineered to alter one particular aspect of fruit ripening: softening. The process of fruit softening is caused in part by the breakdown of pectins-- compounds which give support to the cell walls of fruit. Tomatoes have been engineered to have reduced levels of a pectin-breakdown enzyme called polygalacturonase (PG) using the same "antisense" technique applied to ACC synthase (see above). Not only does this increase shelf-life, but it also improve qualities of interest to tomato processors-- the tomato products are "thicker" (higher pectin to water ratio). This is the technique used in the well-k4). Suppression of ACC oxidase

4) ACC oxidase is another gene involved in ethylene synthesis and has been shown to allow normal development of ripening attributes in attached fruits, even when ACO

was inhibited by 95%. However, it prevented softening, cracking and spoilage normally associated with over ripening. Ayub *et al.* (1996) were the first to demonstrate that the ripening behavior of fruit other than tomato could be genetically modified in melon. These workers used antisense techniques to inhibit ACO. The melon ACC oxidase gene was transferred as an antisense construct to melon using an Agrobacterium transformation and soft vector. The fruits remained green even after ten days and retained their shape whereas controlled fruits shriveled and their rind became yellow known as "Flavr Savr" tomato (developed by Calgene, now owned by Monsanto Fig. 9). A. Role of ethylene in tomato ripening B. Differential regulation of ACS gene expression during the transition from system 1 to system 2 ethylene biosynthesis in tomato. (after Barry *et al.*, 2000).

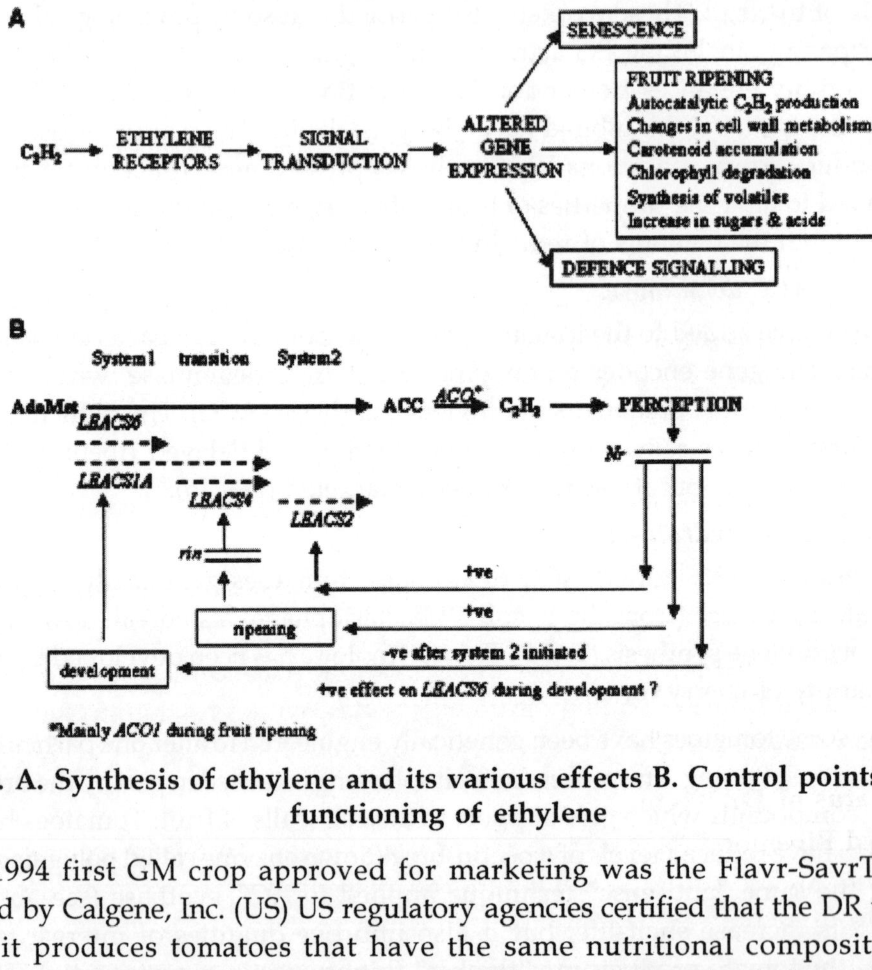

Fig.16. A. Synthesis of ethylene and its various effects B. Control points for the functioning of ethylene

In 1994 first GM crop approved for marketing was the Flavr-SavrTM tomato produced by Calgene, Inc. (US) US regulatory agencies certified that the DR technology is safe, it produces tomatoes that have the same nutritional composition as the conventional ones. GM tomatoes showed no difference in levels of allergens or toxins compared to normal fruit. In addition, field trials have shown that the DR tomatoes do not pose any threat to other plants nor to any non-target organisms.

Other DR tomatoes that followed thereafter have also been granted deregulated status by regulatory agencies in several countries including the US, Canada, and Mexico. In 1996, the UK's food safety regulators also gave their thumbs up to a DR tomato developed by Zeneca Seeds but it is not currently being sold in supermarkets (see box).

A European First

On February 5, 1996 branches of Safeway and Sainsbury's supermarkets throughout the UK started to sell tomato pureé made from genetically-modified tomatoes. This was the first time that food made from a GM organism had been sold in Europe.

Labels on the cans clearly stated that the product had been made with GM tomatoes. Although there was no legal requirement to label the product, both supermarkets adopted an 'open' information policy from the start. For the inquisitive customer there was no shortage of information: leaflets were available describing the product, its benefits to the environment and consumer, the technology, and the regulatory processes through which the product had to pass.

According to the supermarkets, sales in around 80 stores in which supplies were initially available were brisk. Figures indicated that once they bought the product, shoppers came back for more. In November 1997, Safeway Stores announced that they had sold three quarters of a million cans of the product, and that average sales per store of the modified tomato purée exceeded those of the conventional equivalent. One reason might have been the price: the new purée cost 29 pence for 170 grams while the traditional form cost slightly more: 29 pence for a mere 142 grams.

Both supermarket chains pledged that the new product would always be offered alongside its old-fashioned counterpart. This move pleased consumer groups, which had no objection to the purée, provided that it was safe to eat and that consumers were always given a choice.

However, commercial pressures generated by public concern about GM foods early in 1999 forced Sainsbury's to announce that it would withdraw the product from sale. Stocks were exhausted by July 1999.

Current Status of DR Technology

Delayed Ripening (DR) Technology has been applied for use in tomatoes, melons, and papaya. An interesting application of DR technology is in floriculture where experiments are underway to apply the technology to delay the withering of flowers.

In Southeast Asia, DR technology is being applied for use in papayas, a popular subsistence food and part of the general diet in the region. This technology could significantly increase the availability of this nutritious fruit to consumers and to small-scale and mostly resource-poor farmers in the region.

Regulatory approval of countries for crops with the DR trait

Crop	*Countries*	*Types of Approval*
Tomato	Canada	Food
	United States	Environment, Food, Feed
Tomato(Flavr-SavrTM)	Canada	Food
	Japan	Environment, Food
	Mexico	Environment, Food, Feed
	United States	Environment, Food, Feed

Source: Essential Biosafety 2003. Agriculture and Biotechnology Strategies, Inc.

Model systems for fruit ripening

1. Grapes:

In contrast to climacteric fruit, where ethylene is pivotal, the hormonal control of ripening in non-climacteric fruit, such as grape (Vitis vinifera), is poorly understood. Brassinosteroids (BRs) are steroidal hormones, essential for normal plant growth and development but not previously implicated in the ripening of nonclimacteric fruit. Symons *et al.* (2006) have shown that increases in endogenous BR levels, but not indole-3-acetic acid (IAA) or GA levels, are associated with ripening in grapes. Putative grape homologs of genes encoding BR biosynthesis enzymes (z gene was confirmed by transgenic complementation of the tomato (*Lycopersicon esculentum*) extreme dwarf (dx/dx) mutant. Expression analysis of these genes during berry development revealed transcript accumulation patterns that were consistent with a dramatic increase in endogenous BR levels observed at the onset of fruit ripening. Furthermore, we show that application of BRs to grape berries significantly promoted ripening, while brassinazole, an inhibitor of BR biosynthesis, significantly delayed fruit ripening. These results provide evidence that changes in endogenous BR levels influence this key developmental process.This may provide a significant insight into the mechanism controlling ripening in grapes, which has direct implications for the logistics of grape production and down-stream processing.

2. *Strawberry (Fragaria ananassa)*

The strawberry belongs to the genus Fragraria in the rose family, along with apples and plums. The name of the scientific classification was derived from the Old Latin word for fragrant. The largest producing state, California harvests 83% of the strawberries grown in the U.S. on approximately 24,500 acres. And with about 5,000 commercial acres, Florida is the second largest producing state. Ideal temperature for strawberry plants should not exceed higher than 78 degrees or lower than 55 degrees.

Strawberry is widely studied system for non-climacteric ripening, resulting in the identification and characterization of numerous ripening related genes that effects cell wall metabolism, color and aroma(Wilkenson *et. al.* 1995). Characterststics and composition of ripe fruit are the result of biochemical and physiological changes. These changes include a pronounced decrease in fruit firmness that is associated with increased susceptibility to physical damage and disease. Fruit softening is mainly the result of the action of hydrolytic enzymes that modify the carbohydrate components of cell wall. As in all climacteric fruits, ethylene is considered to have a little or no effect on the development of strawberry receptacle (Perkins and Veazie, 1995). Auxins are proposed to be the primary hormone controlling fruit ripening in strawberry. It is synthesized in achenes (true fruit embedded in receptacle) and positively effects the initial growth of the receptacle (Given *et. al.*, 1998). A gradual decline in the supply of auxins from achenes in the latter stages of growth has been proposed to be the basis of ripening. It has also been reported that auxin regulate the transcription of many ripening-related genes (Manning, 1994). Pectin is the most abundant class of macromolecule with in the cell wall of strawberry fruit primarily in the cell wall where it functions in regulating intercellular adhesion between cells. During fruit softening, pectins undergo solubilisation and depolymerization that are thought to contribute to wall loosening and disintegration (Fischer and Bennett, 1995). Several pectin modifying genes have been cloned in strawberry fruit such as pectate lyase, endopolygalactouronase and α- galactosidase. Pectin esterase (PE) activity has also been reported in ripening strawberry fruits (Neal, 1965; Barnes and Patchett, 1976). The cloning of 4 PE genes in strawberry, FaPE1, FaPE2, FaPE3 and FaPE4 is reported. It has been shown that expression of FaPE1 was indused by auxin at the onset of fruit ripening and down regulated by ethylene during fruit senescence. It is proposed that the ethylene-mediated decrease of FaPE1 expression could be an important factor in determining fruit port-harvest quality.

Strawberry genes FaGAST affects plant growth through inhibition of cell elongation:

Strawberry FaGAST gene encodes a small protein with 12 cystein residue conserved in the C-terminal region similar to a group of proteins identified in other spesies with diverse assigned functions such as cell division, elongation and elongation arrest. This fruit is expressed in the fruit receptacle, with two peaks during ripening at the white and the red-ripe stages, both coincident with an arrest in the growth pattern. Expression is also high in the roots but confined to the cells at the end if elongation jone. Exogenous application of gibberellin increased the transcript level of the FaGAST gene in strawberry fruits. Ectopic expression of FaGEST in transgenic Frasaria vesca under the control of the CaMV-35S promoter caused both delayed growth of the plant and fruits with decreased size. The same growth defects was observed in Arabidopsis. In addision, transgenic plants exhibit late flowering and low sensitivity to exogenous gibberllin.

FaGAST plays a role in arresting cell elongation during strawberry Pietro *et al.* (2006) have demonstrated ethylene and carbon dioxide production by developing strawberries and correlated them to indicate STRAWBERRY AS A CLIMACTERIC FRUIT.

The results presented here conflict with notions that stereotype strawberry fruit growth and ripening as a non-climacteric fruit. Unequivocal evidence that C_2H_4 production by developing and ripening strawberry flowers and fruits maturing from DS3 to DS6/24 h+ occurs in a definite and highly reproducible manner is available and that correlates with distinct DSs C_2H_4 production during flower opening and senescence and by fruit developing from DS3 to DS4 is characterized by cyclic diurnal production (that correlates with the growth and expansion of the fruit and chlorophyll loss. Clear evidence that C_2H_4 production by red-ripe fruit is characterized by a non-diurnal cyclic accumulation that coincides with a respiratory climacteric is stated. The in planta findings also contradict reports, gathered from detached and dissected strawberry fruit that suggested strawberry fruit C_2H_4 production to be under negative-feedback control throughout ripening. Our data show that the pattern of C_2H_4 production by strawberry fruit is similar to that exemplified by tomato, where C_2H_4 production is self-regulating via negative-feedback processes until the mature expanded white or pale-green stage, when regulation switches to positive control The approach, to administer STS by injection, took into account the fact that there is no physiological effect on the developmental progress of ripening strawberry when STS, 1-MCP or AVG is administered exogenously and that the levels of C_2H_4 produced from within the seeds are high (as compared with the epidermal output) and may be particularly significant in ripe fruit The physiological response of STS injected into the vascular traces of strawberry flowers and fruits at DS3 suggests that C_2H_4 production is self-regulating via a negative-feedback control mechanism. In contrast, STS caused C_2H_4 to be evolved by positive feedback by fruit at DS4) and DS6/24 h+. In addition, data quantifying the amount of C_2H_4 generated due to tissue damage (removal of achenes) revealed a significant reduction (approximately 20%) in C_2H_4 production in expanded white-fruit peel (DS4), whereas red-fruit peel showed increased (approximately 50%) C_2H_4 roduction. This also highlights the existence of a switch in C_2H_4 self-regulation from negative- to positive-feedback control (respectively) as fruit mature from DS3/4 to DS6 + 48 h.

The switch in diurnal growth-associated C_2H_4 production to one of continuous C_2H_4 production after a 24-h lag phase may reflect a switch from processes governing fruit development and ripening to those that encourage fruit senescence-mediated processes that include fruit softening and microorganism (Botrytis; grey mould) mediated pathogenesis all of which may be encouraged by the elevated levels of C_2H_4, respectively and to which seed derived C_2H_4 contributes significantly.

The results presented here also highlight that the term 'climacteric' may be considered as scientifically 'under discussion', as it can no longer be applied consistently to account for the underlying biological complexity of (at least) strawberry fruit ripening. The data do clearly demonstrate that, by classically physiological definition, strawberry is justifiably a climacteric fruit, because a respiratory increase concomitant with a C_2H_4 increase does occur in ripening strawberry fruit. However, results are correlative and do not prove a causal role for C_2H_4 in strawberry fruit ripening. Despite this, it is highlighted that stereotyping fruit that produce very low levels of C_2H_4 as 'non-climacteric' based on whether C_2H_4 appears to be a prerequisite for fruit ripening may be misleading; this approach fails to consider the underlying biology that will determine the absolute quantity of C_2H_4 (or efficacy C_2H_4 interfering chemicals) to which a physiological response can/ cannot occur. It is suggested that the C_2H_4-sensing mechanism in climacteric fruit may be so sensitive as to preclude the appearance of a phenotypic effect ex post facto. That is, the efficacy with which C_2H_4 inhibitors (such as STS or 1-MCP) can saturate all the available C_2H_4 receptors in strawberry tissues and arrest the C_2H_4 response might be low, whereas the biological efficacy of C_2H_4 reception may be high. In addition, it is demonstrated that C_2H_4 production from strawberry seeds is significant, and this tissue may be considered particularly inaccessible to standard inhibitor applications, compared with receptors of the flesh. This possibility is supported by recently reported molecular data demonstrating that Fa ACOI gene expression correlates with C_2H_4 increases in tandem with the switch in C_2H_4 regulation from negative- to positive-feedback control, postulating a possible role for this gene/mechanism in signaling the progression of the strawberry ripening process. Furthermore, concomitant with this is the increased synthesis of type II receptors (FaETR2), whereby low levels of C_2H_4 may be sufficient to promote ripening-related responses. This opportunity could be exploited further using readily available microarray tools and the genetic studies to dissect a direct analyses on the role of C_2H_4 in the ripening process in strawberry should be targeted towards ACO and ACC-synthase and C_2H_4 receptors expressed in the fruit flesh or (arguably), more importantly, the seed, because our results from dissected fruit showed clearly that the levels of C_2H_4 produced by the 'true fruits' achenes are significant, especially for ripe fruit The possibility that there is a temporal separation of C_2H_4 production initiation between achene and flesh tissue requires investigation. That is, in the same way that declines in achene-derived auxin promotes ripening so to may achene-C_2H_4. This significance of this if proven may explain the inefficiency of attempts to arrest strawberry fruit ripening using C_2H_4 reception inhibitors.

Ultimately, the causal role for C_2H_4 in post-harvest processes must ultimately be assessed in transgenic strawberry plants with blocked C_2H_4 production and sensing, as has been demonstrated for tomato and melon. Samples selected for their physiological state (defined in this instance as C_2H_4 production) will facilitate this full and accurate molecular dissection of C_2H_4-related strawberry fruit ripening. For example, it is shown that fruit at the visual phenotype DS4 may or may not have stopped evolving C_2H_4 in a diurnal pattern cyclically and may or may not have entered the phase where C_2H_4 production in under positive regulation. In this way, fruit at DS4 could have three possible C_2H_4-related physiological phenotypes. The LPD will therefore allow specific physiological stages to be identified, based on the pattern and/or quantity of C_2H_4 that is produced.

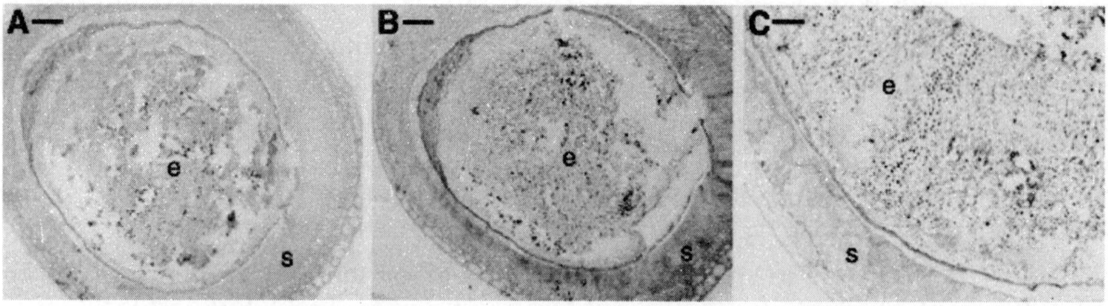

Fig. 17. Silver-enhanced immunogold-labelled light micrographs of strawberry seeds. (A) Developmental stage 4 (DS4) seed negative-control section probed with (pre-immune) normal rabbit serum. (B) DS4 seed and (C) DS6 seed labeled with anti-1-aminocyclopropane-1-carboxylic acid oxidase IgG. e, embryo cells; s, seed coat. Scale bars = 100 μm (A, B) and 50 μm (C). (after Pietro *et al.*, 2006).

Table 6: Ethylene production rates (corrected for response to damage during dissection) by the strawberry fruit samples: epidermal peel minus

	Standardized for fresh weight (pl h^{-1} g^{-1} FW)		Absolute production (pl h^{-1})	
Ethylene prodution rate	Epidermal peel minus achenes	Achenes only	Epidermal peel minus achenes	Achenes only
DS4	280 ± 90	810 ± 90^{a}	34 ± 7^{a}	9.3 ± 0.7
DS6/24 h+	80 ± 20	790 ± 180^{a}	92 ± 0.6^{NS}	9.5 ± 1.6

achenes and achenes only from fruit at developmental stage 4 (DS4) and 6. Data are expressed in absolute (pl h_1) and relative (pl h_1 g_1 FW) terms. A) Samples that are significantly greater (ANOVA; $P < 0.001$). NS, non-significantly different samples.

3. *Mango (Mangifera indica)*

The mango is native to southern Asia, especially Burma and eastern India. It spread early on to Malaya, eastern Asia and eastern Africa. The mango is the apple (or peach) of the tropics, and one of the most commonly eaten fruits in tropical countries around the world. The fruit is grown commercially on a small scale in Florida. The quality of the fruit is generally comparable to Florida mangos, but has other advantages., i.e. the lack of fruit fly and seed weevil populations. Mexico, and to a lesser extent Central America, is a major supplier to U.S. markets today. Mangos were introduced to California (Santa Barbara) in 1880. Mangoes are major fruit crop for export and there are certain exotic but delicate varieties, which do not find place in export market as they perish quickly. Scientists are making efforts to develop such varieties through gene manipulations that have higher shelf lives. Genes for cell wall modifying enzymes such as PG, XET, pectate lyase, expansins and genes for ethylene biosyntheis and perceptionsuch as ASO, ACS, etc. have been cloned and characterized in mango. One of the exotic variety of mango, Dashehari (Mangifera indica), grown mainly in northern India, suffers from rapid and uneven ripening making it unfit for export. Several biochemical and physiological studies have been performed to understand the process of ripening in mango. Isolation and characterization of an alpha-extensin gene, MiExpA1 is correlated with softening in mango. The expression of this gene is under dual control, being triggered by ethylene treatment with in 90 min followed by ripening associated peak in transcript accumulation on third day after ethylene treatment. At the protein level, expression of expansin is detectable from the second day itself and contines throughout the course of softening. Treatment with 1-MCP inhibits both ripening/softening as well as MiExpA1 transcript and protein accumulation. MiExpA1 expression is ethylene dependent and its expression increases with the progression of ripening (Sane *et. al.*, 2005). Genetic engineering in Carabao mango is underway. UPLB-IPB scientist have also developed somatic embryos in that genes will be altered to delay the mango's ripening rate, to prolong the fruit's shelf life. The characterization of one cDNA (pTHMF 1) from M. indica which codes for a protein highly homologous to cucumber, rat and human peroxisomal thiolase (EC 2.3.1.16), the catalyst for the last step in the beta-oxidation pathway has been done. The cDNA for the peroxisomal mango thiolase is 1305 bp in length and codes for a protein of 432 amino acids with a predicted molecular mass of 45,532 Da. Mango thiolase is highly homologous to cucumber thiolase (80%), the only other plant thiolase whose cloning has been reported, and to rat and human thiolases (55% and 55% respectively). It is shown by northern analysis that during fruit ripening THMF 1 is up-regulated. A similar pattern of expression was detected in tomato fruit. Wounding and pathogen infection do not appear to affect THMF 1 expression. There is a possible involvement of thiolase in fatty acid metabolism during fruit ripening. This is the first report of cloning of a plant gene involved in fatty acid metabolism showing an induction during fruit ripening in mango (Bojorqez G *et. al.*,1995). A mango (Mangifera indica L.) cDNA homologue of the ethylene receptor gene ETR-1, referred to as METR1, which codes for a polypeptide of 802 amino acids with a predicted molecular weight of 89 kDa have been isolated (Martínez PG, Gómez RL, Gómez-Lim MA, 2001). The amino acid sequence is highly homologous (over

80 %) to ETRs from other fruits. Genomic Southern blot analysis indicates that two or more ETR homologues exist in mango. RNA blot analysis revealed that the level of METR1 mRNA in the mesocarp increased during fruit ripening. In addition, it was found that the METR1 mRNA increases transiently during wounding of the tissue. This is the first report of an ETR homologue showing an induction during fruit ripening and wounding. Application of an aqueous solution of different concentrations of putrescine (0, 0.5, 1.0, or 2.0 mmol.L-1) containing the surfactant 'Tween-20' (0.01%) were applied and it was found that putrescine treatments significantly reduced ethylene production and response was more pronounced in the postharvest dip application than the preharvest spray. Ethylene production decreased as the concentration of applied putrescine increased. Pre and postharvest application of putrescine increased fruit firmness and decreased sugars, compared with control. Both pre and postharvest application also retarded fruit colour development. After storage, preharvest-treated fruit exhibited higher firmness, TSS and lower fruit rot whilst acidity, total and non-reducing sugars were reduced in fruit treated with both methods, compared with control. So preharvest putrescine spray was more effective than postharvest dip of Putrescine to extend the shelf life and fruit quality of Kensington mango (Malik *et al.*, 2003).

Banana (Musa sp.):

The banana is native of Asia. The banana is of the family Muscaceae, the largest of tree-like herbs grown for fruit, large striking foliage and for fiber (Fig. 19). This family has two genera, Musa and Ensete. The genus Ensete does not bear edible fruit. The genus Musa provides both fruit and fiber. The banana family consists of herbs, some so large as to appear to be trees, reaching several meters in height. Leaves are large, broad, alternate, not toothed, simple, not lobed or divided, but often appearing so because of tattering by wind, spirally arranged and forming a trunk by their sheathing bases. The flowers are bisexual or unisexual with spathelike bracts, in spikes, panicles or heads; sepals 3; petals 3; stamens usually 5; carpels 3;style 1;ovary 3-chambered. The trunk-like cylinder of leaf sheaths supporting the mass of leaves and fruit is called a pseudostem or false stem. The true stem is an underground rhizome commonly called a pseudobulb producing leaves and a fruit stem from a growing point on its upper surface and roots from its lower surface. The blossom stem arises from the central growing point on the pseudobulb and pushes upward through the center of the pseudostem, protected by the encircling leaf sheaths. The time require for the fruit to reach the harvest stage varies form 3 to 4 months after flowering and when the fruit is harvested the pseudostem is cut down. Its place is taken by a sucker that has been allowed to develop from the pseudobulb and it eventually produces a bunch of fruit only to give place to still another sucker as the cycle as the cycle of fruit production continues in the plantation. The large, conspicuous leaves of a mature banana plant are the last of a long series that starts with formation of small scale leaves as the young shoot or sucker emerges above the ground. If the sucker is developing from a deeply buried bud, the scale leaves are followed by long narrow 'sword leaves' and then as the plant grows older by leaves that gradually increase in size and width until the typical mature leaves are formed. On the other hand, if the sucker is arising from a bud at or above the surface of the ground it is apt to produce broad leaves very early in its development. A healthy mature banana plant has from 11 to 14 functioning leaves together with the sheaths of 12 to 16 leaves that developed earlier but those blades have withered and fallen as the plant developed. The last leaf to develop is short but almost as broad as the mature leaves immediately below it. Its sheath encircles the blossom stem as the latter emerges or 'shoots' above the throat of the plant and its blad droops over the blossom bud which bends downward in a few days from the weight of the rapidly developing inflorescence. This 'protecting' or 'follower' leaf gives the blossom bud a temporary degree of protection against the sun. An extensive mass or mat of roots develops from the lower surface of the pseudobulb. They supply water and food materials to the aerial parts of the plant and also anchor and hold the pseudostem upright. The banana varieties grown for commercial use do not produce seeds. New plantations are established entirely from either whole pseudobulbs or portions of them containing one or more lateral buds or 'eyes'. Botanically, the banana fruit is a berry since it meets the definition of being a simple fruit having a pulpy or fleshy pericarp, that is, the wall of a mature ovary, and having one or more seeds in the flesh, but no true stone. Some wild bananas have true seeds in the flesh.

Banana forms an integral part of the farming systems in the humid zones of tropics. World Musa production is around 85.5 million tones annually of which bananas cultivated for export trade account for only 10%. Post harvest technologies that work well in other fruits, extending post harvest life; have not been successful in banana. In part, this is due to imperfect knowledge of the physiological, biochemical and molecular processesassociated with its ripening. Functional analysis of genes through over expression and antiscence expression is underway in banana. Scientist have cloned atleast three promoters from the genes specific for ripening in banana. These are being characterized through transient expression analysis of promoter-GUS fusion construct in order to realize if these promoters can be utilized to express desired genes specifically in fruits. DNA Plant Technology Corporation has been developing banana's with extended shelf life. Similarly, scientists at Zeneca Plant Sciences are developing modified ripening characteristics in banana. Success in this endeavour will improve shelf life in the markets. Significant progress has been made towards the understanding of physiology of fruit ripening and sensscence in the last two decades. In 1969, the first International Symposium on this topic was heldat the Research Station Gorsem In Sint-Truiden, Belgium. Several of the hypothesisraised at the symposium have now been tested and proven. The enzymes of the ethylene biosynthesis, ACC synthase and ACC oxidase, have been identified, purified, cloned, characterized and antibodies developed for analytic reagents.

The feedback regulation of ethylene biosynthesis in banana [Musa sp. (AAA group, Cavendish subgroup) cv. Grand Nain] fruit was investigated to clarify the opposite effect of 1-methylcyclopropene (1-MCP), an ethylene action inhibitor, before and after the onset of ripening. 1-MCP pre-treatment completely prevented the ripening-induced effect of propylene in pre-climacteric banana fruit, whereas treatment after the onset of ripening stimulated ethylene production. In pre-climacteric fruit, higher concentrations of propylene suppressed ethylene production more strongly, despite their earlier ethylene-inducing effect. Exposure of the fruit ripened by propylene to 1-MCP increased ethylene production concomitantly with an increase in 1-aminocyclopropane-1-carboxylate (ACC) synthase activity and ACC content, and prevented a transient decrease in MA-ACS1 transcripts in the pulp tissues. In contrast, in the peel of ripening fruit, 1-MCP prevented the increase in ethylene production and subsequently the ripening process by reduction of the increase in MA-ACS1 and MA-ACO1 transcripts and of ACC synthase and ACC oxidase activities. This suggest that ethylene biosynthesis in ripening banana fruit may be controlled negatively in the pulp tissue and positively in the peel tissue. (Akitsugu Inaba, Xuejun Liu, Naoki Yokotani, Miki Yamane, Wang-Jin Lu, Ryohei Nakano and Yasutaka Kubo, 2007)

Papaya (Carica papaya)

The papaya tree is indigenous to the West Indies and the northern parts of South America, but was quickly introduced to tropical areas of the Old World due to the popularity of its wonderful fruit. The tree is evergreen and has a very distinctive appearance. It has a tall, narrow green or purplish trunk topped by a spiral of large palmate leaves (up to 75 cm across) and is covered in leaf scars further down. The trunk rarely has any side branches, and grows to around 5 m (very old trees have been known to reach 10 m). The papaya is dioecious (some modern cultivars are monoecious also), and therefore several plants need to be grown together to ensure pollination and a good fruit yield. The male flowers are small and grow in long clusters. The female flowers grow on short stalks in the leaf axils and once the plant is mature enough it produces a copious and continuous supply of fruits. The fruits (technically berries) start off green but ripen to yellow with shades of red and orange. They have a thin, smooth skin and a sweet layer of delicious pulp, with a taste and texture similar to peach. At the centre of each fruit is a mass of hundreds of pea-sized dark blue/black seeds. The fruits you commonly find in supermarkets are hand-sized but they do grow to around 50cm and can be very heavy. The fruits are best harvested when 80% coloured and then allowed to finish ripening at room temperature for a few days. Unripe fruit may be eaten as a vegetable after peeling, deseeding and boiling. All parts of the tree, but especially the skin of the fruit, produce a white latex from cut surfaces. The latex contains the proteolytic enzymes papain and chymopapain which are similar to bromelain from pineapples and zingibain from ginger. The latex is commercially harvested, dried and the powder is sold as a meat tenderiser as well as a health food product.

Different ripening stages of papaya (Carica papaya var. 'Sekaki') fruit were evaluated for quality and the activity of softening enzymes. When the skin was 5%, 25%, 50%, 75% and 100% yellow (or YR) stage, quality changes in weight loss, tissue firmness, ascorbic acid, and the softening enzymes activities of á- and α-galactosidases, and pectin methylesterase (PME). During ripening at ambient temperature, tissue firmness decreased rapidly up to 50%YR and then at a more gradual rate until the end of ripening. Weight loss and ascorbic acid level increased slightly during fruit ripening. á-Galactosidase activity increased until 25%YR, whereas, α-galactosidase and PME activities increased linearly with softening until 50%YR. So tissue softening particularly at the initial stages of ripening is related to changes in á-, α-galactosidases and PME. This suggests that these three enzymes play an important role in softening process and contribute to 'Sekaki' papaya fruit quality(International Symposium on Papaya).

Devitt *et al.*(2005)have identified papaya ripening genes. These workers used expressed sequence tags. They generated a total of 1171expressed sequences tags from clones of two independent fruit cDNA libraries derived from yellow and red fleshed fruit varieties. The most abundant gene sequences isolated were those coding for the enzymes chitinase, which breaks down chitin; 1-aminocyclopropane-1-carboxylic acid (ACC) oxidase; catalase, which breaks down harmful hydrogen peroxide into hydrogen and water; and methionine synthase which processes amino acids. Putative genes, contributing to fruit softening, cell wall; hydrolases, cell membrane hydrolases, and ethylene synthesis and regulation sequences. These EST's with gene sequences have been compared with in other plant species, expressed papaya genes which play a role in fruit aroma and color identified.

Chromosomal studies on papaya:

Sex chromosomes in flowering plants evolved recently and many of them remain homomorphic, including those in papaya. The chromosomal location of papaya's small male specific region of the hermaphrodite Y (Y(h)) chromosome (MSY) and its genomic features is determined. The chromosome fluorescence in situ hybridization mapping of Y(h)-specific bacterial artificial chromosomes (BACs) has been done and placed the MSY near the centromere of the papaya Y chromosome. The sequence of five MSY BACs was discovered to examine the genomic features of this specialized region, which resulted in the largest collection of contiguous genomic DNA sequences of a Y chromosome in flowering plants. Extreme gene paucity was observed in the papaya MSY with no functional gene identified in 715 kb MSY sequences. A high density of retroelements and local sequence duplications were detected in the MSY that is suppressed for recombination. Location of the papaya MSY near the centromere might have provided recombination suppression and fostered paucity of genes in the male specific region of the Y chromosome (Qingyi, 2007).

Women in India,Pakistan and Sri Lanka and other parts of the world have long used papaya as a folk remedy for contraception and abortion. Medical research in animals has confirmed the contraceptive and abortifacient capability of papaya, and also found that papaya seeds have contraceptive effects in adult male langur monkeys, possibly in adult male humans as well (Lohiya, 2006) Unripe papaya is especially effective, in large amounts or high doses. Papaya is not teratogenic and will not cause miscarriage in small, ripe amounts. Fruit ripening in Grapes:

Future Research

In-depth understanding of ethylene perception and signal transduction will help in manipulation of fruit ripening. For instance, the molecular characterization of ethylene receptors may permit regulation of ethylene effects or fruit ripening. In the process the role of ethylene in regulating fruit development. Advanced studies of the rin and nor loci and their functional attributes in the ripening of normal tomatoes will be a significant area of future research. Understandably each species of fruit may have different molecular controls of ethylene biology compared with tomato or Arabidopsis. The role of brassinosteroids in the ripening of non-climacteric fruits has been recently brought out in grapes and more fruits need to be screened for ripening.

1-MCP has been a great experimental tool and has aided post harvest physiologists in demarcating various stages in fruit ripening by blocking ethylene perception. Further studies are desired to improve the consistency of quality, post harvest storage and shelf life of many fruits through judicial application of 1-MCP to modulate ethylene responses.

1-MCP could help in identifying individual receptor isoforms. Several new compounds like nitric oxide, lysophosphatidylethanolamine may furnish insights into fruit ripening.One must remember that individual regulatory steps or enzyme is important but ultimately the process of ripening must be visualized as a complex steps of various factors. Identifying genes involved in papaya fruit ripening, researchers generated a total of 1171 expressed sequence tags (ESTs) from clones of two independent fruit cDNA libraries derived from yellow and red-fleshed fruit varieties. Researchers found that the most abundant gene sequences isolated were those coding for the enzymes chitinase, which breaks down chitin; 1-aminocyclopropane-1-carboxylic acid (ACC) oxidase, which controls the release of ethylene in plants; catalase, which breaks down harmful hydrogen peroxide into hydrogen and water; and methionine synthase, which processes amino acids. It was also found that putative genes contributing to fruit softening, among them

5

7

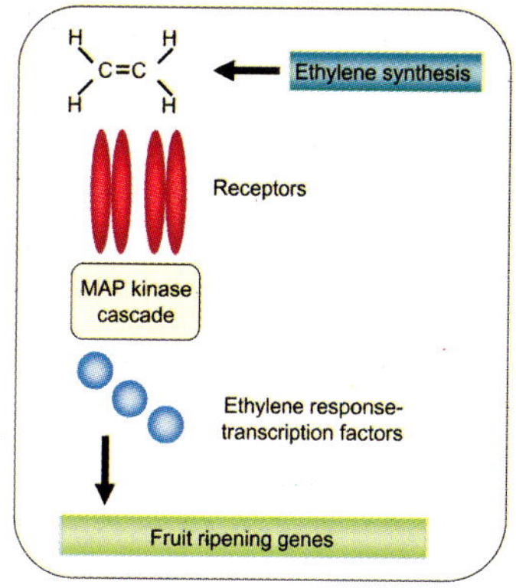

9A

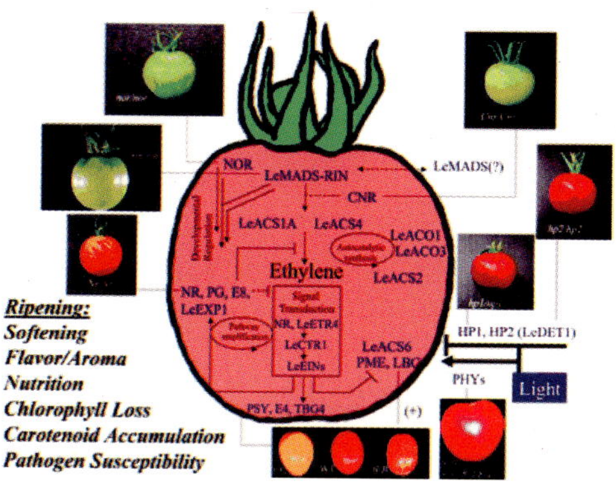

10

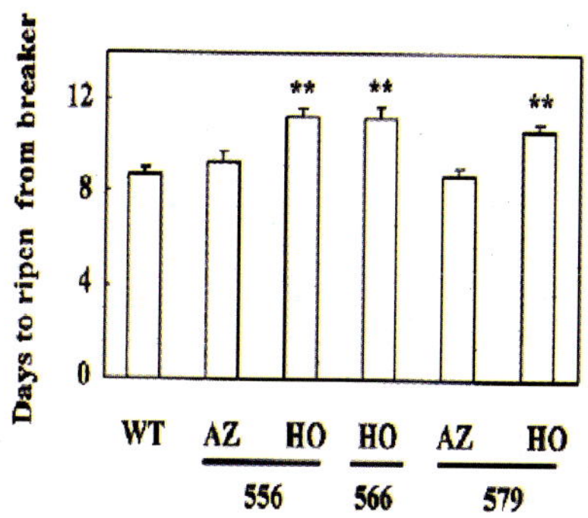

15

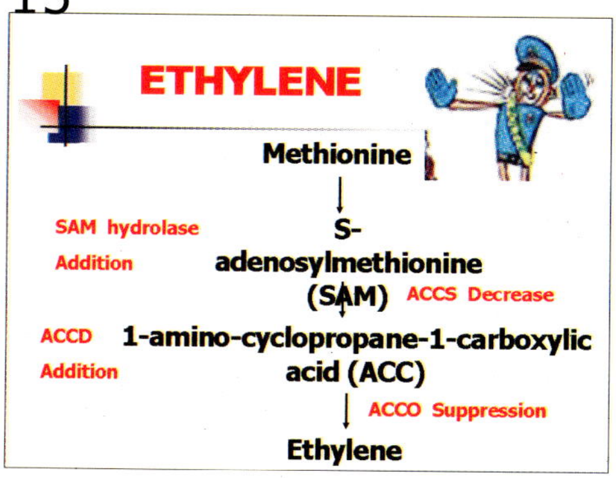

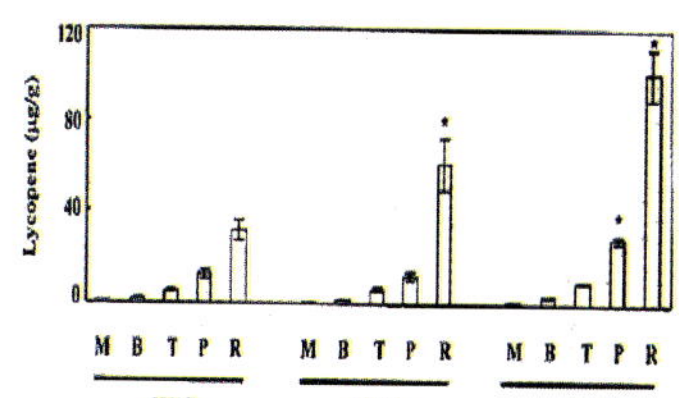

Fig: 5). Pathway for the synthesis of ethylene.
7). Model of ethylene signalling pathway
9A). biochemical changes in tomato fruits at different stages of development
10 B). Quality attributes of SAMdc-transgenic tomato fruit
15). Control points for the regulation of ethylene biosynthesis

18

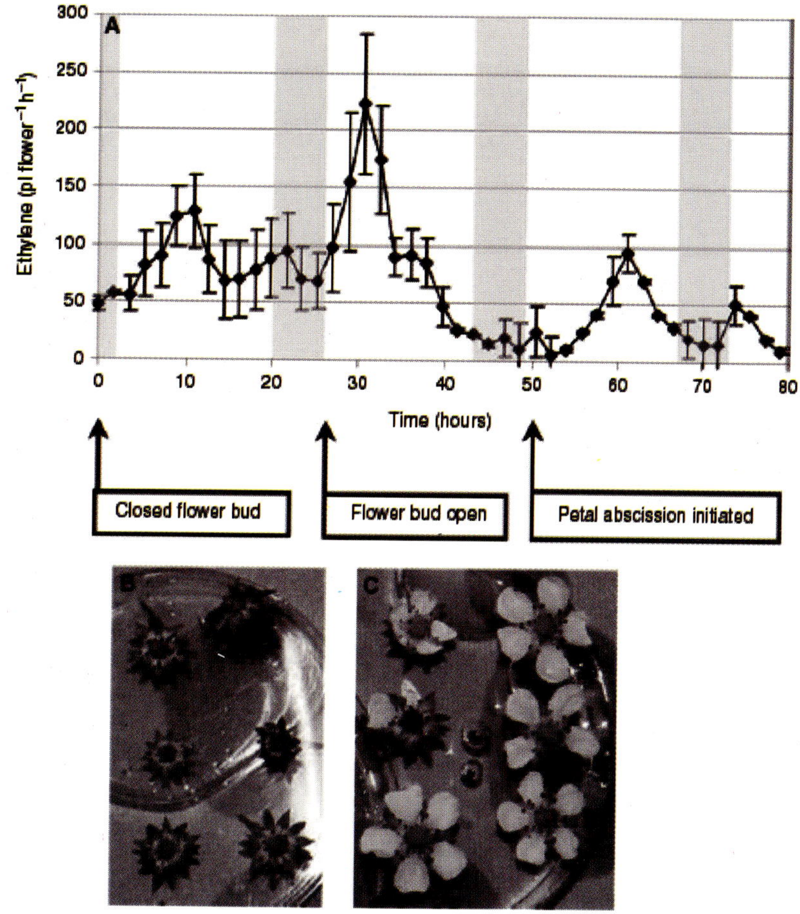

19

Fig: 18 (A) Ethylene production by flower buds as they mature to open flowers in vivo and as they develop towards developmental stage 1 (DS1). Grey bars depict periods of darkness. The response of detached flower buds and open flowers cultured in vitro with their cut pedicels mmersed in either water **(B)** or 10 mM silver thiosulphate **(C)**.(From Pietro et al., 2006)

19). Different types of climacteric fruits

cell wall hydrolases, cell membrane hydrolases, and ethylene synthesis and regulation sequences. By comparing ESTs with gene sequences in other plant species, expressed papaya genes were identified.

REFERENCES

Abeles F B Morgan P W and Saltveit M E 1992 Ethylene in Plant Biology 2nd ed Academic Press INC New York pp 85-273

Adams - Phillips L Barry C & Giovannoni J 2004 Signal transduction systems regulating fruit ripening *Trends in Plant Science*: **9** 331-338

Afta-Aly M A and El-Awady M N 1993 Safety and accuracy of a non-destructive deformation tester manufactured in Egypt for measuring tomato fruit softness *Annals of Agric Sci Ain Shams Univ* Cairo **39**: 307-318

Akitsugu Inaba* Xuejun Liu Naoki Yokotani Miki Yamane Wang-Jin Lu Ryohei Nakano and Yasutaka Kubo 2007 Differential feedback regulation of ethylene biosynthesis in pulp and peel tissues of banana fruit *Journal of Experimental Botany* **58**(5):1047-1057

Alba R Cordonnier-Pratt M -M and Pratt L (2000) Fruit-localized phytochromes regulate lycopene accumulation independently of ethylene production in tomato *Plant Physiol* **123** 363370

Alba R *et al* (2004) ESTs cDNA microarrays and gene expression profiling: Tools for dissecting plant physiology and development *Plant J* **39** 697714

Al-Babili S von Lintig J Haubruck H and Beyer P (1996) A novel soluble form of phytoene desaturase from Narcissus pseudonarcissus hromoplasts is Hsp70-complexed and competent for flavinylation membrane association and enzymatic activation *Plant J* **9** 601-612

Alexander L & Grierson D 2002 Ethylene biosynthesis and action in tomato: a model for climacteric fruit ripening *Journal of Experimental Botany* **53** 2039-2050

Ayub R Guis M BenAmor M Gillot L Roustan JP Latche A Bouzayen M and Pech JC Expression of ACC oxidase antisense gene inhibits ripening of cantaloupe melon fruits *Nature Biotechnology* 1996: **14** 862-866

Balbi V and Lomax T (2003) Regulation of early tomato fruit development by the diageotropica gene *Plant Physiol* **131** 186-197

Barnes MF and Patchett BJ (1976) Cell wall degrading enzymes and softening of the strawberry fruit *Journal of food Sci* **41** 1393-1395

Barry CS McQuinn RP Thompson AJ Seymour GB Grierson D & Giovannoni JJ 2005 Ethylene insensitivity conferred by the Green-ripe and Never-ripe 2 ripening mutants of tomato *Plant Physiology* **138** 267-275

Barr J White W Chen L Bae H and Rodermel S (2004) The GHOST terminal oxidase regulates developmental programming in tomato fruit *Plant Cell Environ* **27** 840852

Barry C Llop-Tous I and Grierson D (2000) The regulation of -1-aminocyclopropane-1-carboxylic acid synthase gene expression during the transition from system-1 to system-2 ethylene synthesis in tomato *Plant Physiol* **123** 979

Baulcombe D C 2004 RNA silencing in plants *Nature* **431** 356-363

Bauchot A Mottram D Dodson A and John P (1998) Effect of aminocyclopropane-1-carboxylic acid oxidase antisense gene on the formation of volatile esters in cantaloupe charentais melon (cv Vendrandais) *J Agric Food Chem* **46** 4787-4792

Biggs M and Handa A (1988) Temporal regulation of polygalacturonase gene expression in fruits of normal mutant and heterozygous tomato genotypes *Plant Physiol* **89** 117-125

Binder BM & Bleecker AB (2003) A model for ethylene receptor function and 1-methylcyclopropene action *Acta Horticulturae* **628** 177-187

Blankenship SM and Dole JM (2003) 1-Methylcyclopropene: a review *Post harvest Biology and Technology* **28** 125

Bisson L (2001) In search of optimal grape maturity Practical Winery & Vineyard July/August Pp 32-43

Bojorquez G and Gomez-Lim MA (1995) Peroxisomal thiolase mRNA is induced during mango fruit ripening Peroxisomal thiolase mRNA is induced during mango fruit ripening *Plant Mol Biol* **28** 811-20

Bramley PM (2002) Regulation of carotenoid formation during tomato fruit ripening and development *Journal of Experimental Botany* **53** 2107-2113

Burg S and Burg E (1962) Role of ethylene in fruit ripening *Plant Physiol* **37** 179-189

DellaPenna D Kates D and Bennett A (1989) Polygalacturonase gene xpression in Rutgers rin nor and Nr tomato fruits *Plant Physiol* **85** 502-507

Brummell DA& Harpster MH (2001) Cell wall metabolism in fruit softening and quality and its manipulation in transgenic plants *Plant Molecular Biology* **47** 311340

Chae HS Kieber JJ & Eto Brute (2005) Role of ACS turnover in regulating ethylene biosynthesis *Trends in Plant Science* 1029-1296

Chen GP Alexander L & Grierson D (2004) Constitutive expression of EIL-like transcription factor partially restores ripening in the ethylene-insensitive Nr tomato mutant *Journal of Experimental Botany* **55** 1491-1497

Coombe B G 1992 Research on development and ripening of the grape berry *American Journal of Enology and Viticulture* **43** 101-110

Dandekar AM Teo G Defilippi BG Uratsu SL Passey AJ Kader AA Stow JR Colgan RJ & James DJ (2004) Effect of down-regulation of ethylene biosynthesis on fruit flavor complex in apple fruit *Transgenic Research* **13** 373-384

Dinesh-Kumar S P Anandalakshmi R Marathe R Schiff M Liu Y (2003) Virus-induced gene silencing *Methods Mol Biol* **236** 287-294

Dixon J & Hewett EW (2000) Factors affecting apple aroma/flavour volatile concentration: a review *New Zealand Journal of Crop and Horticultural Science* **28** 155-173

Fei Z Tang X Alba R Payton P and Giovannoni J (2003) Tomato Expression Database (TED): An interactive management tool for tomato expression profiling data *Proc Comp Sys Bioinform* **203** 424-425

Fischer RL and Bennett AB (1991) Role of cell wall hydrplases in fruit ripening *Annual Rev of Plant Physio And Mol Bio* **42** 675-703

Fraser P Truesdale M Bird C Schuch W and Bramley P (1994) Carotenoid biosynthesis during tomato fruit development: Evidence for tissue-specific gene expression *Plant Physiol* **105** 405-413

Fray R and Grierson D (1993) Identification and genetic analysis of normal and mutant phytoene synthase genes of tomato by sequencing complementation and co-suppression *Plant Mol Biol* **22** 589-602

Fu D Q Zhu B Z Zhu H L Jiang W B Luo Y B (2005) Virus-induce gene silencing in tomato fruit *Plant J* **43** 299-308

Fu D Q Zhu B Z Zhu H L Zhang H X Xie Y H Jiang W B Zhao X D Luo Y B (2006) Enhancement of virus-induced gene silencingin tomato by low temperature and low humidity *Mol Cells* **21** 153-160

Giovannoni J (2004) Genetic regulation of fruit development and ripening *Plant Cell* **16** (suppl) 170-180

Giovannoni J DellaPenna D Bennett A and Fischer R (1989) Expression of a chimeric polygalacturonase gene in transgenic rin (ripening inhibitor) tomato fruit results in polyuronide degradation but not fruit softening *Plant Cell* **1** 5363

Giuliano G Bartley G and Scolnik P (1993) Regulation of carotenoid biosynthesis during tomato development *Plant Cell* **5** 379-387

Given NK Venis MA and Grierson D (1988) Hormonal regulation of ripening in the strawberry a non-climacteric fruit *Planta* **174** 402-406

Hackett RM Ho CW Lin ZF Foote HCC Fray RG & Grierson D (2000) Antisense inhibition of the Nr gene restores normal ripening to the tomato Never-ripe mutant consistent with the ethylene receptor-inhibition model *Plant Physiology* **124** 1079-1085

Hamilton AJ Lycett GW & Grierson D (1999) Antisense gene that inhibits synthesis of the hormone ethylene in transgenic plants *Nature* **346** 284-287

Hiwasa K Kinugasa Y Amano S Hashimoto A Nakano R Inaba A & Kubo Y (2003) Ethylene is required for both the initiation and progression of softening in pear (Pyrus communis L) fruit *Journal of Experimental Botany* **54** 771-779

Herner R and Sink K (1973) Ethylene production and respiratorybehavior of the rin tomato mutant *Plant Physiol* **52** 38-42

Isaacson T Ronen G Zamir D and Hirschberg J (2002) Cloning of tangerine from tomato reveals a carotenoid isomerase essential for the production of b-arotene and xanthophylls in plants *Plant Cell* **14** 333-342

Klee H (1993) Ripening physiology of fruit from transgenic tomato *(Lycopersicon esculentum)* plants with reduced ethylene synthesis *Plant Physiol* **102** 911-916

Klee H & Tieman D (2002) The tomato ethylene receptor gene family: Form and function *Physiologia Plantarum* **115** 336-341

Klee HJ Hayford MB Kretzmer KA Barry GF & Kishore GM (1991) Control of ethylene synthesis by expression of a bacterial enzyme in transgenic tomato plants *Plant Cell*: **3** 1187-1193

Klee HJ (2004) Ethylene signal transduction Moving beyond *Arabidopsis Plant Physiology* **135** 660-667

Klee H (2002) Control of ethylene-mediated processes in tomato at the level of receptors *J Exp Bot* **53** 2057-2063

Lanahan M Yen H Giovannoni J and Klee H (1994) The never-ripe mutation blocks ethylene perception in tomato *Plant Cell* **6** 521-530

Lashbrook C Tieman D and Klee H (1998) Differential regulation of the tomato ETR gene family throughout plant development *Plant J* **15** 243-252

Leclercq J Adams-Phillips L Zegzouti H Jones B Latche A Giovannoni J Peche J and Bouzayen M (2002) LECTR1 a tomato CTR1-like gene demonstrates ethylene signaling ability in Arabidopsis and novel expression patterns in tomato *Plant Physiol* **130** 1132-1142

Lincoln J E Campbell A D Oetiker J Rottmann W H Oeller P W Malik A U and Singh Z and Dhaliwal S (2003) *EXOGENOUS APPLICATION OF PUTRESCINE AFFECTS MANGO FRUIT QUALITY AND SHELF LIFE* in XXVI International Horticultural Congress: Issues and Advances in Postharvest Horticulture Toronto Canada International Society for Horticultural Science *(Lycopersicon esculentum) J Biol Chem* **268** 1942-1943

Manning K (1998) Changes in the gene expression during fruit ripening of strawberry fruit and their regulation by auxin *Planta* **194** 62-68

Martínez PG Gómez RL Gómez-Lim MA (2001)Identification of an ETR1-homologue from mango fruit expressing during fruit ripening and wounding *Journal of Plant Physiology* **158** 101-108

Mehta RA Cassol T Li N Alki N Handa AK & Mattoo AK (2002) Engineered polyamine accumulation in tomato enhances phytonutrient content juice quality and vine life *Nature Biotechnology* **20** 613-618

Nakatsuka A Murachi S Okunishi H Shiomi S Nakano R Kubo Y Inaba A (1998) Differential expression and internal feedback case of conservation *Plant Mol Biol* **26** 1579-1597

Neal GE (1965) Changes occurring in the cell wall of strawberry ripening *J of Science of Food and Agri* **16** 604-611

Pech J and Bouzayen M (1999) Ethylene-regulated gene expression in tomato fruit: Characterization of novel ethylene-responsive and ripening-related genes isolated by differential display *Plant J* **18** 589-600

Perkins-Veazie PM (1995) Growth and ripening of strawberry fruit *Horticultural Reviews* **17** 267-297

Picton S Barton SL Bouzayen M Hamilton AJ &Grierson D (1993) Altered fruit ripening and leaf senescence in tomatoes expressing an antisense ethylene-forming enzyme transgene *Plant Journal* **3** 469-481

Pietro P Lannetta M Laarhobven Luc-Jan Medina-Escobar N James EK McManus MT Davies HV &Harren FJM (2002) Ethylene and carbon dioxide production by developing strawberries show a correlative pattern that is indicative of ripening climacteric fruit *Physiol Plant* **127** 247-259

Qu X & Schaller GE (2004) Requirement of the histidine kinase domain for signal transduction by the ethylene Receptor ETR1 *Plant Physiology* **136** 2961-2970

Ramassamy S Olmos E Bouzayen M Pech JC & Latche A (1998) 1-Aminocyclopropane-1-carboxylate oxidase of apple fruit is periplasmic *Journal of Experimental Botany* **49** 1909-1915

Slater A Scott N and Fowler M (2003) Plant Biotechnology- The genetic manipulation of plants Oxford University Press Oxford

Symons GM Davies C Shavruukov Y Dry IB Reid JB and Thomas MR (2006) Grapes on steroids Brassinosteroids are involved in grape berry ripening *Plant Physiology* **140** 150-158

Tanksley S (2002) Deductions about the number organization and evolution of genes in the tomato genome based on analysis of a large expressed sequence tag collection and selective genomic sequencing *Plant Cell* **14** 1441-1456

Vrebalov J Ruezinsky D Padmanabhan V White R Medrano D Drake R Schuch W and Giovannoni J (2002) A MADS-box gene necessary for fruit ripening at the tomato ripening-inhibitor (rin) locus *Science* **296** 343-346

Wang X Ghosh S and Guo S -W (2001) Quantitative quality control of microarray image processing and data acquisition *Nucleic Acids Res* **29** 75

Watson B (2003) Evaluation of Winegrape Maturity In: E W Hellman (ed) Oregon Viticulture Oregon State University Press Corvallis Oregon pp 235-245

Wilkinson J Lanahan M Yen H -C Giovannoni J and Klee H (1995) An ethylene-inducible component of signal transduction encoded by Never-ripe *Science* **270** 1807-1809

Wilkinson JQ Lanahan MB Conner TW and Klee HJ (1995) Identification of m RNAs with enhanced expression in ripening strawberry fruit using PCR differential display *Plant Mol Bio* **27** 1097-1108

Wu M J Zacarias L Saltveit M E and Reid M S (1992) Al cohols and carnation senescence *Records found* 7

Yen H -C Lee S Tanksley S Lanahan M Klee H and Giovannoni J (1995) The tomato never-ripe locus regulates ethylene- inducible gene expression and is linked to a homolog of the Arabidopsis ETR1 gene *Plant Physiol* **107** 1343-1353

Yokotani N Tamura S Nakano R Inaba A McGlasson WB & Kubo Y (2004) Comparison of ethylene- and wound-induced responses in fruit of wild-type rin and nor tomatoes *Postharvest Biology and Technology* **32** 247-252

Yokotani N Tamura S Nakano R Inaba A McGlasson WB & Kubo Y (2004) Comparison of ethylene- and wound-induced responses in fruit of wild-type rin and nor tomatoes *Postharvest Biology and Technology* **32** 247-252

Yunovitz Hermona Gross Kenneth C (1994) Delay of tomato fruit ripening by an oligosaccharide N-glycan Interactions with IAA galactose and lectins Physiologia *Planta*rum **90(1)** 152-156

Zhao X-C Xiang Q Mathews DE & Schaller GE (2002) Effect of ethylene pathway mutations upon expression of the ethylene receptor ETR1 from Arabidopsis *Plant Physiology* **130** 1983-1991

INDEX

A

ACC deaminase, 389
ACC oxidase, 370, 375, 377, 380, 386, 389, 390, 400, 401, 402
ACC synthase, 361, 377, 380, 386, 388, 389, 395, 400, 407
Acetylsalicylic acid, 384
Achenes, 356, 393, 394, 395, 396
Affymetrix, 1
AGAMOUS (AG), 21, 32
Agrobacterium, 7, 354, 388, 390, 407
AINTEGUMENTA, 23, 30, 33
akinetic mutism, 242-244
Alfa-alfa, 315
Alpha-extensin gene, 397
Altered meristem program 1 (amp1), 17, 18
Alternaria leaf spot, 88
Alzheimer's disease, 230, 275
Ambulatory, 245
Aminoethoxyvinyl glycine, 384
Amylase, 384
Amyloid fibers, 230
ANAC (abscisic acid-responsive NAC), 23
Anthesis, 361
Anthocyanins, 363
Anthrax, 312, 313 , 325, 332, 335, 342, 345, 347,
Antioxidant, 368, 374
Antirrhinum cupuliformis, 23
APETALA 1 (AP1), 32
APETALA 2 (AP2), 31
Arabidopsis response regulator 5, 21
Arabidopsis response regulator, 5, 20
Arabidopsis, 14, 16, 102, 124, 126, 129-131, 339, 357, 361, 369, 370, 374, 393, 402
ARGONAUTE1 (AGO1), 19
ASYMMETRIC LEAVES 1 (AS1), 23
Ataxia, 239, 242, 243, 245, 246, 270
Auxin resistant1 (axr1-12), 22
Auxin resistant6 (axr6), 18
Auxin response factor5 (arf5), 17
Auxin, 15
Axillary meristems1 (rax1), 20
Azygous, 380

B

B chromosome, 9, 10
Backcrosses, 8
Bacterial vector, 320
Bacteriophage, 389
Barcode-tracked drug, 2
Benzaldehyde, 364
beta-secretase, 276
BIN 2, 131
Biotechnology, 1, 4
BLIND gene, 20
Bodenlos (bdl), 18
Bollgard, 156, 165, 170, 178, 184, 196, 197, 205
Bollworm, 147-149, 152, 156, 164, 171-174, 176-180, 182-184
Brassinazole, 392
Brassinolide, 108, 111
Brassinosteroid, 108, 111
Brassinosteroids, 392, 402
Breaker, 361, 380
BREVIPEDICELLUS KNAT6, 18
brevipedicellus (BP), 14, 18, 19
brinjal, 87, 90, 91, 97, 98, 99, 100
bronchopneumonia, 246
Bt, 7, 8
Bt, cotton
Bt, cotton 147, 152, 156, 164, 169, 171-180, 182

C

Caliper life sciences, 1
Cannibalism, 236, 240, 244, 245, 247
Carotenoid, 357, 361, 362, 366, 368, 373, 374, 375
Carrot, 315
Caryopsis, 355, 356, 407
CCAAT, 17, 29
Celera, 1
Cellulose, 375
Chalaza-micropyle, 15
Chaotropic salts, 233
Chloroplast transformation, 333
Choreoathetosis, 243

2- Chloroethan-phosphonic acid, 385
Cinnamoyl alcohol dehydrogenase, 367
Cinnamoyl CoA reductase, 367
Citral, 365
Clavata genes, 24-26, 28
CLV3 18, 24-27, 32
CNA, 26, 27, 30
Co-integrated plasmid, 81
Cointegration strategies, 80
Continuous quality improvement (cqi), 3
Cotton, 101, 103, 108
Cotton fiber elongation, 107-116, 132-136
Cotyledon1 (lec1), 17
cry genes, 151, 152, 154-157, 159, 161, 164, 165, 167, 168, 174, 178
Crystal protein, 151, 159
CUC 18
Cup-shaped cotyledon1 (CUC1) 20 ,31
CYCLIN B1, 20
Cymbopogon, 365
Cytokinin, 14

D

DNA polymerase, 45
DE-ETIOLATED1 (*DET1*) gene, 15
Dementia, 235, 247, 288
Deoxyribonucleases, 233, 234
Diethyl pyrocarbonate, 233
Digitalized radiographs, 2
Dioxygenase, 366
Direct gene transfer 322, 348
Divalent cation hydrolysis, 233
DORNROSCHEN (*drn-D*) mutants, 18
Dossia, 3, 4
Down-stream processing, 392

E

E. coli, 315, 321, 323, 324, 326, 329, 340
early extra petals1 (eep1), 21
Edible vaccines 311-313, 315-317, 319, 321, 323, 325, 327-329, 331-333, 335, 337, 339-343, 345, 347, 349-351
Enantiomer, 265
Engineered minichromosomes, 8
Epitopes, 331,336,337,339,343
Ethaphon, 380
Ethrel, 385, 386
Ethyl methyl phenyl glycidate, 365

F

Fass (fs), 17
Fertilization-independent endosperm (fie), 21
Filamentous flower (fil), 23
Flavonoid, 361, 368
Flavonols, 363
Flavr savr, 390
Floral binding protein2 (fbp2), 33
Fourier transform infra-red spectroscopy, 257, 299
fps2, 21, 33
FUSCA3 (*FUS3*), 17

G

G. arboretum, 101, 110, 131
G. barbadense, 101, 122
GAATTC, 59
Galactanase, 375
GALT, 328, 329
Gea1p and Gea2p, 16
Gel electrophoresis, 229, 233
Gel filtration, 233
Genentech, 1
Genes that regulate *sam*, 28, 29, 30
Genetic engineering, 7
Genetic predisposition, 230, 236
Genetic transformation, 7
Genetically modified, 87, 89, 91, 93, 97, 99
Genome sequencing, 3
Geranial, 365
GFPKN1, 14
GFP-KNAT1/brevipedicellus, 14
Gibberellin pathways, 14
Gibberellins, 358, 372, 407
Gliosis, 236, 239, 243, 254, 268, 278
Glutamate decarboxylase, 375
Glycosyl phosphatidylinositol, 228
Glycosylation, 311, 313
Glycosylphosphatidyl inositol, 255, 269, 270
GNARLEY1, 14
GNOM/EMBRYO DEFECTIVE30, 16, 29
Gossypium, 90
Gossypium hirsutum, 122
Green revolution, 88
Groundnut, 87, 94, 95, 97
Guanidinium thiocyanate, 233
GUS gene, 316, 334
Gynoecium, 13

H

Hamster, 229, 233, 255, 258, 259, 261, 268, 276, 277, 279, 281
HD-ZipIII, 19
Helicoverpa armigera, 149, 171, 173, 177, 178, 187, 212
Helper vectors, 81
Herbicide tolerance, 7, 8, 10
Heterodera glycines, 26
HG-SYV46, 26
His kinase, 370, 371
Histidine decarboxylase, 375
HOBBIT (*HBT*), 18
Homeobox, 15
Homozygous, 380
Hyperhydrosis, 247
Hyperthermia, 247
Hypotonia, 246

I

Iatrogenic, 240, 241, 242, 247
IBM, 3
INNER NO OUTER (*INO*), 33
Input traits', 8
Interoperable personal health, 4
Ionone, 365
IPTG, 50, 60, 61, 79
Isoprenoid, 367
IT miracle model, 4

J

JABBA-1D (JBA-1D) 21

K

Kanamycin 42,78,79,80
KNOTTED1 (*KN1*) 18
KNOX, 14, 15

L

Lactuca sativa, 315
Lateral organ boundaries (lob), 31
LEAFY (*LFY*), 17, 24, 29, 32
Lepidoptera 147, 152, 156, 164, 172, 175, 176, 212, 216
LIPID TRANSFER PROTEIN1 (*ATLTP1*), 16
Lycopene, 354, 368, 373, 374, 375, 377, 378, 379, 380, 386, 387
Lymphoreticular, 284

M

Mahyco, 165, 166, 169-171, 179, 184, 187, 189-191, 193-195, 210, 211, 214, 215, 217
Maize, 315, 316, 319, 322, 325, 326, 333, 341, 343
Mammalian cells, 8
Mass spectrometry, 256
MECH 12, 96
Medical "concierge" system, 3
Medical educational systems, 3
Medical IT, 4
Menthol, 364
Meristem, 14
Methyl butyrate, 365
1-Methylcyclopropane, 384
Methylelinone, 365
Microarrays, 357, 367
Microrna mir164c, 21
Mini-b chromosomes, 10
Minichromosomes, 7, 8, 10
MONOPTEROS (MP)17, 18
Monsanto 156, 164-166, 170, 178, 183-185, 187, 189,192-198, 201-206, 217, 218
Multifocal dementia, 242
Murine zora, 340
Mustard, 87, 91, 94, 95, 97, 98, 99
Myoclonus, 242, 243, 244

N

nac, 21
Nanofabrication, 270
Nanowires, 270
Neral, 365
Neuroinvasion, 284
Neuropathies, 230
Nitric Oxide (NO), 111, 112, 132
Non-climacteric, 358, 359, 360, 363, 369, 392, 393, 394, 395, 402
NOZZLE (NZZ), 33
Nuclear magnetic resonance, 258

O

Oligonucleotide, 229, 277
Omnimedix, 4

Oralvaccines, 313, 314, 315, 327, 337
Organogenesis, 15
oriV, 45
OsNAC genes, 22
Ozone sterilization, 230

P

Papaya 315, 316
Parthenocarpy, 356, 357, 358, 371, 372, 407
PASTICCINO (PAS), 21
pBR322, 44, 46, 47, 48, 49, 50, 67, 69, 71, 75, 80, 81, 82, 85
Pectinesterase, 375
PENNYWISE (PNY), 19
Peroxisomal thiolase, 397
pGVll03, 82
PHABULOSA (PHB) 33
PHANTASTICA (PHAN) 23
Pharmaceuticals, 1, 2
Phenol, 233
Phosphatidylinositol-specific phospholipase C, 256
PhSTM, 33
Phytochemicals, 368
Phytoene synthase, 362, 375
Phytohormones 111, 120
PIN-FORMED1 (PIN1) 16
Plant stem cells, 13, 14, 15
Plaques, 246, 248, 253, 263, 275, 276, 278, 292, 297
plasmid vector 58, 86
Polarity 15
Polyacrylamide gel electrophoresis, 229
Polyamines, 354, 379, 386, 407
Polygalactouronase, 362
Portable memory device, 2
Potato 87, 88, 92, 93, 97-99
Pound-foolish, 19
PRETTY FEW SEEDS2 (PFS2), 33
Preventable medical errors, 2
Prodromal, 242, 245
Prophyletic vaccination 311
Proteases, 230, 250, 263
Proteinase K, 229, 233
Psoralen photoreaction, 233, 234
pUC18, 46, 58
Purkinje cells, 270
Putrescine, 379, 398

R

Rate-zonal sucrose gradients, 233
Rax1, 20
Ribonucleases, 233, 234
RING-H2, 23
RNA-DNA hybridization, 45
Roblastoma-related, 22, 34, 28
ROSULATA (ROA), 27
ROUGH SHEATH1 (RS1), 14
RT-PCR, 27

S

S-Adenosylmethionine, 377, 379, 386
Salicylic acid, 384
Salmonella, 322, 350
SAM, 13, 14, 15
SAM hydrolase, 389
SCREWCROW , 22, 34
Secondary wall deposition, 110, 113, 115, 116
Sedentary, 245
Senescence, 385, 387, 388, 389, 393, 394
Serum transaminases, 242
SET (SU (VAR) 3-9, E(Z), 21
Shoot apical meristem, 13
SHOOT MERISTEMLESS (STM) 17, 27
SHORT INTEGUMENT1 (SIN1) 16
SHORT ROOT 22, 34
Sialoglycoprotein, 229
sin1 gene, 16
Site-specific recombination, 10
Small pox, 312
Sodium dodecyl sulphate, 233
Somatic embryogenesis, 16
Somatic embryos, 397
Spermidine, 354, 379, 386
Spermine, 354, 379, 386, 407
SPLAYED (SYD), 32
Spongiform encephalopathies, 227, 235, 237, 252, 268, 278, 283
Sporadic, 236, 237, 238, 240, 241, 242, 244, 247, 248, 261, 265, 282, 284, 296, 297
Stacked gene varieties, 8
Stenospermocarpy, 357
Stoichiometry, 273, 280
Sulpho-nhs-biotin, 256
Suspensor, 17

T

T-DNA, 79
Tachycardia, 247
Tachypnea, 247
Telomere integration, 9
Terminal transgenes, 9
Terminator 33
Terpenoid, 365
Tetraploid plant, 9
Therapeutics, 1
Ti plasmid, 82
Tobacco, 314, 318, 320, 323, 326, 327, 333, 334, 336, 338, 340, 342, 349
Tomato, 87, 93, 94, 97, 98
Topology, 228
Transgene, 9, 10
Transgenic, 147, 151, 152, 155, 156, 164, 165, 170, 181, 195, 204, 206, 207, 213-215, 219, 220, 361, 375, 376, 377, 379, 380, 388, 392, 393
Transgenic potatoes, 331, 337, 351
Transmembrane, 228, 278
transport inhibitor response 1, 22
Trichome, 101, 102, 124, 126, 128, 129, 131
Trypsin, 233
Tunica corpus, 14
TWN1 AND TWN2, 16

U

ULTRAPETALA (ULT), 33
UNUSUAL FLORAL ORGANS (UFO), 17
Urea, 233

V

VAAMANA (VAN), 19
Valyl-tRNA synthetase, 16
Viral vector 320,322,323,324,342,351

W

WUS gene 15, 18
WUS-CLV3, 33

X

XCL1, 14
x-ray files, 2

Y

Yeast s-adenosylmethionine, 379

Z

zwille (zll), 26